W9-ADR-544

TA
166
.B335
1996

Human Performance Engineering

Designing High Quality, Professional User Interfaces for Computer Products, Applications, and Systems

Third Edition

Robert W. Bailey, Ph.D.

For book and bookstore information
http://www.prenhall.com

Prentice Hall PTR
Upper Saddle River, New Jersey 07458

Library of Congress Cataloging-in-Publication Data

Bailey, Robert W.
Human performance engineering : designing high quality, professional user interfaces for computer products, applications, and systems / Robert W. Bailey.—3rd ed.
p. cm.
Includes bibliographical references and index.
ISBN 0-13-149634-4
1. Human engineering. 2. System design. I. Title.
TA166.B327 1996 95-45147
004'.01'9—dc20 CIP

Editorial/Production Supervision: Maes Associates
Interior Design: Lisa Iarkowski
Manufacturing Manager: Alexis Heydt
Acquisitions Editor: Bernard Goodwin
Cover Design Director: Jerry Votta
Cover Design: Lorraine Castallano

© 1996, 1989, 1982 Bell Telephone Laboratories, Incorporated

Published by Prentice Hall P T R
Prentice-Hall, Inc.
A Simon & Schuster Company
Upper Saddle River, New Jersey 07458

All rights reserved. No part of this book may be reproduced, in any form or by any means, without permission in writing from the publisher.

The publisher offers discounts on this book when ordered in bulk quantities. For more information write:

Corporate Sales Department, Prentice Hall P T R
One Lake Street
Upper River Saddle River, New Jersey 07458
Phone: (800) 382-3429; Fax: (201) 236-7141
E-mail: corpsales@prenhall.com

Printed in the United States of America

10 9 8 7 6 5 4 3 2 1

ISBN 0-13-149634-4

Prentice-Hall International (UK) Limited, London
Prentice-Hall of Australia Pty. Limited, Sydney
Prentice-Hall of Canada, Inc., Toronto
Prentice-Hall Hispanoamericana S.A., Mexico
Prentice-Hall of India Private Limited, New Delhi
Prentice-Hall of Japan, Inc., Tokyo
Simon & Schuster Asia Pte. Ltd., Singapore
Editora Prentice-Hall do Brasil, Ltd., Rio de Janeiro

To my wife Brenda

Contents

Preface

It has been almost 15 years since the first edition of this book was published. The first two editions were enthusiastically received by professors, usability practitioners, and students. With this new edition, we have left the best and changed the rest. The changes are based on comments from readers and an ongoing attempt to keep the book up to date. The book reflects the many new insights gained while **teaching** computer professionals, continuously **reviewing the literature,** and **consulting** with major worldwide corporations.

This book is **written by a practitioner for practitioners.** The book is intended for and has been used successfully by experienced user interface specialists (and other usability professionals), programmers, systems analysts, system designers, project leaders, systems engineers, and students.

This book remains unique. First, it is **user centered,** by including an up-to-date discussion of the strengths and limitations of how users process information. Second, it discusses **major user interface design issues** in sufficient detail for practitioners to make informed decisions. Third, the book deals with **usability** concerns that occur throughout the *entire* development process.

This edition of the book introduces a proven **user interface design model** for helping to ensure the design and development of professional, high-quality user interfaces. The model is **user centered,** promotes an **iterative design process,** and **addresses all major usability categories.**

The book contains major discussions of **prototyping** and **usability testing.** Usability testing research and practice are organized and presented in an understandable way and include discussions of heuristic evaluation, scenario-based evaluation, performance testing, and so on.

The book presents information on traditional **graphical user interfaces (GUIs),** as well as **object-oriented graphical user interfaces.** The book also contains information on designing high-quality **character-based interfaces** and state-of-the-art **multimedia user interfaces,** including discussions of **sound,** high-resolution **images** (photographs), and **full-motion video.**

The critical topic of **task analysis** is discussed in relationship to both traditional graphical user interfaces and **object-oriented** graphical user interfaces. The coverage of **statistics** is sufficient to enable designers to read, understand, and use the three most popular statistical methods: t tests, F tests, and chi square. The book continues to present a strong, up-to-date discussion of major **documentation** and **training** issues.

The focus of the book emphasizes **human performance** issues rather than preference issues. A consistent theme of the book is that well-designed user interfaces help to foster user efficiency. In fact, a professional, high-quality user interface should either reduce the cost of doing business for an organization or increase the sales of a product. In both cases, the outcome is *not* to make things nice for people, but to **achieve greater profitability.**

This is a fact-based book. The major concepts, issues, and proposed solutions (where possible) are based on psychological, usability, or user interface research. Thousands of research papers were read and analyzed to provide source material for the book. This **research-based** approach leads to more fact and less opinion when designers make user interface decisions.

Readers have pointed out that the book provides the perfect balance between being **easy and interesting to read,** and providing adequate, knowledgeable, and informed coverage of major issues. This book was the first and, with the changes included in this third edition, hopefully remains one of the **best sources of usability information** available to practitioners.

Thanks is given to the thousands of computer professionals throughout the world who, while taking my user interface classes, have provided me with countless innovative insights into usability issues. Thanks also to Bernard Goodwin, my editor from the beginning at Prentice Hall, and to Joyce Whiting (Novell, Inc.) for her review and comments on many sections, particularly the chapter on documentation. Finally, a special thanks to all who helped with final production issues, including Brenda Bailey for ensuring accuracy and completeness of tables and figures, Sarah Bailey for helping with the references, and Shauna Bailey for the creative cover design.

Robert W. Bailey, Ph.D.

Part 1

INTRODUCTION AND HISTORY

The following chapter introduces basic concepts related to achieving acceptable levels of human performance and user acceptance in systems. Good human factors (ergonomics) decisions can lead to increased speed, fewer errors, minimal training time, and increased user satisfaction and acceptance. This chapter also helps to establish a historical perspective. Although most computer-related research has been conducted in the past 25 years, studies for improving human performance in other types of systems have been carried out for over 150 years.

1

Human Engineering Acceptable Performance

Introduction

Systems exist to carry out some purpose. They consist of people, computers, and other components that interact to produce a result that the same components could not produce independently. The system may involve one person working with a particular tool to meet a specific goal or a large group of people working with a multiplicity of tools to meet an overall goal.

System designers determine what must be done and the best way to do it. Frequently, the design team deliberately includes members with a variety of different backgrounds and experiences so that each system component is adequately represented by a specialist. In many situations, the most difficult team member to find is the *usability specialist.*

With computer systems, this role has been traditionally filled by system designers or engineers. The reasoning is that system designers are people and should be able to adequately represent potential system users (who are also people). Most knowledgeable system experts would argue that substantial education, training, and experience are needed to be an effective software or hardware specialist. Similarly, it is becoming more apparent that the same high level of skill and knowledge is necessary to be an effective usability specialist.

There are at least three options for adding a usability specialist to a system development team: (1) include a trained and experienced usability specialist as an active member of the team, (2) have a usability specialist act as a consultant to the team, or (3) have one or more of the other specialists on the team become familiar enough with usability technology to become the designated usability specialist.

The best option for any given system depends on many considerations, including the type of system being developed and the personnel available. Guidelines for choosing one or the other specialist will not be presented at this point. Nevertheless, this book takes the position that a usability specialist of some kind will have to be part of most future design teams. This is already being done in many companies, including AT&T, Microsoft, Apple, IBM, and Hewlett–Packard.

The usability specialist's primary responsibility is to ensure an acceptable level of performance for potential users or, in other words, to engineer the human performance in a system. The term *engineering* refers to translating scientific findings into applied technology. Engineering also implies *active participation* by the designer to make things happen the way that he or she wants them to happen. Designers create a product that is exactly as they would have it be (within certain constraints).

Human performance engineering is similar to other computer-related engineering disciplines, such as systems engineering, software engineering, hardware engineering, quality engineering, and knowledge engineering. To be effective as a human performance engineer (or usability specialist) requires much more than simply being "human" and using common sense or intuition. As a minimum, it requires having a working knowledge of many human factors, ergonomics and usability facts, principles, concepts, methods, and techniques. That is, it requires that you understand and be able to appropriately apply the *user interface technology.*

What Is Human Performance?

A first step in making informed "people decisions" is to understand what is meant by *human performance* or the pattern of *actions* carried out to satisfy an *objective* according to some *standard.* The actions may include both observable and nonobservable behaviors (e.g., problem solving, decision making, planning, and reasoning). Things change when people perform.

Table 1-1 illustrates and compares the actions, objectives, and standards of performance for two different activities. These examples emphasize the major differences between behavior and performance.

Table 1-1 Performance Examples

Activity	Actions	Objectives	Standards
	Observable		
Terminal use	(a) Sit at keyboard	Enter data into computer	Accuracy: 99.8%
	(b) Strike keys with finger		Rate: 30,000 keystrokes a day
	(c) Look at source data sheet		Skill development time: 3 months
	Nonobservable		Satisfaction: high
	(a) Emphasize speed over accuracy		
	(b) Try to ignore distractions		
	Observable		
Driving a car	(a) Put key in ignition	Move car from point A to point B	Accuracy: no accidents Rate: within posted speed limit
	(b) Start car		

Table 1-1 Performance Examples (Continued)

Activity	Actions	Objectives	Standards
	(c) Look for obstructions		Skill development time: 3 months
	(d) Press accelerator		Satisfaction: high
	(e) Drive onto highway		
	(f) Watch for other cars		
	(g) Shift gears		
	Nonobservable		
	(a) Consider destination		
	(b) Think about directions		
	(c) Interpret road signs		
	Observable		
Problem solving during a nuclear reactor problem	(a) Observing displays	(a) Determine what is wrong	Accuracy: do not make one wrong decision or take one wrong action
	(b) Observing warnings		
	(c) Talking, swearing, angry remarks	(b) Take appropriate action	Time: solve the problem before the core is damaged and/or radiation is released
	(d) Flushed faces		
	(e) Walking back and forth		
	(f) Waving arms in air		Skill development time: 1 month
	(g) Nodding head		
	(h) Pushing buttons		
	(i) Moving levers		Satisfaction: moderate (do not want discouragement)
	(j) Listening		
	Nonobservable		
	(a) Considering the meaning of display readings		
	(b) Thinking of possible alternative solutions		

Setting Performance Standards

Any performance objective must be measured against some *standard,* particularly against a standard already achieved by a competitive or previous version of the product. The two most common standards are speed and accuracy. To key data into a computer suggests an action (keying) aimed at fulfilling an objective (converting data to computer-readable form). What is missing is a performance standard, some indication of how accurate and fast the keying must be. Without such a standard, there is no way to measure the performance so that improvements can be made. When a good set of standards exists, designers can compare them with the outcomes of user actions and evaluate any differences. In a target-shooting example, finding where the bullets hit or how rapidly the person fired the shots is only meaningful if there is some standard to measure the performance against. The standards should be meaningful and measurable.

Unfortunately, some systems designers allow standards to simply evolve. For example, they do not set requirements for accuracy or the rate at which actions must take place. Under these conditions, there is no way to determine if performance is acceptable; virtually all performance is acceptable. Often it is mistakenly assumed that designers, users, and user management all have a common standard or expectation and that any deviation from this common standard will be quickly recognized and corrected. This naive approach frequently results in considerable disappointment with the human performance levels of new systems.

Measuring Standards

Human performance is commonly evaluated by measuring accuracy, speed, training time, and satisfaction. A more detailed discussion of the characteristics of these standards follows. Later, we will discuss how to convert these standards to *human performance requirements* and then how to use them to guide the development of the user-related portions of a system.

Accuracy

A major goal in almost any system is fewer errors. In fact, many believe that the very essence of acceptable human performance is to allow users to perform activities in a reasonable time with few or no errors. Because faulty design decisions often lead to user errors, designers should try to control errors by making informed decisions. Chapanis (1965a) made this statement:

> Human factors engineers are the first to grant that people make mistakes. But they raise these important questions also: Is some of the blame to be found in the design of the equipment that people use? Do people make more mistakes with some kinds of equipment or vehicles than with others? Is it possible to redesign machines so that human errors are reduced or even eliminated? Research over the past few decades provides us with a resounding "YES" to all these questions. This then is the rationale behind the approach of human factors engineers: they start with the certain knowledge and conviction that people are fallible and careless, and that they have human limitations, but they then turn to the

machine and the job to see whether the *error-provocative features can be eliminated.* (Courtesy of Brooks/Cole Books)

Accuracy standards can be established in almost any system. For keying data into a computer system, for example, the standard can be an accuracy level of 99 percent, or an error rate of 1 percent. An error is defined as nonconformance to a standard, for example, keying a B instead of a V or writing the code 1SK instead of 15K or saying "toy boy" instead of "toy boat."

Measuring the relative proportion of errors made in a given time indicates how closely a user meets the standard set for that work. In addition, inspecting the errors themselves may help explain why a person may not be achieving the required level of performance.

Assume that a person shoots 10 bullets at a target, resulting in holes arranged as in Figure 1-1. If the person aimed for the center and ended up with this pattern of holes, we have good reason to suspect that the performance degradation may be related to faulty gunsight alignment. In this case, the errors themselves suggest a solution to the performance problem.

Another possible pattern of errors is shown in Figure 1-2, which suggests a totally different problem and solution. Lack of steadiness rather than gunsight alignment, appears to be the problem. The person may be jerking the trigger on some shots, or the barrel of the gun may be so heavy that it sways, or perhaps the person does not sight in the same manner each time a shot is made. Whatever the problem, the number and pattern of errors suggest solutions.

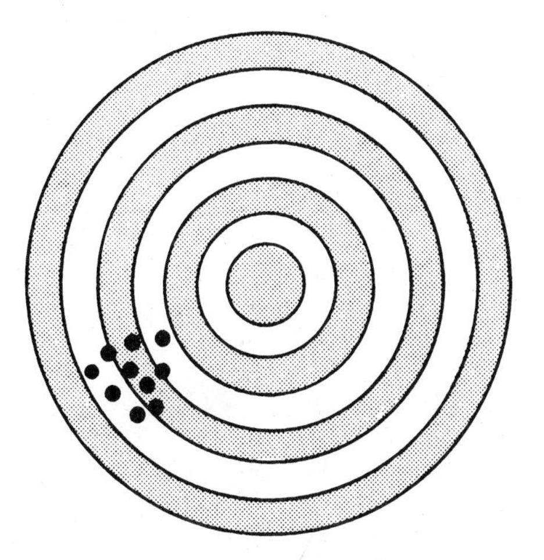

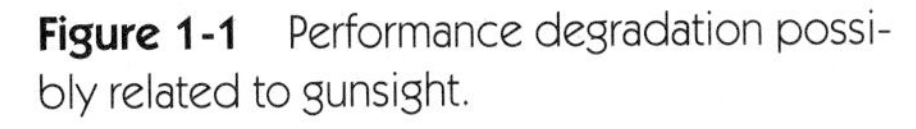

Figure 1-1 Performance degradation possibly related to gunsight.

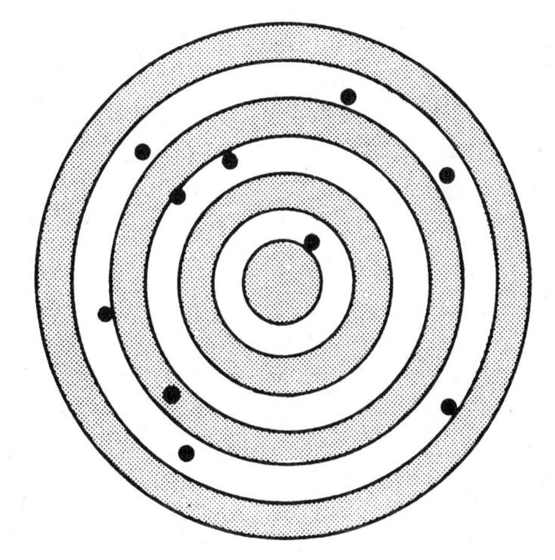

Figure 1-2 Performance degradation possibly related to steadiness.

If the shooter has no idea what the accuracy standards are (i.e., none have been levied), virtually any performance is acceptable. This is like having the person shoot at a blank wall and then adding the target (see Figure 1-3). The shooter is always a winner, and as long as the bullets strike the target at all, the performance is "acceptable."

Figure 1-3 Performance degradation possibly related to accuracy standards.

User Speed

The quantity standard, that is, the rate at which a person works, may be considered in terms such as keystrokes per day, number of shots fired per minute, or crossword puzzles solved per hour. Without having a time standard, it may take two, three, or more times longer to get a job done. As with accuracy standards, if a designer does not set a speed standard, the users will. Some users, recognizing that their early speed (i.e., production rate) will set the standard for years to come, may set a standard so slow that the viability of the new system is seriously questioned.

Achieving an acceptable level of human performance in a system often means reducing to a minimum the amount of time it takes to complete a task. The prediction of efficiency is based on the study of variables affecting the speed of performance in many different activities, such as reading, computing, and checking. Frequently, the designer's ultimate goal is to allow activities to be performed in such a way that more work can be performed by fewer people in less time.

Skill Development Time

Another important standard is the time necessary to develop the skill that is unique to performing a new activity. We refer to this as *skill development time.* This may be a few minutes, or it may be several months or years, as in the case of a skilled surgeon. If no such standard is set, the time taken to develop a skill may, and frequently does, exceed reasonable limits.

To let skills develop haphazardly and to not know when they have matured is a very costly way to operate a system. The major goal of a designer is to find ways of designing activities so that training time is reduced to the minimum and proficiency is maintained after training has been completed.

Generally, the less time it takes to train people, the lower the cost of operating the system. However, there is one additional consideration. Always designing systems so that they are easy to learn may also make systems less efficient to use. There is a definite trade-off. Reducing errors, processing time, and training time (when feasible) can all contribute to a system that costs less to operate.

User Satisfaction

Constructing systems that allow speedy performance and few errors does not automatically guarantee high satisfaction. The system should, *at the same time, satisfy the worker.* User *satisfaction* is usually measured indirectly using interviews or questionnaires.

Creating satisfying work should be a goal of all designers. Unfortunately, too much of the work developed for system users is boring and trivial. Even though pay and friends can provide some satisfaction, the work itself should be rewarding.

Making Trade-offs

Designers constantly face *trade-off decisions.* Deciding to increase satisfaction may lead to more errors or slower processing time. Deciding to reduce errors or processing and training times may lead to a system that is not at all satisfying to use.

For example, providing a slower computer response time may provide more time for a person to review his or her input and thus detect and correct errors. A decision to minimize training time may also lead to longer processing times because the individual must read and digest instructions, rather than automatically perform certain activities. Relying on printed instructions and/or performance aids in lieu of training may also lead to more errors in a system. Thus, improving the results of one performance measure may often degrade the importance of others. The designer must decide which trade-offs lead to the best overall human performance.

Human Performance Model

People performing in systems have in common the fact that they are each *somebody* doing *something, someplace.* Thus, predicting human performance requires an understanding of the *human,* the *activity* being performed, and the *context* in which it is performed (see Figure 1-4). This model of human performance is general enough to serve as model for many, if not all, performance situations. There are different levels of performance.

Figure 1-4 Human performance model.

Perfect (Optimal) Performance

People sometimes seek to attain perfect or near perfect performance. For example, a professional golfer who demonstrates superperformance during a golf tournament can make several thousand dollars; a track star who sets a new world record makes more money. A premium is paid for superior performance in dance, painting, and in many of the crafts, for example, producing well-crafted violins or guitars. Life itself may be on the line in such situations as the Apollo lunar landings, a pilot's communication while landing a commercial aircraft, or people working with high-voltage electricity and nuclear reactors. Near perfect performances, or *optimal* performances, usually have one thing in common: in each, a highly skilled individual performs a familiar (and usually satisfying) activity in a favorable context.

Acceptable Performance

Few designers have the requirement, resources, or know-how to design for near perfect or optimal performance. However, they must at least ensure an *acceptable* level of human performance, particularly when *degraded* performance may cause financial losses. This includes systems in which excessive errors or excessive training time leads to unmanageable costs or boring activities cause high turnover, thereby increasing the costs of recruiting and training.

To achieve a near perfect (optimal) or acceptable level of human performance (or simply to avoid degraded performance), a designer must take into account each of the elements in Figure 1-4:

- The general state or condition of the *human*
- The *activity,* including any required tools or equipment
- The *context* in which an activity is performed

Designers have different degrees of control over these three elements. Designers generally have the most control over the *activity,* but less control over the *people* selected for the performance and the *context* in which the performance takes place. Not always, however. The *Apollo* lunar landing project is unique in that designers had considerable control over all three major elements and could determine to a large extent the final level of human performance.

In most situations, however, maintaining strict control of the elements that affect human performance is limited. For example, an automobile designer can control the activity, but has little control over the potential user or where the activity will be performed. The driver may be inexperienced or intoxicated; he or she may be driving on icy roads at 60 to 70 miles per hour. In this case, designers must usually assume an average driver performing in a normal context.

Even a good understanding of each element (human, activity, and context) is not sufficient to predict human performance. The interaction between elements is also critical. For example, in computer-based activities the interaction between the human and the computer *(i.e., human and activity)* is the critical factor. Interactions between the activity and the *context* may also present design problems. Turning a knob to open a door when both feet are firmly planted on the ground is much different from performing the same act weightless, 200 miles above Earth, or under the pressure of 300 feet of water. Finally, interactions between the *human* and the *context* are also important. A very noisy room may have a different effect on a well-rested person than it has on a person in need of sleep.

The designer should consider the three elements separately and in combination to ensure that the critical elements and their interaction are recognized and dealt with.

The Human

The human is the most complex of the three elements. (Chapters 2 through 6 are directed to an understanding of the human.) Human performance can be affected either positively or negatively by a wide range of conditions or influences that exists within a user, even without considering the nature of the activity or the context. The designer should understand the possible sources of deficiencies in people and take them into account when making decisions. For example, color coding of displays is now common in many systems. However, color coding should not be the sole form of coding because many people are color blind.

The major considerations of the human element of a system are the sensors, brain (cognitive) processing, and responders (see Figure 1-5). People bring a wide range of basic abilities to an activity. These include good vision and adequate hearing (sensors); arms, fingers, and a mouth that function properly (responders); and the ability to think, reason, and make decisions (brain processing). To attempt to design a system without having a good understanding of how people sense, respond to, and process information is like attempting to wire a house without understanding the principles of basic electricity. In both cases it can be done, but the results may leave much to be desired.

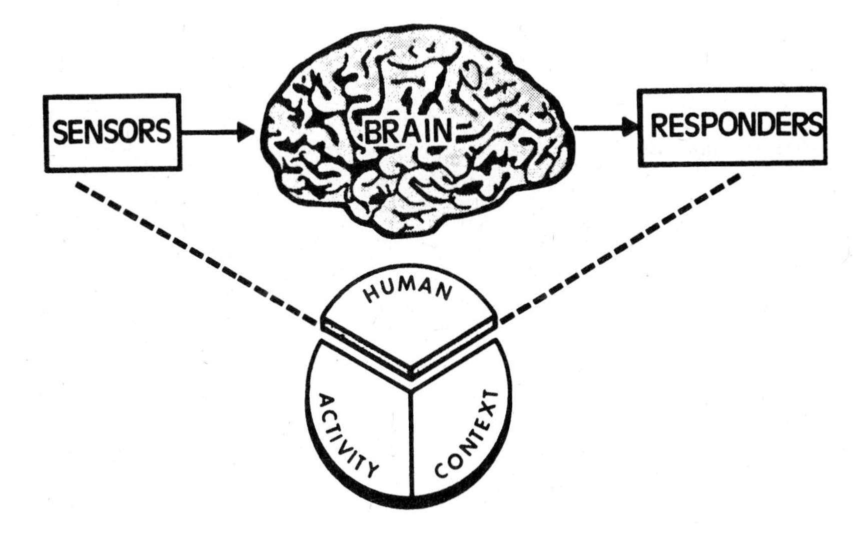

Figure 1-5 The human element.

Degraded performance could result if any of the basic capabilities required to perform an activity are lacking or reduced. We expect degraded human performance if, for example, a person has not had adequate sleep, has less than perfect vision, has not learned certain basic skills, or does not desire to perform the activity requested. Alluisi and Morgan (1976) have suggested that it is also necessary to consider the temporal influences (e.g., biological rhythms, sleep, and fatigue) and organismic influences (e.g., illness, or drug reactions) when considering human performance.

System designers usually make several assumptions about the users of their systems.

- Users have certain skills, such as speaking, listening, writing, and in some cases typing abilities.
- Users have the adequate cognitive skills: the ability to perceive, make decisions, solve problems, and control movement.
- Users desire to perform or be motivated. Even if the person has the basic abilities and has acquired the necessary skills, the question remains as to whether or not the person *will* perform.
- Users are mentally healthy. For example, they are not severely depressed or do not have a chronic high anxiety, do not have a performance-affecting phobia, are not compulsive in their reactions, and do not have perverse or obsessed ideas relating to the performance of an activity.
- All people are the same.
- Individuals do not change over a short period of time.

Designers should be cautioned: these assumptions may not always be true. Some potential users may not have the necessary skills to do a job; furthermore, they may be incapable of learning the needed skills in an amount of time that is practical. Designers must also realize that a depressed or anxious person may "level out," a phobia may be overcome, or irrational thinking may be replaced with clear, rational thought or that the opposite may occur: rational people will become irrational, and so on.

System designers do not usually have much control over the people who will use their system. Ideally, a designer could design for a specific individual, as in the *Apollo* lunar landing missions for which the small select group of astronauts was known to the designers long before critical human performance design decisions were made. Unfortunately, this is rarely the case. The next best situation is for the designer to have a good idea of the characteristics of a potential user population so that design decisions can best accommodate this target population. Because most systems are designed for groups of people, it becomes necessary to deal with the strengths and weaknesses expected in the potential user population.

The Activity

As shown in Figure 1-6, the second major consideration in understanding human performance is the activity performed. (Chapters 7 through 15 examine this element.) Until recently, people designed their own activities; today, a single designer sometimes creates activities for thousands of other people. Cazamian (1970) writes

> For a period measured in millennia, work was carried out in the form of crafts. The craftsman or artist was at one and the same time the organizer and executor of his own works. With a look to agriculture one sees a very slow, from generation to generation, refining of hand tools brought about by the users themselves. Today, we live in a situation where the craftsman's function has, so to speak, split into two parts: those of the organizer (system designer) and those of the executor (system users). In many, many cases the designer is no longer a user and a user has limited or no input to a designer.

Because the designer can control certain conditions relating to the performance of an activity, he or she must know which factors lead to better performance and which tend to degraded performance. For example, designers should know what kinds of work are best done by people (versus the computer) and what training is required to build sufficient skills for an acceptable level of human performance.

Figure 1-6 The activity.

The Context

The final consideration in this human performance model (covered in more detail in Chapter 16) concerns the context in which a human performs a particular activity (see Figure 1-7). It can make a considerable difference if an individual is attempting to connect a cable on top of a telephone pole in shirt-sleeves in Florida or performing the same activity in Minnesota during the winter, wearing a heavy coat, hood, and gloves. While two persons could perform the same activity, the major difference is the context.

Figure 1-7 The context.

There are actually two different context considerations, the *physical context* and the *social context.*

Physical Context

Noise is probably the single most studied factor in the physical context. For example, the performance of a person attempting to communicate over the telephone can be degraded if noise interferes with the perception of speech by one or both parties. Other physical-context-related conditions of interest to researchers include weightlessness, vibration, and insufficient oxygen.

Social Context

Conditions in the social context that may affect human performance include the effects of other people, crowding, and isolation. The effects of the social and physical context are well demonstrated by considering a football player who performs the same task over and over during practice and then in a series of games. The main difference is the context in which the performance takes place. During the football season, the player may perform in snow, rain, sunshine, cold temperatures, hot temperatures, with a noisy partisan crowd or a noisy nonpartisan crowd, or in a practice game with no observers. As the "home court advantage" seems to suggest, the context may make a big difference in human performance.

General User Interfaces Issues

Organizational Barriers

There are many organizational barriers to achieving good user interfaces, including (1) short development schedules (not enough time), (2) other design issues being more important, (3) attempts to keep changes to a minimum, and (4) management not recognizing or rewarding good user interfaces (Gould, 1988). Grudin (1991) has proposed other forces that can be present in development environments that can block the use of good user interface methods. These include

Not motivating developers to spend more effort on user interfaces

Not identifying and/or obtaining access to appropriate users

Late involvement of user interface specialists

Unfortunately, the market pulls products into more complexity. *Every feature has a usability cost* (even if unused). In general, the more features that a system has, the more difficult it can be to learn and use and the more difficult to design a high-quality user interface. As the number of features increases, the cognitive load increases and users become progressively less likely to use the product effectively (Morse and Reynolds, 1993).

Another major problem is having most user interface decisions made by people who have little understanding of user interface technology.

User Interface versus Total System Design

Myers and Rosson (1992) conducted a survey and found that the percent of code devoted to the user interface averages about 48 percent (with a range of 1 percent to 100 percent). The percentage of time spent on the user interface by major process was

Design, 45 percent

Implementation, 50 percent

Maintenance, 37 percent

This suggests that considerable effort is expended in the design and development of user interfaces.

Cost Justifying Good User Interface Design

One major problem in measuring progress with user interface design is that we have no generally recognized usability measures. It is difficult to quantify progress in the quality of user interfaces over the past 10 or 20 years.

Numerous people have attempted to cost justify good user interface design. Mayhew (1990) points out how to conduct a cost–benefit analysis, including making estimates of potential benefits and calculating the expected costs. Karat (1990) reports a study in which she converted reduction in user time on an activity to projected dollar values. She compared data from the first to the third usability (performance) tests. Her results showed a 2:1 dollar savings-to-cost ratio for a small project and a 100:1 ratio for a large project.

Burkhart et al. (1994) report on a study that was conducted while making the transition from a character-based to a graphical user interface. They conducted a series of studies to establish a performance base line. This was done to determine if performance improvements were achieved. They compared performance on 22 similar tasks for the old and new interfaces. They were able to demonstrate an improvement of 56 percent in task time.

Karat (1992) indicates that practitioners must become proficient at communicating that the successful use of human factors is an economic win–win situation for all involved. Imada (1990) notes that successful user interfaces prevent unwanted events from happening. In other words, success creates "nonevents." If done well, slow performance and excess errors do not occur. There are times when designers need to sell good user interfaces as a means for *cost avoidance.*

User Interface Specialists

The user interface specialists that are most successful in the design and development of user interfaces have a certain set of characteristics (Lundell and Notess, 1991). These include

Being competent with user interface issues

Making quick response to problems

Applying techniques associated with favorable outcomes

Discussing user interface issues in team meetings

Ability to do prototyping, conduct walkthroughs, and do performance testing

Mulligan et al. (1991) discuss effective strategies for making good user interface decisions. These include

Abandoning the waterfall method for system development

Becoming familiar with the capabilities and limitations of users

Learning to use a prototyping tool ("our most valuable weapon")

Adopting a user interface standard

Developing product-specific guidelines for issues that come up repeatedly

Some successful user interface designers prefer to specify user interface requirements (Chapanis and Budurka, 1990). They prepare a requirements document that reflects current user interface standards and guidelines. If done well, an application that meets the requirements will not only have a consistent interface, but also one that is highly usable (even if completed by designers with little or no user interface experience).

Kotsonis and Lehder (1990) conducted a study to determine the ideal size for effective user interface design groups. In general, they found that the larger the size of the group, the greater the perceived effectiveness:

Group Size	Percent Feeling Effective
Working alone	27%
Small cluster	41%
Group	51%
Department	79%

They also found that people working alone felt no more integrated into development teams than those in larger user interface groups.

Human Performance Issues

Two important issues should be addressed before proceeding. They both are related to confusion over measuring human performance in systems. The first has to do with the difficulties associated with separating *human* performance from *system* performance; the second, with issues in discriminating between human *performance* and human *behavior.*

Human Performance versus System Performance

Many designers seeking to measure *human* performance actually measure *system* performance. In many systems, human performance is only one consideration. The adequacy of equipment, computers, or even people outside the system boundaries may all partially determine the success of a system. When trying to understand why a system may have problems, the various components must be evaluated *separately.* Too often, people equate poor system performance with poor human performance without taking other variables into account.

While the human is the most complex of all components in any system and rightly deserves to be singled out as the most likely reason an accident occurs or a system falters, human performance is often degraded because of poor design decisions pertaining to the activity being performed, including the tools being used or even the context in which an activity is performed.

One of the most interesting examples of this is the consistency with which human error is considered the cause of many airline accidents. Within two or three days of any major airline accident, the newspapers usually begin to report that human error is responsible. More investigation frequently indicates that other components were equally at fault. Even with many automobile accidents, investigators attempt to attribute the mishap to degraded human performance, frequently ignoring the mechanical adequacy of the automobile and the context in which the automobile was driven (including misleading road signs, rough road surfacing, sharp corners, or unusual weather conditions).

Taylor (1957) provides a good example of problems associated with measuring human performance rather than system performance. Comparing the performance of a boy on a bicycle with that of a boy on a pogo stick helps to illustrate the difficulties. If the main performance measurement is the speed at which the boys travel a quarter of a mile, we should not be surprised to discover that the boy on the bicycle consistently travels the distance in a shorter time. We cannot conclude much about human performance in this situation because the primary variable is not human, but system performance: boy–pogo stick versus boy–bicycle. The boy on the pogo stick may have been doing a better job of pogo-stick jumping than the bicycle rider was doing of bicycle riding. As long as we are dealing with a system-level performance measure, human performance can only be inferred, and the inference in this case could easily be misleading.

This is one reason it is so difficult to judge the adequacy of human performance in a large system. It is much better to measure human performance separately. Unfortunately, in many situations there is no meaningful, uncontaminated way of evaluating human performance within a system. In our example, it is difficult to measure hopping independently of the physical characteristics of the pogo stick or pedaling in the absence of pedals. One cannot pedal a pogo stick or effectively hop through the air on a bicycle.

In a system, human performance is interrelated with all other components working to satisfy the objective of the system. A single component does not act independently of the other components. Even in relatively simple systems (such as the pogo stick and bicycle examples), it is difficult to measure human performance. In more complex systems, it is even more difficult. But it can be done if a designer makes an effort to do so *early* in the design process.

Once the designer has successfully separated the human component from other system components, he or she must evaluate the different elements of human performance. Any measurement of human performance must take into account human characteristics, the activity being performed, and the context in which the activity is performed. For example, we cannot study the human alone and expect to understand a great deal about performance. We cannot study something as simple as a person walking without having some idea of where the walking is being done (i.e., through a muddy field, on an asphalt track, or in 2 feet of water).

Human Performance versus Human Behavior

Performance can be easily confused with behavior. Performance is meeting an objective or producing a result. The actions leading to this result are behavior. This story from Thomas Gilbert's *Human Competence* (1978) illustrates the difference.

> Barton Hogg had achieved a dream he thought might be worthy of Midas. Soldiers had been training at the Fort Jackson firing range for over 9 years. A million GIs had trained in this area, leaving spent bullets, resulting in a half-billion pieces of lead. There must be $100,000 lying there for him, just for the sifting. But he was worried. The 60 laborers he had found by scraping the countryside were not getting the lead out fast enough. But he would have to admit that his regiment looked busy enough, bent over their shovels and sieves in a long line, just as he had deployed them. He had them working in cadence: a shovel of bleached sand into the hardware-cloth box, a sifting of the box, and then the thudding dump into the milk pails he had bought from army salvage. Now, if the 50 college students he had just hired could work as well as the other laborers, perhaps he could [become] a rich man.
>
> The truck arrived annoyingly late, and the platoon of T-shirted students poured off in shouting disarray. Hogg's heart sank as he watched this undisciplined crew. A few of them even carried portable radios, and some had newspapers under their arms. They listened to Hogg's instructions with the same blank inattention that they had learned to give their professors, and they followed his instructions just as poorly. Straggling off into groups (the radios seemed to form the social nucleus), they proceeded to work completely out of cadence. Most were soon on their haunches and shouting blasphemies, radios blaring, with no hint of order. Soon the shovels were discarded, and they were scraping the sieves directly into the sand. Hogg ran from one student to another, shouting each to his feet and inserting the shovel back into his hands. This went on all morning, and to no avail. Derisive hoots chased him into retreat. Defeated, Hogg spent the afternoon in the shade of a truck, visions of Midas shattered. That evening he called them together and he fired them all.
>
> The next morning, buckets of lead, left at odd angles in the sand, attested to the rout of the incompetent students. Hogg found that the unruly gang had sifted out *three times as much lead per labor-hour* as the cadenced crew! (Adapted from *Human Competence*, by Thomas Gilbert. Copyright 1978 by McGraw-Hill, pp. 13–15. Used with the permission of the McGraw-Hill Book Company.)

Like Barton Hogg, many people confuse performance with behavior. Consider our previous example of someone using a rifle for target practice. We watch as the person lifts the gun, sights down the barrel, and pulls the trigger. We observe a set of behaviors that can be measured. For example, we can separately time how fast the person is able to raise the rifle, sight, and fire. We can measure the width of the person's stance, the steadiness of the aimed gun, and the pressure exerted on the trigger. And we can interview the person about the planning and thinking involved in shooting a rifle. But no matter how thorough these measurements are, we still do not have any information on the performance level of the person shooting the rifle (i.e., how many shots are on target). To measure performance, we must inspect the bullet holes in the target. By evaluating the location and pattern of bullet holes, we know something about the performance level of the shooter.

Frequently, we find a designer working hard to ensure that a system measures certain carefully selected behaviors rather than performance. While measuring nonobservable intellectual process does not provide information on performance, behavioral measurements may provide clues to poor performance. For example, to find out why the shooter missed the target on half the shots, we may use a videotape to carefully examine behavioral measures. Perhaps we see the trigger severely jerked on some shots and gently squeezed on others. We now know why some shots missed the target and can make behavior changes that will improve performance.

Sometimes even a careful analysis of a person's behavior may not reveal the reasons for degraded performance, because these reasons are related to nonobservable human conditions (e.g., poor eyesight) or to difficulties with the *activity* and/or the *context.* For example, the difficulties most closely related to degraded performance may be that the resistance of the trigger is too stiff, the gunsight is moved slightly between shots, or the target shooting takes place in a gusting wind. Performance could then be improved by correcting these conditions.

History of Human Performance Engineering

The history of human performance is the history of humankind. The history of *human performance engineering* is the history of humankind's deliberate attempt to improve human performance. Human performance engineering emphasizes the concern of one group of people for the performance of others.

Earliest Human Performance

We do not know when people first became concerned with improving human performance, but even the earliest tools reflect degrees of improvement over the years. These improvements most likely resulted in better human performance. What we do not know is whether the absolute level of individual performance has improved over the last 50,000 years.

Athletic contests provide some of our earliest records. Olympic games began in Greece at least 3500 years ago with Coroebus of Elis, a cook, who won the sprint race in 776 B.C. These early sporting events, like those today, emphasized improving the human element, with the activity (event) and context (weather, number of spectators, etc.) assumed as givens. By 708 B.C., participants began using standard-sized tools, such as the javelin and discus, in some events. When chariot racing was introduced in 680 B.C., the emphasis shifted, possibly for the first time, from preparing only the human element to also preparing the vehicle and horses, items necessary to increase human speed.

People have attempted to improve the performance of others in four basic ways: motivation, training, selection and testing, and human-oriented design. A historical view of each method will be briefly presented.

Motivation

The earliest and probably still most widely used method of improving the performance of another is motivation, either through persuasion or coercion. Typically, one person tries to improve the performance of others by convincing them that improved performance is somehow better. The threat of pain or the promise of a reward sometimes accompanies the discussion.

Sigmund Freud probably made the first systematic attempt to understand motivation as related to human performance. In 1901, Freud published *The Psychopathology of Everyday Life,* essentially a theoretical statement on motivation-related causes of degraded human performance. The chapter titles give an indication of the material covered: "The Forgetting of Proper Names," "The Forgetting of Sets of Words," "Misreadings and Slips of the Pen," and the like.

Also in the early 1900s, Frederick W. Taylor, who will be discussed in more detail later, introduced his methods of "scientific management." He attempted to improve performance by improving motivation. Much of Taylor's early work on motivation led to improved worker performance.

Training

While we know little of early human history, advances in tools, weapons, and shelter suggest that early people passed on to others the knowledge and skill gained in mastering their circumstances. Sometime later, skilled specialties appeared. Direct instruction and experience transmitted the skills and knowledge of these crafts. We know this type of apprenticeship training was being used at least by 2100 B.C., because the code of Hammurabi contains rules and procedures for governing apprentice relationships. This form of training is still used to improve human performance in fields such as masonry, carpentry, and plumbing. Only since the early 1900s has training in the areas of dentistry, medicine, and law moved away from a strong reliance on apprenticeships.

In America, an early form of vocational education appeared in 1745 at the Moravian settlement in Bethlehem, Pennsylvania, with a course in carpentry (Steinmetz, 1967). By the early 1800s, training schools for certain crafts were becoming more popular. For example, the Masonic Grand Lodge of New York, in 1809, established vocational training facilities.

Factory schools for training workers began to appear in the late 1800s. Hoe and Company established one of the first in New York City in 1872. The large volume of business at this manufacturer of printing presses made it necessary to establish a factory school to train machinists. The company found the old-style apprentice approach inadequate for improving the performance of new workers.

Silvern (1970) noted that industrial training increased in popularity in the early 1900s because the public schools were not preparing their graduates for immediate employment. School classrooms presented academic courses, while skills needed for adequate work performance were ignored. To answer this need, academic and skill instruction ("shop" classes) were incorporated into schools around 1930.

Selection and Testing

Selecting those people who can best perform an activity can also improve performance. The first evidence of using tests to select people for a particular performance comes from a Biblical account dated about 1100 B.C. (Judges 7:5–6). The Bible recounts that Gideon was instructed to select a small group of warriors to deliver the Israelites from oppression by the Midianites. Approximately 22,000 people answered his first call for volunteers. Since the battle strategy called for far fewer people, he told all those who were "fearful and afraid" to return to their tents. Twelve thousand returned, leaving about 10,000, still too many. To both reduce the number and retain those specifically needed, Gideon devised a selection test. He had each person who remained go down to the nearest water and take a drink. This simple test divided the remaining people into two groups: (1) those "that lappeth of the water with his tongue, as a dog lappeth" and (2) those "that boweth down upon his knees to drink." Only 300 lapped the water, while the other 9700 bowed down on their knees. Gideon choose those who lapped. It is not clear why.

Dubois (1966) has traced the use of testing to the ancient Chinese (around 1100 B.C.); aptitude tests were used to screen applicants for higher positions in the civil service. The early Greek philosophers also contributed to the development of psychological testing. Plato, in the *Republic,* proposed an aptitude test to select persons who would be suited to a military career.

Modern testing primarily owes its start to Sir Francis Galton, an English biologist. In 1882, he established an anthropometric laboratory in London, where, for a small fee, individuals could have certain traits measured, including vision, hearing, muscular strength, and reaction time. Galton believed that tests of sensory discrimination could gauge a person's intellect. In his *Inquiries into Human Faculty and Its Development* (1883), Galton wrote, "The only information that reaches us concerning outward events appears to pass through the avenue of our senses; and the more perceptive the senses are of difference, the larger is the field upon which our judgment and intelligence can act." Galton had also noted that the mentally retarded tend to be defective in the ability to discriminate heat, cold, and pain, an observation that further strengthened his conviction that sensory discrimination "would on the whole be highest among the intellectually ablest."

James Cattell, one of America's earliest psychologists, used the term "mental test" for the first time in an article written in 1890 (Anastasi, 1963). This article described a series of tests administered annually to college students to determine their intellectual level. The tests included measures of muscular strength, speed of movement, sensitivity to pain, keenness of vision and of hearing, weight discrimination, reaction time, and memory. In his choice of tests, Cattell shared Galton's view that a measure of intellectual functions could be obtained through tests of sensory discrimination and reaction time.

The tests developed by Galton, Cattell, and others (including the famous Binet intelligence tests) were all designed to be administered individually. *Group testing* began shortly after the United States entered World War I. There was a need for rapid classification, according to general intellectual level, of a million and a half recruits. Administrative decisions depended on such information, including assignment to different types of service and admission to officer training camps. For the first time in modern history, tests were used to help match people with work.

The trend toward using group intelligence tests as rough, preliminary screening instruments continued at a rapid rate during the 1930s and 1940s in both the military and private industry. Those who "passed"(i.e., were selected) were then tested for special aptitudes. Among the latter were tests of mechanical, clerical, and managerial aptitudes. During World War II, test psychologists developed specialized test *batteries,* or combinations of tests, for pilots, bombardiers, radio operators, range finders, and several other military specialists.

Human-oriented Design

The final and perhaps most important way of improving human performance within a system is to improve design. The design change may be as simple as putting a longer handle on a broom to prevent back strain or finding a better way for astronauts to cope with weightlessness in space.

One of the most fascinating and well-documented examples of the concern for human performance took place in England around 1800. At the Greenwich Observatory, the Astronomer Royal, Maskelyne, and Kinnebrook, his assistant, were charged with observing the times of stellar transits. The observations were important since the calibration and accuracy of the clock used to establish world standards of time depended on them. Kinnebrook and Maskelyne's times matched, at first, but when Kinnebrook's measurements were as much as eight-tenths of a second off from Maskelyne's, in spite of efforts to make them correspond, Kinnebrook was fired (Boring, 1929).

It is interesting to consider the difficult human performance task that Maskelyne and Kinnebrook were performing. The accepted manner of observing stellar transits at that time, and for at least 50 years after, was the "eye and ear" method. The field of the telescope was divided by parallel cross wires in the reticle. The observer had to note, within one-tenth of a second, the time at which a given star crossed a given wire. This method was accepted and regarded as accurate to one- or two-tenths of a second. In the face of this belief, Kinnebrook's error of eight-tenths of a second was a large one and tended to justify his dismissal and Maskelyne's conclusion that he had fallen "into some irregular and confused method of his own."

About 20 years later, Bessel, a Konigsberg astronomer, sent to England for a copy of Maskelyne's complete observations and, after studying them, decided to see whether this difference between astronomers could be found among other observers. For the next 20 years, he and other astronomers collected and analyzed human performance data in an attempt to find ways to *improve human performance.* However, the negative effect of observer error was not corrected until the chronograph (stop watch) was developed. Like most error-reduction programs, greater accuracy in stellar observation came about by automating an error-prone task, not by improving the human performance.

Others were concerned about human performance in printing. Printers' errors had plagued both writers and readers from at least 1456, the time of the first printed book

(*Gutenberg Bible*). Originally, all detected mistakes were corrected with a pen in each copy. But in 1478 printed errata began to appear, some of them lengthy. For example, one book published in 1507 had 15 folio pages of errata; another author was forced to publish an 88-page volume of errata for his past publications (Wheatley, 1883).

Most printers' errors were considered unavoidable nuisances and were begrudgingly expected and accepted by both authors and readers, except for errors in printed Bibles. The most severe errors were those that actually changed the intended meaning of scriptural messages. For example, in a Bible printed in 1634, the first verse of the 14th Psalm was printed as "The fool hath said in his heart there *is* God"; and in another Bible, 1 Corinthians, verse 9, was printed as "Know ye not that the *un*righteous shall inherit the kingdom of God?" But probably the worst error of all appeared in a Bible published in 1631. In this Bible the word "not" was left out of the seventh commandment, thus leaving "Thou shalt commit adultery." The penalties paid by printers for Biblical errors (i.e., degraded human performance) ranged from heavy fines to excommunication. Printers' errors being such a major concern, it is not surprising to find that another of the first documented, systematic attempts to improve human performance was with manual typesetters (Blades, 1892).

Work Redesign

The early 1900s saw much attention directed toward improving people's performance (usually productivity) by altering the design of their work. Increasing productivity meant increasing people's speed. Much of the original impetus for this early design work came from Taylor (1911, 1947), as well as from influential publications by Munsterberg (*Psychology and Industrial Efficiency,* 1913), Gilbreth (*Brick Laying System,* 1911), and Gilbreth and Gilbreth (*Applied Motion Study,* 1917).

Frederick Taylor began his working career in a machine shop, where he rapidly progressed from laborer to foreman. Meanwhile, he attended college at night and eventually graduated with an engineering degree from Stevens Institute. Two of the most important questions Taylor set out to answer included "Which is the best way to do a specific job?" and "What should constitute a day's work?" The idea was to discover the methods of the best employee, establish a basic time standard for accomplishing work, teach these methods to the less efficient workers, and then pay the worker a premium wage for doing the task as specified. One of Taylor's better known accomplishments in time study was to enable 140 men to shovel the same amount of iron ore that had previously required 600 men.

The Gilbreths, Frank an engineer and his wife a psychologist, focused on identifying and eliminating *wasted* motion. Gilbreth's first work, a study of the motions involved in laying bricks, enabled him to reduce the motions of the bricklayer from 18 to 5, thereby improving individual performance from 120 to 350 bricks per hour.

The Gilbreths worked together for many years and their results demonstrated, early in this century, the advantages of having engineers and psychologists cooperating on projects involving human performance. One of the most interesting of the Gilbreths' contributions was their analysis and breakdown of tasks into basic elements of motion, which they called "therbligs" (i.e., "Gilbreth" spelled backward, with t and h reversed.)

After each therblig was identified, the following six questions were asked to evaluate any human performance improvement:

1 Is each therblig necessary?
2 Can the task be made simpler by having fewer motions?
3 Can there be less motion in performance or degree?
4 Can the steps be combined?
5 Can the sequence be changed?
6 Can more than one be done at the same time?

Lillian Gilbreth is credited with being the first person to develop process charts and symbols. The process charts were the forerunners of the system developer's flow charts, and the symbols are some of the first icons used in business (see Figure 1-8).

Improved Working Conditions (Context)

Murrell (1965) has observed that few organized efforts to study the effect of working conditions on human performance were made until the end of World War I. At that time the Industrial Fatigue Research Board was set up in England. For the first time, a

Name of Therblig	Therblig Symbol	Explanation
Search		Eye turned as if searching
Find		Eye straight as if fixed on object
Assemble	#	Several things put together
Disassemble		One part of an assembly removed
Inspect	0	Magnifying lens
Pre-position		A pin which is set up in a bowling alley
Release load		Dropping content out of hand
Transport empty		Empty hand
Rest for overcoming fatigue		Man seated as if resting
Unavoidable delay		Man bumping his nose, unintentionally
Avoidable delay		Man lying down on job voluntarily
Plan		Man with his fingers at his brow thinking
Hold		Magnet holding iron bar

Figure 1-8 Early icons developed by Lillian Gilbreth.

group of people trained in behavioral science entered industry to study people working. The board's work differed from the contributions of the Gilbreths in that the Gilbreths' principles of motion study were based to a large extent on observation, whereas the board relied more on objective data. For example, by attaching special counters on looms and meters on power motors, the researchers discovered evidence of both daily and weekly fatigue cycles.

In 1927, the famous Hawthorne studies were begun by Elton Mayo at the Western Electric Hawthorne plant (Roethliesberger and Dickson, 1939). These studies extended over a period of 12 years and began with the seemingly simple and straightforward problem of determining the relationship between changes in light intensity and production. The answer proved elusive.

Five experiments were conducted:

1. After the first, researchers concluded that more experimental controls were needed and that they had to eliminate nonillumination factors that affected production output.
2. In the second study, both the control and experimental groups increased their production to an almost identical degree.
3. In the third study, the light level was decreased until the test subjects were working in very dim light (about 3 footcandles); even so, they maintained their level of efficiency.
4. In the fourth study, two volunteers worked in a room until the light intensity equaled that of ordinary moonlight; the subjects maintained their production level and reported no eyestrain and *less* fatigue than when working under bright lights.
5. In the fifth and final experiment, the light was increased daily and the subjects reported that they liked bright lights. When light bulbs were replaced with some that projected the same intensity, the subjects commented favorably on the "increased" illumination. When illumination was decreased, the subjects said that less light was less pleasant.

Despite all these efforts, there was no change in production. Although they have received some criticism in the past few years (cf. Parsons, 1974; Franke and Kaul, 1978), the Hawthorne studies helped to focus on the effects of working conditions on human performance.

Equipment Redesign

World War II was a time of rapid scientific development. It brought about a large number of entirely new and sophisticated types of equipment for human use. Unfortunately, many of these devices were not designed to elicit acceptable levels of human performance. Taylor (1957) has noted that "bombs and bullets often missed their mark, planes crashed, friendly ships were fired upon and sunk, and whales were depth charged" (p. 249).

Until this time, the American, British, and German armed forces had attempted to handle most human performance problems with motivation, training techniques, and selection tests. American and German psychologists had been involved in training and testing since at least 1920 (Fitts, 1946). But in the early 1940s the problems associated with operating many of the new machines increased. Taylor observed that

> Regardless of how much he could be stretched by training or pared down through selection, there were still many military equipments which the man just could not be molded to fit. They required of him too many hands, too many feet, or in the case of some of the more complex devices, too many heads. Sometimes they called for the operator to see targets which were close to invisible, or to understand speech in the presence of deafening noise, to track simultaneously in three coordinates with the two hands, to solve in analogue form complex differential equations, or to consider large amounts of information and to reach life-and-death decisions in split seconds and with no hope of another try. (Courtesy of the *American Psychologist.*)

Of course, people often failed in performing these activities. As a result, psychologists became more active in working with engineers to produce machines that required less of their users while at the same time taking full advantage of people's special abilities. Design began to focus more on the activity to be performed and the tools to be used, with particular attention given to the design of the human–machine interface.

As World War II continued, both American and German psychologists began working on designing military equipment to more closely accommodate the capacities of users. German human performance studies included the shape and color of reticles for gunsights, amount of magnification for telescopic sights, the best positions of the body in relation to a control, the best type of movement for accurate adjustment, and the design of controls (Fitts, 1946).

In 1944, when the war was almost over, the first study primarily concerned with equipment design was conducted in the United States (Parsons, 1972). In that year, a joint project of the Applied Psychology Panel of the National Defense Research Committee and the Armored Medical Research Laboratory investigated the sources of errors in army field artillery. As with the German studies, many of these first studies dealt with evaluating new gunsight scales designed to eliminate errors. Also in 1944, the Applied Psychology Panel established a large field laboratory in Texas to conduct research to improve the design of gunsights for B-29 aircraft artillery.

Shortly thereafter, truly innovative pioneering studies on improving human performance in *combat information centers* were begun at the System Research Laboratory of Harvard University. Combat information centers were complexes where radar and other information were viewed on various display scopes, evaluated, and distributed for weapons and battle direction. Significant early military-sponsored work was also done at Oxford and Cambridge in England and Johns Hopkins University in the United States.

No doubt as a result of much of the work begun during or shortly after the war, a relatively large group of psychologists developed an interest in human performance and continued to work in this field, supported primarily by government funds and working mainly on military problems. Literally thousands of studies on performance in military systems were conducted and reported, both in in-house technical reports and in psychological journals.

In 1949, two of the first books about the subject were published, *Applied Experimental Psychology* (Chapanis, Garner, and Morgan) and *Human Factors in Undersea Warfare* (see Parsons, 1972). Since that time, numerous books have been written on human performance. Also in 1949, an interdisciplinary group of people, including psychologists, design engineers, work study engineers, physiologists, industrial medical officers, and others with a special interest in human performance, met at Oxford to form the Ergonomics Research Society. The new word *ergonomics* was created from the Greek *ergos,* work, and *nomos,* natural laws.

A second major society, the Human Factors Society, later the Human Factors and Ergonomics Society, was founded in the United States in 1957. This society was organized to provide professional and personal interchange of ideas among workers concerned with human performance. To disseminate new knowledge and to promote the application of this knowledge to design, both of these societies publish journals (see Appendix A).

Making Informed Design Decisions

Designing systems for people is serious business. It is not right to hold a designer responsible for human performance problems that are beyond his or her control, but they should be held responsible for *poor* design decisions that lead to degraded performance, particularly if these decisions result from ignorance of human performance technology. Consider an excerpt from the Code of Hammurabi written over 4000 years ago (2150 B.C.): "If a builder has built a house for a man and his work is not strong and the house falls in and kills the householder, that builder shall be slain."

In a similar fashion, perhaps with a slightly reduced penalty, designers should be held responsible for design decisions that lead to less than adequate human performance.

Exercise 1A: Historical Analysis of Printer's Errors

Purpose: To increase your understanding of the extent to which we have progressed in the past 150 years toward improving human performance in printing.

Method: Two lists of printer's errors are shown on page 28. The first is from the first printing of a book published in 1830. The second list is from the eighth printing of a book published in 1976. Both books contain about the same number of words. Compare the number and types of errors appearing in each. How do they differ? Why do they differ? Would you expect them to differ more than they do? Record your observations in a one paragraph report.

Reporting: Many of the exercises in this book will require you to prepare a short report. Most of these written reports will use the same format. Unless told to do otherwise, reports should be no more than one or two typed or handwritten pages in length. This means that you should carefully consider what you write and make the most use of the few words that you are allowed to use. Each report should cover material in the following sections: (a) purpose of the study, (b) method, (c) results, (d) discussion, and (e) conclusion.

1830 Book		1976 Book	
Correct	**Error**	**Correct**	**Error**
about	obout	believed	ebelived
arriven	arrriven	Culture	yulture
befall	befal	Day	Dsy
brethren	brethrren	finally	f8nally
brethren	bretren	glad	gla
Chapter	Ghapter	motor	mrtor
child	chid	save	lave
cloud	clowd	surprised	scriprised
created	crated	that	thlt
daughter	daghter	The	Tne
great	grert	there's	tlere's
highway	higway	twisted	twhsted
homer	horner	We'll	eWe'll
iniquities	iniqities		
judgement	judegment		
known	khown		
meekness	mekness		
multitude	mnltitude		
murderers	rumderers		
rearward	rereward		
the	he		
those	these		
treasures	treusures		
upon	opon		
works	wokrs		

The *purpose* of each study is usually provided. If the purpose of your study is slightly different, or if you can clarify or expand on the stated purpose, do so. In the *method* section, be sure to describe *exactly* what you did and how you did it. Your description of what you did should be so clear and complete that another person could read it and repeat the study doing exactly what you did. Present your *results* as clearly and concisely as possible. Use tables, figures, or graphs if it will help another person to understand what you found.

The *discussion* should include your feelings about the results. How do they relate to the purpose of the study? Were there any surprises? How do your findings relate to real-world problems in systems? What changes could be made in the study or what new study could be conducted to gain even more understanding of what is happening? Finally, the *conclusion* should state specifically one or two points you can conclude as a result of conducting the study.

During the course, you may want to keep your writeups, including the subject's responses, in a loose-leaf binder or special file folder. The results of some early exercises may be used in later exercises.

Exercise 1B: Finding Text Errors

Purpose: To increase your understanding of the progress made in reducing errors, and the difficulties in finding them.

Method: Find at least one typographical error in this book. Make a photocopy of the page with the error, circle the error, and turn it in to your instructor. Include a short, one-page report that discusses the probable reason for the error and the ease or difficulty you had in finding the error. Discuss such things as how long it took to find the error, why text errors are hard to find, and why we have any errors at all in modern-day books.

Reporting: Follow the reporting rules discussed in Exercise 1A.

Part 2

THE HUMAN

Human Characteristics

The following chapters deal with the human, which is the most complex of all system components. Frequently, the people who work with or in new products, applications, or systems are referred to as *users*. We will discuss users in the next five chapters, beginning with their *limits* and *differences* (Chapter 2). To help in understanding how typical users receive information, we will discuss the human *senses* (Chapter 3). Also in that chapter, the users' physical size and strength (*anthropometry*) are discussed. The following three chapters deal with how people process information in their brains. Chapter 4 introduces *cognitive processing* and addresses issues related to perception, intellectual processing, and movement control. Chapter 5 discusses human *memory*, and Chapter 6 reviews human *motivation*. Quality user interfaces require that designers understand and effectively apply the information in these chapters. Being human and using oneself as an example of all people will not lead to professional user interfaces.

User Interfaces

The user interface is the sum total of all design decisions made to enable people (humans or users) to use a computer product. Users can only interact with a product's functions through the computer interface. The user interface allows people (users) access to tremendous computing power, including a near infallible memory, large storage capacities, processing speed measured in billionths of a second (nanoseconds) or faster, and very few (if any) processing errors.

The user interface consists of input devices (including a keyboard, mouse, touchscreen, etc.), output devices (usually a CRT or LCD and speakers), the information input by users (using an input language, such as typed commands, spoken words, or mouse movements), and the information output by the computer (using printed or spoken words, nonspeech sounds, or graphics).

User Interface Designers

User interfaces can be designed by programmers, system analysts, user interface specialists, or even the users themselves. However, most user interfaces are designed and developed by computer programmers. One study suggests that programmers, many with limited (or no) training in user interface design issues, make as many as 90 percent of the user interface decisions in systems (Bailey, 1991). In the following discussions, all who design user interfaces will be referred to as "user interface designers" or, more simply, "designers." On average, the user interface takes about half (50 percent) of the design and development time and usually represents about half (50 percent) the code (Myers and Rosson, 1992).

The User Interface Design Model

It can be difficult to remember all the activities that need to be performed in order to have a quality user interface. In addition, the large number of different ways to design systems makes it difficult to lay out a precise step-by-step user interface design method that fits into all design approaches. The user interface model is shown in the accompanying figure. It is provided to illustrate the major issues associated with designing user interfaces.

User-centered Design

Users as the Nucleus

As the model suggests, the user is truly the *nucleus* of any system development activity. The user must be the focal point because he or she is the most complex and the most difficult component of any human–computer interaction.

In living cells, the nucleus is the kernel that is essential to normal cell functioning. The nucleus of a cell can be clearly identified and is protected by being enclosed in a membrane. When the nucleus behaves properly, the cell usually functions as it was intended. The same is true in interactive computer systems. When users behave consistently with the goals of the system, the product has the best chance of achieving its goals.

In well-designed systems, no user involvement takes place unless absolutely required for system success. The memory, perceptual, intellectual, and/or

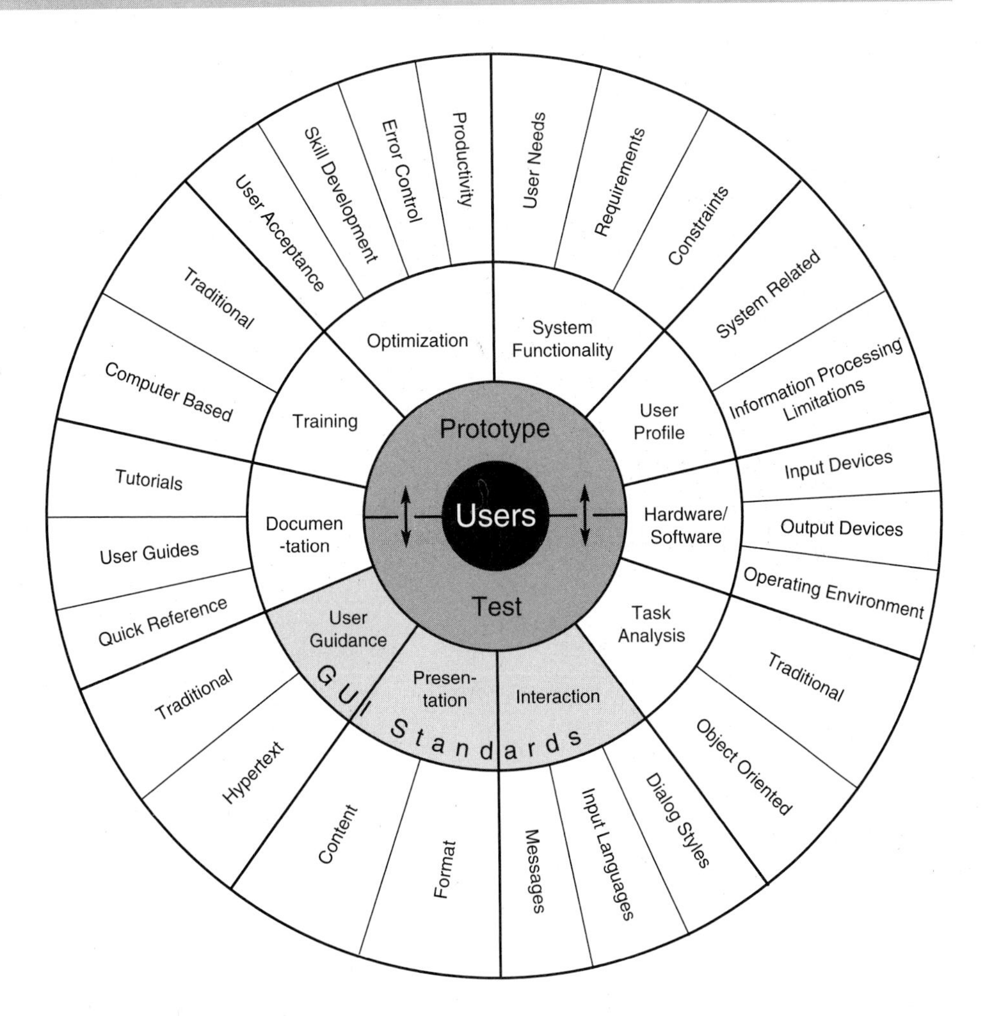

User interface design model.

movement control capabilities of users should be required (not just desired) if a user is to be included as a system component.

Users as Refiners

It is important to note that a user-centered or user-focused design methodology does *not* mean that users should be allowed to make design decisions. They tend to make decisions that are consistent with their *preferences*, which may not help to improve system *performance*. However, once major user interface design decisions are made, users can help in tailoring or refining these decisions. In other words, users are rarely good designers, but they can be good refiners.

Users are critical to the success of computer systems primarily because they can help designers to understand *what* the system should do, that is, by helping to define the functionality of a new system. They are not good at helping designers to decide how the system can best do what needs to be done. The "how" decisions are design decisions and should be made by professional designers. Bridge, building, and airplane builders do not ask their users to make critical design decisions!

In fact, if users are too involved in making design decisions, they tend to design new systems that look very much like their old systems. They are not aware of the advantages of new input or output devices, state-of-the-art operating environments, ways to present ideas, or which interactions are most effective. Users tend to make decisions that focus on improving weaknesses of an existing system, rather than making the work more efficient to perform. Many users focus on having more attractive displays, when the best design decision may be to automate most of what users were previously doing.

Once all functionality decisions are made, the best way to include users *design* decisions is to use them as participants in usability tests. This helps to structure their participation and allows them to suggest refinements, but does not encourage them to make design decisions.

Iterative Design: Prototyping and Testing

One of the most significant design activities is to create user interface prototypes and conduct appropriate usability tests. To date, no other user interface design strategy has proved more effective. Prototypes should be developed to represent only the major components of a new product. Generally, there is little to be gained from prototyping the entire user interface.

Every critical user interface decision should be prototyped and evaluated early enough in the design of a system so that changes can be quickly and easily made. Prototyping software should allow and encourage changes. Once a prototype is ready, the proper level of usability testing should be conducted. Creating good prototypes and conducting appropriate usability testing will make the iterative design process as efficient and effective as possible.

Major User Interface Design Activities

The following 10 major activities should be carried out, generally in the order suggested, during the design of a quality user interface.

Analysis activities

Verify system functionality with users

Develop user profiles

Select input and output devices and the operating environment

User interface design activities

Conduct a task analysis

Design and develop human–computer interactions

Design and develop computer presentations

Design and develop user guidance

Documentation and training design activities

Prepare documentation and performance aids

Develop training

Optimization activities

The order of carrying out the major activities within and between each major grouping is important. For example, within the analysis activities we generally determine and verify the system functionality before attempting to prepare user profiles. Both the product functionality and the user profiles must be determined prior to selecting the best set of input and output devices.

User interface design activities begin with a task analysis. Using information from the task analysis, designers then make the appropriate interaction and presentation decisions. In general, interactions should be dealt with first and then presentation issues, but there is much back-and-forth design work here. For example, radio buttons may initially be selected as the best way to interact with one aspect of the system. However, once the screen presentation is designed, it may be obvious that there is not enough room to use radio buttons. The designer might then elect to use a pull-down selection list instead.

The emphasis next shifts to designing help, documentation, and training. These activities should be started as soon as possible after design decisions begin to be made. The final set of help, documentation, and training should be based on the results of usability tests. For example, help facilities should be developed only when usability testing shows that certain parts of the system require computer-based assistance. In some cases, providing help is not sufficient, and hypertext documentation and/or paper-based documentation or training also may be needed.

Finally, the optimization activity can only be done after *all* other activities are completed. Optimization refers to ensuring that the user interface enable the performance level required by the product, application, or system. This activity provides a final opportunity to evaluate all interacting components of the user interface. Designers generally focus on ensuring that the new product allows performance in the shortest time, with the fewest errors, and the like.

Users as Information Processors

Designers should have a good understanding of how people process information. They should understand human limitations and strengths as well as sensory, memory, and intellectual limits. In addition, they should understand movement limits, including strength, coordination, and flexibility. We will now address many of the most important issues surrounding the humans or users of new products, applications, or systems.

2

Human Limits and Differences

Introduction

This chapter and the four that follow examine how people process information. While practical limitations require that the coverage of each subject be concise, the scientific literature draws on more than a hundred years of research.

The Concept of Limits

Good designers must know three things about the people for whom they are designing:

- What they *can* do: their basic abilities and skills.
- What they *cannot* do: their basic limitations.
- What they *will* do: what they will be motivated to perform.

We tend to think that people can and will do anything, despite much evidence to the contrary. The reality is that people's abilities and skills have very definite limits. For example, not one person on Earth can run 100 miles an hour, jump 25 feet high, or lift 10,000 pounds.

Designers must take these strengths and limitations into account as they design. It is generally easier to focus only on people's *strengths* and attempt to expand these abilities. Galileo's telescope improved our ability to see long distances, Ford's automobile enabled us to travel more quickly, the Wright brothers' flying machine enabled us to move even faster, and the computer enabled us to make computations more quickly. New ways of improving human abilities will no doubt come to pass in the future.

Designers must also recognize people's *limitations* and should not expect people to perform beyond their ability. For example, any high jumper will have great difficulty jumping higher than 9 feet. This type of physical limit seems inherent and fairly obvious. If a designer built a system that required people to jump, unassisted, over a 9-foot wall, the system would be inoperable and fail. The failure might even be blamed on faulty human performance. But we cannot expect people to reliably perform beyond their limits.

Figure 2-1 shows a hypothetical set of extreme limits, the widest possible boundary that could be placed around a system. If the system required all users within the system to

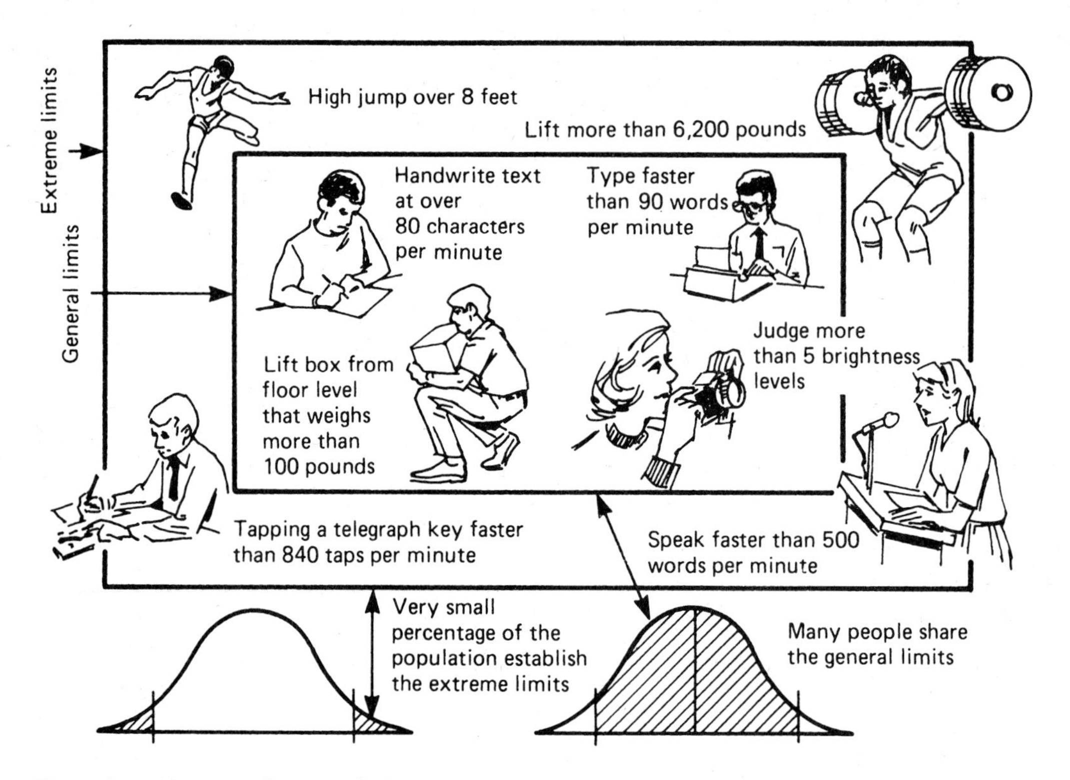

Figure 2-1 Human performance limits.

lift 6200 pounds, send Morse code at 840 taps per minute, or speak at a rate of 500 words per minute, it would certainly fail since fewer than five people (in some cases only one) can perform any *one* of the four extremes, let alone several. A designer must become familiar with human performance limits, particularly those of the people who will be using the new system.

Occasionally, people are able to extend these extreme limits, but any increase, even for those on the extreme outer boundaries, is very slow. Instead of extreme limits, most designers should deal with a set of limits more representative of their entire user population. In fact, they must usually identify and deal with the subset that applies to their user population.

For example, the specific limits of each astronaut involved in the *Apollo* lunar landing missions were known and well defined because the user population was small. On the other hand, a designer attempting to construct a map of the New York subway system must consider a set of limits covering a much wider user population, one that is not at all well defined. The latter case includes people who can read English and those who cannot; those who can see different colors and those who are color-blind; those who ride the subways daily and those with no experience. Even in this case, however, designers can make certain assumptions about human limits. Table 2-1 shows the number of people in the United States with disabilities. These disabilities obviously illustrate considerable limitation for a large part of the population in their ability to perform with many modern-day systems.

Table 2-1 Number of People with Disabilities in the United States

Visual	
Blind	580,000
Severely impaired	1,800,000
Audition	
Severe loss	2,400,000
Moderate loss	6,500,000
Orthopedic	
Impairments	23,400,000
Cognitive	
Learning disabilities	18,700,000
Speech impairments	2,000,000

From Elkind, 1990.

One assumption is that not all people can turn in optimal performances all the time. Therefore, if a system that requires typing is being designed, designers do not plan for people who can type 500 words per minute, but for people who can type 60 words per minute. Systems are not designed for people who can lift 500-pound boxes (without help) constantly for 8 hours a day, but are designed for people who can consistently lift 50 pounds.

The real secret to making good design decisions that do not compromise human limitations is to understand the full range of human limits in the user population so that system users are not required to sense, process, and respond to information faster or with greater accuracy than humanly possible. This includes appreciating that some people have limits that others do not have. Designers must also avoid erroneously assuming their own characteristics are also those of the users. Being human does not make one typical or representative of the user population.

Types of Human Limits

Human limits can be divided into three major areas: sensory, responder, and cognitive processing limits.

Sensory Limits

Sensory limits include the basic sensory *thresholds,* as well as sensory *deficiencies.*

Thresholds

Thresholds include the least amount of light that can be perceived, the smallest lettering that can be reliably read, and the faintest noise that can be reliably heard. We will discuss specific thresholds in Chapter 3. Now, it is sufficient to recognize that a set of basic stimulus characteristics must be present for the human sensory apparatus to function at all.

Deficiencies

Not all people have a full complement of senses. Even those that can, for example, see, hear, touch, and smell have these abilities in different amounts. Sensory deficiencies are common.

Considering the number of people that have seeing defects of one type or another probably best shows the varying levels of sensing. For example, almost 50 percent of the population in the United States wears glasses. Obviously, if someone's deficient vision is not corrected, performance on a vision-related activity could be degraded.

Another visual deficiency, color blindness, is present in some form in about 8 percent of the male and less than 1 percent of the female adult population. A designer who frequently uses color coding and who is not in a position to screen out people with defective color vision may find degraded performance.

Responder Limits

Two common, performance-related responder limits are the user's reach and strength. Some automobiles today have a hood latch that can only be released by exceptionally strong people. The designer has not recognized the strength limit of the user population. Not being able to reach a car's headlight switch while strapped in by a seatbelt illustrates the overlooked reach limit.

Human sensory and responder limits, one would think, would be the most obvious. We can easily pick out the people who wear glasses or use crutches, as well as those with obvious coordination problems or those with speech defects. Even so, these limits are frequently overlooked. Less obvious characteristics such as defective color vision can be fairly accurately determined through a quick test.

Cognitive Processing Limits

Probably the most difficult limits to identify are those associated with the brain-cognitive processing limits. When considering human performance, two sets of limits are especially interesting: *response time* and *accuracy.*

Response Time

Response time can be conveniently separated into (1) the time it takes to recognize a signal and decide on the appropriate movement (*reaction time*) and (2) the time it takes to move (*movement time*). For example, consider the time required for you to stop an automobile when a child runs into the street after a ball. Your response is broken down as follows. You become aware of the situation and make the decision to move the foot; this is the *reaction* component. The *movement* component consists of actually moving your foot from the gas pedal to the brake pedal and pushing down.

Designers need to account for both of these factors. For example, if a new system is being developed and the designer allows one-tenth of a second (100 milliseconds) for a particular user task, can the user respond quickly enough to be successful? Consider the following. The reaction portion of response time (the time to initiate a movement) consists of a series of delays. Wargo (1967) has suggested that the delays look like those in Table 2-2. The total delay ranges from 113 to 528 milliseconds (i.e., about one-tenth of a second to about one-half of a second).

Table 2-2 Reaction Times

	Typical Time Delays (msec)
Sensory receptor	1–38
Neural transmission to brain	2–100
Cognitive-processing delays (brain)	70–300
Neural transmission to muscle	10–20
Muscle latency and activation time	30–70
Total	113–528

In general, "fast" people, under ideal circumstances, can react to a visual stimulus in about 200 milliseconds. To expect people to react in a shorter time will lead to disappointment when the new system is operational. This is a good example of a cognitive limit. It takes time to react. In answer to our question, a design requirement of 100 milliseconds will most certainly result in degraded human performance.

Reaction time varies with the sensor that is used. The different reaction times associated with different sensors have been known for well over 100 years. For example,

To *hear* a signal and make a simple response, on the average, takes 150 milliseconds.

To *see* a signal and respond, takes closer to 200 milliseconds.

To *smell* a stimulus and respond takes 300 milliseconds.

To sense *pain* and respond takes as long as 700 milliseconds (Swink, 1966).

The fastest response occurs if people are able to *hear, see,* and *touch* the stimulus all at the same time.

These reaction time limits can be shortened if the person is well practiced in the activity, is alerted shortly before the signal occurs, or if the stimulus is increased in size or intensity. The reaction time is lengthened when a person is fatigued, is using a depressant drug, or must make a very complex movement in response to the signal.

If a system requires reactions that are close to or exceed the cognitive limits, performance will be degraded. For example, it was once considered possible for pilots of two supersonic planes flying in a head-on collision course to alter their respective courses after seeing the other aircraft. After calculating the time required for responding once the other plane was sighted, it was determined that neither of the pilots would have sufficient time to move. As a result, both the detection and movement processes were automated.

In communicating with one another, we have definite limits on the speed with which responses can take place. Table 2-3 shows estimates of many of these limits.

Table 2-3 Inherent Speed Limits

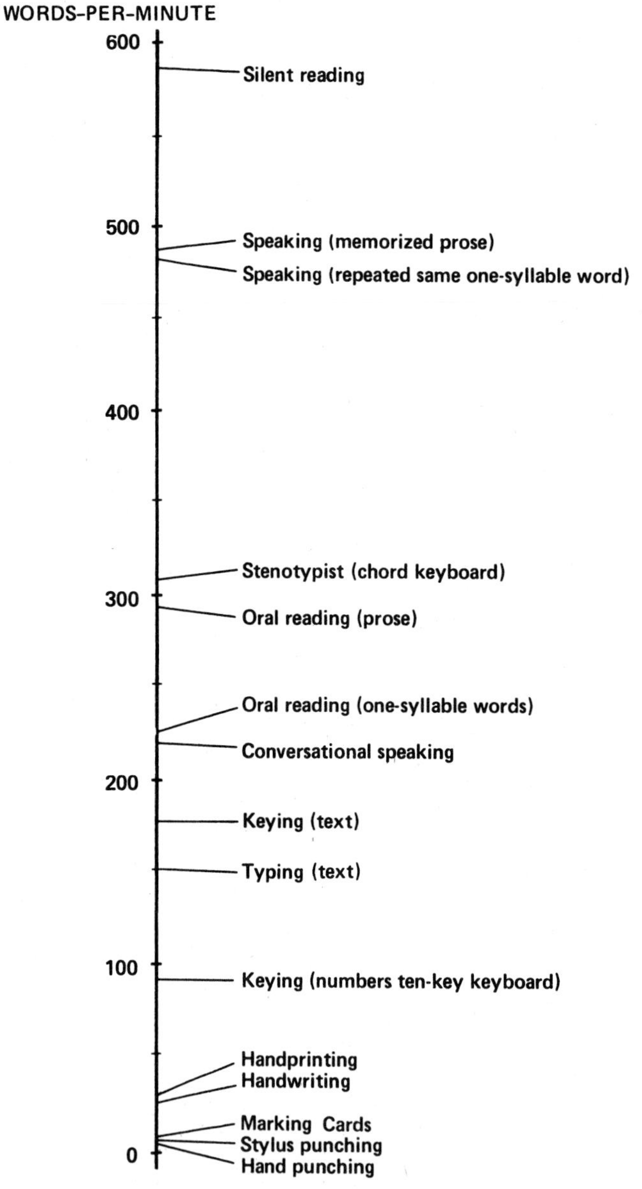

(Derived from Shackel, 1979; Turn, 1974; Newell, 1971; Deininger, 1960; Devoe, 1967; Hershman and Hillix, 1965; Seibel, 1964; Pierce and Karlin, 1956)

Accuracy

Unlike reaction time, which is physiologically limited, accuracy seems to be a little more under a person's control. It appears that people establish their own accuracy level on a task-by-task basis and attempt to meet it. With experience, a person seems to arrive at a level that is the most comfortable in terms of achieving the activity's objective. For example, the same person may have an error rate for reaction-time experiments (a person sees and responds to a signal) of about 1 percent, for hand printing of 0.5 percent, and when keying of 0.03 percent. The accuracy level is very much dependent on the activity being performed, including any penalties associated with making errors.

People seem to set more stringent accuracy criteria for some activities than for others. The accuracy criteria for driving a car is more stringent than that for typing. The acceptable accuracy level for dialing telephones (97–98 percent accuracy) is more stringent than for pushing buttons to make a telephone call (95 percent accuracy). Certain critical activities, once learned, can be performed with near perfect accuracy for long periods of time.

As a general rule, accuracy is activity related and may vary considerably with people performing the same activity, and even slightly for one person performing the same activity. Unfortunately, some activities are so poorly designed that they encourage errors.

Speed

In some activities there seems to be a relationship between speed and accuracy: the faster an activity, the higher the probability of errors. In these cases, slowing the activity reduces errors. However, in other activities, speed has little relationship to the number of errors. And in still others the fastest performers actually make fewer errors. One such activity is keying by experienced operators. Klemmer and Lockhead (1962) reported a difference among experienced key operators in both speed and accuracy. The fastest key operators were about *twice as fast* as the slower key operators; however, the slower key operators tended to make *10 times more errors.*

Number and Size of Stimuli

The number and size or amounts of stimuli also effect accuracy. Designers should know that people can discriminate among only a small number of different sizes, brightnesses, line lengths, and the like. In fact, when stimuli are presented separately, people can usually only discriminate among five to nine different categories. For example, if loudness is used as the signal, most people will only be able to discriminate among five different loudnesses; if brightness is used, most people will only be able to discriminate among five different light brightnesses.

There is even evidence that people cannot readily tell the difference between a quarter and the Susan B. Anthony dollar (introduced in the United States in 1978). People complained that the Anthony dollar was too easily mistaken for a quarter. U.S. Mint officials brushed this aside as a matter of "perception" rather than "reality." They pointed out that the size difference between a quarter and an Anthony dollar is the same as the difference between a nickel and a quarter. In addition, the dollar is 43 percent heavier than the quarter. However, many people still found it difficult to tell the difference, and the Anthony dollar is little used.

Cognitive Processing and Accuracy

Making discriminations also involves cognitive processing. People frequently have difficulties making accurate judgments even in rather straightforward, simple situations.

Estimating

The human ability to estimate has very definite limits. People do not do all things as well as most designers think they should. For example, people tend to overestimate time when they are passively involved and underestimate time when they are actively involved. Thus, if a person sitting in a chair is asked to say when 5 minutes have passed, he or she will tend to overestimate the time. The estimate will be that 5 minutes have elapsed when perhaps only 4 minutes have actually gone by. On the other hand, if a person is totally involved in building a model airplane and is asked to judge when 5 minutes have passed, he or she will tend to underestimate; when the person judges 5 minutes, 7 minutes may have already gone by. People do not accurately represent time, and time estimations depend considerably on the activity being performed during the time interval.

When people estimate physical quantities, their judgments, though more predictable, still vary. People tend to underestimate distance. They tend to overestimate vertical height when looking down (e.g., from the Empire State Building) and underestimate when looking up (from the street to the top of the Empire State Building). People tend to overestimate temperature when it is hot and underestimate it when it is cold. Weight is usually overestimated if bulky and underestimated if compact. Someone asked to estimate a number of items without counting will consistently underestimate. These examples all show people's limits in performance requiring judgment.

Multitasking

Most designers assume that people can perform only one task at a time. This may not be the case. It may only appear so because people usually *learn* to do one thing at a time. People can *do* two or more things at the same time. Take, for example, driving a car and talking to someone in the seat next to you or walking and whistling a tune. Certainly, as the skills are first being learned, it is difficult to do two or more at the same time, but once a skill has been developed people can do amazing things at the same time. For example, Neisser (1976) reported on people who can listen and speak and listen and read at the same time. Expecting users to perform only one activity at a time, and at the most two, seems to be more habit than reality. Nevertheless, when dealing with human limits, requiring people to perform more than one task at a time means that each activity must be well learned and practiced until a high skill level in both (or all three) has been attained. Otherwise, people are limited to the performance of one activity at a time. If forced to do more than one without sufficient skill, degraded human performance will result.

Most people cannot do everything well. Their ability to sense, process, and make appropriate responses in different situations has limits. All designers should appreciate these limits. It is the designers' responsibility to know and understand the extent of these limits prior to making design decisions that would be affected by limitations. If designers do this, it is much more likely that their systems will elicit an acceptable level of human performance.

Speed and Accuracy of Fingers

Lachnit and Pieper (1990) reported that the thumb and little finger were significantly faster than the other three fingers. The little finger had the fewest errors. Participants who were pianists were reliably faster than participants who were typists. When errors occurred, certain fingers were more likely to be involved than others. High probability fingers were (1) those in closest proximity to the intended finger, and (2) those located on the thumb side. For example, if the middle finger was the finger that was supposed to press a key, and a different finger was moved, the most likely finger to be moved would be the pointer finger.

Individual Differences

No two people are the same, and people continually change as well. Even the members of a small group of potential users that appear to be the same are, on closer inspection, different. There are no identical twins as far as human performance is concerned. The *differences among people* must be recognized as another way of considering limits. Some differences are inherent and long term; people are born with a certain set of characteristics that last a lifetime. These long-term differences include sex, as well as certain basic physiological capabilities, such as the existence of eyes, ears, arms, legs, and fingers.

Psychological Changes

Other long-term differences include certain psychological capabilities that people are born with, including the ability to perceive, reason, and remember. These capabilities are not the same for all people; and even though they may vary slightly throughout a lifetime, their basic existence or nonexistence seems to remain fairly constant. Perceptual, reasoning, and verbal skills are frequently measured and reported as intelligence quotients (IQ). It is doubtful whether all that is measured by an IQ test is inherent. In fact, much of what is measured is probably learned. The fact remains that certain basic psychological abilities are inherent, including the ability to perceive, reason, and remember.

Physiological Changes

As far as movement control is concerned, there are numerous examples of people who are born with an impaired ability (see Table 2–1). Again, it is very apparent when an impairment is extreme, as in cases of total paralysis. It is less apparent when the ability to control movement is only partially impaired. Some people are born with the potential for movement control that results in an experienced ballerina, professional athlete, or eloquent speaker. With others, it seems that no matter how hard they try, they cannot become professional golfers, Olympic swimmers, or world-class tennis stars.

Physiological changes also take place in the body over time. Many of these changes are due directly to the aging process. The body structure obviously becomes larger as an individual grows from infancy through childhood to adulthood. Later in life the physical structure actually begins to shrink slightly. Sensory capacities also change with age. This can be seen readily by observing the larger proportion of eyeglasses and hearing aids among older people.

Reaction Time and Age

Fozard et al. (1990) reported that there is a relatively constant rate of slowing over the adult life span. On a choice reaction-time task, people slow at about 20 milliseconds per decade or 2 milliseconds per year. Men are reliably faster than women at every age.

Age-related Cognitive Slowing

Some studies suggest that older people appear to experience a *generalized slowing* across all information processing stages (Vercruyssen et al., 1989; Cann, 1990). Other studies suggest that age-related slowing is most apparent in the *perception* or *response selection* stages (Vercruyssen et al., 1989). Other studies suggest that the slowing is most apparent in the *decision making* and *response preparation* stages (Diggles-Buckles and Vercruyssen, 1990; Fozard et al., 1990).

Age and Errors

Rabbitt (1990) conducted a study to determine the number of errors made by people in four different age groups (19–30, 50–59, 60–69, 70–79). He found that members of all age groups made the same percentage of errors. The only error-related difference was the ability to remember errors after the test, which declined with age beginning at age 50.

Differing Capabilities

People begin with basic capabilities that definitely differ. This means that some people are more limited than others. As changes take place, whether due to new learning, skill development, or physiological changes, they tend to widen the differences among people. Most designers are concerned with systems that involve human performance by adults and recognize that these adults can range in age from 18 to 80. The multitude of differences among people must be taken into account, and the better they are taken into account, the better the likelihood of a successful system.

Design decisions should also reflect people's numerous short-term differences. These short-term differences include fatigue, stress, illness, and drug effects. Recognizing the existence of these differences is important, because the limits of people will change as they are put under additional stress, become more fatigued, or perhaps develop a cold.

Human limitations, then, are not stationary and are directly related to both long- and short-term differences. Human limits not only change as a person grows older, but they may change on a day-to-day, hour-by-hour basis. Good designers take into account that these differences exist. Their designs reflect this knowledge.

Summary

This chapter has introduced the human component of the performance model with a discussion of human limits and differences. If designers are to produce systems that optimize human performance, they must first know what people *can* do (basic abilities and skills), *cannot* do (their limitations), and *will* do (motivation). The next factor that designers must take into account with regard to limits is that people differ widely in their capabilities and limits. Finally, system designers ought to be aware that users will change over time. The system designer who takes these three concepts to heart will have found some of the initial ideas related to the edict "Know thy user." This knowledge helps to prevent the most common problem in system design, designing for oneself.

Exercise 2: Determining Memory Limitations

Purpose: To demonstrate one of the most restricting of all human limitations: the ability to remember and record a relatively small number of items.

Method: The *memory span* is defined as the maximum number of digits (or other items) retained after a single reading or hearing. This exercise will illustrate the number of digits most people can remember for a short period of time.

You will need four adults as subjects. The people can be tested individually or as a group. Have each participant sit across a table from you. They should number their papers from A to T (there are 20 separate numbers).

When the subjects are ready to begin, read the numbers shown in the list that follows. Read one series of numbers at a time (e.g., 6 4 1 7). Instruct the subjects to listen carefully to the numbers and, after you have finished reading all digits, to write the numbers on their papers. For example, if you read 6417, they should wait until you have finished and then write 6417. Continue this process until all 20 codes (sets of numbers) are presented.

Be sure to read the digits clearly and slowly (about one digit per second). Do not group them (e.g., 435–257–720) or provide other cues while reading the numbers.

A. 7524
B. 63927
C. 38472
D. 97381
E. 261947
F. 195382
G. 825146
H. 9724635
I. 6925138
J. 1739265

K. 74621835
L. 58273149
M. 31842796
N. 536184972
O. 173825694
P. 829351476
Q. 5028419673
R. 9281037465
S. 8375319206
T. 35829174605

Based on your results, (1) on average, about how many characters can people hear, store, and accurately write down? (2) In this exercise, what code length do most people get right most of the time? (3) For those people who were able to accurately report any of the codes that had 9, 10, or 11 characters, how does this benefit them in the use of a new system? Do people make most errors in the beginning, middle, or end of the codes? How do these results relate to the design of new computer systems?

Reporting: Write a report using the standard report format discussed in Exercise 1A. What does the memory limit in this type of situation seem to be? In other words, what is the number of digits that most people are able reliably to remember? How can knowledge of this limit be applied in the design of a new product, application, or system?

3

Sensing and Responding

Introduction

Human information processing frequently begins with *sensing* (see Figure 3-1). The sensing is done by specialized nerves called *receptors.* Some receptors are spread throughout the body (e.g., pain receptors), while others are concentrated in one location and function as part of a sense organ (e.g., the eyes). Receptors begin functioning long before birth. Absence of sensory functioning (e.g., lack of a reaction to pain) is one test used to determine whether a person is deceased.

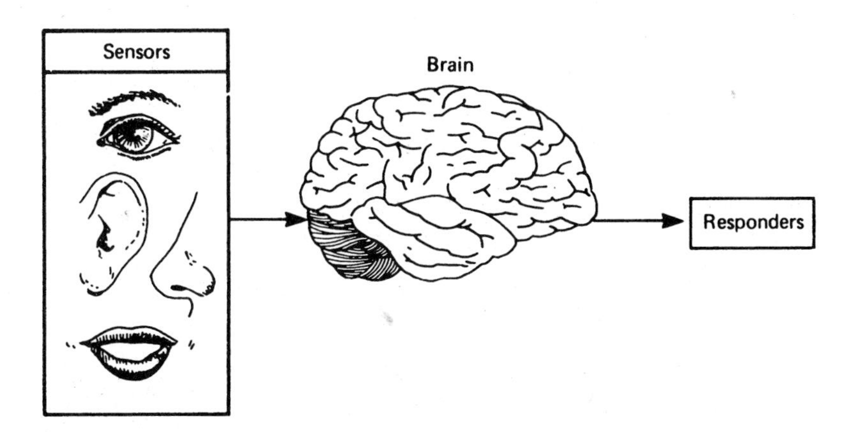

Figure 3-1 Simplified model of human information processing with the sensors emphasized.

Children are usually taught that people have five senses: vision, hearing, smell, taste, and touch. It is with some difficulty that older students acknowledge the existence of another five senses: cold, warmth, pain, kinesthetic, and vestibular. Input to the brain of information coming from these 10 senses is frequently responsible for initiating human information processing.

Sensing is not the same as perceiving. The two terms are often interchanged. Sensing is the capture and transformation of information required for the process of perception to take place. Although the sensors sometimes can be positioned to sense better (e.g., cocking the head to better position an ear), they are not improved with continued use. It is perception (in the brain) that is enhanced with experience.

Stimuli

Information comes to each sensor as a *stimulus.* If strong enough (but not too strong), it is *sensed* and passed to the brain. A stimulus is a physical event, or a change in physical energy, that causes physiological activity in a sense organ. The stimulus for the ear is sound; for the fingertips, pressure; and for the nose, odor.

A stimulus may activate a sensor and yet not be the appropriate stimulus for the specific sense organ involved. For example, if you press hard on your eyeball, you will experience a visual sensation, but not the sensation that is characteristic of normal vision. Thus, in the eye, pressure works as a stimulus, but it is an inappropriate stimulus; the appropriate stimulus is light. When the sensors are properly functioning, they are flooded with stimuli. And while they continually receive a tremendous amount of information, much of it is filtered out by cognitive processes in the brain.

Sensors and Receptors

Most people can sense a wide range of stimuli from both external and internal sources. Each sensor is designed to respond to a specific type of stimulus. The major function of sensors is to receive and transform stimulus energy into a form that the brain can recognize and process.

One way sensors can be classified is based on the source of the stimulus and the location of the sensor. According to this method of classification, there are four types:

1. Those located in the eyes, ears, and nose, which give us information concerning changes that take place at a distance from the body
2. Those located in the skin, which give us information concerning changes immediately adjacent to the body
3. Those located in the visceral organs, which tell us about changes in our internal organs
4. Those found in muscles, tendons, joints, and the labyrinth (inner ear), which give us information concerning movements of the body and the position of the body in space

The senses may also be classified according to the type of energy that is the proper stimulus (see Table 3-1).

All the senses are selectively receptive to certain types of stimuli. For example, vision responds to radiant energy, but only to wavelengths between 400 nm (nanometers or one-billionth of a millimeter) and 700 nm. The temperature sensors respond to infrared wavelengths. The tactile sense may respond to relatively slow pressure changes, while the

Table 3-1 Various Forms of Energy and the Senses Each Stimulates

Stimuli	Sense
Electromagnetic	Vision
Mechanical	Hearing
	Touch
	Pain
	Vestibular
	Kinesthetic
Thermal	Cold
	Warmth
Chemical	Taste
	Smell

Adapted from *Human Behavior: A Systems Approach* by N. W. Heimstra and V. S. Ellingstad. Copyright © 1972 by Wadsworth Publishing Company, Inc. Reprinted by permission of the publisher, Brooks/Cole Publishing Company, Monterey, California.

ear responds to very rapid pressure changes. The sense of smell is especially sensitive to chemical stimuli in gaseous form, while the sense of taste is most sensitive to chemicals in liquid form.

Sometimes the sensitivities overlap. For example, both the tactile and auditory senses respond to pressure changes (oscillations) in the 20-Hz (hertz) to 1000-Hz range. Even so, under most circumstances there is a restricted set of stimuli that activates each of the different sensors.

Sensory Limits

A light may be so dim that we cannot see it or a sound so quiet that we cannot hear it. The simplest definition of a *threshold* is that it is a point on an intensity scale below which we do not detect the stimulus and above which we do. Both thresholds and upper limits are important to human performance. The threshold represents the smallest amount of stimulation necessary to produce a sensation. Table 3-2 shows some approximate thresholds for five senses. Galanter (1962) used familiar stimuli to help in understanding how sensitive the senses are.

In addition to a threshold, each sense has an *upper limit,* a point on the physical stimulus-energy continuum above which the sensation will not become more intense no matter how much the stimulus increases. For some sensations, this limit is not known

Table 3-2 Some Approximate Sensory Thresholds

Sense	Detection Threshold
Sight	Candle flame seen at 30 miles on a dark clear night
Hearing	Tick of a watch under quiet conditions at 20 feet
Taste	Teaspoon of sugar in 2 gallons of water
Smell	Drop of perfume diffused into a three-room apartment
Touch	Wing of a bee falling on your cheek from a distance of 1 centimeter

Adapted from Galanter, 1962.

because stimuli intense enough to reach this point would damage the sensors. For example, sound pressure intense enough to establish an upper limit for loudness would probably cause deafness.

Another type of threshold is the *difference threshold.* This threshold is the minimum physical difference in the amount of stimulation that produces a perceptible (just noticeable) difference in the intensity of a sensation. For example, if a person must rely on a change in light or noise intensity as a warning signal, the amount of decrease or increase in intensity necessary for an observer to detect a change in the brightness of a light or loudness of a sound is the difference threshold. A problem with difference thresholds is that they vary with the overall intensity. In general, the size of the difference threshold increases as stimulus values increase (i.e., it is harder to tell the difference between 100 and 105 pounds than between 10 and 15).

Sensing and Performance

People have many senses, including vision, hearing, cold, warmth, pain, touch, smell, taste, kinesthesis, and the vestibular sense. We will discuss the senses most closely related to human performance.

Vision

More is known about vision than any of the other senses. The light that stimulates the eye is a form of electromagnetic radiation. As shown in Figure 3-2, it belongs to the same class of physical phenomena as radio waves, radar waves, and x-rays. The eye converts light into a form that can be used by the brain. For human performance, vision may be the most important sense.

As shown in Figure 3-2, visual stimuli come only from a narrow band in the electromagnetic spectrum, a band that covers wavelengths ranging from approximately 400 to 700 nanometers (Riggs, 1971). All other sources of radiation cannot be seen.

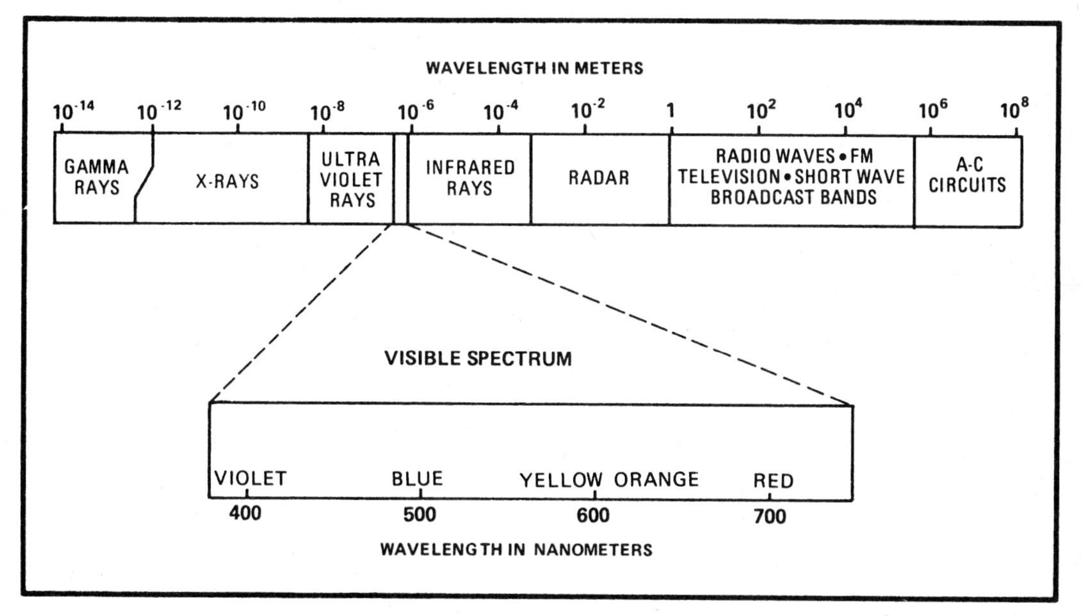

Figure 3-2 Electromagnetic spectrum showing the visible wavelengths (adapted from Riggs, 1971).

The Human Eye

Important Features

Five of the most important performance-related features of the human eye are shown in Figure 3-3. Light enters the eye and passes through the *cornea, pupil, lens,* and *humor,* on its way to the *retina.* The cornea and lens help to bend and focus the light on the retina. Unlike the cornea, however, the shape of the lens is continually being modified to change the focus. The curvature of the lens is increased for distant vision and decreased for near vision. The size of the pupil helps to control the amount of light that enters the eye and reaches the retina. The pupil widens (dilates) in dim light and constricts in bright light.

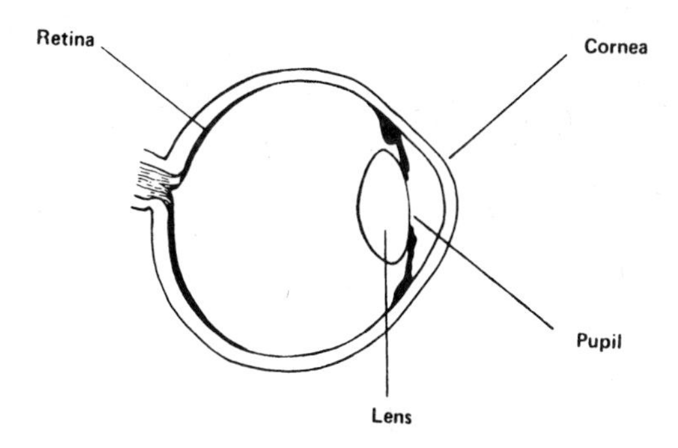

Figure 3-3 Some important features of the human eye.

The retina is the part of the eye that contains light-sensitive receptors. It contains two types of receptors: cones and rods. Cones are primarily used for seeing when the light is bright. They are *color sensitive,* which means that they can differentiate among different colors. Problems associated with color blindness or color weakness are usually related to defective cones. It is the cones that allow people to see shapes and colors clearly in well-lighted conditions.

The rods are receptors that are sensitive to dim illumination and are not sensitive to color. There are about 6 million cones and 125 million rods in the human eye.

Sensitivity

The sensitivity of the eye at a given moment depends on many things, including the size of the stimulus, the brightness and contrast of the stimulus, the region of the retina stimulated, and the physiological and psychological condition of the individual. The most important considerations will be discussed in the following pages.

Size of Stimulus

The measurement of size is the *visual angle.* This is the angle formed at the eye by the viewed object (see Figure 3-4). Usually, this is given in degrees of arc, where 1 degree = 60' (minutes of arc), and 1 minute of arc = 60" (seconds of arc).

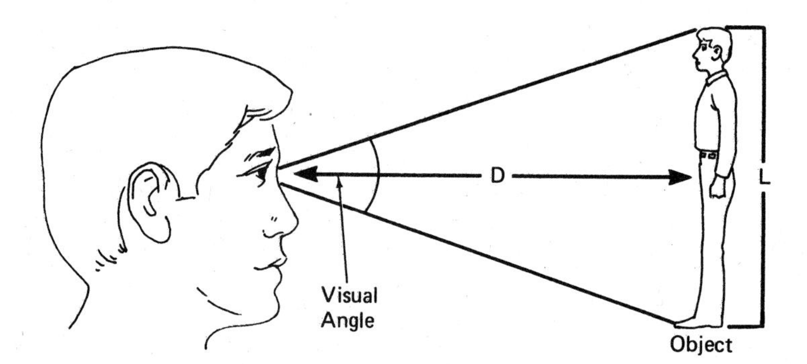

Figure 3-4 The meaning of visual angle.

The visual angle of an object subtended on the retina is calculated using the formula

$$\text{visual angle (minutes of arc)} = \frac{(3438)(\text{length})}{\text{distance}}$$

where

length = size of the object measured perpendicular to the line of sight
distance = distance from the front of the eye to the object

The visual receptors are actually about 7 mm behind the foremost point of the eye, but this distance is so small that it has little effect on most calculations (see Graham, 1965).

Generally, the minimum perceptible visual angle is approximately 1 second of an arc (e.g., a thin wire against a bright sky). The threshold for character recognition is about 5 to 6 minutes of arc. Some comparative visual angles are shown in Table 3-3.

Table 3-3 Comparison of Visual Angles

Object	Distance	Visual Angle
Sun	93,000,000 miles	30 minutes
Moon	240,000 miles	30 minutes
Quarter	Arm's length	2 degrees
Quarter	90 yards	1 minute
Quarter	3 miles	1 second

Adapted from Cornsweet, 1970. *Courtesy Academic Press.*

The following visual angles are recommended for reading tasks (Human Factors Society, 1988):

Where reading speed is important
- Minimum: 16 minutes of arc
- Preferred: 20 to 22 minutes of arc
- Maximum: 24 minutes of arc

Where reading speed is *not* critical, the visual angle can be as small as 10 minutes of arc.

Characters should never be less than 10 minutes of arc and never larger than 45 minutes of arc. If trying to determine the *smallest size* for a stimulus target so that it can be reliably read, the formula becomes

$$\text{length} = \frac{(\text{visual angle})(\text{distance})}{(3438)}$$

For example, consider the following question: How large must a set of characters be so that they can be easily read at a distance of 20 feet? If we assume a visual angle of 16 minutes of arc, at 20 feet (240 inches) away, the smallest acceptable size (length or height) for characters would be a little over 1 inch (1.1 inches or 2.8 centimeters, cm).

If trying to determine a farthest *distance* a person can be from a stimulus target so that it can be reliably read, the formula becomes

$$\text{distance} = \frac{(3438)(\text{length})}{\text{visual angle}}$$

For example, consider the following questions: What is the maximum distance a person can be from a set of characters so that they can be reliably read? If we assume a visual angle of 21 minutes of arc and characters that are ½ inch tall, the maximum distance for reliable reading is about 82 inches (205 cm).

Brightness

A second important consideration is the brightness of the stimulus. If a stimulus is not sufficiently bright, it will not be sensed clearly. Strictly speaking, brightness is the subjective sensation when viewing an object. Luminance is the measure of light per unit area that is emitted by a surface. Unfortunately, luminance can be expressed in terms of millilamberts, footlamberts, candelas per meter squared, or nits. Table 3-4 shows some luminance conversion factors.

Table 3-4 Luminance Conversion Factors

		Candela per Meter Squared (cd/m^2)	Footlambert (ftL)	Millilambert (mL)	Nit (nt)
1 Candela per meter squared	=	0.2919	0.3142	1	
1 Footlambert	=	3.426		1.076	3.426
1 Millilambert	=	3.183	0.9290		3.183
1 Nit	=	1	0.2919	0.3142	

Approximate luminance values for a variety of commonly experienced conditions are shown in Figure 3-5.

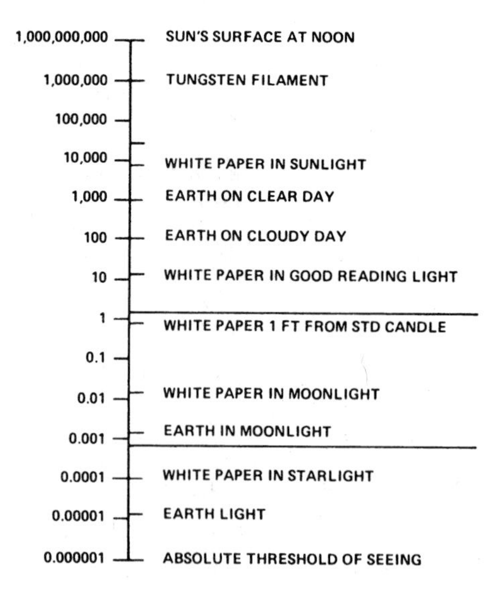

Figure 3-5 Examples of various luminance levels.

Visual Field

The most restricted visual field is that area that can be seen when the head and the eyes are motionless. Figure 3-6 shows that the monocular field (for one-eyed vision) extends from 104° from the line of sight on the temporal side to some 60° to 70° on the nasal side. The visual field can be enlarged by rotating the eyes, with the horizontal field for each eye alone being approximately 166°. The visual field obviously can be further extended by movements of the head or body.

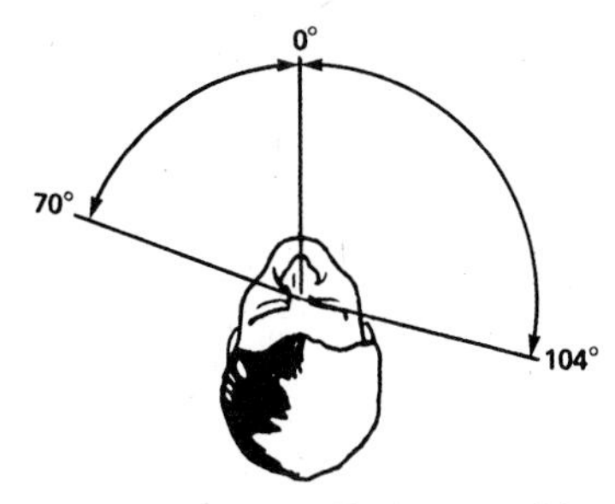

The visual field for one-eyed vision with the eye motionless.

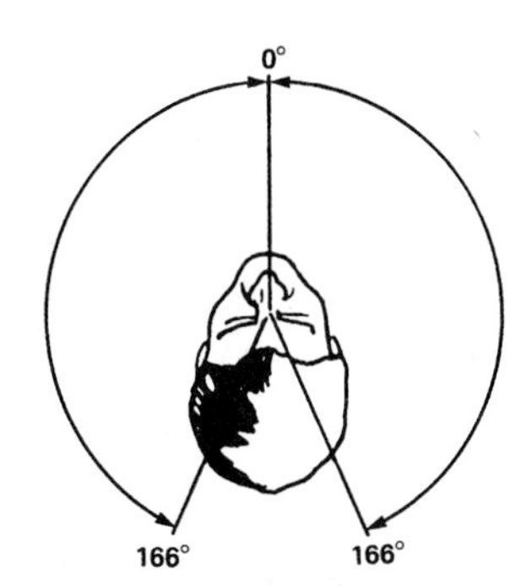

The visual field for each eye when the eyes are allowed to rotate, but not the head.

Figure 3-6 The visual field (adapted from Woodson and Conover, 1964).

Region of Retina Stimulated

Most people try to have the visual image of most interest strike about the center of the retina (on the fovea). This is the area of most sensitivity on the retina. Images that are viewed outside this area may not be as clear or their color so vivid. This leads to two important considerations: those related to visual field and those having to do with peripheral color vision.

Peripheral Color Vision

Not all areas on the retina are equally sensitive to color. Toward the periphery, objects (particularly those that are moving) can still be distinguished, while their colors cannot. Figure 3-7 shows the visual regions in which the various colors can, under normal illumination, be correctly recognized. The area where color is perceived extends for about 60° on either side of a fixed point when the eyes are motionless. Color perception occurs from 30° above to 40° below the horizontal line of vision. With the eyes motionless, blue can be recognized at about 60°, with the colors yellow, red, and green recognizable as we move progressively closer to the fixed point.

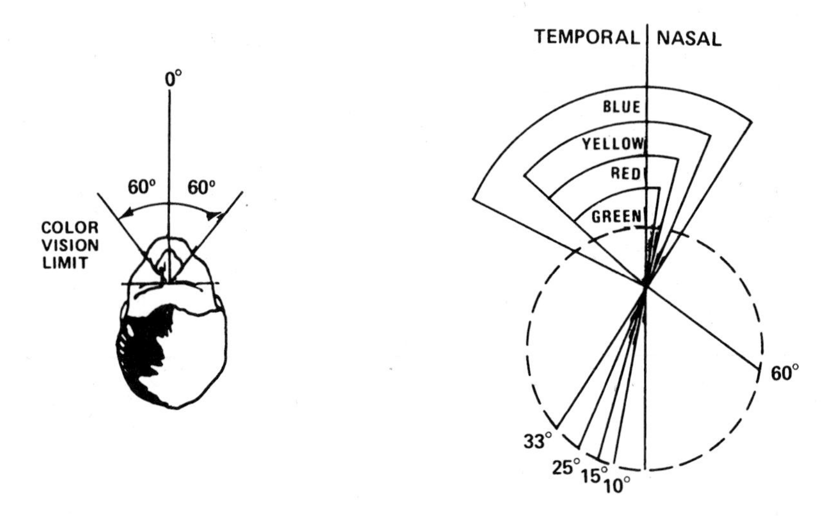

Figure 3-7 The visual field for color with eye motionless (adapted from Woodson and Conover, 1964).

Visual Impairments

In many, if not most, computer systems, users will have considerable difficulty if the visual sensors (the eyes) are not functioning properly. Some eye problems can be easily corrected (e.g., nearsightedness), some can be corrected only with great difficulty (e.g., cataracts), and others cannot be corrected at all (e.g., color blindness).

As will be discussed in more detail later, eye problems can be particularly troublesome for people who spend a good part of their day using visual display terminals (VDTs). Unfortunately, many system designers tend to assume that all users will have perfect vision. In some cases, this erroneous assumption has proved to be very costly.

Hyperopia, Myopia, and Night Blindness

We will discuss only three of several visual deficiencies that could exist and that are correctable if recognized. Two of the most common correctable visual defects, hyperopia and myopia (nearsightedness), are usually due to shape abnormalities in the eyeballs. Figure 3-8 illustrates the structure of normal, hyperopic, and myopic eyes.

In hyperopic people, the eyeball is shorter than normal, so the refracted light rays from near objects focus somewhere beyond the sensory surface at the back of the eye. The surface is stimulated adequately only by images of distant objects. Hardening of the lens with age often brings on increasing problems with hyperopia. As the lens loses elasticity, the individual finds it more difficult to focus clearly, because the lens cannot contract to allow nearby objects to come into proper focus. Generally, hyperopic people experience deterioration of vision for this reason in the years following age 40. By about 60, the hyperopic person must view an object from at least 39 inches away to even have a chance of seeing it clearly.

The reverse is true for myopic persons. In nearsightedness, the abnormally long eyeball causes light rays to focus best at a point somewhere just short of the sensory surface so that the surface is not sufficiently stimulated. The myopic person usually has little trouble in seeing close objects clearly. Both of these conditions can be fairly easily corrected with the use of eyeglasses or contact lens.

Another visual impairment that can be corrected is night blindness, a condition in which a person has less than normal vision in dim light. It may be due to a vitamin A deficiency, which can be corrected in as short a time as 30 minutes with an intravenous injection of vitamin A. An excess of vitamin A, however, will not enhance sight. In fact, too much vitamin A can cause a toxic condition leading to nausea, fatigue, and *blurred vision,* all of which could degrade human performance.

Lens, Pupil, and Humor Impairments

Some conditions can be at least partially corrected with modern surgical techniques. For example, the lens can become clouded, obstructing the passage of light and making it difficult to see clearly. When the lens (or its capsule) clouds, it is termed a *cataract.* Cataracts can be removed surgically.

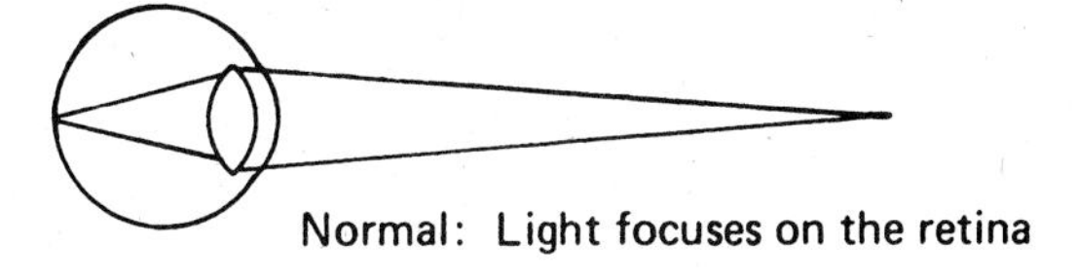

Hyperopia: Light focuses in back
of the retina for near objects

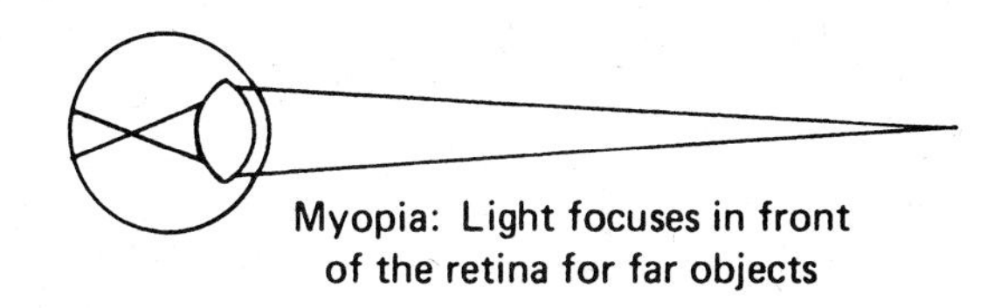

Myopia: Light focuses in front
of the retina for far objects

Figure 3-8 Structure of normal, hyperopic, and myopic eyes.

Other impairments can be much more difficult to correct, if at all. The lens can lose its elasticity, causing its focusing to become sluggish. This can be a difficult problem for people who must read from a source document and CRT interchangeably, when both are at different distances from the eyes (requiring constant refocusing).

The size of the pupil tends to shrink with age and also opens and closes slower. This can cause a chronic reduction in the amount of light that reaches the retina. The humor, or jellylike fluid, in the inner eye can also become cloudy. Among other difficulties, the clouding of the humor tends to substantially increase the problems associated with *glare.* Glare is caused by the scattering of light in the humor. This tends to reduce visual contrast because scattered light is superimposed on the retina, which reduces its sensitivity. The more clouded it becomes (usually with age), the more the incoming light is scattered.

Blindness and Color Defectiveness

Two types of visual deficiencies usually not correctable are blindness (partial or total) and color defectiveness. A person is defined as *blind* if vision (after correction) is 20/200 in his or her better eye. If a person with impaired vision recognizes letters at 20 feet that a person with normal vision would recognize at 200 feet, then he or she is said to have 20/200 vision. This degree of visual acuity is generally inadequate for meeting day-to-day visual demands. There would certainly be human performance problems if a blind person was expected to perform in a system where the possibility of blindness had not been taken into account when it was designed. A person is defined as being *partially sighted* if the vision in his or her better eye (after correction) is less than 20/70 but better than 20/200. Such people need special considerations if expected to perform adequately in vision-oriented systems. In some cases, the existing design may be adequate, but in other cases, special material, equipment, or procedures may be required to produce a satisfactory level of human performance.

Two major defects are associated with seeing colors: color weakness and color blindness. The most common defect is *color weakness.* These people are capable of seeing all colors, but tend to confuse some of them, *especially under dim light.* Their ability to distinguish different colors tends to be less acute than for people with normal color vision.

The second most common defect is *color blindness.* Most people who are color-blind tend to confuse red, green, and gray. Many more men than women have defective color vision. Approximately 8 percent of men are color defective, but less than 1 percent of women (Hsia and Graham, 1965). Other studies have shown the incidence among black and Indian males to range from 2 to 4 percent and among Chinese from 5 to 7 percent (Kherumian and Pickford, 1959). Goldschmidt (1963) lists percentages that range from about 4 to 10 percent across 11 groups of different national origins. Thus, in some places in the world, it is possible to have an all-male user population in which one out of every ten has defective vision. You can easily see the implications for using color in systems. Color coding should not be used without some form of backup coding.

Finally, it should be noted that only a very small percentage of people (about 3 in 100,000) see no color or only one color. Some of these people see everything as gray; others see everything as the same color.

Age

Age usually causes physiological changes that bring about a decline in vision. Also, older people's vision can be improved by increased illumination. Visual acuity is relatively poor in young children, but improves as a person reaches young adulthood. From about the middle twenties to the fifties, there is a slight decline in visual acuity and a somewhat accelerated decline thereafter. The use of glasses or contacts helps to compensate for this decline.

Drugs

Drugs may temporarily affect vision. Most people are familiar with the warning against driving if certain prescribed drugs are used. Although driving is probably the most hazardous activity affected, on-the-job performance can also suffer. Green and Spencer (1969) have listed about 200 commonly prescribed drugs known to affect vision.

Audition

General

The sense of hearing is probably second only to vision as the most important sense associated with human performance. Hearing is particularly important in spoken communication. In addition, it provides sound information, ranging from loud warning signals to soft, soothing music. Our ears probably allow the greatest amount of annoying stimulation. People can close their eyes if they wish to eliminate visual sensations, but cannot as easily avoid auditory stimulation.

Sound is the result of pressure fluctuation generated by vibrations from some source. Hearing is the phenomenon of sensing these vibrations. Before any auditory sensations are perceived, acoustic energy must set off a series of mechanical and neural events, beginning with sound waves moving through the air and striking the eardrum, which transmits them to receptors in the inner ear. Nerve impulses are then transmitted to the brain.

Frequency and Intensity

Sound is measured in terms of frequency and intensity. We define *frequency* as the number of cycles of pressure change occurring in 1 second. If the sound pressure changes from positive to negative and back to positive 256 times a second, the frequency is 256 cycles per second or 256 hertz (Hz). The psychological correlate of frequency (i.e., the auditory sensation we experience) is *pitch.* The rate at which the eardrum vibrates determines what one hears, and the rate of movement is determined by the frequency of the sound wave. Familiar frequencies associated with singing and playing the piano are shown in Figure 3-9.

We generally accept that the human ear responds to frequencies from about 20 to 20,000 Hz. Although very rare, hearing has been reported for sounds from 5 Hz up to 100,000 Hz. The normal person is most sensitive to frequencies in the region from about 1000 to 4000 Hz. After age 50, the ability to perceive tones at higher frequencies gradually declines. Few people over the age of 65 can hear tones with a frequency over 10,000 Hz. This loss of perception of high frequencies interferes with identifying others by their voices and with understanding conversation in a group.

The *intensity* of sound depends on the pressure of the sound wave that strikes the eardrum. The psychological correlate of intensity is *loudness.* The human ear can hear sounds so weak that they are almost impossible to distinguish from the random movements of air molecules. The loudest sound that people can hear without experiencing discomfort has a pressure about 1 million times greater than the weakest. The range of sound intensities is so large that a special number scale has been developed to measure it. Sound intensities are measured in decibels (dB). To get an idea of how decibel levels are associated with some familiar sounds, see Figure 3-10.

Hearing sensitivity varies greatly among individuals, by as much as 20 dB or more. Even an individual's hearing sensitivity may vary 5 dB within a very short period of time. At the lower frequencies, a tone must be much louder to be heard than a tone in the mid-frequencies. Frequencies between 1000 and 8000 Hz require the least intensity to be heard.

Hearing sensitivity differences are most frequently associated with age (older people tend to have more difficulty) and past exposure to loud noises (the more exposure, the greater the probability of having a hearing difficulty). Extended exposure to high sound-pressure levels (over 100 dB) may result in permanent damage to the ear. Prolonged exposure to high-intensity sound can result in the diminishing of the ear's sensitivity to all frequencies, but especially to the higher frequencies. The 1985 *Statistical Abstract of the United States* indicates that 19 percent of working-age Americans have a hearing impairment.

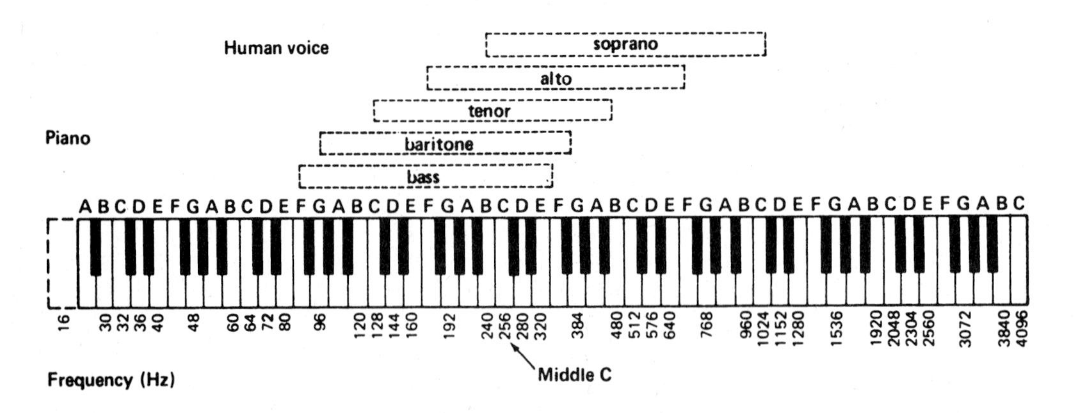

Figure 3-9 Familiar frequencies associated with singing and piano playing (adapted from Gilmer, 1970).

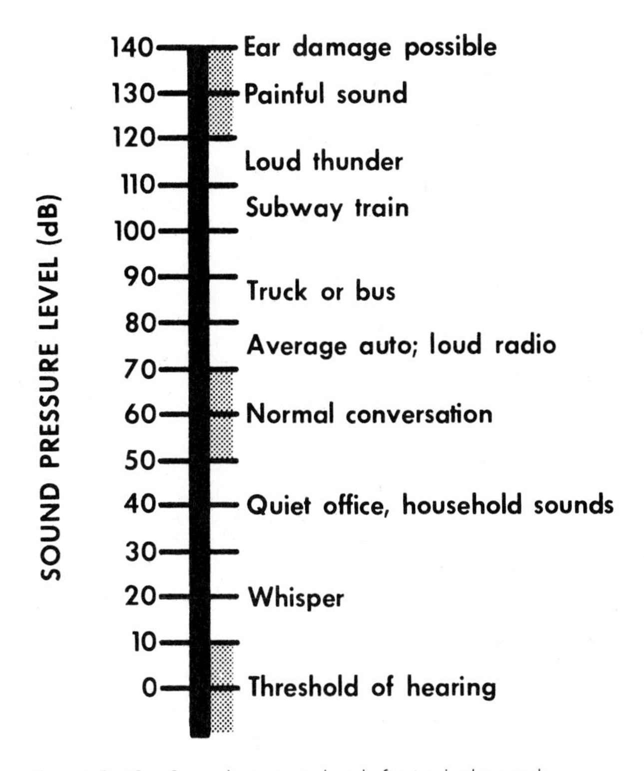

Figure 3-10 Sound pressure levels for typical sounds.

Speech Communication

Introduction

In many systems, the performance of people depends heavily on speech communication. Communication among people is most frequently by means of speech. The speech communication chain is shown in Figure 3-11. In general, speech communication can be degraded if there are difficulties with the speaker, transmission of sound waves, or the listener. Errors made by a speaker are sometimes considered amusing and given such names as "bloopers" or "Freudian slips." Currently, there is little a designer can do about many speech errors. Even in systems for which the speech content has been reduced to the words yes and no, eventually a speaker could accidentally mean one and say the other. However, the designer does have some control over the likelihood that the speech content will be audible, distinguishable, and understood.

A Major Disaster

A KLM Royal Airlines jumbo jet roared down the fog-shrouded runway at Tenerife in the Canary Islands. Suddenly, a taxiing Pan American World Airways jumbo jet loomed

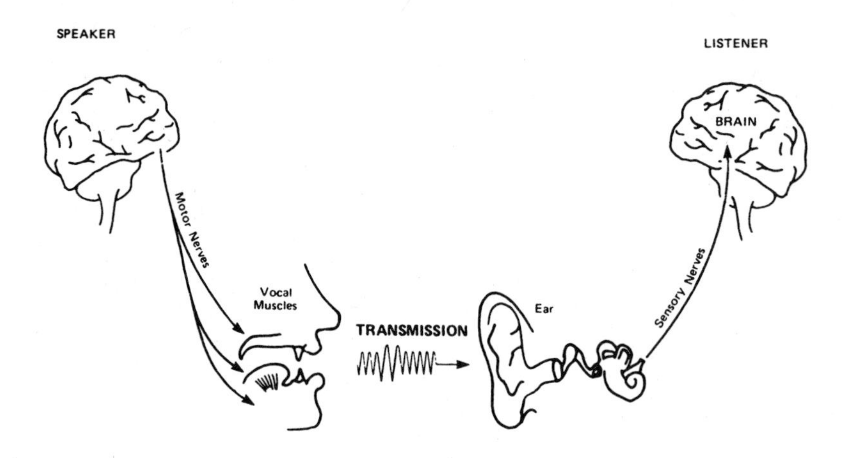

Figure 3-11 Speech communication chain.

in the fog. The KLM captain yanked his control column back in a desperate attempt to take off and fly over the Pan Am plane. But it was too late.

It all began when the Pan Am jet left Los Angeles with 373 passengers on board. About the same time the KLM jet left Amsterdam with 234 passengers. While both planes were en route to Las Palmas, a bomb exploded in that terminal and several planes, including the Pan Am and KLM jets, were diverted to nearby Tenerife. The KLM and Pan Am planes arrived early in the afternoon of May 27, 1977.

The KLM pilot, who had been on duty for more than 9 hours, was under considerable pressure. If he stayed at Tenerife much longer, the 747 would have to remain overnight in the Canary Islands. Flight schedules and passengers' plans would be disrupted.

While the KLM and Pan Am planes were waiting, the weather at Tenerife deteriorated. Fog rolled in, decreasing visibility. When it finally came time to leave, the KLM plane could not use any taxiways to get to the end of the runway since all were jammed with other planes diverted from Las Palmas. To reach the end of the airport's only runway, the KLM plane had to taxi down *that* runway.

The Pan Am plane was cleared to taxi down the same runway behind the KLM jet. The Tenerife controller ordered the Pan Am plane to taxi partway down the runway to a turnoff and then to exit to a clear taxiway. But the fog became so thick that the Pan Am crew missed their assigned turnoff and continued taxiing down the runway behind the KLM jet.

Meanwhile, the KLM plane reached the end of the runway and turned around. The captain still needed two separate clearances for takeoff: the first from air traffic control (ATC) and the second from the tower. Without waiting for the clearances, the captain began to advance the throttles. The following conversation was heard from the cockpit:

KLM copilot: "Wait a minute, we don't have any ATC clearance."

KLM captain, retarding the throttles: "I know that, go ahead, ask."

The KLM copilot requested and received clearance from ATC. He then read ATC clearance back to the tower, but even before he finished, the pilot advanced the throttles again. The copilot told the tower hurriedly, "We are now at takeoff." (That commonly means "takeoff position.")

Tower: "Okay, stand by for takeoff and I will call."

But only the word "okay" was clear. The Pan Am crew radioed the tower to announce their position on the runway. That transmission caused a squeal that partly blocked the rest of the tower's orders to the KLM crew to stand by. Saying, "We go!" the KLM captain gunned the 747 down the runway. The next two transmissions were heard in the KLM cockpit.

Tower: "Roger Papa Alpha (Pan Am). . . report the runway clear."

Pan Am: "Okay, we'll report when we're clear."

KLM engineer (in the KLM cockpit): "Is he not clear, that Pan American?"

KLM pilot: "Oh, yes."

In the dense fog, the Pan Am crew could not see the KLM jet roaring down the runway toward them. But the radio transmissions alarmed them.

Pan Am pilot: "Let's get the hell out of here!"

Pan Am copilot: "Yeh, he's anxious, isn't he?"

Pan Am pilot: "There he is—look at him."

Pan Am copilot: "Get off, get off, get off!"

The Pan Am pilot tried to steer his jumbo jet off the runway. But it was useless. The KLM jumbo was going about 130 miles an hour when it struck the Pan Am plane. Most of the 583 who died were killed on impact.

In this situation, adequate communication between the aircraft, tower, and air traffic controller could have saved many lives. The example is unique only because of the large number of lives lost. In system after system there are misunderstandings and missed messages due to degraded speech communication. To avert the possibly disastrous consequences of some breakdowns and guarantee an acceptable level of human performance in all systems that depend on speech, designers should ensure that the important aspects of speech communication are carefully considered.

It is the designer's responsibility to know where in the system speech communication must occur, the frequency and types of those exchanges, why they will occur and their likely content, the conditions under which they will occur, and the consequences of failed communication at any point. Designers also must be aware of the issues that relate to speech communication, the conditions that have the potential to degrade speech communication, and the techniques that may be used to circumvent or overcome problems in speech communication.

Normal Speech

The frequency of a normal voice lies almost entirely within the range from 100 to 8000 Hz. The loudness of average speech at a distance of 1 meter is usually in the range from 60 and 75 dB. Maximum intelligibility is attained when speech intensity is about 40 dB. Intelligibility decreases slightly when the sound is above 100 dB.

Signal-to-Noise Ratio

In human terms, noise is unwanted sound. Speech performance can be degraded under noisy conditions. Listener performance depends on the ability to distinguish a desired sound (signal) from an undesired sound (noise). In a system whose purpose is the transmission of speech, the voice of the speaker is "signal" and all else is noise.

The *signal-to-noise ratio,* abbreviated as S/N ratio, is a measure of the relative sound intensity of the signal (speech) to the background noise. Because a decibel is a log scale, to compute the S/N ratio, *subtract* the number of decibels in the noise from the number of decibels in the speech signal. For example, if the average intensity of speech is 75 dB and the average intensity of the noise in which the speech is spoken is 70 dB, we have an S/N ratio of +5 dB. If the average intensity of the speech is 60 dB and it is heard in an environment with an overall noise level of 60 dB, the S/N ratio is 0 dB. Finally, if the average intensity of the speech is 50 dB and it is heard in an environment with noise at 80 dB, the S/N ratio is –30 dB.

To illustrate how speech sounds can be confused, Miller and Nicely (1955) conducted a study in which subjects listened to consonants against a background of random noise. As the noise increased, it soon became impossible for people to distinguish the different letters. Kryter (1972) reviewed their results and made the following observations:

At S/N = –18 dB, all consonants are confused with one another.

At S/N = –12 dB, the consonants such as *m, n, d, g, b, v,* and *z* are confused with one another, and the consonants such as *t, k, p, f,* and *s* are confused with one another, but the consonants in the first group are seldom confused with those in the second group.

At S/N = –6 dB, although the m and n are confused with each other, they are clearly distinguishable from the other consonants.

At S/N = 0 dB, certain individual consonants are still confused.

At S/N = 12 dB, all the consonants are readily distinguished.

For satisfactory communication of most voice messages in noise, the speech level should exceed the noise level by at least 6 dB.

Improving Hearing

The most effective way to protect speech from noise interference is to control and isolate noise at its source. When noise is a problem, the effects of noise on speech communication can be reduced somewhat by using ear-protective devices, better-quality microphones, automatic gain controls, high-pass filters, peak limiters, loudspeakers, and headsets. Chapanis (1965a) provides a brief list of recommendations for designing systems requiring speech communication.

When noise levels are high (about 80 dB or more) or when speech levels are high (about 85 dB or more), ear plugs will usually improve speech intelligibility in face-to-face communication.

The most effective microphones are those that have high sensitivity to acoustic speech signals and reject other acoustic signals and noises at the speaker's location (noise-canceling microphones).

Loudspeakers rather than headsets should be used when (1) environmental noise levels are low, (2) listeners must move around or would otherwise be hampered by wires and cables, or (3) many listeners must hear the message.

Headsets rather than loudspeakers should be used when (1) environmental noise levels are high, (2) different listeners must hear different messages, (3) reverberation in the room is a serious problem, (4) the listener must wear special equipment (such as an oxygen mask), or (5) the power output is too low to operate a loudspeaker.

Face-to-Face Speech Audibility

When a talker is speaking to someone in a quiet office and the listener is about 1 meter away, the level of conversational speech is usually about 65 dB. Given a good estimate of the expected background noise level in a new system, the designer can evaluate the estimated speech signal (65 dB) against the estimated noise level using the S/N ratio to help determine if there will be difficulties in speech communication. For satisfactory communication of most voice messages in noise, the speech level should exceed the noise level by at least 6 dB. Keep in mind, however, that as the noise level increases most users can also increase their voice level (within certain limits).

Language Considerations

Language Issues

There are at least 3000 different languages in the world. Of these languages, 130 are each used by more than 1 million people. Some systems are designed to accommodate the fact that users will speak different languages, and other systems require that all communication be done using only one language.

The use of English in many international systems such as air traffic control is a good example of using only one language. English is the world's most common first and second language. For that reason (and others), English is the language of international air traffic control. Some years ago, air traffic controllers in Quebec fought to be allowed to use French for flights originating and terminating within Quebec. This was opposed because there would still be a danger to other aircraft who must comprehend the instructions being given to all aircraft.

When a person speaks a second language, the problem of accent arises. Also, the same language spoken in different countries will display variations in word usage (e.g., compare English spoken in the United States, Ireland, and Australia). Finally, there are regional dialects within countries.

Matching Expectations and Message

Speech communication is influenced by a set of expectations that exists merely by knowing what activity is being performed. A physician returning a telephone call usually means that the listener expects some specific medical advice. The same person receiving a call from a dentist's receptionist is likely to expect a message concerning an upcoming appointment or an overdue bill. After we drive into a gas station, we expect the attendant to ask, "How much do you want?" In any given situation, we call up the "educated expectation" or schema that we think is most appropriate for the particular activity. This schema helps us to understand the message. If the schema is not appropriate or if the message is totally unexpected or bizarre, it may hinder understanding of the message. Consider trying to understand the message from the physician while thinking that it is the dentist's receptionist calling. To help prepare the listener to receive the message, a designer should provide as many cues as possible *before* the message begins.

Vocabulary Size

One important way to increase the intelligibility of a message (particularly in the presence of noise) is to limit the size of the vocabulary. For example, if the number of words to be transmitted over a noisy communication channel is kept small, and if the entire list of these words is known to both the listener and the talker, the chances of accurately communicating are greatly increased. The smaller the vocabulary set, no matter whether it is made up of sentences, words, or letters, the better the communication. Keep in mind that to take full advantage of limited set size both talker and listener must know the set thoroughly. The difference between very small (2 word) and moderate-size (1000 word) vocabularies can change by 18 dB the signal-to-noise ratio required for a given level of intelligibility (Howes, 1957).

Familiarity and Length of Words

Other things being equal, the more frequently a word occurs in everyday usage, the more readily it is correctly identified when transmitted using speech. Kucera and Francis (1967) provide a frequency listing of over 50,000 present-day American English words.

The length of the word also influences its intelligibility; the longer the word, the more readily it is correctly identified. The listener is able to identify a long word by hearing portions of it, particularly a familiar, highly probable word, whereas missing one syllable of a short word is more likely to prevent the identification of the entire word. Both familiarity and length factors can change by 10 to 15 dB the signal-to-noise ratio required for a given level of intelligibility (Howes, 1957).

Word Context

Another factor influencing the intelligibility of speech in noise is the context in which the words are heard. For example, a word is harder to understand if it is heard in isolation than if it is heard in a sentence. Listeners can identify words contained in meaningful messages at signal-to-noise ratios that would be unacceptable if the message consisted of an equal number of unrelated words. A temporary employee from Texas once asked for permission to fly home early from New York to attend a "fire." When asked to provide more information, she replied, "Fire, fire, fire—don't you like to go to fires?" Further questioning determined that she was not a pyromaniac who enjoyed watching buildings burn, but that she wanted to attend the Texas State *Fair.* Whenever possible, designers should provide for having words always contained in a meaningful context.

Speech Sound Confusions

Situations may arise with a noise level so great that certain speech sounds may be confused. When individual characters are to be communicated verbally in noisy environments, a designer may want to have users use the international word-spelling alphabet shown in Table 3-5.

Table 3-5 International Word-spelling Alphabet

A	Alpha	N	November
B	Bravo	O	Oscar
C	Charlie	P	Papa
D	Delta	Q	Quebec
E	Echo	R	Romeo
F	Foxtrot	S	Sierra
G	Golf	T	Tango
H	Hotel	U	Uniform
I	India	V	Victor
J	Juliet	W	Whiskey
K	Kilo	X	X-Ray
L	Lima	Y	Yankee
M	Mike	Z	Zulu

Auditory Defects

Frequently, designers erroneously assume that all people working in a new system will have adequate hearing and, because of this, place much weight on spoken communication and other auditory signals. Actually, hearing impairments are relatively common, and the chances of a user having difficulty in hearing are fairly high. Designers should take into account the basic abilities of *all* potential users.

Two kinds of deafness may figure in human performance problems. *Conduction deafness* involves roughly the same hearing loss at all frequencies; that is, the person suffering from it has about the same difficulty in hearing at one frequency as at another. The term *conduction deafness* is used because the deafness originates in deficiencies of mechanical conduction in the ear. The ear may be stopped up, the eardrum may be broken, or organs of the middle ear may be damaged. The effect of conduction deafness is much the same as putting earplugs in one's ear.

The second kind of hearing loss is *nerve deafness.* As the name suggests, in this type of deafness something is wrong with the auditory nervous system. Either the nerves themselves have been damaged or damage has been done in the inner ear. It is characteristic of nerve deafness that hearing loss is much greater at higher frequencies. This means that the nerve-deaf person can hear low-pitched sounds reasonably well, but hears high-pitched sounds rather poorly or not at all. Such a person has a great deal of trouble understanding speech because some of the higher frequencies are very important in speech comprehension. They are unable to distinguish easily between word sounds. The curves for each type of deafness are shown in Figure 3-12.

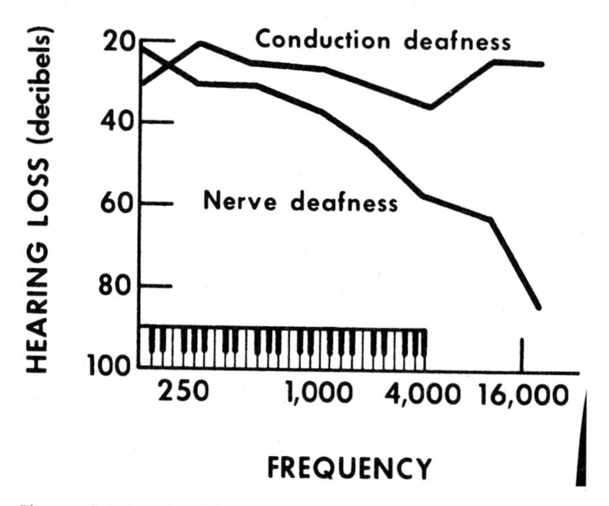

Figure 3-12 Graphic comparison of hearing loss due to conduction versus nerve deafness.

People with hearing impairments should be accommodated as users in a system by making use of their other senses. On the other hand, the *hard of hearing* usually have sufficient hearing to benefit from most of the regularly used decisions for enhancing human performance. This assumes that those people who are hard of hearing are equipped with the necessary amplification devices, such as hearing aids. If they are not so equipped or if they have adjusted the volume to reduce the noise level around them, then the hard of hearing represent a user group that requires special attention. For example, auditory alarms as the sole source of alerting people will not work well for some people who are hard of hearing and are of no use for the deaf. A designer should build in redundant means of alerting people.

A hearing problem that receives very little mention is the phenomenon of "ringing in the ears," more properly known as *tinnitus.* For most people, tinnitus is a transitory event, sometimes triggered by an understimulation or absence of sound in the environment. Tinnitus is also known to be a complaint of individuals under emotional stress and disappears when those circumstances change. For a smaller number of people, tinnitus is a permanent problem resulting from diseases of the ear or from head injury. Individuals suffering from tinnitus must overcome noise generated by their own sense organ in order to accurately perceive stimuli.

One of the most difficult problems for a designer is temporarily impaired hearing. Being exposed to loud noises for extended periods of time, whether from aircraft noise or loud music, can temporarily impair a person's hearing. A much more subtle effect on hearing comes from the use of certain drugs. These drugs can produce temporary or even permanent impairment of hearing.

For example, with the discovery of the antibiotic streptomycin came the finding that the drug could cause impairment of hearing, as well as vestibular (balance) disturbances. A large number of other drugs, some of which are very commonly used, such as aspirin, can temporarily or permanently impair hearing. Gignoux et al. (1966) describe a case involving a 16-year-old girl with normal hearing who became totally deaf after taking 40 aspirin tablets in a suicide attempt. She partially recovered her hearing but, for reasons unknown, a loss of 65 to 80 dB persisted in one ear, as compared to a 20-dB loss in the other. Jarvis (1966) reports an example of partial deafness in a patient who took two or three aspirin tablets every two hours for three days after a tooth extraction.

Accommodating Hearing Limitations

During the basic design stage of system development, designers should produce a *user profile* of the people who will perform in the system. When hearing may degrade performance, they should be addressed in the user profile.

Designers should realize that systems designed to operate adequately only with people who have good hearing are destined to exhibit human performance problems. People will continue to expose themselves to loud noises (e.g., rock music) and to take drugs, prescribed or otherwise, both of which are known to affect hearing. Most designers are not in a position to restrict the outside activities or the drug intake of people who will be working in their systems and, consequently, should design a system that functions well even if the hearing of some users is impaired.

Kinesthetic Sense

The kinesthetic sense is probably the third most important sense related to human performance in most systems (behind vision and hearing). It is used when people are positioning, making movements, controlling forces, judging weight, and so on. For people to control actions, they need to know the position of body parts both before and after a movement. The primary source of this information is the kinesthetic or muscle sense. For example, people can close their eyes and still walk, sit down, and get up. Kinesthesis provides information on the position of the limbs, how far they have moved, and the posture of the body as a whole. A unique feature of kinesthesis is that stimulation comes from within the organism itself rather than from the outside world. Kinesthetic stimuli are always present, although most of the time we are not aware of their presence, and sensitivity varies with different parts of the body.

Sacks (1987) describes a woman who totally lost her kinesthetic sense:

> Christina was 27, a self-assured woman who enjoyed hockey and riding. She had two young children, worked as a computer programmer, and had scarcely known a day's ill-

> ness. After an attack of abdominal pain, she was scheduled to have her gall bladder removed. Christina was admitted to the hospital and placed on an antibiotic to reduce the chances of complications. Later in the day, she found herself very unsteady on her feet, with awkward flailing movements, and dropping things from her hands.
>
> As the time for surgery came closer, her symptoms became worse. Standing was impossible unless she looked down at her feet. She could hold nothing in her hands, and they "wandered" unless she kept an eye on them. When she reached for something, or tried to feed herself, her hands would miss, or overshoot wildly. Her face was oddly expressionless and slack, her jaw fell open, even her vocal posture was gone. "Something awful has happened," she mouthed, in a ghostly flat voice. "I can't feel my body. I feel disembodied."
>
> Tests indicated that she had an almost total kinesthetic deficit, going from the tips of her toes to her head. The sensory receptors in the muscles, tendons or joints were not working at all. She observed that, "I 'lose' my arms. I think they are in one place, and I find they're in another. It is like my body is blind. My body cannot see itself."
>
> There was no neurological recovery a week, or even years later. Christina learned to use her other senses (primarily vision) to provide feedback about her body. She was eventually able to walk, talk and return to work using her computer. (pp. 45–50)

One main function, perhaps the most important function, of the kinesthetic sense is to enable people to control their voluntary muscular activities without the aid of vision. When the eyes are used in executing positioning movements (e.g., shifting gears in a car and keying on a typewriter), kinesthetic cues are not used too much.

However, this type of visual positioning requires more time to accomplish than when more reliance is placed on the kinesthetic sense (i.e., "feeling" where to shift the gears or which key to depress on a keyboard). While most of us think of music as an auditory phenomenon, the ability of pianists to memorize and reproduce passages of music has much more to do with the kinesthetic sense.

When considering the large number of movements made daily by people performing different activities, the kinesthetic sense may be as important to successful performance as the visual or auditory senses. Design decisions should be made that expedite the shifting of dependence from visual cues to kinesthetic cues as quickly as possible when new skills are being learned. In addition, designers need to develop activities that provide clear and unambiguous opportunities for the kinesthetic sense to function properly.

Cutaneous Senses

The surface of the body can be classified as a sense organ in the same way as the eyes or ears. Because the skin covers the entire surface of the body, it provides widespread contact with the immediate environment. The skin's several kinds of nerve endings respond to mechanical, thermal, electrical, and chemical stimuli. The sensations produced by these different stimuli take the form of pressure, pain, cold, and warmth.

Touch (Pressure)

Touch or pressure is experienced when a depression is formed on the skin by some mechanical stimulus. The skin sensor indicates that an object is touching the body and, within certain limits, where it is touching, what size and what shape it is, and whether it is moving, still, or vibrating.

The threshold for pressure varies with the area of the skin and depends to a large measure on the concentration of nerve fibers and skin thickness at each location. Among the most sensitive areas of the body are the tip of the tongue and the fingertips; among the least sensitive are the kneecaps.

The great sensitivity of certain areas of the skin, including the fingers and hands, has enabled designers to develop numerous different types of controls that can be identified by touch. In many aircraft, for example, the control used for lowering the wheels when landing is in the form of a small wheel. The use of touch can also provide some advantage when the visual and auditory channels are overloaded. The touch sense is able to receive and respond to stimuli every bit as quickly as the auditory sense and in many cases faster than the visual sense. In high-noise areas or in cases where visual and auditory detection may be impaired (for example, in early stages of hypoxia), touch warning signals appear to offer significant advantages.

The possibility of using the touch sense on areas of the body other than fingers and hands has been considered as a way of enhancing human performance in systems. There are now available a variety of systems designed to provide optical information to skin areas using a matrix of artificial photoreceptors and a corresponding matrix of skin stimulators. These aid the blind with reading and recognition of pictorial information or improve mobility in their surroundings.

Geldard (1960) developed a touch-communication technique by which subjects were able to learn a touch code for letters and numbers. He used vibrators located at different positions on the chest to send signals consisting of changes in duration and amplitude of vibrations. His best subject was able to learn the touch code and receive communications at the rate of 38 words per minute.

Linvill designed a reading aid for the blind with the stimulus array on the skin of the fingertip. With an improved model of this device, blind subjects can read at rates of over 60 words per minute (Linvill and Bliss, 1966; Bliss, 1971). Collins and Bach-y-Rita (1973) have developed a technique for transmitting pictorial information via the skin. This scanning technique uses both mechanical and electrical stimulation of the skin. It uses a camera with a zoom lens controlled by a subject. The person aims the camera in any desired direction, and the video signal is converted to a touch stimulus that is picked up by touch receptors on the back or the trunk.

Perhaps the greatest disadvantage of using the touch sense is that pressure sensations remain only as long as the rate of movement into or out from the skin continues. With constant pressure and no movement, sensation soon ceases.

Pain

Pain can motivate and deteriorate human performance almost more than any other stimulus. Sudden unexpected pain is a sensation that usually signals immediate action. Prolonged or chronic pain is an entirely different phenomenon, with potential profound negative effects on performance.

People's sensitivity to pain has received little recognition by system designers. Although overstimulation of virtually any sensor will produce the sensation of pain, its use as a warning or in other ways in systems has been minimal. Pain can occur with either mechanical, thermal, electrical, or chemical stimuli. Receptors are spread through almost all tissues of the body. There are actually three kinds of pain: superficial or cutaneous pain; deep pain from muscles, tendons, and joints; and visceral pain. The threshold for pain varies with the place stimulated. Probably the most sensitive place of the body is the cornea of the eye, and the area that is least sensitive to pain is the sole of the foot.

People have a considerable ability for adapting to pain. If, for example, a needle is held to the skin with a steady, unvarying force, the pain aroused gradually loses its intensity and eventually disappears altogether, leaving a sensation of pressure, which then gradually declines and also disappears. As with touch, if pain is used in a system, the effect of the pain must be capitalized on quickly. Unfortunately, people react relatively slowly to pain stimuli. Subjects in one study, for example, were able to respond to visual stimuli on the average of about 200 ms, whereas they required about 700 ms to respond to pain stimuli.

Pain may be aroused by stimuli such as radiant heat or cold that are not in direct contact with the skin. The sensation of heat or cold is aroused first, then a sensation of pain, followed by adaptation and a return of the sensation of heat or cold. This confusion may lead to ambiguous sets of responses to the stimuli, which could lead to unpredictable and degraded performance.

Temperature

The skin also has receptors that are sensitive to temperature changes. Thermal receptors are stimulated by raising and lowering the temperature of the skin. Normal *skin* temperature is about 91.4°F (32° or 33°C), and stimuli at this temperature elicit no thermal sensation.

Note that the normal skin temperature of 91.4°F is considerably below the average deep body temperature of 98.6°F (37°C). At any one time, the different regions of the skin can be at different temperatures. And although different regions of the skin may be at different temperatures, adaptation can exist simultaneously at all regions. As shown in Figure 3-13, thermal adaptation will only occur when the skin temperature is between about 60°F (16°C) and 105°F (40°C). Below this range, the skin will transmit a cold feeling and, above it, a feeling of warmth.

Again, we have the problem of relatively quick adaptation to temperature changes. This suggests that the temperature receptors may be used as sources of information in a system, but it would require a relatively quick interpretation because the receptors tend to adapt so quickly.

Taste

A person can normally identify four distinct qualities of taste: sweet, sour, salty, and bitter. All taste sensations are combinations of these four primary taste qualities. The most sensitive to stimuli is the bitter taste sense.

Actually, in many cases taste is the result of the interaction of several senses. Pepper tastes as it does because it stimulates both taste and pain receptors. The sense of smell is

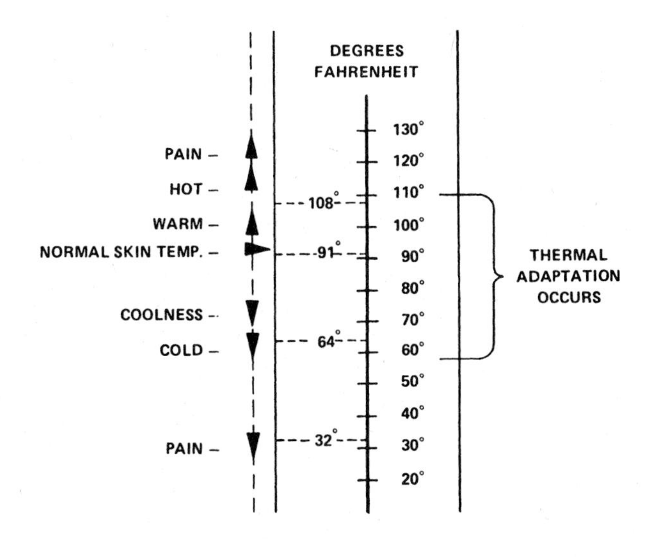

Figure 3-13 Effects of temperature on skin receptors.

particularly important in determining what we taste. Food can be almost tasteless to someone suffering from a head cold mainly because of a loss of the sense of smell. Thus, much of what we call taste is actually smell. In fact, odors from food may pass upward into the nasal area, stimulating the olfactory system as much as thousands of times as strongly as the taste system. The taste of food also may be affected by its temperature and its appearance.

Reduced taste sensitivity is associated with the atrophy and loss of taste buds from the tongue as people approach the age of 70 and beyond. A 1990 study of 10 of Chicago's top chefs found that 70 percent of them had a reduced sense of taste (30 percent had a reduced sense of smell). Sensitivity may be affected by temperature, internal factors such as a salt deficiency, adaptation, and individual differences. In addition, many people are taste-blind for certain substances. A substance used frequently by psychologists for demonstrating taste blindness is phenylthiocarbamide, for which 15 to 30 percent of the subjects exhibit taste blindness.

This leads us to conclude that there would be great difficulty in using taste as a source of information in many, if not most, systems. However, a creative designer may find ways of capitalizing on the unique characteristics of the taste sense for providing information to help achieve an adequate level of human performance. In one system, a telephone company employee touched the end of wires to his tongue. If he tasted salt, he rightly concluded that the wire was live.

Smell

The sense of smell is activated when an odorous stimulus, in the form of vaporized chemical substances, comes into contact with specialized areas in the upper region of each nostril. Some smells may also involve mechanical and chemical stimuli.

The primary qualities of smell are not as well known or understood as those of vision, hearing, or taste. There are probably at least six qualities of odor: spicy, fruity, burnt, resinous, flowery, and putrid, although there may be as many as 50 or more primary sensations of smell. Smell stimuli are almost always mixtures of basic smells, resulting in complex odors.

From a designer's point of view, one of the most important characteristics of smell is the minute quantity of the stimulating agent required to effect a smell sensation. It is estimated that smell is 10,000 times as sensitive as taste. For instance, the substance methyl mercaptan can be smelled when only 1/25,000,000,000 mg is present in each milliliter of air. Because of this low threshold, this substance is mixed with natural gas to give the gas an odor that can be detected when it leaks from a gas pipe. Similar odors are also used because they smell like decaying flesh. These odors are able to attract scavenger birds. Workers can more easily find leaks by watching for birds circling around certain areas of a pipeline.

In some systems it may be useful to use odor as a warning, rather than bells or flashing lights. For example, most of us have been involved in fire drills while at school and have noted the lack of urgency in these drills as people left the building. One wonders if the drills are actually helpful since they represent such a low-fidelity simulation. If smoke and flames were actually part of the drill (making a higher-fidelity situation), would the people perform in the same manner? It is possible that they would not. Using a noxious odor that is poured into the air rather than a bell that rings would greatly improve the motivation of the people to leave the building. This would make the fire drills more like an actual situation where a fire exists. Thus, noxious odors could be used in situations where people are expected to leave an area quickly, as tear gas is already used. In these cases the sense of smell would become every bit as important as sight or hearing and would be very useful to help to achieve a particular system objective.

One main problem with using the smell sense is that smell discrimination is influenced by other odors in the environment. Also, the existence of *odor blindness* in many people should be recognized by designers who may want to use certain odors as warning stimuli in a system. The simple blocking of the nasal passages with the common cold can eliminate the ability to smell. Finally, like many other senses, receptors in the nose adapt very quickly and cease to respond to stimuli. Thus, even a strong odor will gradually become imperceptible.

There are some substances that elicit differences in the sense of smell between men and women. The most extraordinary compound tested is exaltolide. Most men and most children have difficulty smelling this substance, but most women find it very strong and the strength seems to vary with the menstrual cycle and in pregnancy (Kling and Riggs, 1971).

Vestibular

Located in the inner ear, this sense provides people with information about their position in three-dimensional space and their movement through space. The vestibular receptors sense body posture and positions in relation to the vertical and act something like a gyroscope that helps people stay reasonably upright. They are also sensitive to acceleration and deceleration. This sense is closely related to motion sickness.

Sensory Adaptation

All senses adapt, but, as we have seen, some adapt much more quickly than others. The sensitivity of each sensor is modified by the continuous presentation of stimuli to that receptor or adaptation. With adaptation the receptors become less effective. Few people realize that there is one process of adaptation that *heightens* the performance of the receptors. When the eye adapts to dark places, certain visual receptors become *more sensitive* during the course of adaptation.

Sensory Interaction

The senses have been discussed as though they function independently in the information-gathering process. Normally, the senses combine to produce an integrated experience. When eating, for example, the sight of food arranged on the table, the conversational tones and background music, and the tactile sensations, aromas, and taste of the food all combine to enhance the experience. So when considering the adequacy of human senses, designers need to be aware that they are not used individually but in combination.

The Body and Performance

Introduction

In designing systems for optimal human performance, we need to understand not only the senses but also the capabilities of the body that enable individuals to respond (see Figure 3-14). This understanding helps a designer to envision in greater detail the potential users of a system.

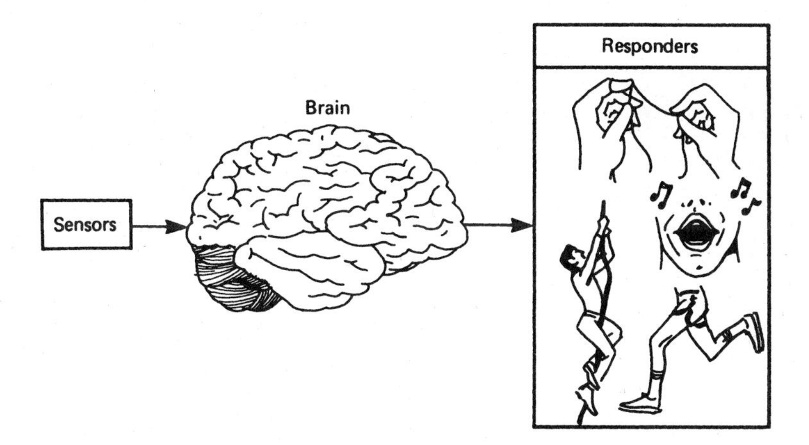

Figure 3-14 Simplified model of human information processing with the responses emphasized.

Body Dimensions

A designer must consider the body's relationship to performance in at least two ways. First, certain body dimensions seem to be related to optimal performance in some activities. It seems that the taller professional basketball players are, the more success they have on the court. Second, body dimensions must be considered in designing work areas so that a user will fit into the space provided and will be able to make the required *reaches* or *movements*. An automobile, for example, should comfortably accommodate larger drivers, yet have the controls readily accessible to smaller drivers.

Because they vary greatly, human body dimensions can pose many problems for designers. Fortunately, all people fall within certain limits. If these extremes are defined, they can be considered in any design for human performance. For example, the smallest group of people on record are the Pygmy people of central Africa. Gusinde (1948) found a mean stature of 4'8" (143.8 cm) for adult males and 4'6" (137.2 cm) for females. Conversely, the tallest group of people are the Nilotes of the Sudan, where the mean was 5'11" (182.9 cm) for males and 5'6" (168.9 cm) for females (Roberts, 1975). The tallest human reliably recorded was Robert Wadlow (1918–1940), 8'11", while the shortest was Pauline Musters (1876–1895), at 1'11" tall. This range, from about 2 feet to 9 feet (61 cm to 274 cm), establishes the extreme outer limits for human adults.

Outer limits have also been established for other body dimensions. Studies of body dimensions, or *anthropometry,* have been conducted throughout history. In ancient Egypt, the length of the middle finger was considered to be a third of the height of the head and neck and a nineteenth of the whole body. Leonardo da Vinci followed the rule that the head, from crown to chin, is contained eight times in the stature. The systematic study of individual differences in size probably derives from the pioneer work of the Belgian astronomer Quetelet, who published his *Anthropometrie* in 1870. Since that time, much anthropometric literature has been written, particularly in the last few years, and details of body measurements are available for over 10,000 population samples throughout the world. Several such studies were conducted in the 1920s and 1930s, but it was not until World War II that well-controlled studies of body measurements were begun.

Based on the findings of these studies, we now have information concerning many body dimensions. One interesting fact that comes from these results is that there seems to be a worldwide increase in height. American soldiers of World War II, for example, were an average of 0.07 inch (2 cm) taller than those of World War I (Davenport and Love, 1921; Newman and White, 1951). Dreyfuss (1967) estimates that the average height of men in the United States from 1900 to 2000 could increase as much as 3 inches (7.6 cm).

Full growth in males occurs at about age 20 and in females at about age 17. Around age 60, height begins to decrease. Old age, however, is not the only time humans lose height. In fact, although temporary, the height of an individual usually decreases as the day progresses because of the compression of the intervertebral discs. Body heights are greatest immediately after rising and least before retiring. The difference averages about 0.95 inch (2.5 cm) in adult males (Backman, 1924).

The design-related problems associated with the variations of body dimensions in different populations have begun to emerge as a result of the increase in world trade and the attempt to establish worldwide standards. International studies have focused on the extent of differences in body dimensions and their importance for design, particularly in systems that will have international use.

Average Person Fallacy

Designing for the body dimensions of the so-called *average person* is usually a mistake. The result of such a design is that the smaller 50 percent of the users may be unable to reach the controls comfortably or read the displays, and the larger 50 percent will not have sufficient room to move about comfortably. No one is average in all dimensions, and few people are average in even a few dimensions.

To test the concept of the average person, Daniels and Churchill (1952) categorized 4063 men according to 10 measurements used in clothing design. For each of the 10 dimensions, the men making up about the middle 30 percent were considered average. The results are as follows:

Of the original 4063 men, 1055 were of approximately average stature.

Of these 1055 men, 302 were of approximately average chest circumference.

Of these 302 men, 143 were of approximately average sleeve length.

Of these 143 men, 73 were of approximately average crotch height.

Of these 73 men, 28 were of approximately average torso circumference.

Of these 28 men, 12 were of approximately average hip circumference.

Of these 12 men, 6 were of approximately average neck circumference.

Of these 6 men, 3 were of approximately average waist circumference.

Of these 3 men, 2 were of approximately average thigh circumference.

Of these 2 men, 0 were of approximately average inseam length.

Of the original 4063 men, not one was average in all 10 dimensions, and less than 4 percent were average in even the first three. The average person simply does not exist.

Anthropometric Measures

Instead of using an average person as a design model, designers are encouraged to use tables of anthropometric measurements. These measurements provide information on body dimensions that are of value in a wide range of design situations. Good design practice uses this information to accommodate at least 95 percent of the potential users in a system. Designing for 95 percent of the potential user population eliminates only a few people and at the same time avoids costly problems encountered in dealing with extreme dimensions.

For example, in decisions associated with the distance a person can reach when secured by a seat belt, a designer should use the dimensions of a person at the 5th percentile. This would mean that the headlight switch would be easily reached by 95 percent of the people who would drive the automobile. On the other hand, decisions associated with head, shoulder, and leg clearance should be based on the height and other dimensions of the largest users (i.e., those at the 95th percentile). Percentile values are needed for each critical dimension. Thus, the designer must first determine which dimensions directly relate to human performance in a particular activity.

Static and Functional Measures

Two basic kinds of measurements are of value to a designer. *Static* measurements are those taken when an individual is in a nonmoving position, for example, standing back against a wall for measuring height. *Functional* measurements are those taken with people assuming common movement positions in which the position itself helps to determine the dimension of the body parts involved. Functional measurements are used, for example, for determining leg clearance at a desk or reach from a seated position.

When dealing with either static or functional dimensions, the type and amount of clothing that a person is wearing, including the height of the shoes, headgear (e.g., helmet), and the type of equipment that an individual might be carrying or wearing (e.g., a parachute, weedsprayer, or gun), must be taken into account.

Human Dimensions

The measurements given in the following figures and tables are for adult workers. This set of dimensions was selected from among hundreds of different possibilities because it tends to be the most useful in the design of computer systems. Keep in mind that a large number of other measurements are available for specialized applications (cf. Garrett and Kennedy, 1971).

These measurements were compiled from numerous sources (see Israelski, 1977). To help make them more useful, male and female data that were originally reported separately were combined and new percentile values calculated. In addition, the dimensions were evaluated against historical data and, where appropriate, increased in size to ensure that they truly represented the larger users of the late 1980s and 1990s. One good example of the need for this type of adjustment was mentioned earlier in this chapter, when it was pointed out that stature is increasing at the rate of about 0.3 inch (8 mm) per decade.

Basically, there are three types of measurements: (1) those that help in making decisions concerning *reach,* (2) those that are useful in making decisions concerning *clearance,* and (3) those for making decisions where *ranges* of dimensions are important. In all three cases, the design goal is to accommodate 95 percent of the system users. This is done in slightly different ways depending on which type of measurement is being used.

The reach dimensions focus on the smaller users of a system. To ensure that at least 95 percent of all users can reach a certain height (bookshelf), easily reach a set of controls, or are tall enough to reach or see over a cubicle wall, the 5th percentile measurements are typically used. This is illustrated graphically by the bell-shaped curve in Figure 3-15. The few people that have measurements falling below the 5th percentile will have more difficulty making reaches in these systems, but the other 95 percent should have no trouble at all.

The clearance dimensions help to ensure that workplaces are large enough to easily accommodate 95 percent of their users. In this case, the focus is on the larger measurements, those at the 95th percentile. A bell-shaped curve is illustrated in Figure 3-16. Note that the tail of interest is now on the other side of the curve. If people with the larger measurements are adequately dealt with, then those with smaller measurements are automatically accommodated. For example, if sufficient clearance is left under a visual display terminal (VDT) for a person with long legs (95th percentile), then those with shorter legs will also be comfortable.

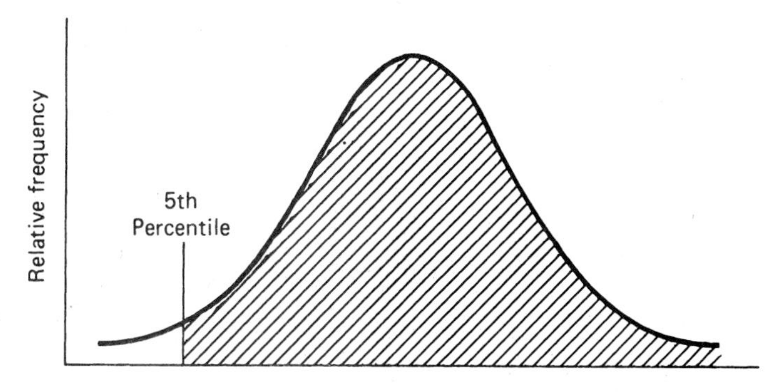

Figure 3-15 Accommodating 95 percent of users by focusing on the 5th percentile reach measurements.

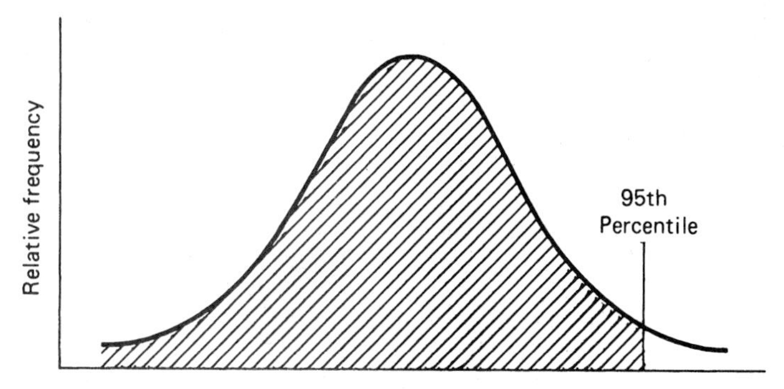

Figure 3-16 Accommodating 95 percent of users by focusing on the 95th percentile clearance measurements.

Finally, several measurements are most useful only when considering a full range of values. As shown in Figure 3-17, in this case we are most interested in accommodating the *middle* 95 percent. One of the most commonly used examples of this is the adjustable chair. When purchasing furniture for a group of people of varying sizes, it is usually a

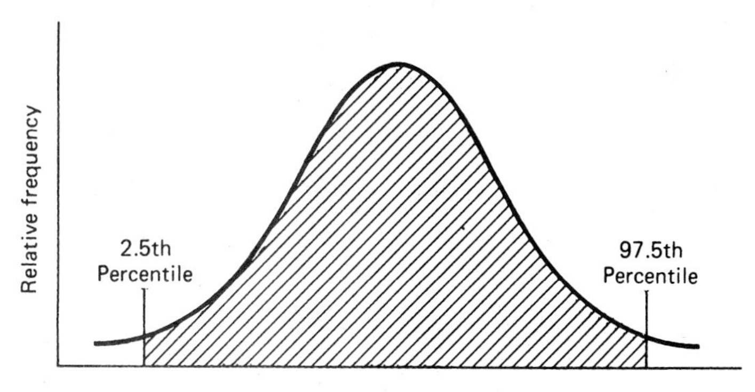

Figure 3-17 Accommodating the middle 95 percent of users.

good idea to find chairs that can easily adjust in height from 3 to 4 inches (8 to 10 cm). This will accommodate 95 percent of the users by considering the full range of measurements from the 2.5th percentile to the 97.5th percentile.

Anthropometric Measurements

Table 3-6 shows common anthropometric measurements used in office systems. Keep in mind that the measurements in Table 3-6 and Figure 3-18 do not take into account

Table 3-6 Body Dimensions for Office Workers in Inches (cm). To Be Used with Figure 3-18

		Estimated percentiles				
Legend	**Dimensions**	**2.5th**	**5th**	**50th**	**95th**	**97.5**
A–1	Vertical reach	73.5 (187)	75.0 (190)	80.5 (204)	86.0 (219)	87.5 (222)
A–2	Stature	61.5 (156)	62.0 (158)	66.5 (168)	70.5 (179)	71.5 (181)
A–3	Eye height, standing	56.5 (144)	57.5 (146)	62.0 (157)	66.0 (167)	67.0 (169)
A–4	Head circumference	21.5 (54)	22.0 (55)	22.5 (57)	23.0 (58)	23.5 (60)
B–1	Thumb tip reach	27.5 (69)	28.0 (71)	31.0 (78)	33.5 (86)	34.5 (87)
B–2	Shoulder height	50.0 (128)	51.0 (130)	55.0 (139)	59.0 (149)	59.5 (151)
B–3	Elbow to floor	39.5 (100)	40.0 (101)	42.5 (108)	45.0 (115)	45,5 (116)
B–4	Foot length	9.0 (23)	9.5 (24)	10.0 (25)	10.5 (27)	11.0 (28)
B–5	Foot width	3.0 (8)	3.0 (8)	3.5 (9)	4.0 (10)	4.0 (10)
C–1	Head to seat height	32.5 (83)	33.0 (84)	35.5 (90)	37.5 (96)	38.0 (97)
C–2	Eye height, sitting	28.0 (71)	28.5 (72)	30.5 (78)	32.5 (83)	33.0 (84)
C–3	Shoulder breadth	16.0 (40)	16.0 (41)	18.0 (46)	19.5 (50)	20 (51)
C–4	Hip breadth	13.0 (33)	13.5 (34)	15.0 (38)	17.0 (42)	17.0 (43)
D–1	Hand length	6.5 (17)	6.5 (17)	7.5 (19)	8.0 (21)	8.0 (21)
D–2	Hand width	3.0 (7)	3.0 (7)	3.5 (8)	3.5 (9)	4.0 (9)
E–1	Knee height	19.0 (48)	19.0 (49)	21.0 (53)	23.0 (58)	23.0 (59)
E–2	Popliteal height	15.0 (38)	15.0 (39)	17.0 (42)	18.0 (46)	18.5 (47)
E–3	Buttock to popliteal	17.5 (45)	18.0 (45)	19.5 (49)	21.0 (54)	21.5 (55)
E–4	Buttock to knee	21.5 (55)	22.0 (56)	23.5 (60)	25.0 (64)	25.5 (65)
E–5	Elbow to wrist	9.0 (23)	9.0 (23)	10.5 (26)	11.5 (29)	12.0 (30)
E–6	Thigh clearance	5.0 (12)	5.0 (13)	6.0 (15)	6.5 (17)	7.0 (18)
E–7	Shoulder to elbow	12.5 (31)	12.5 (32)	14.0 (35)	15.0 (38)	15.0 (39)
E–8	Elbow rest height	8.0 (19)	8.0 (20)	9.5 (24)	11.0 (28)	11.5 (29)
E–9	Shoulder to seat height	22.0 (56)	22.5 (57)	24.5 (62)	26.5 (67)	27.0 (68)

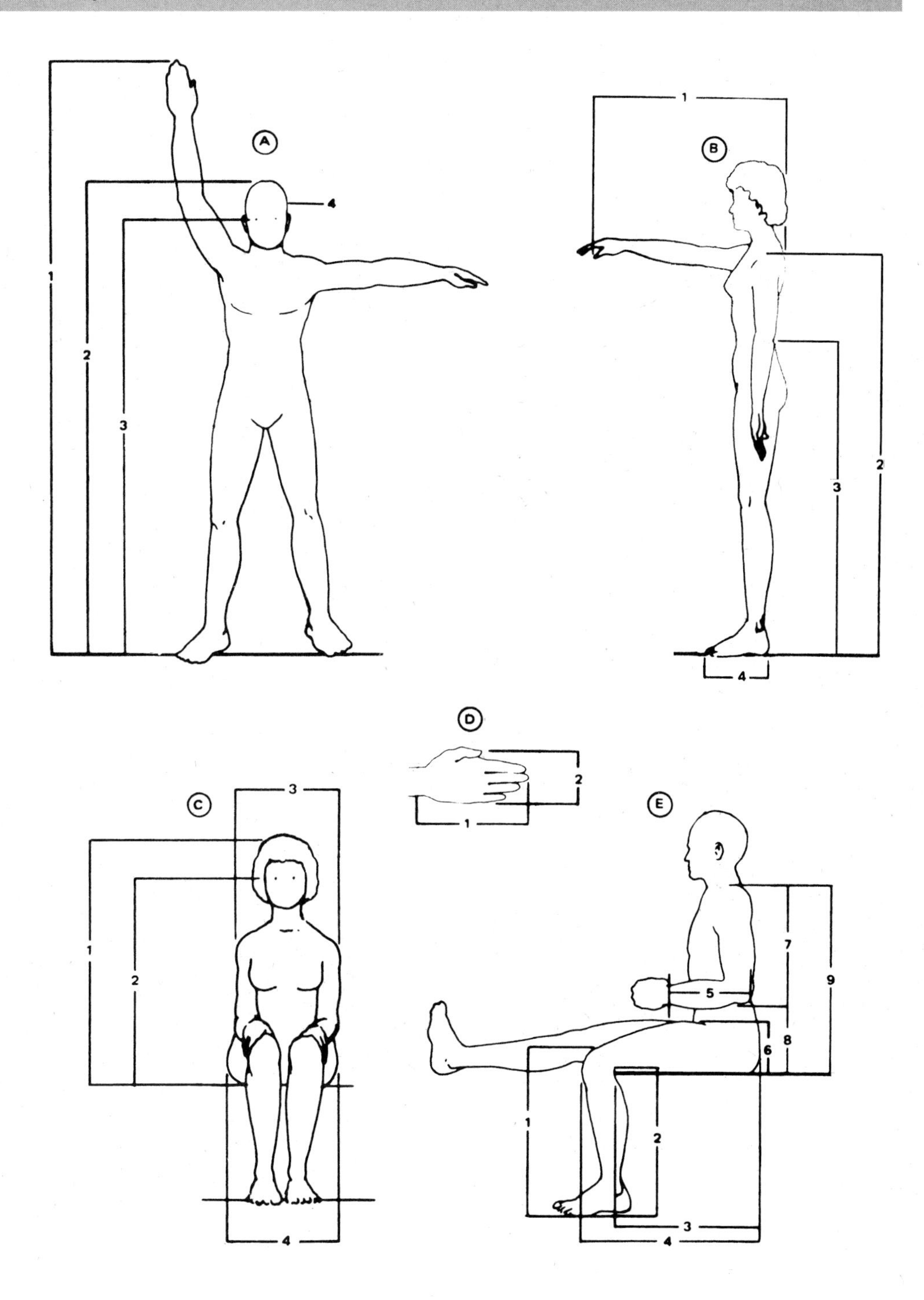

Figure 3-18 Pictures showing locations of measurements (adapted from Woodson and Conover, 1964). Use with Table 3-6.

increases for clothing. It is usually a good idea to add at least 1 inch, where appropriate, for shoes. If work is being done outside, where coats and hats are being worn, then this also must be taken into account.

Obviously, certain dimensions relate more closely to specific types of work than to others. The sets of measurements shown in Tables 3-7 and 3-8 (see also Figures 3-19 and 3-20) are most useful for helping to establish the reach characteristics previously discussed. All these measurements represent smaller 5th percentile users. Consider, for example, the measurement shown for vertical reach in Table 3-7. If users with the shortest (5th percentile) arms can reach the distance shown in the table, then the 95 percent of users that are larger can reach that far and farther. Thus, about 95 percent of users can reach the distance shown in Tables 3-7 and 3-8.

Table 3-7 Useful Body Dimensions for Reach (Accommodating 95% of Users)

Dimensions	5th Percentile	Example of Use
Vertical reach (A–1)*	75 in. 190 cm.	Placement of highest book shelf
Eye height, standing (A–3)	57.5 in. 146 cm.	Allowing most workers to look over the side of their cubicles
Thumb tip reach (B–1)	28.0 in. 71 cm.	Maximum distance for placing CRT controls

*See Table 3-6.

Table 3-9 presents several clearance-related body dimensions, one from each of the major groups shown in Table 3-6. Examples of how these dimensions can be used are also shown. In these cases, the 95th percentile values are most appropriate.

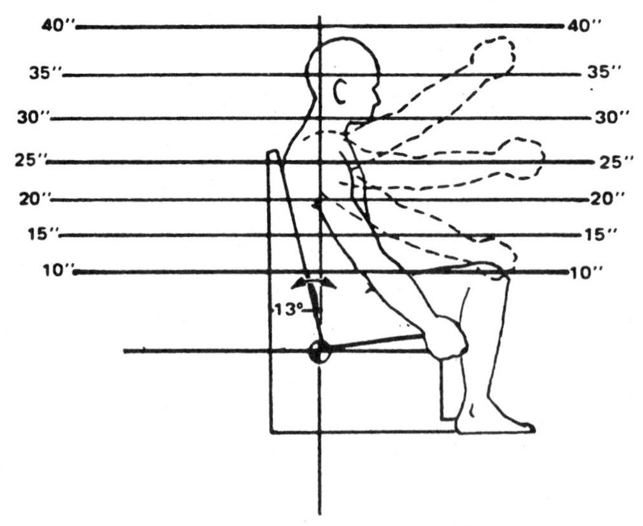

Figure 3-19 Illustration of height above seat surface (use with Table 3-8).

Table 3-8 Useful Dimensions Showing Portions of a Reach Envelope To Be Used with Figures 3-19 and 3-20

Desk Top: Angle to Left or Right	Height above Seat Surface Using Right Arm/Hand in Inches (cm)						
	10	15	20	25	30	35	40
L 120						9.5 (25)	10.0 (26)
L 105						11.0 (28)	10.5 (27)
L 90						12.5 (31)	11.0 (28)
L 75						13.5 (34)	11.5 (29)
L 60			16.0 (40)	16.5 (42)	15.5 (39)	14.5 (37)	12.0 (30)
L 45		17.0 (43)	17.5 (45)	18.0 (46)	17.0 (43)	15.5 (39)	12.5 (32)
L 30		19.5 (50)	19.5 (49)	20.5 (51)	19.5 (49)	17.5 (44)	14.0 (35)
L 15		21.0 (53)	21.0 (54)	21.5 (55)	21.5 (54)	19.0 (48)	15.0 (39)
0		22.5 (57)	23.0 (58)	23.5 (60)	23.0 (58)	20.0 (51)	17.0 (43)
R 15		24.0 (61)	25.0 (64)	25.5 (65)	24.5 (62)	22.5 (57)	19.0 (48)
R 30	24.5 (62)	25.5 (65)	27.0 (69)	27.0 (69)	26.0 (66)	24.0 (61)	20.5 (52)
R 45	25.5 (65)	27.0 (69)	28.0 (71)	28.0 (71)	27.0 (69)	25.5 (65)	22.0 (57)
R 60	26.0 (66)	28.0 (71)	29.0 (73)	28.5 (72)	28.0 (71)	26.0 (66)	23.0 (58)
R 90	26.5 (67)	28.0 (71)	29.0 (74)	29.0 (74)	28.0 (71)	27.0 (68)	23.5 (60)
R 105	26.0 (66)	27.5 (70)	28.5 (73)	28.5 (72)	28.0 (71)	27.0 (68)	24.0 (61)
R 120	25.0 (63)	26.5 (67)	27.5 (70)	27.5 (70)	27.0 (69)	26.0 (66)	23.5 (60)
R 135	23.5 (60)						

Some frequently used body dimension ranges are shown in Table 3-10. These dimensions are used to help ensure that the majority of users are accommodated in situations where the use of range information is most appropriate. Note that, in order to accommodate 95 percent of people, designers need to use the 2.5 and 97.5 percentile values.

Calculating Percentile Ranks

In any distribution, the percentile rank of any specific value is the percent of cases out of the total that falls *at* or *below* the specific value. For example, out of a group of 345 people, we wish to find the percentile rank that 115 fall at or below. This would be figured as follows: (100)(115/345) = (100)(0.33) = 33. Thus, the percentile rank is 33, or the 115th score is at the 33rd percentile.

The maximum score must have a percentile rank of 100, and the minimum score must have a percentile rank of $100(1/N)$, where N equals the total number of people in the group under study. The percentile rank of the median must be 50, since 50 percent of all observed values in a frequency distribution must lie at or below the median.

Table 3-9 Useful Body Dimensions for Clearances (Accommodating 95% of Users)

Dimension	95th Percentile	Example of Use
Stature (A–2)	70.5 in. (5' 11") 181 cm	Eliminate overhead obstructions (without shoes or hat)
Head to seat height (C–1)	37.5 in. 96 cm	Privacy for taller users seated in a cubicle
Foot length (B–4)	10.5 in. 27 cm	Footroom under a typing table (without shoes)
Shoulder breadth (C–3)	19.5 in. 50 cm	Determining adequate room for three passengers in the backseat of a car
Hand width (D–2)	3.5 in. 9 cm	Ensuring a large enough space for reaching in a small area
Thigh clearance (E–6)	6.5 in. 17 cm	Ensuring sufficient room for placing knees under a visual display terminal (VDT)

Performance and Body Dimensions

Efficiently performing certain tasks seems to be easier for persons with certain anthropometric combinations (cf. Roberts, 1975). Numerous studies have been conducted in an effort to associate body measurements with human performance. After reviewing the available literature, Carter (1986) concluded that champion performers of a particular sport exhibit similar patterns of body size, with patterns tending to become narrower as levels of performance increase.

Table 3-10 Useful Body Dimensions for Ranges (Accommodating 95% of Users)

Dimension	Percentiles	Example of Use
Eye height, standing (A–3)	56.5 to 67 in. 144 to 169 cm	Approximate minimal vertical window opening for standing users (without shoes)
Eye height, sitting (C–2)	28 to 33 in. 71 to 84 cm	Approximate minimal vertical window opening for viewing while seated
Popliteal height (E–2)	15 to 18.5 in. 38 to 47 cm	Approximate range of chair adjustments (without shoes)
Elbow rest height (E–8)	8.0 to 11.5 in. 19 to 29 cm	Approximate range of arm rest adjustments
Head circumference (A–4)	21.5 to 23.5 in. 54 to 60 cm	Ensuring full range of sizes for safety hats

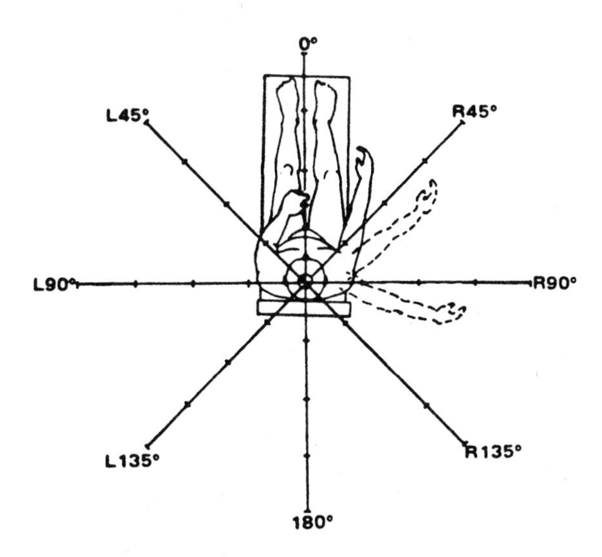

Figure 3-20 Illustration of angle to left or right (use with Table 3-8) (adapted from Kennedy, 1964).

Differences in performance levels have also been observed between male and female athletes. For example, in most sporting events the top man will outperform the top woman. This disparity in performance, however, does not seem to be a result of gender-related *sensory* or *cognitive* abilities. Rather, it is because their bodies are different—men are usually stronger than women.

As the performance of an individual in a particular activity becomes more cognitive, that is, depends more on decision making and reasoning and less on movement, the importance of anthropometric measurements diminishes (as do the differences between male and female). Studies suggest that the speed with which an untrained individual can voluntarily react to a visual stimulus has little relationship to body size (cf. Pierson, 1961). For these activities, the speed of performance seems to be more closely related to cognitive processing than to the size of body parts.

Personalizing Work Areas

Personalizing work areas for users helps to promote physical health and increased comfort. A carefully planned work environment may actually increase productivity.

Chairs

Providing chairs that are easily adjustable in height is a good place to start. They should be comfortable and provide firm support to the lower back (lumbar region). Users should adjust the chair so that their forearms form approximate right angles with their upper arms. Their feet should rest flat on the floor, or they should use a footrest that is high enough so that their thighs are parallel to the floor when seated.

Desks

If at all possible, place the keyboard and monitor on a desk designed for a computer. Most traditional writing desks are too high for comfortable computer use over a long period of time. Having at least 1 inch of clearance between the top of the user's thighs and the bottom of the desk is also important.

Displays

Place the display screen directly in front of users, and locate the monitor at a comfortable viewing distance. When users are sitting in their chairs, the top of the display should be no higher than eye level. Make sure that users do not see glare on the screen (antiglare filters help), the keyboard, or the mouse. Also try to encourage users to keep their screens clean and dust free.

In some work assignments, it can be important to frequently look away from the visual display. Some professionals believe that it is a good idea to focus on an object about 20 feet away several times each hour.

Keyboard and Mouse

Position the keyboard directly in front of users. When users are typing, encourage them to keep their shoulders relaxed, letting their upper arms hang freely. Allow enough room on a desk for easy access to the keyboard and unhindered movement of the mouse. The forearms should be nearly parallel to the floor when typing. The mouse should be positioned at the same height as the keyboard. If a mouse pad is used, make sure it is not so thick that it substantially raises the height of the mouse. Users should normally hold the mouse using their preferred hand (for most people this is the right hand). Provide sufficient work space to allow users to use the entire arm to move the mouse around.

The mouse is designed to allow a loose grip when being used. This enables users to relax their hand and helps to minimize excess tension. Also, by periodically varying the way the mouse is held, the same motion is not repeated over a long period of time.

Making Workstation Adjustments

Romero et al. (1993) suggest a way for users to quickly estimate a starting reference point for arranging a computer workstation. To use the formulas, all that is needed is the person's stature (height) measurement. The stature measurement and resulting calculations are assumed to be in inches. For example, to calculate the height of the top of the seat from the floor for a person who is 64 inches (163 cm) tall,

$$0.271 \times 64 \text{ inches} - 1.7 = 15.6 \text{ inches (39.6 cm)}$$

The other formulas are as follows:

Location of the sitting user's eyes from the floor:

$$0.726 \times (\text{stature}) - 2.17 =$$

Horizontal location of the user's eyes from the screen:
$0.412 \times (\text{stature}) - 0.24 =$

Location of the center of the screen from the floor (assumed to be about 10 inches below the person's line of sight):
$0.605 \times (\text{stature}) - 3.89 =$

Location of the top of the (compressed) seat from the floor:
$0.271 \times (\text{stature}) - 1.71 =$

Location of the center of the lumbar support from the top of the (compressed) seat pan:
$0.148 \times (\text{stature}) - 3.73 =$

Location of the lumbar support from the front of the seat pan:
$0.081 \times (\text{stature}) + 10.02 =$

Height of the underside of the work surface:
$0.295 \times (\text{stature}) + 4.21 =$

Knee clearance under the work surface:
$0.331 \times (\text{stature}) - 5.31 =$

Height of the keyboard from the floor:
$0.331 \times (\text{stature}) + 3.14 =$

Breaks and Exercises

A well-designed work area can help users work more comfortably. It also may help to minimize the chances of users experiencing some physical discomfort. Long periods of repetitive motion, coupled with an improper work environment and incorrect work habits, may be related to physical discomfort or even injury. Two of the most commonly reported problems include tendinitis and carpal tunnel syndrome.

To help to increase comfort and reduce health problems, organize the work of users so that they alternate using their computer with other activities. This encourages the use of different muscle groups throughout the day. Getting up and walking around several times a day are also beneficial.

Frequent breaks can play an important part in staying alert and being more comfortable while working. Users should be encouraged to take periodic breaks to rest their eyes, move their bodies, and enhance blood circulation. Doing exercises during the day can be helpful. Some simple exercises include having users gently press their hands against a table, stretch, and hold for 5 seconds. Another is to have users stretch and massage their fingers, hands, wrists, and forearms. Also, they can be encouraged to gently shake their hands and fingers to relieve tension and help to increase blood flow. At least twice a day, they can rotate their shoulders in a full forward circle several times; then roll them backward several times.

Range of Movement

Movement is considered to be a composite, an integration of both structure and function. The *structure* of human movement involves the complex interaction of the skeleton and the muscles. The skeleton supports the human body and provides the system of links and hinges that form body levers. Muscles are the source of power.

There are over 200 bones in the skeleton. Most are connected by joints that permit movement. With the exception of those used in speech, the long bones of the legs and arms and the miniature long bones of the fingers and toes most directly affect human performance. The bones of the joints are held together by ligaments and muscles. The ligaments are inelastic and in general *limit movement* as they tighten at the end of a movement. In joints where movement is restricted, ligaments are found on each side of the joint, remaining taut throughout the entire range of movement. The muscles operating the joint are paired, with one set flexing the joint and the other extending it.

The movable joints of the body are of several different types. The three most important are hinged joints (such as in the finger), pivot joints (such as in the elbow), and ball-and-socket joints (such as in the shoulder and hip). Designers tend to assume that the range of motion for joints is fairly uniform within a user group. In reality, however, ranges of motion vary considerably and are usually determined by the joint's bony configurations; the attached muscles, tendons, and ligaments; and the amount of surrounding tissue. Joint mobility tends to decrease only slightly in *healthy* people between the ages of 20 and 60. The incidence of arthritis, however, increases so markedly beyond age 45 that any older population usually reflects considerably decreased average joint mobility (Morgan et al., 1963).

On the average, women exceed men in the range of movement at all joints but the knee. Slender men and women have the widest range of joint movement, while obese individuals have the smallest. Average and muscular body builds have intermediate ranges. Physical exercise can increase the range of motion of a joint, but excessive exercise can result in a muscle-bound condition that decreases the range of motion. The range of movement of one part of the body is affected by the position or movement of neighboring parts. In addition, movements made while prone are not necessarily the same as those made while standing. Numerous range-of-movement limitations can be found in Damon et al. (1966).

Strength

In some activities, the strength of an individual is as important as body dimensions and range of motion (cf. Kroemer, 1970, 1974, 1975). Strength increases rapidly in the teens, more slowly in the early twenties, reaches a maximum by about the middle to late twenties, remains at this level for 5 to 10 years, and thereafter declines slowly but continuously. By about the age of 40, muscle strength is approximately 90 to 95 percent of the maximum attained in the late twenties; this reduces to about 85 percent by age 50 and about 80 percent by 60.

However, not all muscle strength declines with age at the same rate. For example, hand grip seems to remain relatively strong in later years. Conversely, the strength of the back muscles decreases more rapidly with age than either the hands or arms. As one would expect, body build is related to strength. In general, the larger the muscle, the stronger a person is. Body position is an important factor affecting strength. When large force must be exerted, people usually assume the position in which they can best exert their maximum strength.

Designers should ensure that the maximum resistance for controls and the limits for lifting or carrying are based on the strength of the *weakest* potential user. In addition, resistance levels for controls should not require the application of maximum strength. These resistance levels should be low enough to prevent fatigue or discomfort, but high enough to prevent inadvertent operation of the controls. They should also provide clear and concise kinesthetic cues.

It is interesting that the apparent maximum strength exerted by most people seems to be far less than their potential. In certain situations, one's capability can be astounding. Several years ago, a 123-pound woman was reported to have lifted a 3600-pound station wagon off her son who was trapped beneath it. More recently, news accounts told of a 12-year-old boy who rescued his father in the same manner (Morehouse and Gross, 1977). The enormous strength summoned by such life-and-death emergencies suggests that we normally use only a portion of our actual capacity.

Exercise 3A: Reading Large versus Small Print

Purpose: To determine if there is a performance difference when reading text that is printed in very large versus very small letters.

Large Letters

Many people feel the standard typewriter keyboard is not designed as well as it could be. Some think that if the letters were arranged in alphabetical order the keyboard would be easier to remember. Such an arrangement of keys probably would not reduce movement time for skilled typists. The keyboard might be arranged to take advantage of the frequency with which letters appear in written text. Letters that tend to follow each other could be next to each other on the keyboard. Or the keys could be arranged to reduce the distance the fingers must travel while typing. With the existing typewriter format, some of the most frequent letters require long finger movements from one key to another.

Small Letters

Speed is of secondary importance when dialing a telephone. More critical is the ability to dial without making errors. With some new numbers, by the time the first few digits are dialed, the last ones are forgotten. Many telephone numbers consist of a three-digit prefix and a four digit suffix. Despite the fact that the prefix is usually more familiar and less likely to be forgotten, it is always dialed before the less familiar suffix. One study reported fewer errors and faster dialing when the prefix digits were dialed last rather than first. The order in which numbers are listed in telephone directories is still correct, with the less familiar digits being the last read but the first dialed.

Large Letters

Speed is of secondary importance when dialing a telephone. More critical is the ability to dial without making errors. With some new numbers, by the time the first few digits are dialed, the last ones are forgotten. Many telephone numbers consist of a three-digit prefix and a four digit suffix. Despite the fact that the prefix is usually more familiar and less likely to be forgotten, it is always dialed before the less familiar suffix. One study reported fewer errors and faster dialing when the prefix digits were dialed last rather than first. The order in which numbers are listed in telephone directories is still correct, with the less familiar digits being the last read but the first dialed.

Small Letters

Many people feel the standard typewriter keyboard is not designed as well as it could be. Some think that if the letters were arranged in alphabetical order the keyboard would be easier to remember. Such an arrangement of keys probably would not reduce movement time for skilled typists. The keyboard might be arranged to take advantage of the frequency with which letters appear in written text. Letters that tend to follow each other could be next to each other on the keyboard. Or the keys could be arranged to reduce the distance the fingers must travel while typing. With the existing typewriter format, some of the most frequent letters require long finger movements from one key to another.

Method: This exercise will require four subjects. Test each subject one at a time. Have the subjects sit at a desk or table. Make sure the large print and small print are each read from *exactly* 20 inches. Do not let your subjects move their head or the book.

Before they begin, read the following instructions to them:

> In just a moment I will give you a paragraph to read. Please read the paragraph out loud. I will be timing you and also listening for reading errors. Read as quickly and as accurately as you can.

After reading the instructions, give the paragraph to the subject and begin timing when the person begins reading. You may want to use photocopies of the paragraphs. Follow along on your copy and mark any words or phrases where they make reading mistakes.

Have each subject read one paragraph with small print and one with large print. Make sure that they do not read the same paragraph each time. Also, make sure that you give two of your subjects the paragraph with large print first and two of your subjects the paragraph with small print first. Again, make sure that you keep the reading material exactly 20 inches away from their eyes for both the large and small print.

Reporting: Prepare a report using the format discussed in Exercise 1A. Were there performance differences (either time or errors or both) when reading the two different paragraphs? Did subjects try to increase the visual angle of the smaller characters? Did your participants prefer either the large or small print?

Exercise 3B: Calculating the Visual Angle

Purpose: To determine visual angle of the large print versus the small print.

Method: Assume a reading distance of 20 inches (51 cm). Measure the height of the lowercase *s* character in both the large- and small-print paragraphs. Calculate the *visual angle* of the *s* (which also represents the visual angle of most of the other lowercase characters) in the two paragraphs. Also determine the ideal size (in inches or centimeters) of the lowercase characters for a visual angle of 15 minutes of arc. If one *point* is equal to $\frac{1}{72}$ inch, how many points would be the ideal character height?

Reporting: Prepare a report using the format discussed in Exercise 1A.

Exercise 3C: Determining and Comparing Body Dimensions

Purpose: To demonstrate the wide range of body dimensions—even within one person.

Method: Use yourself as the subject in this exercise. You may need someone to help to take some of the measurements. Using a measuring tape, take your own measurements for those dimensions shown in this chapter. Compare your measurement of each dimension with those shown for the 2.5th, 5th, 50th, 95th, and 97.5th percentiles in Table 3–6. Record which percentile value is closest to your actual measurement.

Reporting: Write a report using the format shown in Exercise 1A. Include a table showing your measurements and how each one compared with the percentile values in the book. Do your measurements reflect only one percentile value, or do they reflect numerous different values? What does this mean? How do your measurements relate to the average-person fallacy?

Exercise 3D: Determining the Anthropometric Adequacy of Cars

Purpose: To demonstrate how car designers use anthropometric data when making design decisions.

Method: Select three cars (compact, mid-size, and luxury) and take the following measurements: (1) the width of the backseat, (2) the distance from the top of the front seat to the bottom of the roof, and (3) the distance from the back of the front seat to the volume control on the radio. Compare your measurements with the appropriate ones in C-3 (for three people), C-1, and B-1 of Table 3–6.

Reporting: Write a report using the format shown in Exercise 1A. Did the car designers properly use the anthropometric data? What adjustments would you make (if any)?

4

Cognitive Processing and Performance

Introduction

The senses present the brain with information about the world. The brain interprets and processes this information and may send appropriate messages to responders (e.g., fingers, legs, or mouth). The brain processes that take place between sensing and movement, called *cognitive processes* or *cognition,* are the essence of human performance. Cognition refers to all the processes by which sensory input is transformed, reduced, elaborated, stored, recovered, and used. The cognitive processes may also operate in the absence of external stimulation, as in internally produced reasoning, decision making, and problem solving. Cognition is involved in all human performance.

Understanding human cognitive processes is similar to discovering how a computer is programmed. If a program stores and reuses information, by what routines or procedures are these done? It does not matter whether the computer stores information on magnetic tape or electronic cores. Designers are usually more interested in the program (psychology) than the hardware (physiology). Computer programs have much in common with human cognition because they are both descriptions of how information is manipulated. The computer program analogy helps us to understand cognitive processing in people.

The Brain

Neurons

The human brain is a 3-pound mass of tissue comprised of trillions of cells. About 100 billion of the brain cells are *neurons* linked in networks. These networked cells cause human performance to take place. Neurons convey information to other neurons by generating impulses (action potentials). These impulses propagate like waves down the length of the cell and are converted into chemical signals at the point where one neuron seems to touch another neuron (the synapse). The nerve impulses travel at speeds up to 100 yards per second, or about 200 miles per hour. At this rate, a nerve impulse can travel from the toe to the brain in about 20 milliseconds. When compared with the speed at which an electrical signal moves in a copper wire, about 1 million yards per second, nerve impulses are relatively slow.

Cerebral Cortex

The cerebral cortex is the deeply convoluted surface region of the brain that enables perceptual and intellectual processing. The cortex is less than $\frac{1}{8}$ inch thick. If flattened, the human cortex would cover four pages of typing paper, a chimpanzee's would cover only one, a monkey's would cover a postcard, and a rat's would cover a postage stamp (Calvin, 1994).

The cortex consists of four lobes that make up the left and right hemispheres of the brain (see Figure 4-1). For example,

- The *frontal* lobe assists in planning, controlling movements, and producing speech.
- The *temporal* lobe assists in hearing and interpreting music and languages.
- The *parietal* lobe enables the reception and processing of information from the senses.
- The *occipital* lobe specializes in vision.

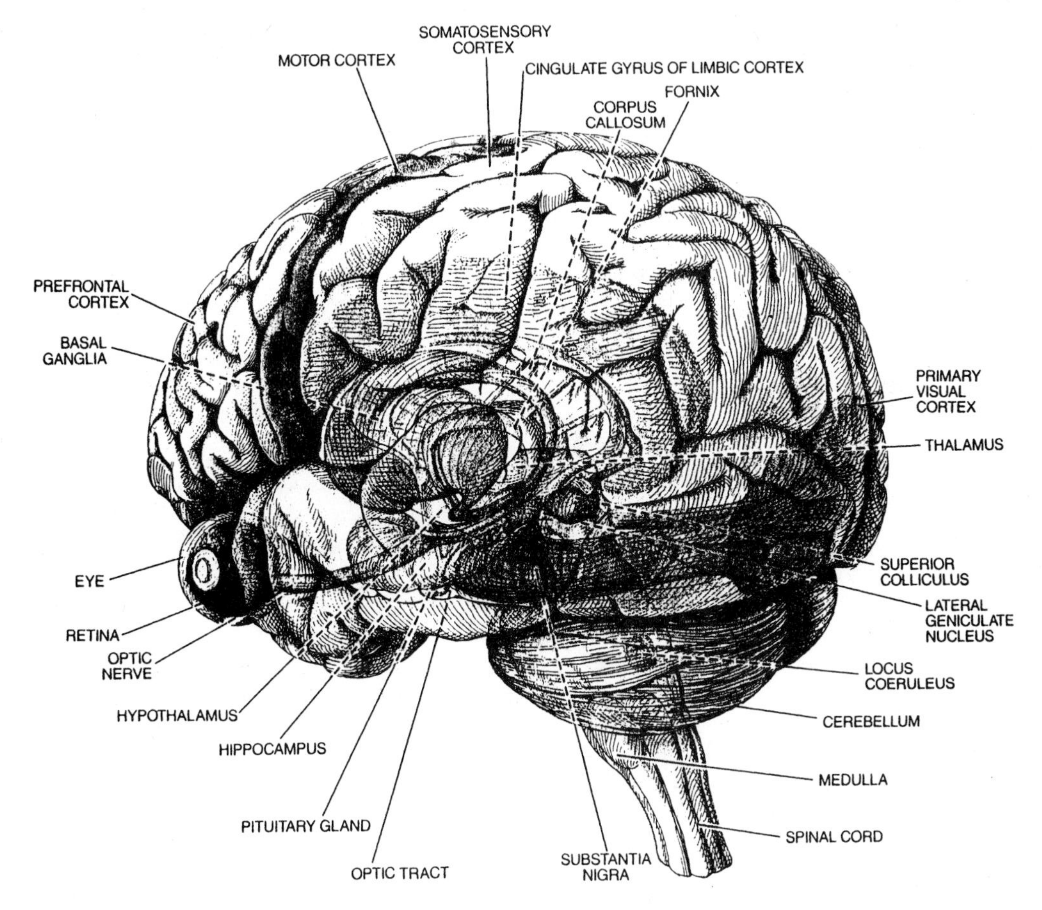

Figure 4-1 The human brain.

The two hemispheres are connected by a nerve bundle called the corpus callosum (the largest databus in the brain). Certain areas of the cortex clearly deal with specific human performance processes. For example, the motor cortex (located toward the rear of the frontal lobe) is associated with human movements. Within the motor cortex, smaller areas are identified that deal specifically with mouth movements, motor programming, and so on. Obviously, the auditory cortex is associated with hearing, and the visual cortex (at the very back of the brain) is associated with seeing.

Other areas of the brain that are closely associated with human performance include

- The *hippocampus,* which helps to consolidate recently acquired information from short-term memory into long-term memory
- The *thalamus,* which takes sensory information and relays it to the cortex
- The *cerebellum,* which helps to govern muscle coordination and the learning of rote movements
- The *limbic system* (amygdala), which generates emotions from perceptions and thoughts

Research on brain processing is very incomplete in terms of helping designers be better designers. As significant new discoveries are made, it will enable designers to make more informed design decisions. The better that we understand how people process information, the better will be our ability to match human processing capabilities with computer processing capabilities. Brain research currently reports discoveries at the level of the following two examples (Begley et al., 1992):

Example 1

The brain hears loud sounds in a totally different place from quieter sounds. Also, the areas that hear tones are laid out like a piano keyboard. The physical distance in the brain between areas that hear low C and middle C is the same as the distance between areas that hear middle C and high C (just like a piano).

Example 2

When learning a complex game, the brain uses much more energy than is used to play the game once the game-playing skill is developed. The greater the drop in the energy level between learning and playing, the higher the person's IQ. Intelligence may be a matter of neural efficiency. Smart brains may get away with less work because they use fewer neurons or circuits, or both.

Stage Processing

In the past few years, it has become popular to consider the processing of information by the brain in terms of *stages.* Information is believed to come from the sensory receptors to a *perceptual stage.* After processing, the information passes to a *translation stage* (intellectual) where it is translated from perception to action. The response is then selected and passes to a *movement control stage.* Each stage is thought to have access to a memory store. The important components of cognitive processing are shown in Figure 4-2.

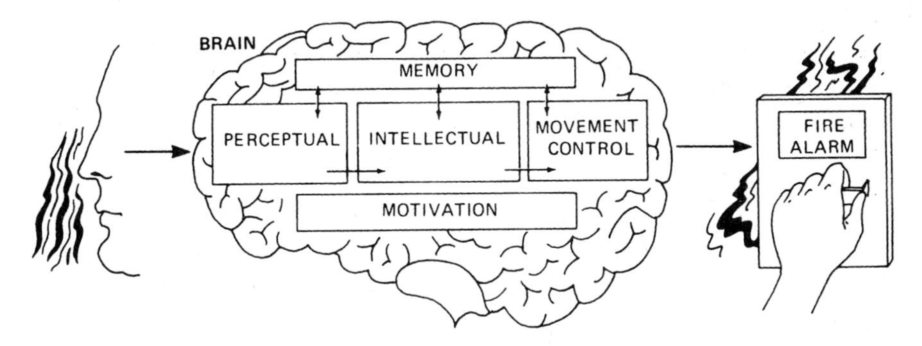

Figure 4-2 Major components in human cognition.

Human Information Processing Model

This book will assume that at least three brain processing stages take place in most activities: *perception, intellection,* and *movement control* (Welford, 1976). The stage model shown in Figure 4-2 represents the cognitive processes of most interest to system designers. This general approach to relating cognition to human performance originated with Broadbent's (1958) model of attention.

The human information processing model is useful for studying and understanding cognitive processing. However, the stage model carries with it numerous difficulties. For example, when the three hypothetical stages are examined closely, it becomes very difficult to distinguish one stage from the next. Also, processing often does not seem to be strictly serial. Nevertheless, this stage model approach is a useful heuristic and will be used in our discussion of cognitive processing. It is used to give designers an insight into the processing that goes on in the brain to aid them in making design decisions that facilitate cognitive processing.

The stage model in Figure 4-3 also suggests that each of the three stages of processing makes use of *memory.* Human memory is discussed in Chapter 5. In addition, Figure 4-3 shows *motivation* to be a major consideration related to human performance. Performance motivation is presented in Chapter 6.

The *sensors* shown at the extreme left of Figure 4-3 fall into two groups. The *external* group consists of sensors that receive data from sources such as the eyes, ears, nose, mouth, and skin. The *internal* group can be further divided into two categories. First, internal sensors in muscles, tendons, and joints supply data concerning the control of movement. Second, a number of less understood internal sensors measure the state of the blood, body dehydration, and other bodily conditions.

The *responders,* shown at the extreme right of Figure 4-3, also fall into two groups. One includes the hands, feet, vocal organs, and other voluntary muscles, which

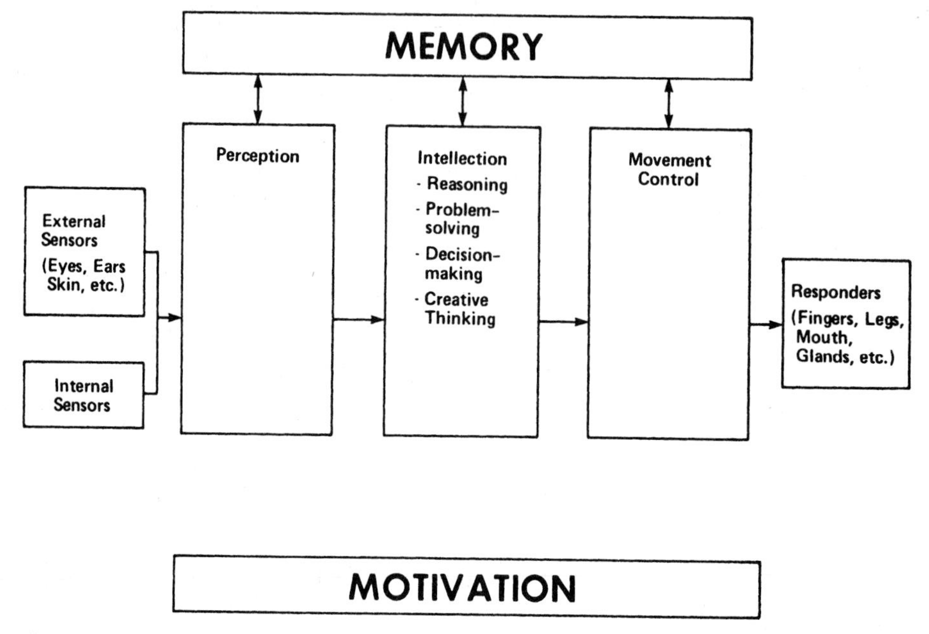

Figure 4-3 Stage model of major human performance processes.

are considerably important in human performance. The other group consists of various reactors in the autonomic system, such as the action of glands, heart, and lungs, that have less obvious effects on performance. In recent years, there have been substantial advances in bringing the autonomic, nonvoluntary body processes under voluntary control. Through the use of *biofeedback,* some people have been able to control severe pain (including headaches), blood pressure, anxiety, and so on, without using medication.

If you were to pull open your desk drawer and see a snake coiled and ready to strike, it would be one thing to *recognize* it, quite another to *decide* what to do about it, and still another to *carry out* the required action. Thus, the process begins with data from the various sense organs flowing into the perceptual mechanism and then to the intellecting mechanism where a decision is made. Finally, after an appropriate response is determined, the orders are passed on to the movement control mechanism, which carries out and monitors the response to ensure proper execution.

We have considered the preceding stages in terms of the *time* taken for processing to occur in each, but they can also be seen in terms of *errors* that occur in each stage. An incorrect response—an error—could be made if there were a failure in any one of these three cognitive stages. For example, if the snake were perceived as a piece of old rope, the response would be to throw it away. But if it were a poisonous snake, that response would obviously be incorrect. The error could be attributed to *faulty perception.* Assume, however, that the snake was perceived as a snake, but the decision was made to reach in with the hand and remove the snake. If the snake were poisonous and did strike as the person was reaching for it, this would also be considered an incorrect response. This error could be attributed to an *inappropriate decision.* Finally, assuming that the snake was appropriately perceived and that a decision to remove the snake with a stick was made, the execution of this decision may still be carried out, but in a jerky, awkward manner. If the person's movements are not smooth and precise, the snake might still be able to strike. In the latter case the incorrect response could be associated with *faulty movement control.*

Using another example, consider a typist looking at an X on a sheet of paper and then hitting the X key on a keyboard. The X is first perceived, a decision is made on how to translate what is perceived into an appropriate action, and then the orders for striking the X key are carried out by the movement control mechanism. Again, it is fairly easy to see how errors could occur at each point in the process.

As noted earlier, different amounts of time could be spent in each of the three cognitive stages. To use an extreme example, typists can perceive an X, decide the appropriate response, and then strike an X on a keyboard in less than a second. In this case the decision was made within milliseconds because it was well practiced and there were few alternatives. It is also possible that a business executive when asked for a decision on an important policy matter could ponder the question, consider a large number of alternative decisions, and after 2 or 3 hours write down a response. In this case, the question was perceived by listening and responded to (hours later) by writing. The interim cognitive processing took a considerable amount of time, measured in hours rather than milliseconds. Even though the situations were different and the time involved was different, it appears that the order of cognitive processing was the same (perception to intellecting to movement control).

Processing Stages and Skilled Performance

Ability and Skill

Fleishman (1972) provides a clear distinction between ability and skill. According to his definition, *ability* refers to a basic trait of an individual. Many of these abilities are the product of previous learning. Abilities are what an individual brings to a new situation. These abilities become the basis for learning a new activity. It is assumed that the individual who has many highly developed abilities can more readily become proficient in performing a wide variety of other (similar) activities.

The term *skill* refers to a level of proficiency obtained in a specific activity. When we talk about proficiency in wiring a house, driving a car, or keying at a terminal, we are referring to specific skills. Designers are usually interested in taking full advantage of existing abilities for developing specific user skills as quickly as possible.

Skilled Performance

The development of the specialized concept of skill took place during and shortly after World War II and was stimulated by the study of highly skilled people performing real activities. Bartlett (1943) observed that skilled responses were not a series of unrelated simple movements but *coordinated actions.* The smooth sequence of movements needed to kick a ball or the integration of information required for playing a game of chess constitutes skilled responses. Skill refers to the seemingly automatic execution of those cognitive processes that produce rapid and accurate performance. This definition may be applied to all skills, even though different skills may be dominated by one of the three stages (perceptual, intellectual, or movement control). Welford (1976) argues that all skills involve all three of the major cognitive mechanisms, but that different types of activities emphasize different processes. When we speak of perceptual or intellectual or movement skills, we are classifying in relative rather than in absolute terms.

Perceptual Skill

Welford suggests that *perceptual skill* consists of giving coherence to the sensory data that pour in through the sense organs and linking these data to material already stored in memory. Individuals may differ in both their basic perceptual ability and the level of perceptual skill that they have developed. Some people are able to recognize and readily identify large numbers of different people, rocks, or birds. Others have great difficulty in recognizing even a small number of different objects. The development of perceptual skill seems to require continuous exercising of the perceptual mechanism. Neisser (1976) has observed that perceptual skill differs from movement control skill (e.g., sculpting or playing tennis) in that the perceiver's effects on the world are usually negligible. A person does not change objects by looking at them or events by listening to them. Although perceiving does not change the world, it does change the perceiver (this idea will be discussed in more detail in Chapter 4).

Intellectual Skill

Many skills in the arts and trades probably have less to do with the ability to execute particular responses, such as making a brush stroke or connecting a pipe, than in deciding what colors to use in the painting and how best to install plumbing in a house. A highly skilled musician transcends the mere playing of an instrument and makes ingenious new interpretations of the music. These are all forms of *intellectual skill* that are analogous to the skills of an administrator, manager, politician, or military officer (Welford, 1976). For all these individuals, the input data may be perfectly clear, and the actions needed to effect their decisions (e.g., writing or speaking) may be straightforward, due to extensive experience in perceiving and in movement control. Intellectual skill lies in the efficient linking of perception to an appropriate action based on reasoning, decision making, or problem solving.

Movement Control Skills

Once data have been perceived and an appropriate action chosen, the emphasis shifts to making a response, in many cases a movement. These are called *movement skills* and include such things as riding a bicycle, eating food with a fork or chopsticks, pronouncing words clearly, hitting a golf ball, and typing. As these skills develop, the movements become highly coordinated. Well-practiced movements are noted for their *lack* of intellectual control. For example, after the skill of bike riding is acquired, we no longer have to think about moving the handle bars to maintain balance.

Fitts and Posner (1967) observed that the organization or patterning of movement control skill involves both spatial and temporal factors. They suggest that

> The simple act of picking up a pencil involves skill in that the movement must be precise in amplitude and the fingers, in order to grasp the pencil, must move in a coordinated way at the right time in the reaching sequence. Similarly, speaking one's name requires the modulation of amplitude and pitch of the voice in a complex temporal pattern. The writing of a name may involve the coordinated activity of as many as twenty different small-muscle groups in the arm and hand. These simple acts of reaching, speaking, and writing become so highly overlearned and automatic in an adult that it is easy to forget the laborious way in which they were originally learned as a child.
>
> We marvel at the execution of the soloist and the timing of the supporting symphony orchestra, or at the control of the quarterback as he throws a pass to the end who is running at full speed down the field. No less remarkable, however, are the linguistic skills that we employ every day in communicating with other people.

Many people feel that movement control skills are relatively simple in comparison to perceptual and intellectual skills. However, Fitts and Posner have rightly pointed out that programming a computer to hit a pitched baseball is as complex as programming it to play chess. In both cases, all three cognitive processes are involved, although in the baseball example the emphasis is on the movement control stage and in the chess example the emphasis is on the intellectual stage.

Developing Skills

The acquisition of a skill may not increase basic abilities, but does result in improved efficiency of the cognitive processes. For example, the exertion of force is not a skill in itself, although the controlling and appropriate application of force do require skill. Similarly, visual and auditory reception of stimuli is not skilled until it is organized by the perceptual process. Thus, the reception of stimuli by our senses and the carrying out of actions by muscles are not included in what is being defined here as skill. *Skill is exclusively cognitive.* That is, skill refers to changes that take place in the brain, even though skilled execution relies to a large measure on the condition of the sensors and responders.

Senses do not seem to improve the more they are used (i.e., the constant use of the eyes does not improve their ability to see). The cognitive processes *do* improve with use. The responder's size, strength, and extent of flexibility are finite and for most adults change very little. To increase strength, a person can build up the fibers in a muscle by exercise; however, new fibers are not added, and the building up of existing fibers has limits that are soon attained. In addition, improved strength is not necessarily related to improved skill. A person may become strong without becoming more skilled.

Skills appear to improve indefinitely as long as they are practiced; however, the rate of improvement slows considerably after a person has had extensive experience. Skills that are developed and not practiced tend to deteriorate.

Improving Performance

We have discussed how skill can be related to the three processing stages. A designer should (1) consider each activity being performed and (2) decide which of the three processing stages dominates. The designer can then concentrate his or her efforts on making decisions that will lead to improved performance.

There are some cautions related to skilled performance that are of interest to designers. First, it is possible to design an activity that is impossible to perform no matter how skilled the individual. Second, although it is possible for people to become skilled in many activities, they frequently do not perform in one activity long enough to acquire all the necessary skills. As a result, many users never develop a complete set of even the minimum skills necessary to reach an acceptable level of performance.

By considering the dominant stage in each activity, the designer can make decisions concerning the best way to conduct training to develop the appropriate skills. For example, for movement-control-dominated activities, the potential user should be provided considerable practice in making the necessary movements. In an intellectual-dominated activity, practice should minimize the perceptual and movement control processes and place emphasis on the appropriate intellectual processing (e.g., practicing making critical decisions or solving problems).

The designer can expect different types of errors from each stage, and the cognitive processing time in each stage can be affected by certain design decisions. For example, barely perceptible stimuli, such as nearly illegible print, can require substantially more perceptual processing time per item. Over a large number of items, this slowness of processing can mean that more work time and possibly more people will be required to perform a given amount of work.

Processing Levels

The designer should also consider that people can process performance-related information on at least two levels: the automatic and the conscious (see Figure 4-4). For example, driving home from work in a car pool with two other people, an experienced driver (i.e., a driver *skilled* in operating an automobile) is able to find the proper roads and respond appropriately to road conditions, signs, and other cars. All this is done *at the same time* that the driver is participating in a discussion with others in the car. In this example, both processing levels are being used. The *automatic* level controls performance related to the skill of driving. At the same time, the *conscious* level controls much of the performance related to listening, thinking, and speaking. Both types of performance can occur at the same time as long as one of the activities is performed automatically.

Automatic performance is usually considered as skilled performance and can only take place after the skill for that particular activity has been developed. Skilled performance is made up of actions that were originally under conscious control, but that have become automatic and relatively inflexible.

Miller et al. (1960) reported an interesting comparison of automatic performance in people and machines. Consider a stereo that has a system for playing and changing records. Whenever the appropriate stimulus conditions are present, for example, when the arm is near enough to the spindle or when a particular button is pushed, the routine for changing the record is executed. There are sensors that discriminate between different-sized records, and there are responders that push the next record into place and lower the tone arm gently into the groove of the record. The entire performance is obviously automatic. This routine for changing records is built into the machine and requires only the proper stimulus to elicit the performance. For our discussion, the main difference between

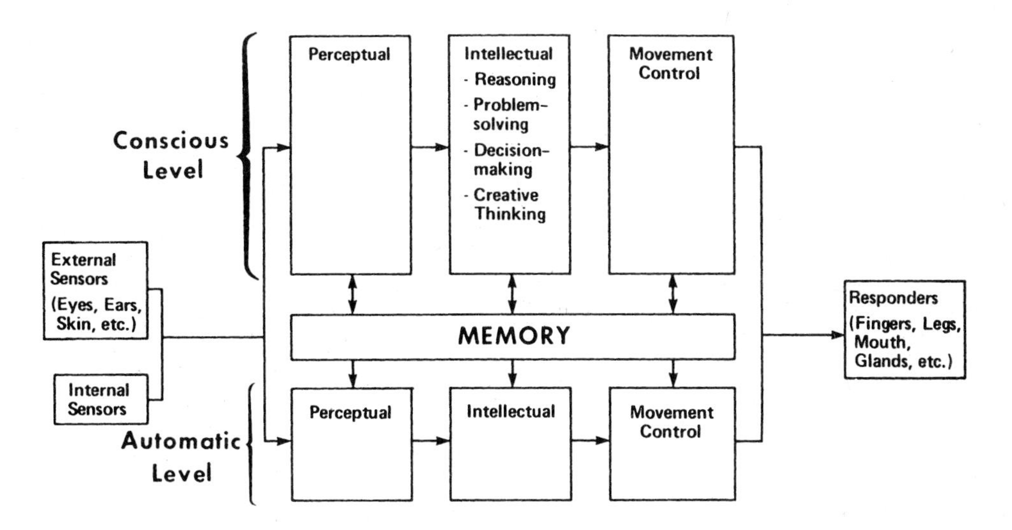

Figure 4-4 Two-level model of human performance processes as related to the stage processing model.

this machine and people is that people develop their own routines, which we frequently refer to as skills.

Keele (1968) suggests that a shift from conscious to automatic control of performance provides at least three advantages to a user. First, the degree of conscious attention required to perform an activity is reduced. Second, successive stimuli are anticipated so that appropriate movements can be planned in advance to coincide with the stimuli, rather than lag behind. And, third, it is possible for movements to be made at a faster rate—much faster than when conscious thought processes are required.

Most psychologists agree (cf. Woodworth, 1899; Lashley, 1951; Fitts, 1966; Pew, 1966; Welford, 1976) that skilled performance generally begins with strict conscious control, eventually reaching a level of automatic actions with occasional conscious monitoring, and finally reaching a point where it is almost exclusively automatic.

At the turn of the century, Book (1908) wired a typewriter to record the time of occurrence of successive key strokes and then collected data while people learned the touch method of typing. His study represents an interesting example of the shifting from conscious to automatic control in a real-world activity. People first tried to consciously memorize the locations of different letters on the keyboard. Then they went through several discrete steps: look at the letter in the material that was to be typed, locate this letter on the keyboard, strike the key, and look to see if it was correct. Each character was typed with considerable conscious effort. After a few hours of practice, these conscious actions began to fit together into small routines of automatic actions centered around individual letters. At this point we would say that the learner had acquired a *letter skill.*

With practice, people began to anticipate the next letter and to develop new automatic routines for dealing with familiar sequences like "the" and "ing." The letter routines were combined into word routines and a *word skill* was developed. Eventually, the experienced typist was able to read the text several words ahead of the letters being typed, thus developing a *phrase skill.* The end result of the skill-development exercise was to have almost totally automatic typing performance with little conscious thought involved.

Conscious processing seems to rely more heavily on visual and auditory sensory feedback than does automatic processing. As the transition from conscious to automatic control continues, a person becomes less dependent on peripheral feedback for performance and needs less time to check on the progress of the activity. In some activities the emphasis can be observed to shift from feedback-dependent movements to smooth, feedback-free execution. MacNeilage and MacNeilage (1973) observed that "the need for peripheral sensory feedback can be thought of as inversely proportional to the ability of the central nervous system to determine every essential aspect of the following acts" (p. 424).

As conditions change or as novel circumstances occur, people must change their automatic performance or errors could occur. Unfortunately, some designers do not recognize that changes in people do not come about as readily as changes in software or hardware. People have a tendency to deal with new situations in old, automatic ways. For example, some radios in new automobiles have the tuning knob on the left and the volume control on the right (with one labeled TUNE and the other labeled TONE). This layout is just the opposite of how radios have been designed for well over 30 years. It is interesting to observe users as they "stop and think" before adjusting the volume or changing stations. Quick automatic motions without thought frequently result in turning the wrong knob, even after several months of experience with the new radio.

Another example of how automatic performance can be a problem was reported recently in the newspaper. A policeman was suspended and ordered to undergo a psychiatric evaluation because he wrote the number seven with a line through the downstroke. He had made sevens that way for 30 years, since he was in the seventh grade. The police chief said the policeman's writing was confusing to typists, and he had been told to change the way he made sevens. The policeman said he tried to break the habit, but forgot while writing some reports.

Conscious Awareness

When people attain a high level of skill, they can perform one or more actions without *conscious awareness.* In fact, the concept of performing *automatically* suggests that the performance is taking place without conscious control and in many cases without conscious awareness (Welford, 1976). Highly skilled people that are put in situations where considerable demands are placed on the efficient execution of the skill experience moments of such intense involvement that there is no awareness of the passage of time. Only after the difficult situation has passed is there awareness of what has taken place, and even then the details may be fuzzy (Dreyfus and Dreyfus, 1979).

Conscious awareness is lost and automatic performance is gained when it is possible to dispense with monitoring. This is likely when the predictability of a situation is such that actions are always, or virtually always, accurate. In such cases, the performance does not require much checking for possible modification. It is just this kind of accuracy that seems likely to be attained and maintained by considerable practice on the same activity, especially the intensive practice that is an inevitable feature of highly repetitive tasks such as speaking, typing, keypunching, or driving an automobile. In addition, when performing automatically with little or no monitoring, a person is free to consciously carry out actions related to other activities.

Frequently, automatic processes continue uninterrupted once they begin, and it is very difficult for people to self-analyze the components involved. This is one reason that knowledge acquisition is so difficult when designing expert systems.

Performance that takes place automatically is not usually remembered. Automatic performance seems to make use of memory (obviously, or the driver would not know where to drive, which roads to use, or what the signs mean), but does not appear to feed new information into memory, at least not in a form that can be readily retrieved. This may be why when a person drives home after work there are few, if any, memories associated with the trip itself. The driver may remember a conversation that took place and a child that ran into the road (as well as the sick feeling associated with the "near miss"), but may have difficulty remembering other cars, taking the usual curves in the road, the landscape, or even the time spent. In one study, when asked to think back and estimate the time it took to perform a work activity, people had a tendency to believe that performance of a well-learned activity took considerably longer (up to 40 percent) than it really did (Butler et al., 1979).

Errors

The two-level model of human performance can be of help in making design decisions that will reduce human errors in a system. We have already discussed that different types of errors could be made in each of the three processing stages. In addition, different types of errors are characteristic of conscious and automatic performance.

People are frequently the *victims* of unwanted releases of automatic performance, just as animals are for their instinctive behaviors. James (1914) observed, "Who is there that has never wound up his watch on taking off his waistcoat in the daytime, or taken his latchkey out on arriving at the door-step of a friend?" Norman (1980) has suggested that these kinds of errors involve the following principle: pass too near a well-formed habit and it will capture your behavior. In other words, if a habit is strong enough, even cues that only partially match the situation can activate the incorrect program that leads to incorrect performance and thus errors. Norman gives an example of automatic performance as being triggered and carried out without conscious awareness of why the performance was even needed. A colleague reported to him that before starting work at his desk at home he headed for his bedroom. After getting there he realized that he had forgotten why he had gone there. "I kept going" he reported, "hoping that something in the bedroom would remind me." Nothing did. He finally went to his desk, realized that his glasses were dirty, and returned to the bedroom for the handkerchief he needed to clean them.

A designer should seek to have few errors made as users develop skill on an activity. Kay (1951) and Von Wright (1957) have shown that errors made in the first few trials of performing a new activity tend to become ingrained. In fact, Kay has argued that one major problem in achieving an acceptable level of performance lies in getting rid of, or in other words *unlearning,* these early errors. It seems that for those activities that eventually will be automatic a designer would do well to seek for *errorless learning* (cf. Terrace, 1963).

Speed and Errors

Pew (1969) suggested that there is a "robust" *speed–accuracy trade-off* that is very much apparent in most choice reaction-time studies. Swensson (1972), on the other hand, found the speed–accuracy trade-off to be very elusive. It is probably fair to say that the precise relationship of speed and accuracy is still not well understood. Consider, for example, two studies in which subjects were given instruction to perform either (1) as fast or (2) as accurately or (3) as fast and accurately as possible (Howell and Kreidler, 1963; Fitts, 1966). A comparison of the accuracy levels for the three conditions in each study is shown in Table 4-1. Both studies found little difference between any of the groups with respect to *speed* of responding. The *speed instruction* resulted in no significant increase in speed, but a very large reduction in accuracy.

Table 4-1 Comparison of Accuracy Levels

	Instructions		
Study	**Speed–Accuracy**	**Accuracy**	**Speed**
Howell and Kreidler (1963)	95.9%	97.8%	87.3%
Fitts (1966)	86.4%	89.7%	77.4%

There is a common misconception that as people gain experience on an activity they always tend to make fewer errors. Designers should note that in activities for which performance is primarily automatic the proportion of errors will remain fairly constant, but the *speed* with which the activity is performed will increase with practice. Thus, experience gained on automatic activities affects speed of performance more than accuracy of performance. For example, with handprinting, people initially learn to write faster and make fewer errors. Once the basic skill is developed, the speed continues to increase, while the proportion of errors remain about the same (and may even increase). When learning to type, as typists gain several months and eventually years of experience, they type more words per minute with the same proportion of errors to key strokes. People reach an acceptable error rate relatively soon after beginning to perform many activities, but continue to increase their speed of performance for a longer period of time. In fact, Crossman (1959) has shown that the time taken by workers to make a certain product was still decreasing after 7 years.

A study by Howell and Kreidler (1963) illustrates this phenomenon. They had subjects perform a choice reaction-time activity over a period of four blocks, with each block containing 5 trials, for a total of 20 trials. As subjects gained experience, their response speed significantly improved. But, as far as accuracy was concerned, the authors reported "no significant trend of improvement... over the complete 20 trials" (p. 42). Accuracy remained at a fairly constant level throughout the experiment, and nearly all improvement in performance came about as a result of increased speed.

A second example comes from a field study (Bailey and Koch, 1976). In this study, the performance of several newly trained clerks was monitored daily for four consecutive weeks. For comparative purposes, the performance of experienced workers was also monitored. It was observed that the average time new clerks spent per customer became shorter as they performed over the 4-week period. That is, they worked *faster* with practice. Although the new clerks were not able to process customers as quickly as experienced clerks at the end of 4 weeks, the trend was certainly in that direction. But, for the same period of time, the error rate after only 1 week's experience for the new clerks was already at the same level as experienced clerks, and it remained at that level for the remainder of the study.

It is interesting to note that the majority of errors occurring during automatic processing in systems are likely contributed by the *most experienced* people. For example, as key operators gain experience with a computer system, they learn to perform faster, but their *rate* of making errors remains fairly stable. Because they process more data, they have more opportunities to make errors; and because they continue to make about the same proportion of errors, they may actually have a greater number of errors.

The number of self-detected errors increases slightly as people become more experienced (cf. Conrad and Longman, 1965; West, 1967; Schaffer and Hardwick, 1969). The increased capability for self-detection does not appear to keep pace with faster performance. Even with more errors being self-detected, the absolute number of undetected errors also continues to increase as people gain experience (i.e., work faster). Thus, the term experience often reflects the ability of a person to do things faster. People seem to reach an accuracy level as soon as the basic skill begins to develop and for the most part retain that accuracy level over a long period of time.

Speed

In most well-learned activities, the fastest performance comes from performing totally automatically. Disruptions to this automatic processing will slow down the overall processing time. A designer should guard against unwanted interruptions in order to have activities that are under automatic control performed in the shortest time possible. This may mean that a set of activities be designed so that the automatic processing can run its course, with consciously controlled processing taking place before or after the activity. For example, a form that requires filling-in information familiar to the user, except for two or three unfamiliar items, should have the unfamiliar items grouped at the end of the form. In this way the automatic performance is allowed to run with a minimum of interference.

The fastest performance is only possible if conscious thought can be avoided. Consider an example about tennis by Gallwey (1974):

> I asked students to stand at net in the volley position, and then set the machine to shoot balls at three-quarter speed.... At first the balls seem too fast for them, but soon their responses quicken. Gradually I turn the machine to faster and faster speeds, and the volleyers become more concentrated. When they are responding quickly enough to hit the top-speed balls and believe they are at the peak of their concentration, I move the machine fifteen feet closer than before. At this point students will often lose some concentration as a degree of fear intrudes.... Soon they are again able to meet the ball in front of them with the center of their rackets. There is no smile of self-satisfaction, merely total absorption in each moment. Afterward some players say that the ball seemed to slow down; others remark how weird it is to hit balls when you *don't have time to think about it.*

Morehouse (in Morehouse and Gross, 1977) reports on his successes as a psychologist in achieving super performance from world-class athletes. One of his major points is to "*think* when you need to, and *do* when you need to, but make it a rule to keep thinking and doing separate" (p. 65). A person cannot effectively plan and act at the same time. Planning requires analysis. Consciously analyzing performance while performing detracts from quickly accomplishing the activity. Movements are slowed.

Processing Levels and Skill Development

Initially, almost all performance may require considerable conscious thought. When learning to drive, practically every move is consciously considered before it is initiated. For many activities, one of the main objectives of a designer is to realize a shift of as many tasks as possible from conscious to automatic control in the shortest time possible.

In relatively simple tasks, a user is not told what to do, but is shown and then guided through the proper actions until the skill begins to be developed. In these cases, knowledge is gained as the skill develops, not before. But for complex activities, a considerable amount of knowledge must be accumulated *before* the skill-building exercise begins. A designer should determine the type of activity to be performed, consider all reasonable alternatives for building the skill in the shortest time, and then decide the best way to proceed. For each task or series of tasks, the best way may be different. In some cases the knowledge needed can be obtained by selecting people who already have it. In other cases the needed knowledge is so minimal that a person can begin developing skills with some brief verbal instructions.

Summary

A human information processing model was presented and discussed. Major considerations include stage processing, skilled performance, processing levels, and a more detailed understanding of speed, errors, and speed–accuracy trade-offs. It was proposed that (1) designers should consider each activity being performed in relation to the three-stage model, decide which stage dominates, and then use that knowledge to improve performance, and (2) designers should be aware of the fact that different types of errors occur in each processing stage.

Perception, Problem Solving, and Decision Making

Introduction

The major human information processing components will be briefly addressed (see Figure 4-5). The discussion will focus on issues related to achieving acceptable human performance in systems. Experience has shown that the better a designer understands how people truly process information, the better will be his or her user-related design decisions.

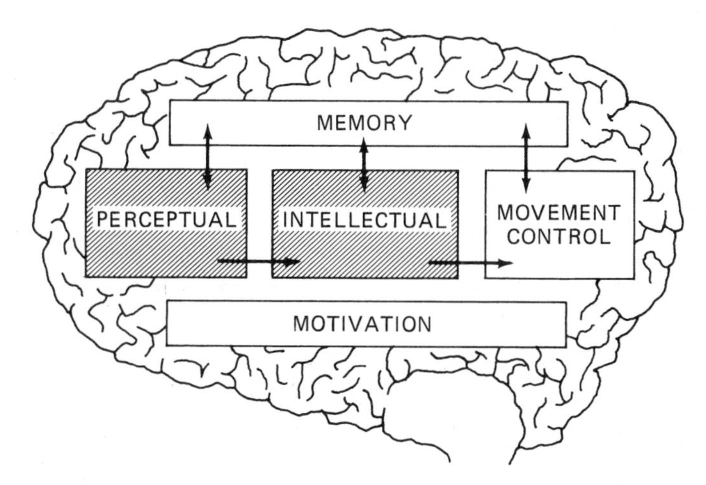

Figure 4-5 Major components in human information processing.

Perception and Perceptual Skills

The word *perceive* is often used as a synonym for *see;* however, perception may involve any sense, including hearing, tasting, smelling, or feeling. Another person's perception cannot be observed directly and is usually inferred from observations of performance.

Perception involves the interaction of two sources of information available to the perceiver. The first is the information available through our senses, and the second is the accumulated knowledge of the perceiver stored in memory. The whole process of perception hinges on being able to relate new experiences with old experiences in some meaningful way. Perceptual skill, in fact, may be defined as developing ways of quickly and efficiently combining new experiences with old.

Perceptual Flexibility

The perceptual process as a whole seems quite flexible. For example, Kohler (1962, 1964) reported a series of experiments in which subjects wore special prism goggles that reversed the image on the eye, transforming the entire visual world into a mirror image. The subjects wore these prisms for several days or weeks. At the beginning, the subjects functioned clumsily and would, for example, see someone apparently on their left, move to the right, and bump squarely into the person. After a while, however, the subjects adapted to this new way of looking at the world and were able to function quite well. One subject learned to ride a bicycle with ease while wearing the prisms.

In another study, people were asked to reach for a target while looking through similar prism goggles that displaced their visual world several inches to the left or right. After watching their reaching hand for several minutes through the prisms, they were soon able to recorrelate vision with touch and reestablished normal performance despite the optical distortion. When the prism goggles were removed, the people would reach for the same targets and miss in the opposite direction. Other research has shown in these types of situations that, contrary to all appearances, *visual perception* does not change at all. It is *kinesthetic perception* that is affected (Harris, 1980). Instead of correcting the distorted visual input, people automatically adjust their kinesthetic perceptions to match the visual input.

It is the persistence of the altered kinesthetic perception that requires the readjustment period once the prism goggles are removed. This readjustment period can last for quite a long time (Harris, 1980a). People spent 15 minutes a day for 4 days drawing pictures and doodling while watching their hand through mirror-reversing prism goggles. Afterward, when they tried to write letters and numbers correctly while blindfolded, they often wrote them backward without being aware of their errors. It is interesting to note that only the hand that is observed through the prisms makes the kinesthetic adjustment; the other hand is unaffected by the visual distortions.

These and other similar findings have practical application in interactive computing. For example, if a visual movement is linked to a hand movement in another location (as when one uses a mouse to control a CRT), the arm's kinesthetic sense may become distorted. The distortion could happen without the user's awareness and could induce errors in movements made without visual guidance, such as reaching for familiar control buttons without looking.

Designers should proceed cautiously when developing systems that use spatially separated controls and displays. To help avoid these kinds of perceptual difficulties, designers should develop systems with coincident controls and displays such as using a light pen in computer graphics.

Matching Patterns

One of the most remarkable features of the perceptual processes is the capacity to respond to a wide range of differing patterns. For example, the size of type on a page has little effect, over a considerable range, on the speed or accuracy with which the page is read. We can view the printed page, the face of a friend, or any object from various distances and yet perceive it as constant in size. The ability to recognize spoken words is even more remarkable. Within limits, a speaker can vary his or her rate of speech or loudness; another voice with very different frequency characteristics can even take over and the listener will continue to understand.

Perception is more than just sensing isolated lines or patches of color. The overall arrangement is crucial. For example, the line segment shown at the top of Figure 4-6 is identified more accurately when it is part of a picture that looks like a three-dimensional object (item a) than when it is within a less meaningful context (items b to f) (Weisstein and Harris, 1974).

In proofreading, we sometimes may fail to detect errors (distortions) because the context leads us to expect a certain pattern. This is illustrated in Figure 4-7. Quickly read the short statement and see if it is correct. If someone reads Figure 4-7 quickly and is asked to point out the error, he or she would likely suggest that Jack and Jill fetched a pail of *water,* not milk. This is true. However, the number of errors in this short passage goes far beyond the substitution of milk for water. The words *went* and *a* are repeated. People frequently overlook this because we learn to ignore certain types of errors in favor of recognizing a consistent and meaningful whole.

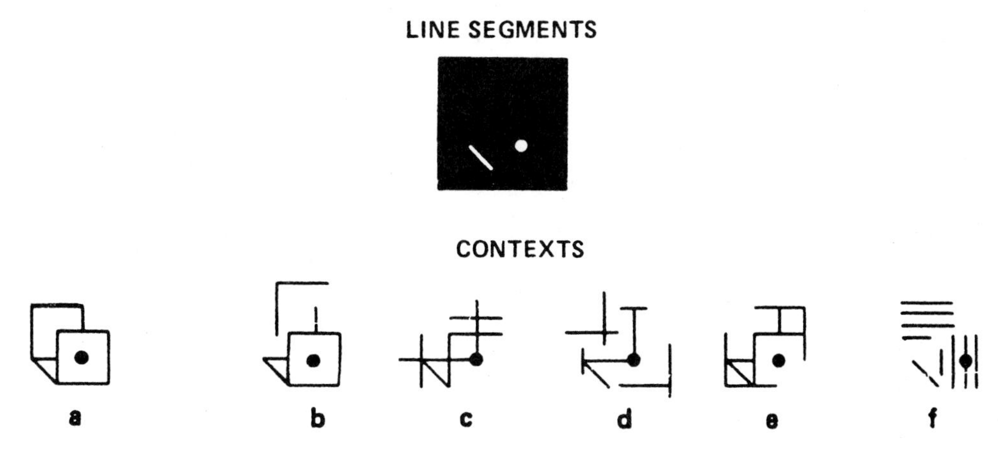

Figure 4-6 Visual detection of line segments.

Perception can be thought of as having the immediate past and the remote past brought to bear on the present in such a way that the present makes sense. The *skilled* perceiver detects features and structures of which a naive viewer or listener is not even aware because he or she lacks past experience (or knowledge of what is task relevant). A younger child, for example, sometimes ignores information that older children and adults recognize quite effortlessly.

Schema

Neisser (1976) proposed a perceptual cycle, which is illustrated in Figure 4-8. The cycle begins with a *schema* that directs exploration and samples the environment; this in turn modifies the schema, which directs new exploration and so on. The term *schema,* used in slightly different ways by Bartlett (1932), Woodworth (1938), Piaget (1952), and Posner (1973), is defined here as the portion of the perceptual cycle that is internal to the

Jack and Jill went

went up the

hill to fetch a

a pail of milk

Figure 4-7 What's wrong with this sentence?

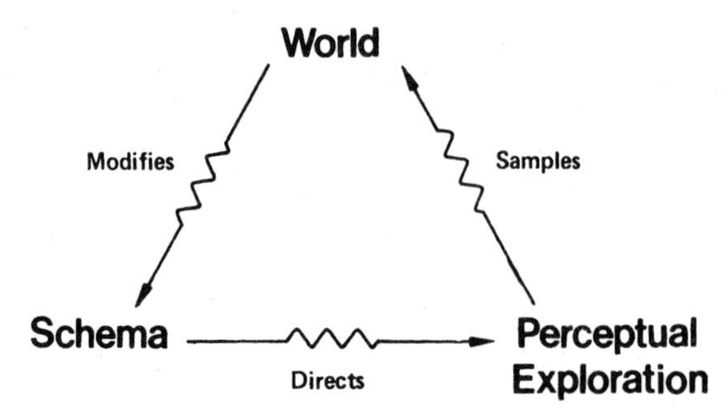

Figure 4-8 The perceptual cycle (adapted from Cognition and Reality: Principles and Implications of Cognitive Psychology, by Ulric Neisser, W. H. Freeman and Company. Copyright, 1976).

perceiver, modifiable by experience, and somehow specific to what is being perceived. The schema accepts information from the senses and is changed by that information. When viewed as an information acceptor, a schema is like a *format* in computer programming language. Formats specify that information must have certain characteristics to be interpreted coherently. Other information could be ignored or lead to meaningless (perhaps confusing) results. A schema also functions like a *plan* (cf. Miller et al., 1960) for finding out about objects and events and for obtaining more information to meet the format requirements.

Perceptual Cycle

We see perception as an active, constructive process; at each moment the perceiver constructs anticipations of certain kinds of information that enable him or her to accept them as they become available. *Perceptual explorations* are directed by the schema. Because we perceive best what we know how to look for, it is the schema (together with the information actually available in the world) that determines what will be perceived. Anticipatory schemas then are plans for perceptual action. These anticipations enable a person's eyes to move, head to turn, and hands to explore, all to be in a better position to sample the world. The outcome of the explorations, the information picked up, modifies the original schema. Thus modified, it directs further exploration and becomes ready for more information.

The perceptual cycle is probably best illustrated with an example by Neisser (1976):

> It is no evolutionary accident that babies are born with a tendency to look toward the sources of sounds, nor that the outer parts of the retina are sensitive to motion and change although they are poorly endowed for pattern vision. The sound of a footstep, like the first peripheral glimpse of a movement, is an effective guide to further exploration. In its own right, it indicates only that somebody is moving in a certain region of the environment. Nevertheless, it allows the perceiver to anticipate what a glance in that direction might reveal. This "anticipation" is not a deliberate and conscious hypothesis, of course, but a general readiness for information of a particular kind. If the perceiver actually executes

the exploratory glance, he embarks on the perceptual cycle; otherwise, he fails to perceive the object.

> To see my visitor properly, then, I must swivel my head and eyes around for a better look. In that better look, the visitor's face will probably be imaged on the central fovea of my eye. But perception is not complete in that moment either; during the next few seconds I will shift my gaze repeatedly as I look at him. Each exploratory eye movement will be made as a consequence of information already picked up, in anticipation of obtaining more. I will not be aware of the fixations or their sequence; only of the visitor himself.

Seeing, listening, and feeling are all perceptual skills that develop over time. *Expectations* appear to direct perception. For example, the old joke that the optimist sees the doughnut while the pessimist sees the hole does not imply that either is mistaken. Each will be confirmed by what he or she has seen. People can identify a picture or word far more easily when it is anticipated or plausible (i.e., matches expectations) than when it is rare or out of context. People can identify a letter better when it is part of a word than when it occurs in a meaningless string of letters, even when more than one letter is equally plausible at that position in the word (Reicher, 1969; Johnston and McClelland, 1974).

Bruner and Minturn (1955) report a study in which people were briefly shown either a letter (e.g., L, M, Y, or A) or two numbers (e.g., 16, 17, 10, 12). Afterward they were briefly shown a broken B in which the curved part of the figure was separated from the vertical line (see Figure 4-9). The subjects indicated whether the broken B was a B or a 13. As expected, the subjects showed a tendency to perceive a B when expecting a letter and to perceive a 13 when expecting numbers.

Consider the items in Figure 4-10. The answer to the first item is "sandbox." The time taken to perceive and answer each of the remaining items is due in large measure to appropriate schema already existing in memory. To understand the message in some problems requires the development of new schema. However, once an answer is given or a rule is established and the same type of situation occurs, existing schema can be used and recognition occurs much faster. Most people can eventually sort out, organize, and arrive at a

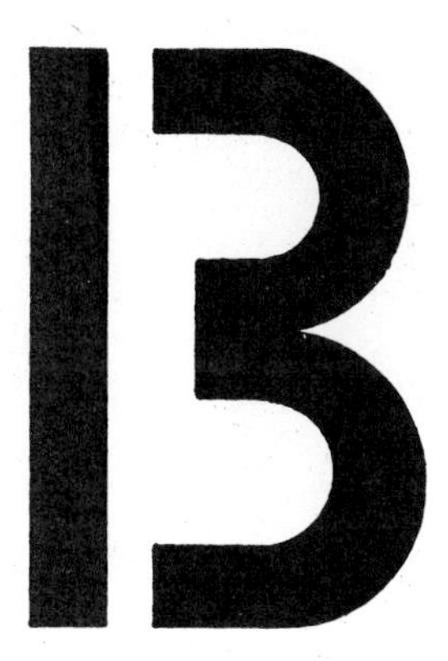

Figure 4-9 This figure could be perceived as a B or 13, depending on your expectations.

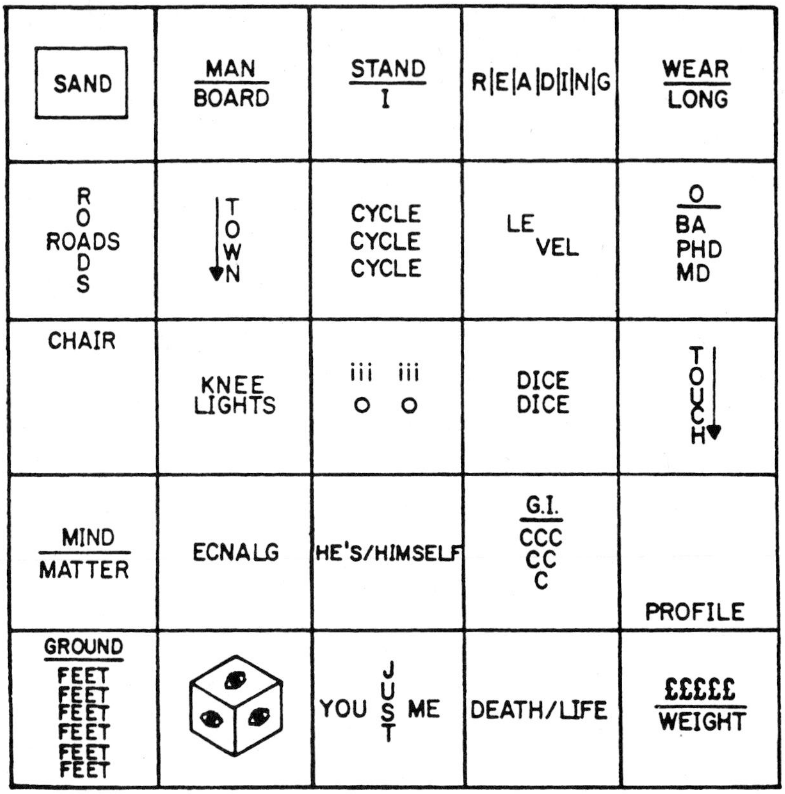

SAND	MAN BOARD	STAND I	R|E|A|D|I|N|G	WEAR LONG
R O ROADS D S	T O W N	CYCLE CYCLE CYCLE	LE VEL	O BA PHD MD
CHAIR	KNEE LIGHTS	iii iii o o	DICE DICE	T O U C H
MIND MATTER	ECNALG	HE'S/HIMSELF	G.I. CCC CC C	PROFILE
GROUND FEET FEET FEET FEET FEET FEET		YOU J U S T ME	DEATH/LIFE	£££££ WEIGHT

Figure 4-10 What is the message in each box?

meaningful solution to each of the problems by making the necessary schema changes. Because of the lack of appropriate schema, solving these problems is very difficult for computers.

Combining Sensory Input

We must not believe that people deal with each bit of sensory information that comes to the perceptual mechanism separately. People perceive total situations, not isolated sensory inputs. The perceptual process "puts it all together" in a way that helps people to understand what is occurring in the world. For example, we see someone walk and also hear the footsteps; we hear someone talk and also observe movements of the facial features. When eating, we touch, taste, and smell as well as chew. When viewing an argument, we see the gestures and attitudes of the participants and hear their words and tones of voice. The act of perceiving is a composite of experiences, and it is this composite that is stored and becomes the schema to which new experiences are related.

Developing Perceptual Skills

Gregory (1966) reports the experiences of a man who, blind from the age of 10 months, had his sight restored at the age of 52. When he was shown a simple lathe, he could not recognize it although he knew what a lathe was and how it functioned. When he was allowed to touch it, he closed his eyes and ran his hands over the parts of the lathe. Afterward, when he stood back and observed it, he said, "Now that I've felt it, I can see it." This man had developed a perceptual skill that allowed him to see the world better by touch than by sight. Without knowing what to look for or how to look for it, the object when first seen could not be perceived. Even the simplest perceptual skills must be learned.

Each time we reread a good book we inevitably gain new insights and new information. Each reading of a good book provides new information not because the words have changed, but because the *reader* has changed. The schema used for the second reading is much different from the schema used in the first reading. Our understanding of the world builds line upon line and precept upon precept.

Assume that we are interested in teaching people to use a new computer system. In the language of cognition, one of the designer's tasks is to create a set of experiences that will enable each user to build schema consistent with the quick and accurate use of the new system. One way to approach this problem is to give each person free access to the system, as well as whatever time is necessary to become proficient (this may take some users days, others months). A second approach, and one that is usually more efficient, is to develop a means of assisting the person in *developing appropriate schema.*

The latter approach is best known as *training* and is discussed in more detail later. Training has the advantage of guiding a potential user through the important conditions of an activity in some controlled manner, building schema as the user has each new experience. Leaving a person to explore a new system in some unstructured trial-and-error manner is usually a very inefficient way of constructing new schema.

It also tends to be more efficient to select users who already have a particular set of desirable schema. The schema-related requirements of an activity can be clearly outlined in a user profile. These requirements could include a knowledge of keyboard operation, a general knowledge of the activity to be performed, and perhaps other specific capabilities. A period of formal training will then follow with the express intent of building, in some systematic way, the existing schema related to performance with the new system. Once the basic schema (which will include experience with the system in a variety of different contexts) has been formed, the person is then considered minimally proficient and is able to proceed with performing the activity.

Illusions and Errors

Although people are usually quite good in judging shape, size, and distance, they may be fooled by illusions. Illusions are the result of errors made in the perceptual process. Illusions may be produced when we make assumptions about how things usually are (i.e., we rely too heavily on inappropriate schemata), which causes us to distort or misunderstand the information we actually receive. Many automobile, airplane, and other accidents are no doubt contributed to by illusions.

An interesting illusion, closely related to the real-world activity of proofreading, is the one shown in Figure 4-11. Look at the figure and as quickly as possible count the number of *f*'s. People usually respond by saying that there are three *f*'s. In actuality, there are six *f*'s. The difference probably occurs because the perceptual process, in an effort to make the statement meaningful, is unable to ignore the sound of the words. It overlooks those *f*'s that sound like v's, the ones in the "of's."

FINISHED FILES ARE THE RE-
SULT OF YEARS OF SCIENTIF-
IC STUDY COMBINED WITH THE
EXPERIENCE OF MANY YEARS.

Figure 4-11 How many *f*'s are in this sentence?

Certain types of perceptual errors that commonly occur in processing information may be related to illusions. This type of error is particularly easy to identify in computer systems where people are dealing with letters and numbers. If, for example, a person looks at a hand-printed 5 and thinks that it is an S or at a 2 and believes that it is a Z, or hears an H and keys an A, we would say that he or she has been fooled into misreading or mishearing the correct character.

Conclusions

A designer's primary objectives in making decisions that will ensure the adequate operation of the perceptual process should be to ensure that perceptions are accurate and to provide means for quickly building perceptual skills. The latter objective is obviously related to providing training and other experiences to assist a person to build the necessary schema. Keep in mind that accurate perceptions are most likely when a person encounters data, information, or conditions that are familiar and consistent with past experiences.

Incidentally, the answers to the messages in Figure 4-10 are (left to right, top to bottom): sandbox, man overboard, I understand, reading between the lines, long underwear, crossroads, downtown, tricycle, bilevel, three degrees below zero, high chair, neon lights, circles under your eyes, paradise, touchdown, mind over matter, backward glance, he's beside himself, G. I. overseas, low profile, 6 feet underground, icebox, just between you and me, life after death, and 5 pounds overweight.

Intellectual Processing

Intellectual processing is the second stage in the human information processing model. Effective processing in this state is highly dependent on having accurate perceptions. People are continually making decisions, solving problems, and reasoning in one form or another. This includes daily decisions such as what to wear, as well as momentous decisions, such as what college to attend. Good decision making, problem solving, and reasoning come with experience and constitute what we will refer to as intellectual skill.

Systems should be designed to encourage the development and use of the intellectual skills related to the activities to be performed. Two of the most important intellectual processes are problem solving and decision making. We will first present some information related to problem solving, or creative thinking, and then the processes involved in decision making. These two major intellectual processes are closely related to having successful systems.

Problem Solving

Problem solving is one of the highest and most complex forms of human mental processing. Problem solving is defined as the combining of existing ideas to form a new combination of ideas. It requires, then, a considerable accumulation of knowledge as a basis for recombining ideas and forming solutions.

A *problem* is a situation for which the human does not have a ready response (Davis, 1973). Being confronted with a problem makes an individual uncomfortable and therefore frequently provides motivation to find a solution. In other situations, users may attempt to solve problems because they provide a challenge, an opportunity to carry out a fantasy or resolve curiosity (Malone, 1980). In addition, a person with more experience may have a ready response for a situation that confuses someone with less experience.

Wallas (1926) lists the best known set of steps for problem solving:

Preparation: clarifying and defining the problem, along with gathering pertinent information

Incubation: a period of unconscious mental activity assumed to take place while the individual is doing something else

Inspiration: the "aha!" experience, which occurs suddenly

Verification: the checking of the solution

Solving problems in the real world frequently follows this sequence (Shulman et al., 1968):

Problem sensing: A person initially detects, to his or her discomfort, that some kind of problem exists.

Problem formulating: The person subjectively defines a particular problem and develops his or her own anticipated form of solution.

Searching: The person questions, gathers information, and occasionally backtracks.

Problem resolving: The person becomes satisfied that he or she has solved the problem, thus removing the discomfort.

Three Mile Island

The activities just described can be seen in the following description of a nuclear power plant accident. This situation was essentially a 2 hour and 22 minute problem-solving exercise. While reading the description, notice the iterative nature of the *problem formulation* and *searching* activities.

At 4:00 A.M. on March 28, 1979, a serious accident occurred at the Three Mile Island (TMI) nuclear power plant near Middletown, Pennsylvania. Mechanical malfunctions caused the initial problems, and the inadequate response of plant operators made things much worse. In the minutes, hours, and days that followed, events escalated into the first major crisis ever experienced by the United States nuclear power industry (Kemeny, 1979).

The control room at TMI, even under normal conditions, gives the impression that much is going on. The plant's paging system continuously spouts messages, and panel upon panel of red, green, amber, and white lights and alarms sound or flash warnings many times each hour.

Besides the nuclear reactor itself, probably the most important element in the TMI system is water. Water in the form of steam runs the turbine to produce electricity, and water also keeps the fuel rods from becoming overheated. On the morning of March 28, the water pumps stopped working and the flow of water to the steam generators halted. At this point, three emergency feedwater pumps automatically started.

There were two operators in the control room when the first alarm sounded, followed by a cascade of alarms that numbered 100 within minutes. One of the operators later reported, "I would have liked to have thrown away the alarm panel. It wasn't giving us any useful information." Fourteen seconds into the accident, one of the operators noted that the emergency water pumps were running. He did *not* notice two lights that told him a valve was closed on each of the two emergency water lines and that no water could reach the steam generators. One light was covered by a yellow maintenance tag. No one knows why the second light was missed.

At this point a relief valve should have closed to deal with ever increasing steam pressure. It remained open. However, a light on the control room panel indicated that the valve had closed. With the relief valve stuck open, the pressure and temperature of the reactor coolant erroneously dropped. About 1 minute and 45 seconds into the incident, the steam generators boiled dry because their emergency water lines were blocked.

Five minutes and 30 seconds into the accident, steam bubbles began forming in the reactor coolant system, displacing the coolant water in the reactor itself. The operators still thought that there was plenty of water in the system. With more water leaving the system than being added, the core was on its way to being uncovered.

Eight minutes into the accident, the operators finally discovered that no emergency feedwater was reaching the steam generators. One operator scanned the lights on the control panel that indicate whether the emergency water valves are open or closed. He checked a pair of emergency water valves, which are always supposed to be open except during a specific test of the emergency water pumps. The two were closed. He opened them, allowing water to rush into the steam generators. The major effect of the loss of emergency water for the first 8 minutes was that it *confused and misled the operators as they sought to solve their primary problem.*

Later investigation would show that throughout the first two hours of the accident the operators failed to recognize the significance of several things that could have warned them that they had an open relief valve and a loss-of-coolant accident. For example, one set of emergency instructions states that a pipe temperature reading of 200°F indicates that the relief valve is open. But the operators later testified that the pipe temperature normally registered about this high because the valves leaked slightly. Recorded data during the accident showed that the pipe temperature actually reached 285°F.

At 4:11 A.M., an alarm signaled high water in the containment building's sump, a clear indication of a leak or break in the system. The water had come from the open relief valve.

At 4:15 A.M., a disc on the drain tank burst as pressure in the tank rose, sending more radioactive water onto the floor and into the sump. At 4:20 A.M., instruments measuring radioactivity inside the core showed a count higher than normal, another indication, unrecognized by the operators, that cooling water was being forced away from the fuel rods. During this time, the temperature and pressure inside the containment building rose rapidly from the heat and steam escaping from the open relief valve.

Shortly after 5:00 A.M., TMI's four reactor coolant pumps began vibrating severely. This was another unrecognized indication that the reactor's water was boiling into steam. The operators feared the violent shaking might damage the water pumps, so they shut down two of the pumps. Twenty-seven minutes later, they turned off the two remaining pumps, stopping the forced flow of cooling water through the core.

At about 6:00 A.M. someone observed that the block valve, a backup valve that could be closed if the relief valve stuck open, had *not* been closed. The block valve was shut at 6:22 A.M. The loss of coolant was finally stopped. *It had taken 2 hours and 22 minutes to solve the problem.* Even so, the crisis continued for several more days, causing considerable mental stress for people living close to the facility. The direct financial cost of the accident was estimated to be over $2 billion.

Barriers to Problem Solving

One main difficulty faced by the problem solvers at Three Mile Island was their natural reliance on doing things that worked in the past. This is consistent with past research findings, which show that the two main barriers to effective problem solving are habit and pressure to conform. Habit and conformity are also referred to as rigidity, fixation, mental set, predisposition, resistance to change, and fear of the unknown. Perhaps the most difficult challenge for a designer is to develop ways for people to overcome the rigidity that causes them to use the incorrect, but habitual solution to a problem. One of the best illustrations of the existence of this rigidity or fixation with traditional solutions is the water jar problem (Luchins, 1942).

Table 4-2 shows a number of problems that a subject can solve by using available empty jars to obtain a prescribed volume of water. For example, consider the first problem

Table 4-2 Water Jar Problems

Problem No.	Jars Regarded as Given A	B	C	Required Amount (quarts)
1	32	4		20
2	100	20	3	74
3	14	163	23	99
4	18	43	10	5
5	9	42	6	21
6	20	59	4	31
7	23	49	3	20
8	15	39	3	18
9	28	76	3	25
10	18	48	4	22
11	14	36	8	6

shown in the table. The subject has a 32-quart jar, a 4-quart jar, and all the water that he or she needs. The problem is, "How can a person measure exactly 20 quarts of water?" How would you solve the problem?

The solution for this first problem requires a person to fill the 32-quart jar, then fill the 4-quart jar three times from the large jar: 20 quarts will then remain in the large jar. The second problem, to measure exactly 74 quarts, can be solved in an equally simple way. The 100 quart jar is filled and, from it, the 20-quart jar is filled once and the 3-quart jar twice. This will leave 74 quarts in the large jar. All the remaining problems can be solved using this same general method.

After working some of the problems, most people develop a fixed way of solving them and continue to solve all the problems in the same way. Only a few people recognize that for a couple of the problems a better solution is available. For example, consider the seventh problem on the list. Again, the problem can be solved by the method just described: the 49-quart jar is filled and from it the 23-quart jar is filled once and the 3-quart twice. However, by studying the problem for a moment it is apparent that there is a simpler solution. One can fill the 23-quart jar and then fill the 3-quart jar from it. This also gives 20 quarts. The existence of this type of fixation or rigidity hampers much problem solving.

The common expressions made by people who have adopted a set of rigid, long-standing habits and who resist the introduction of new ways to solve problems include "It has been done the same way for 20 years so it must be good," "We have never used that approach before," "We are not ready for it yet," "Somebody would have suggested it before if it were any good," "I just know it won't work," "You'll never sell that to management," or "You don't understand the problem."

Without concluding that all problem solving is the same, we can recognize some commonalities because of several important and identifiable dimensions of problem solving. The designer should focus on these commonalities when attempting to design a system that will enable an acceptable level of problem-solving performance.

Dimensions of Problem Solving

Davis (1973) has suggested three problem-solving dimensions. The first of these attempts to answer the question, "Is the problem really a problem?" Recall the definition of a problem as a situation to which a person does not have a ready response. Simple arithmetic "problems" and simple questions such as "Who flew the first airplane?" are not problems at all to most adults. Remember, some situations that represent true problems when a person first begins working in a system may not be problems after he or she gains experience.

One of the most difficult and most critical features of problem solving lies in defining problems. The designer should develop a system that makes problems as simple as possible. This sometimes requires that a large, single problem be broken into subproblems, each a subject for problem solving. For example, thinking of ways to improve the design of an automobile may be too complex a problem. Broken into subproblems, users could focus on improving seating, the driver's station, ways of entering and exiting, and engine performance. Whenever a problem is simplified, however, it must be defined broadly enough to allow *totally new* approaches to appear. For example, if the problem were stated as "How can we build a better bus?", we would limit our view to one approach, that of transporting people on a bus. More broadly stated, the real problem is "How can we best move people from one place to another?", which opens new ways of dealing with the problem.

The second dimension addresses the question "Does the activity require some type of systematic, organized approach to problem solving, or are problems solved through trial-and-error searching?" From a designer's point of view, the problem-solving approach makes a difference. Trial-and-error solutions are usually unplanned and random. Problems that are best solved using trial-and-error methods should be accompanied by design decisions to facilitate that kind of performance. On the other hand, if the problem-solving behavior is to be more systematic and organized, then the activities should be designed to assist *that* approach. Human performance problems may arise, for example, when a designer assumes that a solution will be systematically arrived at, but because of inadequate training the user relies totally on trial and error, or in situations where a designer has given little thought to the best means of solving certain types of problems, which may force many users into trial-and-error behavior. A designer should know which general approach is best and design the system to facilitate that approach.

Consider the anagram problem shown in Figure 4-12 (Smith, 1943). Can you solve it by rearranging all the letters to form one word? The solution to this problem can be brought about in a variety of ways depending on the design of the system. In the typical case, a person usually rearranges the problem letters mentally in some fairly systematic

LOVE TO RUIN

Figure 4-12 Anagram problem.

way to find the solution word. However, if a designer elects to provide a user with pencil and paper, it may encourage the anagram-solving activity to become much more trial and error. Frequently, a problem that can be solved systematically will be solved using trial and error if a designer supplies the appropriate material and the time to do so. A basic design decision must be made to determine whether the best human performance can be obtained systematically or through trial-and-error problem solving. In one case, the systematic solution may come more quickly but with more errors. However, if a trial-and-error solution of the problem is attempted, the time to arrive at a solution may be much greater.

Larkin et al. (1980) observed that limited short-term memory capacity constitutes one of the most severe constraints on problem solving. Their suggestion is to always consider providing a computer, paper and pencil, or some other means to extend the problem solver's short-term memory. Of course, the cost associated with keying or writing down information is that it will probably take more time to solve the problem. Even so, using notes is usually much faster than taking the necessary time to memorize (i.e., put in long-term memory) all relevant information. The solution to Figure 4-12 is "Revolution."

A designer should decide whether the short-term memory capacity of a user is adequate for the kinds of problems to be solved or if a means for assisting memory should be provided. Consider, for example, the time required to solve the problem in Figure 4-13 with and without writing.

Systematic problem solving usually requires much training and a considerable amount of experience. In some cases, a system designer may be willing to accept trial-and-error problem-solving performance initially, but as the user becomes more experienced, the designer may accept only systematic problem-solving performance. In this case, the design of the system should reflect the original trial-and-error performance and then make provision for a smooth transition to systematic problem-solving performance as the necessary skill is developed. Conditions, materials, and facilities that enhance trial-and-error performance will not necessarily enhance systematic problem-solving performance. How about a hint on solving Figure 4-13. The following has the same unique characteristic: "Madam, I'm Adam."

The third question is "Does the activity require one correct solution or many?" A distinction exists between a simple problem-solving task that requires only a single correct solution and more complex activities that require many original ideas or solutions. A person working in this system may be required to come up with a single solution in some circumstances and multiple solutions in others. A designer must accommodate both situations.

For example, a user (telephone engineer) attempting to determine the best layout of telephone equipment and facilities (e.g., lines, cables, and poles) for a new subdivision

What is unique about this statement?

Name no one man

Figure 4-13 Palindrome problem (adapted from Smith, 1943).

will think of a number of possibilities, but will ultimately select and implement only the solution that best meets the telephone requirements of the new subdivision. The system designer should provide all that is necessary to increase the probability that the user (the telephone engineer) will arrive at the best solution. Designers must determine ahead of time the best means for solving different types of problems and not leave this decision to users; the system should facilitate user decisions. People can be trained to apply different problem-solving strategies. But this is only possible if the best strategy is determined in advance and made part of the design of the system.

Problem-solving Techniques

Brainstorming

One or more of the following techniques can improve users' problem-solving skills. *Brainstorming* is probably the best known form of *forced* problem solving. The designer can provide for brainstorming as part of system performance. Brainstorming can be done by an individual writing out a large number of possible solutions or a small group of people contributing several ideas for a solution.

If brainstorming is to be used, the following rules should be kept in mind (Davis, 1973). First, criticism is ruled out, and adverse judgments of ideas must be withheld until later. The goal of a user is to produce a large number of ideas or possible solutions. Designers can provide systems that encourage individuals to brainstorm by requiring that they generate a long list of ideas as possible solutions to a system-related problem. Second, originality is desired. In fact, the wilder the possible solution, the better. Reducing or eliminating ideas is easier than coming up with good ones in the first place. Creative, even wild ideas can become imaginative problem solutions. Of course, many of the really impractical solutions will not be used, but a small percentage could justify the exploration. Third, the greater the number of ideas, the better. Ideas seem to become progressively more original when more and more are listed.

Finally, a designer should provide for a situation in which users can combine and improve proposed solutions. In addition to improving ideas of their own, users in small groups can suggest how ideas of others can be turned into even better ideas or how two or more ideas can be joined into still a better solution. If designers provide systems that encourage originating and combining previously unrelated ideas, the chances of having creative problem solutions seem to improve.

Attribute Listing

Attribute listing, another idea-finding technique, also yields novel idea combinations or problem solutions. Crawford (1954) has indicated that each time an improvement is made in a product or system, it is done by changing an attribute. Original invention occurs by improving the attributes (parts, qualities, or characteristics) of an object or by transferring attributes from one situation to a new situation. For example, a pencil's attributes are size, color, and shape. Each of these attributes can be altered to develop new kinds of pencils. A pencil can be skinny or fat, one color or striped, circular or hexagonal.

By changing one or more attributes, the item itself changes. This also applies to systems. A system may be improved by changing one or more of its attributes. Designers should develop systems that encourage users to critically consider attributes of situations when solving problems.

Checklists

Whenever we read through the Yellow Pages to locate, for example, automotive repairs, we are using an *idea checklist.* Checklist strategy amounts to examining some kind of list that suggests solutions suitable for a given problem. Once an auto mechanic is located, the problem is solved. Designer-provided checklists can provide possible solutions directly or indirectly. Paging through the Yellow Pages, a catalog, or a thesaurus is a direct means of finding a problem solution. We can use checklists indirectly to stimulate production of new ideas beyond those provided in the list itself. Osborn (1963) devised the following list to help to inspire solutions:

Put to other uses? New ways to use as is? Other uses if modified?

Adapt? What else is like this? What other idea does this suggest? Does past offer parallel? What could I copy? Whom could I emulate?

Modify? New twist? Change meaning, color, motion, sound, odor, form, shape? Other changes?

Magnify? What to add? More time? Greater frequency? Stronger? Higher? Longer? Thicker? Extra value? Plus ingredient? Duplicate? Multiply? Exaggerate?

Minify? What to subtract? Smaller? Condensed? Miniature? Lower? Shorter? Lighter? Omit? Streamline? Split up? Understate?

Although a checklist may not always be appropriate for system users, printed or computer-generated checklist performance aids often can facilitate problem solving in many systems. These techniques should shorten the time taken to arrive at solutions. Incidentally, the answer to Figure 4-13 is that each phrase can be spelled backward and have the same meaning.

Designing for Effective Problem Solving

We should design systems that encourage efficient problem solving. All the problem-solving techniques discussed can be used in systems to help ensure effective and imaginative problem solving. A designer's responsibility is to be aware of these and other techniques and to incorporate them as an integral part of the system. This means that problem solvers should have the necessary instructions, performance aids, training materials, or computer outputs to aid acceptable problem-solving performance. A problem that helps to illustrate the ideas just discussed is shown in Figure 4-14. See how long it takes to reorganize your own ideas and arrive at the solution. What could a designer have done to help you arrive at the solution sooner?

When considering problem solving, designers need to have some idea of the extent of knowledge users have concerning a particular problem situation. Another and equally

How quickly can you find out what is so unusual about this paragraph? It looks so ordinary that you would think that nothing was wrong with it at all and, in fact, nothing is; but, it is unusual. Why? If you study it and think about it you may find out, but I am not going to assist you in any way. You must do it without coaching. No doubt, if you work at it for long, it will dawn on you. Who knows? Go to work and try your skill. Par is about half an hour.

Figure 4-14 "Not so easy" problem.

important issue is how efficiently other relevant information can be organized and accessed so that it can be brought to bear easily on specific problems. This may require, for example, a separate computer database or an extensive library of relevant documents. The answer to the problem in Figure 4-14 is that there are no *e*'s in the paragraph.

Decision Making

The reader should be cautioned that the contrasting of problem solving and decision making as two separate processes is not always possible. Obviously, in some situations decision making may be a special case of problem solving, and in other situations problem solving may involve multiple decisions. The problem solving–decision making distinction is used here to help to identify and emphasize many of the different types of design decisions associated with intellectual processing. For our purposes, *problem solving* usually involves the discovery of a correct solution in a situation that is new to an individual. *Decision making* involves the weighing of known alternative responses in terms of their desirability and then selecting one of the alternatives.

Uncertainty

Decisions are usually made along a continuum that ranges from absolute uncertainty at one end to absolute certainty at the other. Uncertainty, here, refers to the uncertainty of the consequences of the decision. Purchasing stock in the stock market is an example of decision making with uncertainty. The future state of the world is not known when the decision is made; thus, the consequences of the decision are not known and there is high uncertainty.

Examples of making decisions with less uncertainty concerning the consequences include deciding whether to walk up two flights of stairs or to take an elevator. The consequences of the decision are more or less known, and the issue is focused more on which is the best decision. Designers must deal with the full range of decision uncertainties. Where uncertainty is low, a designer should make provisions for a correct decision based on available information. Frequently, this can be accomplished by using decision tables, logic trees, or contingency tables. A simplified decision table is shown in Table 4-3.

In general, a decision is easier to make if the decision maker knows or has a fairly good idea what the outcome will be. Suppose that we have to decide whether or not to

Table 4-3 Decision Table Example

If This Condition Exists:	Take This Action:
A	1
B	2
C	3

carry an umbrella when leaving home in the morning. There are two possible actions: take an umbrella or leave it home. Likewise, there are two states of the world that are of most interest: either it will rain or it will not rain. If the person knows for certain that it will rain, then he or she will carry an umbrella. If the person knows for certain that it will not rain, then he or she will leave the umbrella at home. In both of these situations, an individual makes a decision under conditions of little uncertainty. However, the world is not known for certainty. On some mornings we may not be at all certain whether or not it will rain that day. Even after listening to a weather report on the morning news, looking out the window at the sky, and considering what happened during the night, there may still be some uncertainty as to whether it will rain. After collecting all this information, an individual can establish a personal probability for rain during the day. The person may reason that the probability of rain is 75 percent.

In general, uncertainty adds complexity to decision making. If a designer determines that a decision situation contains uncertainty, she or he should provide instructions, performance aids, additional training, and any help to reduce the uncertainty as much as possible in a form that aids decision making. The designer may require the use of a computer to help decide among alternatives.

Decision-making Skill

Individuals tend to differ greatly in their decision-making capabilities. One major reason seems to be that some people have developed their decision-making skills to a greater degree than others. To ensure an acceptable level of decision making in a system, the designer should rely on both selection and training: select those with the most developed decision-making skill and then provide skill-building practice for them.

Since people can be trained to be better decision makers, designers should make proper training materials available to users to help to ensure adequate training and practice. If good materials are provided, user decisions will be made more quickly, more accurately, and with less training, and the decision-making performance itself will be satisfying.

Computer-aided Decision Making

A designer should provide all the information and tools necessary to make good decisions. In computer-based systems, the computer should be used to provide decision alternatives so that a decision maker can make use of as much information as is available.

People can integrate only a limited amount of information to make a meaningful and reasonable decision. Systems that require people to integrate large masses of data before making a decision will have degraded performance.

Experienced users can make good estimates of the likelihood of different events occurring if given sufficient information to do so. The *computer* can then take these likelihood estimates and, using Bayes's formula (Winkler, 1972) or some other appropriate means, combine them. This enables the final decision maker to decide among alternatives based on the probability of each alternative under different sets of circumstances and relieves the decision maker of a considerable amount of assimilation and storage of data.

In some systems, decisions are made sequentially, with later decisions being contingent on the consequences of earlier decisions. Generally, in these cases, the user is more concerned with the outcome of a series of sequential decisions rather than the outcome of any single decision. People are fairly good at evaluating a series of events occurring as a result of decisions and accumulating this knowledge in order to make later decisions. However, a designer can assist users by providing a computer for storing and adequately presenting the results of earlier decisions in a way that facilitates future decisions.

Risk

When people make risky decisions, they usually want to know what is at stake and what are the probabilities for certain outcomes. The designer can help an individual evaluate both stakes and the probabilities of outcomes by providing as much information about these two items as possible. Generally, people can assess what is at stake quite accurately, but do not assess outcome probabilities well, even when considerable information is available.

In decision making one assumes that a person weighs the stakes against the probability of occurrence. If the stakes are low (i.e., there is little to be gained or a small payoff) and the probability of an undesirable outcome is high, the individual will select an alternative course of action, if one is available. For example, a person may be driving a car with limited visibility due to a heavy rain. Passing a car on a narrow road without a clear view ahead means that the possibility exists that the driver may be killed. The gain, in this case, would be a few minutes saved. Most people would elect not to pass, but some people would make the decision to pass the car. Each decision is very personal.

To adequately weigh the stakes against probable outcomes, one must have a true appreciation of the actual risk involved in deciding on one course of action over another. It appears that people are not equally equipped to make this evaluation. In fact, the evaluation of risk may, in itself, be a skill that is developed with experience. A child may not perceive a situation as risky, while an adult may recognize the risk.

Other Considerations

Designers should keep in mind that early decisions tend to restrict the range of choices for future decisions, and, consequently, many users are tightly bound up due to decisions they have made in the past.

Finally, there are at least four decision-making characteristics that the designer should recognize when developing a system:

Users usually wait longer to decide than needed (overaccumulate information).

Users tend not to use all available information.

Users tend to be hesitant in revising original opinions, even if new information warrants revision.

Users usually consider too few alternatives.

Exercise 4A: Proofreading for E's

Purpose: To demonstrate how the perceptual process can affect proofreading.

Method: You will need four people to serve as subjects. Have the subjects sit at a desk or table with a pen. Make four photocopies of the paragraph entitled "Salt versus Sugar" that follows. Give a copy of the paragraph to each subject and read these instructions: "Go through this paragraph and put a line through every *e*. Perform this task as quickly and accurately as you can. Ready? Begin!"

Time the subjects to help to encourage completion in the shortest possible time. However, the time information will not be used when analyzing the results. What is of most interest is the number of *e*'s that were *missed.*

SALT versus SUGAR

In March, 1962, a shocked nation read that six infants had died in the maternity ward of
the Binghamton, New York, General Hospital. The deaths had occurred because the
infants had been fed formulas prepared with salt instead of sugar. The error was traced to a
practical nurse who had inadvertently filled a sugar container with salt from one of two
identical, shiny, 20-gallon containers standing side by side, under a low shelf, in dim light,
in the hospital's main kitchen. A small paper tag pasted to the lid of one container bore the
word "Sugar" in plain handwriting. The tag on the other lid was torn, but one could make
out the letters "S . . . lt" on the fragment that remained. As one hospital board member put
it, "Maybe that girl did mistake salt for sugar, but if so we set her up just as surely as if
we'd set a trap."

To help in your analysis of the results, consider the following: (a) the paragraph contains 68 *e*'s; (b) the number of *e*'s per line is as follows: line 1 has 5, 2 has 10, 3 has 9, 4 has 6, 5 has 6, 6 has 7, 7 has 6, 8 has 9, 9 has 7, and 10 has 3; and (c) there are 12 occurrences of the word "the."

Reporting: Prepare a report of your findings. Use the format discussed in Exercise 1A. Are silent *e*'s much more likely to be missed than pronounced *e*'s? Does the position of an *e* in a word have any effect? Is there any word that has an *e* that is more likely to be missed? Even though you told your subjects to *visually scan* the paragraph, does it seem like they "heard" some of the words while they were reading? What does this study suggest about the perceptual process?

Exercise 4B: Developing a Skill

Purpose: To evaluate the skill development process.

Method: Use yourself as the subject. You will need a hand-held mirror, a pen, and a timing device (e.g., a wall clock, wristwatch, etc.). Sit at a desk or table. Hold the mirror with one hand and your pen with the other. Draw with the hand you usually use for writing.

When the exercise begins, you will watch what you are drawing by looking *only in the mirror.* Use the mazes given (you may want to make a photocopy).

Before you begin using the mirror, time yourself on how fast you can do Maze 1 *without* using the mirror. This will be a good estimate of how fast you should be able to draw after the mirror-drawing skill has been well developed.

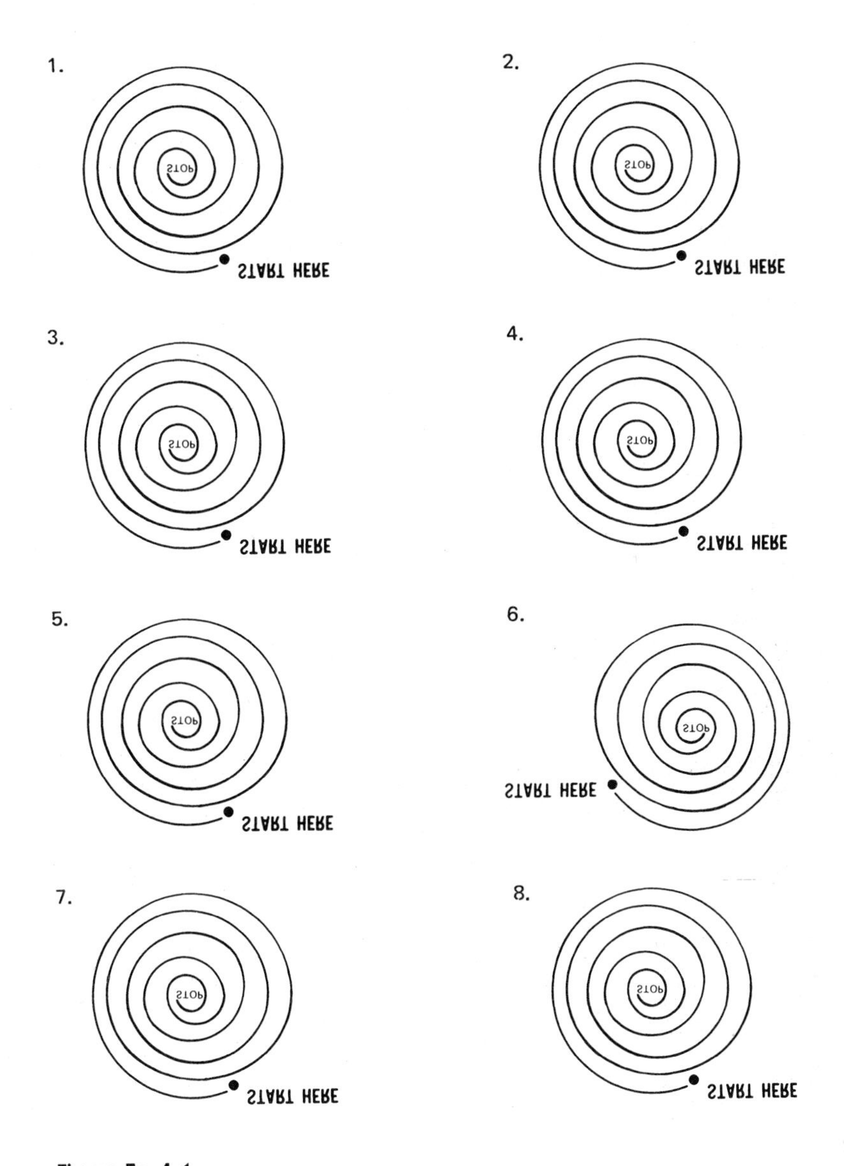

Figure Ex. 4-1

Next, go to Maze 2. Put your pen where it says "Start Here" and move in the white space to where it reads "Stop." Do not raise your pen from the paper once you have

RECORDING PERFORMANCE

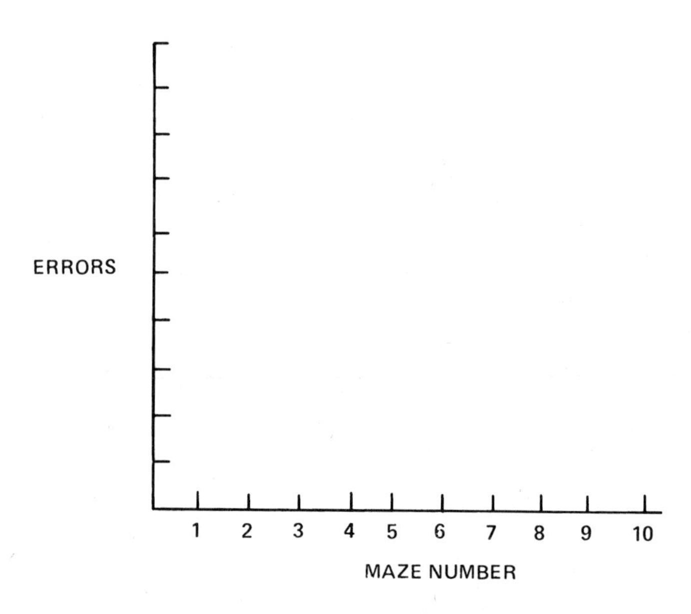

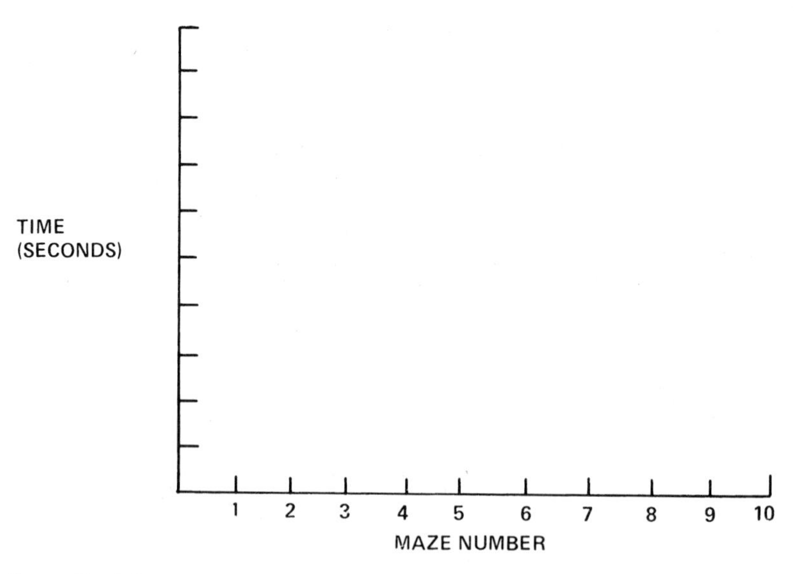

Figure Ex. 4-2

started. Complete each of the mazes in the correct order (2 through 8) as quickly and as accurately as you can.

After you finish each maze, record your time. After finishing all the mazes count your errors. An error should be counted each time that your pen touches a boundary line.

Reporting: Write a report using the format discussed in Exercise 1A. Include in your *Results* section Figures Ex 4-1 and Ex 4-2. Plot your own results in the figures, and discuss the skill development process.

As the mirror-drawing skill developed, did you get faster or make fewer errors or both? Was the skill development process smooth and consistent? How can this skill be best characterized: as a perceptual skill? or as an intellectual skill? or as a movement control skill?

5

Memory

Introduction

Human performance is frequently degraded because people forget. As shown in Figure 5-1, memory is one of the five brain processes that is important when trying to understand how people process information. The central processing stages of perception, intellection, and movement control all require memory to function properly.

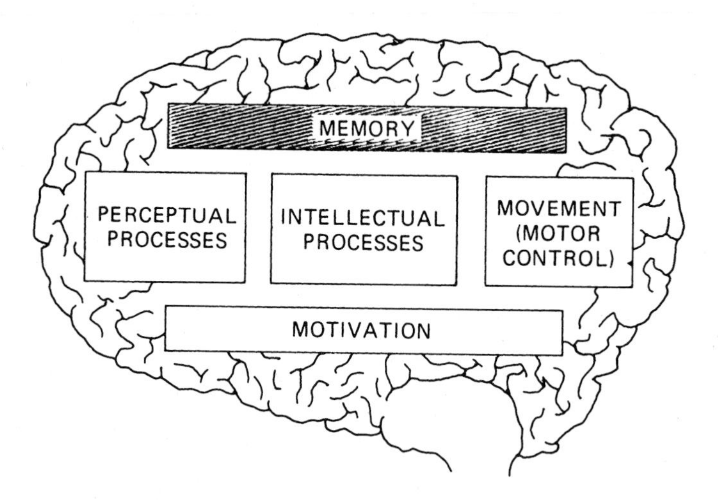

Figure 5-1 The human information processing model.

The fallibility of human memory is well described in a conversation between the King and Queen in Lewis Carroll's *Through the Looking-Glass:*

> The King was saying, "I assure you, my dear, I turned cold to the very ends of my whiskers!"
>
> To which the Queen replied, "You haven't got any whiskers."
>
> "The horror of that moment," the King went on, "I shall never, *never* forget!" "You will, though," the Queen said, "if you don't make a memorandum of it."

Designers should be aware of at least three types of human memory: sensory memory, short-term memory, and long-term memory.

Sensory Memory

There do not seem to be many design-related issues associated with sensory memory. However, it is important for designers to be aware of its existence and to make full use of its capacities when needed.

Sensory memory can be demonstrated in a number of different ways. Move your finger rapidly back and forth in front of your eyes; you should see it in more than one place at a given time. Or note the trail that a Fourth of July sparkler or lighted cigarette leaves when waved in the dark. These and other more controlled demonstrations show that an image persists (is briefly stored) after the stimulus is no longer present. This persistence of information makes it available for further processing even after the stimulus has moved or terminated. In the case of vision, the persisting information seems to be stored in the sense itself rather than in the brain (Sakitt, 1975). This also may be the case for the other senses. We will refer to this persistence of a stimulus that probably takes place in the sense itself as *sensory memory* (Sperling, 1960; Averbach and Coriell, 1961).

Sensory memories are known to exist for vision, audition, and (possibly) touch. They are characterized by being very brief and, at least in the case of vision, as being a literal representation (i.e., a more or less photographic image) of the stimulus. Items in sensory memory quickly fade or are erased by new inputs. In the case of vision, an image usually persists for about a quarter of a second or slightly longer (Averbach and Coriell, 1961; Dick, 1969, 1970; Haber and Hershenson, 1973; Dick and Loader, 1974). In his original studies, Sperling (1960) reported a duration of about 1 second. Others have observed visual sensory memory durations in excess of 2 seconds where the stimulus–background contrast was great (Averbach and Coriell, 1961; Mackworth, 1963). This means that a visual stimulus is available for cognitive processing for at least ¼ second up to about 2 seconds after the stimulus is removed.

The auditory sensory memory appears to last for at least ¼ second (Massaro, 1972) and may last as long as 1 to 5 seconds (Glucksberg and Cowan, 1970; Kubovy and Howard, 1976). In addition to vision and audition, there is some evidence for a sensory memory for touch that lasts for about 0.8 second (Bliss et al., 1966). Similar memories may exist for other senses as well.

It is still not altogether clear what role sensory memory plays in cognition. Glass et al. (1979) suggest that, since auditory information such as speech is spread over time, it is necessary to preserve brief segments so that they can be processed as units. For example, in English a person can signal a question simply by raising the intonation pattern, as in "You are tired of *studying?*" Here it is necessary to preserve segments long enough to tell that the intonation is in fact rising. The auditory sensory memory makes this possible. The function of the visual and touch sensory memories is even less clear. Jonides (1979) has proposed that visual sensory memory gives people time to detect, switch attention to, and process events that occur off the center of the field of vision. Massaro (1975) believes that the visual sensory memory assists in performing activities such as reading.

Designers should note that the duration of sensory memory may be somewhat under their control. At least for vision, the duration of sensory memory can be lengthened by optimizing the stimulus–background contrast. Difficulties with visual sensory memory would likely show up as an increase in errors. These errors would be characterized either by a lack of pattern (Miller and Nicely, 1955) or an excess of visual confusions (Keele and Chase, 1967). If a typist or key operator were typing codes, say seven-character codes such as DLTRVSA, we would expect that there would be more errors in the rightmost positions and fewer in the left positions (Bryden, 1960; Bryden et al., 1968; Mewhort and Cornett, 1972; Heron, 1957) and more errors in the center positions than in the end positions (Merikle et al., 1971; Merikle and Coltheart, 1972; Coltheart, 1972).

Short-term Memory

A second memory store, one that has more obvious human performance implications, is *short-term memory.* Short-term memory is sometimes referred to as *working memory* (Baddeley, 1992). People use short-term memory to hold information temporarily, usually for a few seconds. Hundreds or even thousands of tasks performed each day require this type of remembering. Most of the information dealt with daily is "throwaway," actually meant to "pass in one ear and out the other." We would soon become overburdened if every sight, sound, touch, smell, and pain were stored in our memory. A temporary memory store, what we are calling a short-term memory, is therefore convenient.

Encoding

The exact visual, auditory, or kinesthetic image or message is not directly stored in short-term memory. Rather, the information stored must first be *encoded.* The exact form of encoded memory is still being debated. Information is converted into a form that is consistent with human physiology and that aids further processing. The precise physiological form of the code is not important for our purposes, but understanding that the information has been encoded and therefore is in a form much different from the original stimulus image *is* important.

Evidence suggests that some of the *visual* information received by an individual is encoded in short-term memory in *auditory* form (Sperling, 1960; Conrad and Hull, 1964). These studies have shown that some errors made in visual tasks are errors that would more likely occur in an auditory task when sounds, not shapes, are confused. For example, an A substituted for a K or a B substituted for a C. The issue of encoding, then, has at least two dimensions that interest us: The information is encoded into a form that can be conveniently stored in human memory, and, probably, some visual information is transferred into auditory form (possibly to assist rehearsal). The important point is that the brain deals with information in an encoded form that is much different than the original stimulus.

Capacity and Duration

Probably the two best-known characteristics of short-term memory are capacity and duration. The short-term memory store is small and appears to hold from four to six units of information for most adults (Broadbent, 1975; MacGregor, 1987). Some people have slightly fewer and some slightly more units available to them. Those with more available units tend to perform better in

Reading comprehension

Drawing inferences from text

Learning technical information

Reasoning skill (Baddeley, 1992).

A unit is any organization of information that has previously become familiar, such as familiar words or a familiar configuration of pieces on a chessboard. For example, if someone looks up a telephone number and stores this information in short-term memory for a few seconds while dialing, he or she would usually store and correctly recall the seven characters as long as there were some familiar combinations. If, however, the person looks up an unfamiliar 10-character telephone number and tries to remember it while dialing, his or her performance would almost surely have errors. In the latter case, the short-term memory store was expected to hold more than its capacity.

Thus, short-term memory units may be made up of one, two, five, ten, or more items. A familiar telephone number prefix (e.g., 543) may use up only one unit, whereas an unfamiliar telephone number may use up all available units. Shepard and Sheenan (1965) conducted a study in which subjects reproduced from memory eight-digit numbers of two different kinds. In the first case, the last four digits of each number were selected at random, while the first four digits were selected from two familiar four-digit prefix sequences. In the second kind, this order was reversed so that one of these two familiar sequences *followed* the four random digits as a suffix. They found decreases of 20 percent in response time and 50 percent in errors when the last four digits were familiar rather than the first four. It was better (faster, with fewer errors) to have the familiar characters at the end, rather than the beginning. This finding has implications for codes that have some parts that are more familiar than others. To improve human performance in activities that use short-term memory, one must take into account the capacity limit of short-term memory and ensure that it is never exceeded.

The relationship between the number of items to be remembered and the length of time that they will persist in short-term memory is shown in Figure 5-2. In one case, a single unit of memory (a word) was used. In a second case, three units of memory (three-consonant codes, such as HZB or CFW) were used. In both cases, people performing a difficult irrelevant task during the recall interval minimized rehearsal. Obviously, the single unit was retained more accurately and longer than the three units. In fact, almost all the three-consonant codes had completely disappeared from short-term memory after only 18 seconds. If the designer needs to have information remembered, he or she should probably keep the messages as brief as possible and ensure little interference. People forget longer messages sooner. Therefore, when working with short-term memory, *the shorter the code, the better.*

There are some who feel that it is not a matter of units, but the time required in rehearsal. For example, the number of words that can be stored in short-term memory is inversely related to the spoken duration of the words. This is most likely related to the sub-

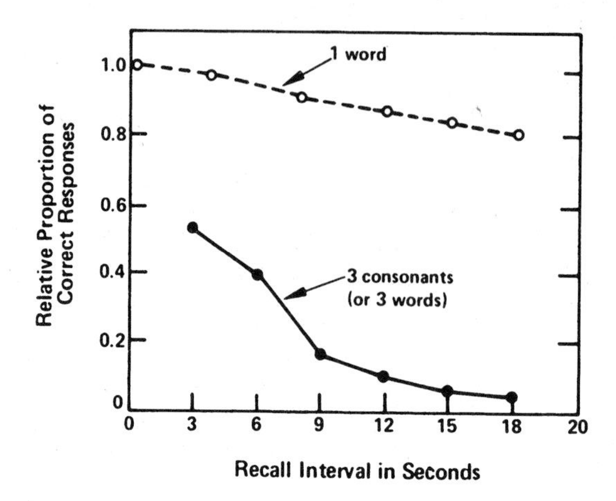

Figure 5-2 Short-term memory studies.

vocal rehearsal process. People can remember about as many words as they can say in 2 seconds. Short-term memory seems to reflect sheer speed of processing.

Rehearsing

Rehearsal retains information in short-term memory. When a designer must use longer strings of characters (sometimes referred to as *codes*), he or she should divide them into groups of three or four to help rehearsal. For example, the code 217964831 becomes 217 964 831. Designers also should keep in mind that during rehearsal of a code or other information other intellectual activities should be limited. If an individual is rehearsing in an attempt to remember long enough to key something, or to speak to someone, or to write down an item number or other code, other activities should not interfere with this rehearsal process. Rehearsal tends to cause material to be stored in the practiced form. The most likely mistakes are those related to the rehearsed sensory features (Conrad and Hull, 1964; Conrad and Rush, 1965; Locke and Locke, 1971).

Patterns

By rehearsing we ignore new inputs and may even determine *new patterns* and possible rhythms to help maintain the items already in short-term memory. Consider the code 427947247. This may be recalled by establishing a pattern of characters in which the last digit of each three-character set is the same: 427 947 247. Such creative rehearsal, a skill that can improve with practice, becomes very important in situations when material must be remembered for a short time. Designers can also facilitate remembering this kind of information by building patterns into codes and then teaching users how to work with the patterns.

The series of letters NTH EDO GSA WTH ECA TRU can be extremely difficult for most people to remember, even for a very short period. In fact, it appears to require 18 units of memory when only about 6 units are actually available. Some people may try to remember each of the 6 three-character units and may be able to do so successfully. Other people may stare at the 18 characters and look for a pattern or ways to make the code more familiar. Believe it or not, this particular code can be organized to the point where only one or two units of memory are required. By changing only one letter, it actually contains six words: "The dog saw the cat run." Once this pattern is recognized, the load on the short-term memory is drastically reduced and this particular code can be recalled easily with few errors.

Ericsson et al. (1980) report on a person who was able to use patterns to increase his short-term memory from 6 to 79 digits, but only after more than 230 hours of practice (1 hour a day, 3 to 5 days a week for about 1½ years). He was read random digits at the rate of one per second; he then recalled the sequence. If the sequence was reported correctly, the next sequence was increased by one digit; otherwise, it was decreased by one digit. This individual categorized each number into groups of three or four and then applied a mnemonic that was usually (90 percent of the time) based on running times and ages. For example, the number 3492 was remembered as "3 minutes and 49 point 2 seconds, near world-record mile time"; and 893 was "89 point 3, very old man."

Short-term Memory for Numbers

Chapanis and Moulden (1990) report the results of a study using short-term memory to recall individual numbers. They found that some numbers were more easily remembered than others. The relative difficulty for individual numbers (from fewest errors to most errors) was 0, 1, 7, 8, 2, 6, 5, 3, 9, and 4. The characters that were substituted for each other the fewest number of times were 0 and 1. The characters that were substituted for each other the most were 2 and 3, 3 and 4, 4 and 5, and 5 and 6.

Two-character combinations were remembered in the following order. The table should be read from left to right and top to bottom (e.g., 00 had the fewest errors and 64 had the most errors).

00	11	10	20	01	30	70	80	90	50
05	60	02	40	09	12	07	77	08	03
21	88	04	06	81	22	71	17	99	33
18	15	19	14	55	31	44	41	51	89
91	13	16	98	61	25	27	56	28	66
87	35	85	96	58	67	52	29	97	23
48	24	38	65	62	42	37	78	83	57
84	69	82	86	46	63	26	68	54	75
32	53	76	95	36	45	79	74	72	34
43	39	73	92	59	94	93	49	47	64

About 88 percent of the easiest two-character combinations contained a 0 or a 1 or contained identical digits (00, 11, 22).

The easiest to remember three-character combinations tended to contain a 0 or 1 and/or repetitive numbers, such as 000, 111, 222. For example, the easiest 5 percent of three-character combinations (those with the fewest errors) were the following:

000	777	111	800	222	888	040	100	333	999
005	200	444	002	009	600	022	090	110	201
666	020	210	211	300	555	006	011	500	700
900	810	001	004	220	400	890	008	007	016
030	070	080	112	311	610	910	060	103	630

The hardest to recall tended to contain a 4 and/or a 9. For example, the 20 most difficult to remember three-character combinations (those with the most errors) were the following:

947	759	564	675	968	823	645	939	739	472
439	683	724	576	574	549	495	483	264	673

Serial Position

Many conditions related to short-term memory could be considered. We will briefly discuss only one, those related to serial position errors. This means that, if a seven-character code is to be remembered for a short time and then written down, dialed, or keyed into a CRT, errors will tend to occur in certain character positions more frequently than in others. For example, in a seven-character code, most errors tend to occur in the fifth position and the fewest usually occur in the first position. Figure 5-3 shows a set of serial position curves. It is interesting that whether an item in the seventh-character position will be recalled depends on whether the original stimulus was auditory or visual (Conrad and Hull, 1968). With auditory stimuli, the last character tends to be recalled about as well as the first character, but with a visual stimulus, the last character tends to be mistaken more often (see Figure 5-4.)

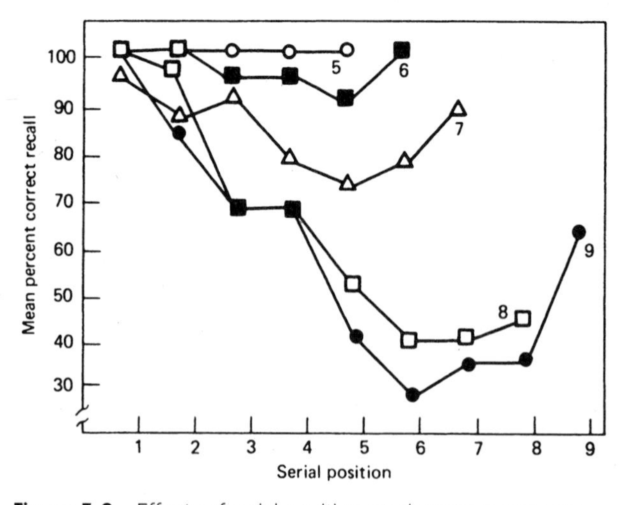

Figure 5-3 Effects of serial position on short-term memory.

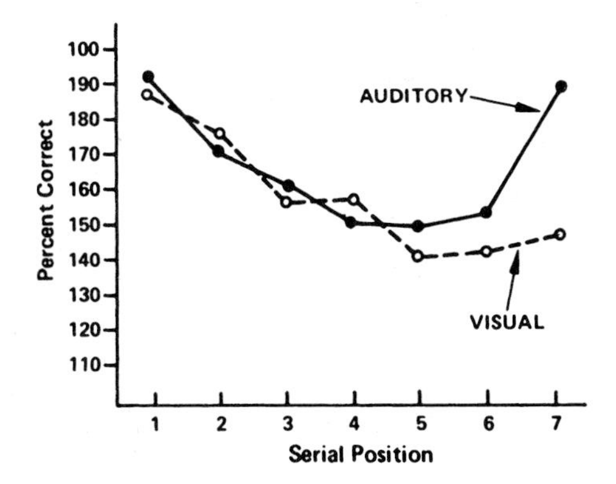

Figure 5-4 Illustration of serial position curves (adapted from Conrad and Hull, 1968).

Conclusions

Obviously, designers should not require users to retain even a small amount of information for a short time (5 to 10 seconds) if rehearsal is not possible. And they should not require users to retain even easily rehearsed information much beyond 20 seconds. Note in Figure 5-2 how the proportion of correct responses begins to decline after even the first few seconds.

Primarily, we lose material from short-term memory by replacing it with new information. New information seems to push out old information because of the restricted number of units that short-term memory can hold at any one time. One of the best ways for a designer to ensure that information in short-term memory persists is to provide for no new information while the old information is needed and being rehearsed.

Designers should particularly keep in mind problems associated with short-term memory, such as limited capacity (four to six units), relatively short duration (usually less than 10 seconds), and the requirement of continued rehearsal without interference to maintain information in short-term memory.

Long-term Memory

Long-term memory is a more or less permanent memory store. The concepts of learning and memory are closely related; the results of learning must be remembered for experience to accumulate. Long-term memory is essentially unlimited; at least it seems capable of storing all memories that come about in the lifetime of an individual. Unlike items stored in short-term memory (which could disappear in 20 seconds or less), items stored in long-term memory appear to last forever.

Items tend to make it into long-term memory if the item can be easily "hooked up" or linked with something that is already there. Thus, long-term memory relies heavily on *organization* to build and maintain content. Although we cannot be sure that information is ever actually lost from long-term memory, some stored memories may become less accessible or less easy to retrieve.

When most people refer to learning, remembering, and forgetting, they are really thinking of long-term memory. Designers should be interested in knowing what already has been learned and stored in the memories of users and what types of information should be added to memory (what must be learned by users) for users to perform adequately. Ensuring that information is *properly stored* and *easily accessible* is important also. We usually say a person has forgotten information when it would be more accurate to say that he or she *has lost access to it.*

A discussion of memory is also a discussion of forgetting since the terms are complementary: The amount forgotten equals the amount learned minus the amount retained. Designers deal constantly with users who forget. If people did not forget, a designer would need to present information to system users only once to ensure that appropriate knowledge was gained for performing the activity. Those who have tried to train people know that hardly ever is once enough.

Two of the most persuasive lines of evidence underlying the belief that people do not lose information stored in long-term memory are related to brain stimulation and hypnosis. The first of these relates to some experiments done in the 1940s in which electrodes were used to penetrate a small portion of the inner brain. The patients were awake during this procedure. When these small areas were stimulated, about 3.5 percent of the patients reported having vivid recollections of situations, feelings, and odors relating to experiences that they once had but supposedly had forgotten (Penfield, 1958, 1969; Penfield and Perot, 1963).

Twenty-four patients claimed to have had an auditory experience; they "heard" a voice, music, or some other meaningful sound. Nineteen patients claimed to have had a visual experience; they "saw" a person, group of people, a scene, or a recognizable object. Only 12 reported a *combined* auditory and visual experience. Furthermore, detailed examination of these few patients' reports suggests that they could have been reconstructions or inferences rather than memories of actual events (Loftus and Loftus, 1980).

The second form of evidence involves the use of hypnosis. Frequently, hypnotized people can recall vividly experiences that occurred when they were much younger. People seem to have difficulty recollecting these experiences without hypnosis. Barber (1965), however, pointed out that no evidence exists to support the view that recall under hypnosis is superior to recall under ordinary waking conditions. Maybe people report more *willingly* rather than more ably when under hypnosis. Although memories are not always accessible, the evidence is unclear whether these memories are lost.

Why then do we seem to forget material from long-term memory? To provide even the briefest answer, one must understand that at least three operations take place related to remembering and forgetting: encoding, storage, and retrieval.

Encoding

Consider the following example. We have a filing system organized, in part, according to the categories shown in Figure 5-5. We file new articles pertaining to these categories in the appropriate section. A new article on handprinting errors, for example, would be placed in a file cabinet under the handprinting category. From time to time an article

HUMAN ERROR
History of Human Error
General
Theory
Detection (Manual and Computer)
Correction
Costs
Reporting
Error Rates

HUMAN INFORMATION PROCESSING
Processing Capacity
Short-Term Memory
Perception
Memory
Speed-Accuracy Tradeoff
Bayesian Processing/Decision
Problem Solving

DATA ENTRY
Handprinting/Writing
Keyboard/Keying
Optical Recognition
CRT (Displays)
Forms Design
Character Design
Code Design

Figure 5-5 Filing system categories.

appears that does not fit neatly into one of the existing categories. We must decide where to file the new report. For example, an article on whether or not to use "boxes," "tic marks," or "underlines" on forms to reduce handprinting errors could be filed under either "handprinting" or under "form design." This categorization process must be consistent to work. Problems may arise later if the new article is placed under "form design" and another similar article is placed under "handprinting." In human memory, we refer to this process of deciding how to classify information as *encoding*. It requires an individual to perceive the information and determine one or more essential characteristics of the information. As discussed in the short-term memory section, part of the encoding process involves changing the form of the information prior to storing it.

Generally, what actually gets stored is not the same as what was originally perceived. Usually, only the essence of what was sensed and perceived will be encoded. This is a basic difference between the filing system in our example and human memory. What is in our physical filing system is the exact article that was originally received and not an abstract or other approximation.

Possibly the encoding process becomes more efficient with experience; some people learn how to learn better (or accumulate many more associations from which they can retrieve). This means that being exposed to numerous different types of learning experiences may enable people to encode more effectively.

Storage

The second necessary operation in long-term memory is *storage*. In our example, we assume that a new article is stored in the filing cabinet and not placed in a desk drawer or accidentally thrown out. Furthermore, we hope that the print on the paper will not disappear with time so that the information, once put into the file cabinet, stays in the same form with the same amount of clarity as when it was filed. We assume that the same can be said for human long-term memory, although we have little information on the true form of a memory at filing time (after encoding).

The storage process may last over a relatively long period of time, even while we sleep. There are at least two studies suggesting that sleeping and dreaming may be part of this process (Horgan, 1994). In one study, researchers placed electrodes into the hippocampus of several rats. As the rats learned to navigate a maze, their neurons fired in certain, clearly defined patterns. For several nights after the rats' maze exercises, their hippocampal neurons displayed similar firing patterns. It seemed that the rats were playing back their memories of running the maze. The major difference was that the firing was more rapid, as if the memories were being run on fast-forward. The firing occurred during slow-wave sleep, a phase of deep sleep in which dreams can occur.

A second study was done with people who were learning to detect symbols hidden in images flashed at the periphery of their vision. The researchers observed improvements in performance over a 10-hour period following a training session. To determine whether sleep played a role in this phenomenon, they disrupted the sleep of volunteers after they had their training session. Interfering with the subjects' slow-wave sleep had no effect. But an equivalent disruption of rapid eye movement (REM) sleep, which is marked by vivid dreaming, kept the subjects from improving overnight. Is it possible that dreams represent "practice sessions"? Possibly, slow-wave and REM sleep allow different types of "sleep-practice" to take place.

Retrieval

The third long-term memory operation is called *retrieval*. Retrieval is the opposite of storing. If we wished to find the reports on the use of boxes with handprinting, we would have to decide exactly where to start looking in the file system. Is it filed under "handprinting" or under "form design"? Or has a new category been established that covers in greater detail this particular topic of interest? There are many potential locations for this article, and it may not be clear exactly where the article has been stored. The person who originally filed the article may be able to retrieve it quickly, but only if not too many other articles have been filed in the meantime. Someone unfamiliar with the system might have to search each category until he or she finds the article. People usually are not aware of searching in different locations when trying to remember something stored in long-term memory, but it seems that this is precisely what we do. In some cases, people try to remember the storage location by recalling how the information was originally filed, including the circumstances surrounding the original filing. In many cases, if the original means of storing can be recalled, the appropriate storage area can be identified quickly.

Forgetting

Forgetting may be due to a failure of any of these three operations (and possibly others) in long-term memory. Original encoding may be incorrect (information may be stored under another category), information may be in some way degraded during or even after storage, or information may be difficult to retrieve because the search process takes place in the wrong part of the file.

For many years it was thought that forgetting was primarily a result of disuse or decay. This became known as the "leaky-bucket" description of forgetting. This concept of forgetting suggested that learning was the result of practice or use and that, when information was not used, forgetting occurred. Therefore, *disuse* was considered as the main cause of forgetting. Disuse was related to time and, consequently, the longer the time interval, the more information was supposed to have leaked out or decayed. Forgetting in long-term memory, according to this view, was due to a failure of storage, with neither encoding nor retrieval playing a major role.

In fact, most forgetting from long-term memory usually does not occur in this way. It is possible to demonstrate that forgetting is most affected by what a person does between learning and performance. For example, people forget less if they sleep after learning than if they are awake and engaged in some activity (Jenkins and Dallenbach, 1924; Ekstrand, 1972). Forgetting varies according to the nature of the activity that takes place after information is stored and before it is used. The leaky-bucket or disuse idea of forgetting as a complete explanation is not acceptable.

As we discussed previously, some forgetting may take place because it never got stored properly (possibly due to inadequate sleep or dreaming). This suggests that there has not been ample opportunity for new information to be *consolidated* (or connected) with past learning (Muller and Pilzecker, 1900; Hebb, 1949). The idea is that normal activity produced by learning tends to continue (i.e., perseverate) after the end of the presentation of new material. Again, this concept (lack of consolidation) may account for some forgetting, but it is inadequate to account for all forgetting.

Interference

Another concept pertaining to forgetting has to do with interference. Two kinds of interference are of interest to designers: proactive interference and retroactive interference. *Proactive interference* suggests that material learned *prior* to the learning of new material may interfere with the use of the new material in a performance situation. In studies of proactive interference, a group of people learn to perform task A, then learn task B, and then actually perform on task B. Another group of people do *not* learn A first, but learn B, and then perform on B. The performance of the two groups is then compared. The people who learned A and then B did not perform as well as those who learned only B. Certain kinds of material (A) seem to interfere with future learning and performance of similar material (B).

Occasionally, people regress to using information learned in A rather than information learned in B. To use a simple example, people learn to turn on a house light by flipping a switch up (situation A), and many people have learned this over a period of years. Some new light switches require the depression of a button (situation B), and these same people then learn to use the button. People may press the button the majority of the time, but occasionally (particularly in times of boredom or stress) may attempt to flip the button up like the switch in situation A. This suggests a regression to earlier learning. This, then, is the idea of proactive interference.

Designers must be alert when determining the "best way" for people to accomplish an activity. People will always bring with them a set of previously learned ways of doing things. Designers must find out what kinds of responses have already been learned and then incorporate as much as possible the same kinds of responses in the new system. This will decrease the effects of proactive interference.

Retroactive interference, on the other hand, occurs when one learns to perform task A, then learns task B, and is then expected to perform task A. This is usually the easier of the two to understand because the interfering condition is introduced after the subject has learned to perform the original task. In some systems, designers take great pains to train people to perform a certain task and then, before these people have an opportunity to develop skills in performing that task, train the people on a second task. At the end of this dual training session, the trainees are expected to perform primarily the task for which they were originally trained. Frequently, they do not do well. The interference from learning the second task may affect the long-term memory of an individual, and one of the predictable outcomes is an increase in errors due to forgetting.

Recall versus Recognition

Before we discuss problems related to the retrieval of information, the designer should be aware that there are two commonly used ways of measuring remembered information: *recall* and *recognition.* In a typical recall situation, a person is presented with a list of instructions on how to perform on a certain type of terminal. After these instructions have been presented, the individual is asked to *recall* as many of the instructions as possible. This procedure places considerable emphasis on the actual retention of the instructions originally presented.

Another way of measuring retention is *recognition.* With recognition, an individual must select from among two or more alternatives, such as on a multiple-choice test or when using a menu. Recognition is usually considered the most sensitive of the direct measures of retention, since it often demonstrates retention even when a person is unable to recall material. Recalling information is much more difficult than recognizing information. A designer should make full use of this fact and reduce the number of times people need to *recall* information or instructions and increase the number of opportunities for them to *recognize* one or more available alternatives. By doing this, the amount of forgetting is less.

Inappropriate Retrieval Strategies

What is stored determines what retrieval cues are effective in providing access to what is stored (cf. Tulving and Thomson, 1973). When we forget something, it does not necessarily mean that the memory is lost; it may be merely inaccessible because the context in which we are trying to remember it does not permit retrieval strategies consistent with the strategies originally employed at the time that the information was learned. Thus, in many situations, forgetting may be understood as being due to a lack of appropriate retrieval cues. There is evidence to suggest that people can remember better if they are expected to do so in the *same context* as that in which learning originally took place.

One frequent problem when trying to remember is the apparent *blocking* of the correct response in favor of no response or an incorrect one. There are many who feel that probably the main cause of forgetting is not a failure of storage, but competition between alternative responses when an item in storage is trying to be remembered. This type of remembering difficulty is a very common experience for most people. They may find themselves trying to remember a name or a fact while "hanging on" to some obviously incorrect response. In situations such as this, a stimulus may cause a person to seek a particular memory, but take the wrong route and end up at an incorrect storage place. The erroneous information is fed over and over again into consciousness, and an individual continually rejects it. The difficulty comes in trying to reroute or rechannel toward a different place in the memory store.

Other Types of Forgetting

Some information may be intentionally forgotten, particularly things that are painful to remember. Freud (1901) suggested that some memories are intentionally (though unconsciously) blocked, and this blocking inhibits recall. He called this *repression.* Also, numerous conditions generally related to disturbances of the brain can lead to forgetting. Forgetting that is due to tumor, head injury, disease, or old age is frequently referred to as *aphasia* or *amnesia.* Generally, designers are not concerned with these effects.

Conclusions

This brief discussion of forgetting should help designers make decisions that reduce the amount of forgetting that occurs by people working in a system. We have considered the "leaky-bucket" theories that suggest that in some way and for some reason information gradually disappears from memory. More likely, however, some form of the material placed in long-term memory is retained throughout a lifetime. Interference, whether by something that was learned before or something that was learned after the pertinent material, is a problem in many systems. Both complete and partial inability to retrieve can be dealt with by making design decisions that emphasize recognition over recall. Also, to enhance retrieval of information, a designer may have people work and, consequently, retrieve information in the same or similar setting in which information was acquired originally.

Sensory-related Characteristics of Memory

For memory that begins with visual stimuli, the evidence suggests a division into three different memory stores, sensory, short term, and long term, each with a specific set of unique characteristics. Similar distinctions are not quite so clear for stimuli received through other sensors. The characteristics of human memory are closely related to the type of material (e.g., visual, auditory, touch, and smell) being stored.

Auditory stimuli seem to follow much the same pattern as visual stimuli. There is evidence of a sensory memory that is slightly shorter than visual sensory memory and that lasts for about 100 milliseconds (ms) (Baddeley, 1976). Estimates for this memory also exist that range from 50 ms (cf. Loeb and Holding, 1975) to 250 ms (Massaro, 1975). In addition, the short-term memory associated with the auditory sensor seems to be shorter than for the visual sensor. Auditory short-term memory appears to last for only about 5 seconds (cf. Pollack, 1959; Wickelgren, 1969; Glucksberg and Cowan, 1970). Forgetting in short-term memory seems to occur by displacement of existing items by new auditory items rather than decay.

Auditory long-term memory clearly exists. Without it we could not recognize voices on the telephone or songs heard on the radio or identify the animal associated with a bark or meow. Two very different fields have contributed to most of what is understood about auditory long-term memory. The military has been training people for years to make auditory discriminations when listening for sonar signals that designate a submarine (cf. Corcoran et al., 1968). The other application is in the field of music, where researchers have studied, among other things, the ability of people to learn and remember specific combinations and sequences of tones (cf. Deutsch, 1973).

Two other senses have received some study in terms of memory: touch and smell. The touch sense seems to have a sensory memory that lasts about 800 ms (Bliss et al., 1966). Evidence for a touch short-term memory indicates that forgetting occurs fairly rapidly over about 45 seconds (Gilson and Baddeley, 1969). Together, these studies suggest that the sensory and short-term memories for touch are both longer than those generally associated with vision or hearing.

Again, the existence of a tactile long-term memory is widely accepted because of familiar experiences that we all have with it. Most people are able to recognize in the dark or with their eyes closed the feel of such things as velvet, metal, wood, a favorite vase, or the headlight switch in an automobile. One interesting idea that is emerging from research in this area is that the right hemisphere of the brain seems to be much more important than the left hemisphere in terms of tactile memory (Milner and Taylor, 1972). For example, one study has shown that blind subjects appear to be more efficient at reading Braille with the left hand (and hence the right hemisphere) than with the right hand.

It is commonly believed that learned smells are never forgotten. Baddeley (1976) has noted that smells also appear to be extremely good retrieval cues for apparently forgotten events. In Victor Hugo's words, "Nothing awakens a reminiscence like an odor." What little research is available suggests that there are no sensory or short-term memories associated with this sense. In addition, there is no evidence of forgetting, no suggestion of a difference between short- and long-term memory, and there appears to be nothing equivalent to rehearsal (Engen and Ross, 1973).

Memory Skill

Mnemonics

Remembering names, dates, events, formulas, and so on, can be *improved* with proper training and practice. This suggests that there is a remembering skill. People have been interested in improving this skill since ancient times. As long ago as 500 B.C., the Greek poet Simonides was teaching how to develop a trained memory. Today, this memory skill is developed most dramatically by the use of *mnemonics.*

Mnemonics are conscious ways for helping to ensure the retention of material that would otherwise be forgotten. They range in complexity from a string tied around a finger to complex visual imagery schemes. In essence, mnemonics are cognitive performance aids. Most mnemonics require a person to either *reduce* or *elaborate* on information being received (Baddeley and Patterson, 1971).

Reductions

Reductions frequently take the form of acronyms. When trying to remember the names of the Great Lakes, a person may think of the word *homes,* which contains the first letter of lakes Huron, Ontario, Michigan, Erie, and Superior. Or when trying to remember the colors of the spectrum and the order in which they appear, a person may use the acronym ROYGBIV (red, orange, yellow, green, blue, indigo, violet).

Tying of a piece of string around a finger to help one remember to make a phone call is a way of reducing the memory load. Another technique that is used to help people remember the number of days in different months is shown in Figure 5-6 (Lorayne, 1976). Close both hands into fists and place the fists side by side, back of hands up, as shown. Starting on the knuckle of your left little finger, moving from left to right and including the valleys or spaces between the knuckles, recite the 12 months. Thumbs are not counted. The months that fall on the knuckles are months with 31 days. The valleys represent the months with 30 days or less.

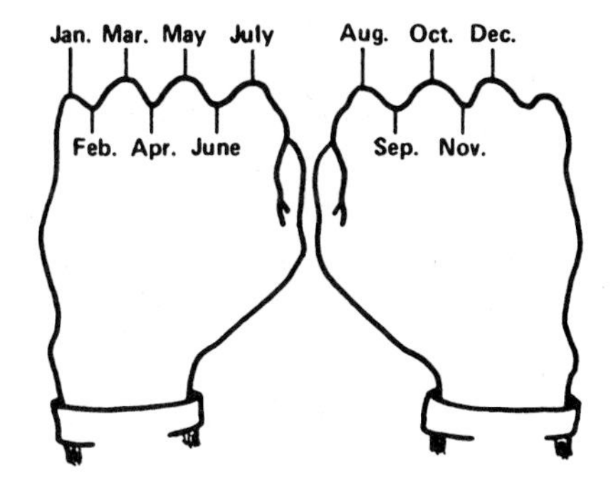

Figure 5-6 Mnemonic for remembering the number of days in a month.

Another commonly used memory aid is the use of rhymes like "I before E except after C," which helps people remember how to spell such words as "believe" and "receive." The alternative to remembering that rhyme is to memorize all words that contain "ie" or "ei." Thus, reducing the memory load from several hundred words to six words helps to have an acceptable level of performance without extensive training and memorization.

In the case of reduction, there is a danger of reducing the information so much that it is no longer possible to reconstruct the original. Even so, people seem to have a remarkable ability to make use of relatively simple memory aids to assist performance. We use this ability frequently in remembering to perform certain activities. It is not a foolproof method, as evidenced by the many times we still forget. But it is interesting that we more often than not *do* remember what the memory aid means and perform accordingly.

Elaboration

Elaboration, on the other hand, involves *adding* information to make the material easier to remember. This can be done in at least two ways: (1) by taking greater advantage of meaning that already exists in remembered words, phrases, or concepts and (2) by using visual imagery. In other words, there are two basic forms of elaboration: the first depends on *verbal cues* and the second on *visual imagery.* In many cases the two are combined.

Verbal Cues

The use of existing words, phrases, or concepts is based on a basic rule of long-term memory. To remember any new piece of information, it must be associated with something the person already knows. Association, as it pertains to memory, simply means the tying together or connecting of two or more things. A person beginning to study music may be told by the teacher to learn the lines of the treble clef, E, G, B, D, and F. The teacher helps the student remember these letters by thinking of the sentence "Every good boy does fine." The teacher is doing nothing more than helping the student to remember something new and meaningless by associating it with something the student already knows.

Baddeley (1976) gives us another example of making use of words, phrases, or concepts to remember. He provides a rhyme to help in remembering pi to the first 20 decimal places.

Pie.

I wish I could recollect pi.

Eureka cried the great inventor

Christmas Pudding Christmas Pie

Is the problem's very center.

The rule here is very simple: one simply counts the number of letters in each word and comes up with the answer 3.14159265358979323846.

When considering pi, it is interesting to note that pi has been evaluated many times to see if there is a pattern to the numbers. There is none. One of the most important testimonials to the extent to which the memory skill can be developed is that there are a few people that have used elaboration mnemonics to memorize and accurately recall pi to over 10,000 decimal places.

In some cases, elaboration requires a person to remember a considerable amount of material. This is well illustrated in a "peg system of memory" described by Lorayne (1976). The method is based on a technique published as early as 1849 by an English schoolmaster named Brayshaw (Hunter, 1957). The technique involves associating each of the digits 1 through 0 with one or more consonants. Lorayne proposes that the digits and consonants be matched as in Table 5-1.

Table 5-1 Lorayne's Technique for Improving Memory

Digit–Consonant	Hints for Remembering the Digit–Consonant Matching
1 = T (or D)	A t has one downstroke.
2 = N	A small n has two downstrokes.
3 = M	A small m has three downstrokes.
4 = R	The word *four* ends with the R sound.
5 = L	The Roman numeral for 50 is L.
6 = J (or SH, CH, soft G)	A capital J is almost a mirror image of a 6 (J).
7 = K (or hard C, hard G)	A capital K can be formed with two 7's (K).
8 = F (or V)	A handwritten small f and an 8 have two loops, one above the other (f).
9 = P (or B)	The letter P and the number 9 are almost exact mirror images (P).
0 = S (or Z, soft C)	The first sound in the word *zero* is Z.

The consonants are then incorporated in a key word. All this is necessary because it is difficult for most people to remember numbers. But by knowing the sounds of the consonants associated with each number, it is possible to combine the sounds in words in order to remember them. If a person wants to remember the number 39, all that he or she must do is come up with a word that uses both M and P, such as "Mop." The M is for 3 and the P is for 9. Knowing the 10 digit–consonant combinations gives a way to make any number meaningful; that is, it gives a way to associate the number with the set of familiar verbal materials already in memory.

A person can convert numbers to letters as needed or else have a specific word associated with one- and two-digit numbers. Lorayne suggests that the best performance comes from memorizing the following peg words, each of which is constructed using the following digit–consonant combinations:

1. tie
2. Noah
3. Ma
4. rye
5. law
6. shoe
7. cow
8. ivy
9. bee
10. toes
11. tot
12. tin
13. tomb
14. tire
15. towel
16. dish
17. tack
18. dove
19. tub
20. nose
21. net
22. nun
23. name
24. Nero

25. nail
26. notch
27. neck
28. knife
29. knob
30. mouse
31. mat
32. moon
33. mummy
34. mower
35. mule
36. match
37. mug
38. movie
39. mop
40. rose
41. rod
42. rain
43. ram
44. rower
45. roll
46. roach
47. rock
48. roof
49. rope
50. lace
51. lot
52. lion
53. limb
54. lure
55. lily
56. leech
57. log
58. lava
59. lip
60. cheese
61. sheet
62. chain
63. jam
64. cherry
65. jail
66. church
67. chalk
68. chef
69. ship
70. case
71. cot
72. coin
73. comb
74. car
75. coal
76. cage
77. coke
78. cave
79. cob
80. fez
81. fat
82. phone
83. foam
84. fur
85. file
86. fish
87. fog
88. fife
89. fib
90. bus
91. bat
92. bone
93. bum
94. bear
95. bell
96. beach
97. book
98. puff
99. pipe

This list can be used in several ways. One of the most common would be to remember a long list of items with associated numbers. Assume that there are 31 steps to be done to complete a preventive maintenance routine. Assume also that the maintenance needs to be done in a place where having a written performance aid is not possible. The user must memorize the maintenance steps and their order.

The user memorizes the peg words during training, along with their associated maintenance activities. For example, if three of the mid-steps are

1.
2.
.
.
.
14. grease rear joints
15. replace filter
16. reset dial
.
.
.
30.
31.

and if the three peg words are 14 = tire, 15 = towel, and 16 = dish, the maintenance steps could be remembered as "a tire covered with slippery grease," "a towel with sand being sifted through it," and "a dish with a digital readout in the middle." The associations between tire and grease, towel and filter, and dish and dial are established.

The user would work through the peg words until the maintenance was completed. If someone later were to ask what step number 15 was, the user would immediately envision the towel with the "filtering sand" and reply, "replace the filter." This is one way that has been proposed to use the verbal form of elaboration to improve performance.

There are other techniques for establishing peg words. One that is frequently used requires that the peg word rhyme with the number it represents, as, for example, in the following list:

Peg Words That Rhyme

One is a bun
Two is a shoe
Three is a tree
Four is a door
Five is a hive
Six is sticks
Seven is heaven
Eight is a gate
Nine is wine
Ten is a hen

This technique requires, first, the memorization of the 10 words rhyming with numbers 1 through 10. Suppose that the first item to be remembered is telephone. Then one might imagine a bun with a telephone in it. Bun is always used for the first item, since it rhymes with one. If the second item to be remembered is dog, one might imagine a dog standing in some jogging shoes ("two is a shoe"). When trying to remember, one thinks of the number, then the associated rhyming word, and finally the associated image. Using this technique, and if given enough time to construct images, people are able to remember better (Bugelski et al., 1968).

Baddeley (1976) points out that one important advantage to using a verbal mnemonic lies in the fact that it is multidimensional. He uses as an example his semantic concept of a cat, which involves not only the word and its various verbal associates (dog, mouse, cream, etc.) but also what cats look like, sound like, the texture of their fur, what it feels like to pick up a cat or to be scratched by one. Baddeley suggests that this multidimensional coding has two major advantages: It allows greater discriminative capacity, and it allows redundant coding.

The story is told that many years ago, before people kept carefully written records, witnesses for major events such as births and weddings were young men and women. The younger, the better, so that the memory of the event would last for more years. The witnesses were often beaten to make the day even more memorable. This represents another, albeit crude, way of providing or building into the system multidimensional cues for memories that must later be recalled.

Information that is coded along several dimensions is less likely to be forgotten than information coded along only one dimension. The greater the number of dimensions, the greater the probability that at least one cue will survive to provide an available retrieval route (cf. Bower, 1967). Consider the last time you tried to remember the name of a person that seemed to be "on the tip of your tongue." You may have tried alternative means of retrieval such as thinking of the color of hair, height, facial expressions, or voice. The name was probably remembered only after reviewing several dimensions of the person.

It is also possible that high-imagery words are easy to learn at least partly because they have both a verbal and a visual content. The greater the number and range of encoded features, the greater the probability that one of these will be accessed, hence allowing the item to be retrieved.

Imagery Cues

Another form of elaboration is to use *visual imagery,* where a mental picture is created and "viewed." In this case, the images are created without the benefit of direct sensory data. It is a cognitive experience with the new image constructed from an item or items already in memory. Probably, the main value of using imagery in remembering is to have an object that a person can "see" and that can be cognitively manipulated to be linked to other objects or cues. This process helps in the storage and retrieval of the information being manipulated. Imagery requires the memorizer to remember a great deal of additional information about the location representing the item, the elaborate image signifying it, and the links between location, image, and information.

To illustrate the extent that imagery can be developed, consider some of the phenomenal memory feats described by Luria (1968) for a person referred to simply as **S.** Luria began his study of S in the mid-1920s. S was working at the time as a newspaper reporter. His editor noted that he never took notes but was able to remember a full day's worth of assignments. The editor tried to embarrass him one day by having S write down the assignments after they were given. S responded by writing down *exactly* what the editor had said. The editor encouraged S to visit Luria to give the psychologist an opportunity to study this remarkable memory.

S's capacity for memorizing material appeared to be unlimited either by the number of items or by the time span over which the items were remembered. For example, S could remember perfectly material he had memorized as long as 15 years earlier. Not only was S able to remember the test material provided by Luria, but he was also able to describe what the psychologist was wearing that day, where he was sitting, and so on.

One problem given to S was to remember a list of 50 numbers like this:

6680542616
8479354237
3803470283
6095050176
2732573504

S required about 3 minutes to study this list. He could then exactly reproduce the table in about 40 seconds. It would take him about 30 seconds to reproduce only the second column in reverse order. The same set could be recalled in order several months later or up to 15 years later.

The key to S's memory feats was his use of imagery. As he was memorizing he might imagine, for example, a familiar street in his hometown. While walking along the street, he would place images of objects to be recalled in various places along the way. Usually, the objects, such as names, numbers, or events, were placed on the front porch, a gate, in the window, on the lawn, or another conspicuous place. Each object was placed at only one house and in the proper order. To recall items, S merely had to "walk" along the street and observe what had been placed at each house. This type of visual imagery served to improve his memory considerably.

The few errors that S made are very interesting. Most of the errors were omissions. The errors usually could be attributed to him doing such things as placing a white object on a white background, such as a bottle of milk in front of a white house or placing a dark object in a darkened passageway. The error then occurred when he did not "see" the item during recall. Eventually, S learned to avoid such errors by being more careful where he placed items.

S's skillful use of imagery to improve his memory enabled him to remember considerably more material than he could before the skill was developed. However, the vivid imagery that he learned to use so well tended to interfere sometimes with his ability to understand the meaning of a message. His use of imagery often led him to think about irrelevant details of images, rather than the message that they conveyed.

Bower (1970, 1972) has identified the following principles for use with imagery:

1. Both the cues and items must be visualized.
2. The images must be made to interact so as to form a single integrated image. When subjects are told to imagine two items on opposite walls of a room, no facilitation occurs.
3. The cues either must be easy for the subject to generate himself or must be provided by the experimenter during recall.
4. Semantic similarity among the encoding cues impairs performance.
5. Contrary to popular belief, there is no evidence that bizarre image combinations are more effective than obvious relationships.

Imagery mnemonics seems to provide a retrieval plan that directs a person to the appropriate storage location. Having a good, familiar set of locations provides a learner with a clearly labeled set of files in a well-organized file cabinet. Visual imagery then provides a powerful link between the cue and the item.

Keele (1973) has noted that imagery alone is not sufficient for improving recall. Imagery must be accompanied by *organization.* Atwood (1969) conducted a study in which subjects were led into his laboratory, noting various places (such as doorways and desks) that could later be used for memorization. Subjects in one group were then asked to mentally put images of various objects to be recalled in the different places in the laboratory—a knife on the desk, perhaps. Presumably, the subjects would be able to recall the information stored by mentally retracing their steps through the laboratory and recalling the objects in the different places.

Another group of subjects was also told to construct images, but the places in which they mentally put the objects were not organized in any particular way; for example, they imagined a knife sticking in the ground. The group that put the images in places in the laboratory were able to remember much better. Atwood also reported that if people were asked to imagine printed, abstract words (such as "metaphysics") in different locations instead of imagining objects, recall was very poor. Good organization of poor images is not sufficient for good recall.

Conclusions

Designers can take steps to improve the memory of users in at least three ways. First, they should ensure that people are not expected to perform functions that put an unreasonable demand on human memory. As memory limitations are better understood, they should be reflected in this early design activity. Second, designers should develop *interfaces* that support and encourage an acceptable level of remembering. Shakespeare's Globe Theater, for example, was called a "memory theater" because there were special cues in the architecture that served as pegs to help actors remember their lines. Designers can do similar things when developing a system. Cues to assist human memory should be everywhere. They should be built-in as part of displays, controls, workplaces, human–computer dialogs, codes, computer command languages, and so on.

Finally, the memory of users can be improved through the design of good *facilitators,* that is, providing training, instructions, and performance aids that make full use of the mnemonic concept. The designer can assist users in creating and organizing meaningful images to help in remembering certain items over a long period of time. In other words, a design should help users develop their memory skills. (Some users may not even be aware that this is possible.) And after a user constructs and practices appropriate images, the designer should ensure that performance aids will complement and support the initial imagery training.

Exercise 5: Using Imagery to Memorize Digits

Purpose: To demonstrate the potential of improving human memory by using imagery techniques.

Method: Use yourself as the subject. After studying the material in the textbook on using imagery, memorize the first 100 digits given in the accompanying table. The numbers represent pi taken to 1000 decimal places (i.e., 3.14159 . . . etc.). Using computers, this number has been carried out to 1 million digits without finding a pattern to the numbers.

If you really want to determine the potential of your memory with mnemonics, see how many digits over the first 100 you can remember. Can you memorize all 1000? Can you memorize enough to set a new world's record (over 100,000)?

You will be expected to write down the first 100 digits in the correct order in class.

Reporting: There is no report required for this exercise.

Pi to 1000 Decimal Places									
3.14159	26535	89793	23846	26433	83279	50288	41971	69399	37510
58209	74944	59230	78164	06286	20899	86280	34825	34211	70679
82148	08651	32823	06647	09384	46095	50582	23172	53594	08128
48111	74502	84102	70193	85211	05559	64462	29489	54930	38196
44288	10975	66593	34461	28475	64823	37867	83165	27120	19091
45648	56692	34603	48610	45432	66482	13393	60726	02491	41273
72458	70066	06315	58817	48815	20920	96282	92540	91715	36436
78925	90360	01133	05305	48820	46652	13841	46951	94151	16094
33057	27036	57595	91953	09218	61173	81932	61179	31051	18548
07446	23799	62749	56735	18857	52724	89122	79381	83011	94912
98336	73362	44065	66430	86021	39494	63952	24737	19070	21798
60943	70277	05392	17176	29317	67523	84674	81846	76694	05132
00056	81271	45263	56082	77857	71342	75778	96091	73637	17872
14684	40901	22495	34301	46549	58537	10507	92279	68925	89235
42019	95611	21290	21960	86403	44181	59813	62977	47713	09960
51870	72113	49999	99837	29780	49951	05973	17328	16096	31859
50244	59455	34690	83026	42522	30825	33446	85035	26193	11881
71010	00313	78387	52886	58753	32083	81420	61717	76691	47303
59825	34904	28755	46873	11595	62863	88235	37875	93751	95778
18577	80532	17122	68066	13001	92787	66111	95909	21642	01989

6

Motivation

Introduction

As used in the human information processing model shown in Figure 6-1, *motivation* is any influence that gives rise to performance. Assuming equal ability, skill, and knowledge, high motivation usually results in high levels of performance, and low motivation results in low levels of performance. Given two people with equal ability, skills, and knowledge, but substantially different motivation levels, the person with the higher level of motivation will perform best.

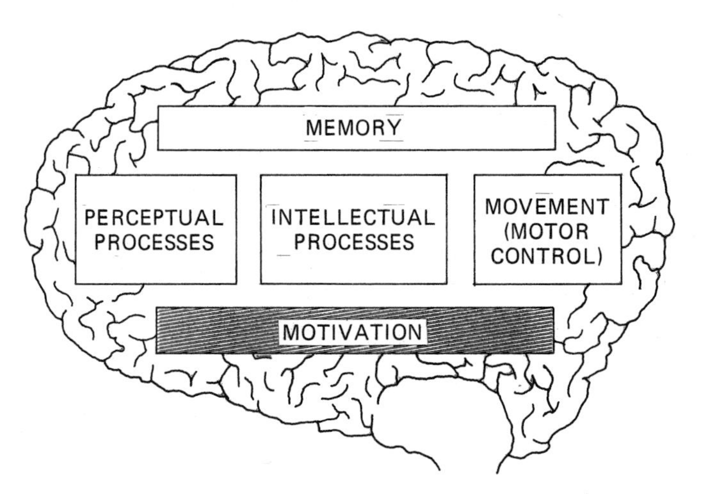

Figure 6-1 Human information processing model with *motivation* highlighted.

The strength of motivational influences can range from virtually none to very high. At virtually none, a user could have the skills and knowledge necessary to perform an activity but not do anything. He or she might not show up to work at all (go fishing or stay in bed) or come to work but not feel like performing. At a very high level of motivation a user may arrive early, work hard all day (take no breaks and only 10 minutes for lunch), and then take work home to do in the evening. People in this latter group are sometimes referred to as *workaholics*. Most users are motivated to perform at a level somewhere between these two extremes.

Traditionally, designers have not taken much of an interest in user motivation. This aspect of performance was left to the management of the new system to develop and implement. Recently, however, we have discovered that the nature of the work in a system can affect the motivational level of many users. The better the work design, the higher is the motivation level of many users.

Internal versus External Motivation

The reason that the design of work affects some users more than others appears to be related to whether internal or external motivational influences most affect a user. Internal motivational influences are closely related to performing the work itself. External motivational influences are related to conditions outside the actual performance of the work. Probably the best-known external influence is money.

With many useful classification schemes, the distinction between internal and external motivation can sometimes become fuzzy. In general, an activity is labeled internally motivating if its completion brings no obvious external reward. Conversely, an activity is labeled externally motivating if it leads to external rewards like food, money, or social reinforcement. Generally, another person or machine dispenses an external reward (cf. Condry, 1977). Some problems with this distinction are immediately apparent. For example, there may be potential future rewards for an activity that are not obvious to an observer—learning a skill may lead to external rewards in future situations where that skill is valuable.

Instead of defining internal motivation as the absence of external rewards, we can also define it in a positive way as a need for competence and self-determination (White, 1959; deCharms, 1968; Deci, 1975) or as a search for an optimal amount of psychological incongruity (Hunt, 1965). Still another way of defining internal versus external motivation is to let the user make the distinction, that is, to allow users to determine whether their actions and consequences are largely under their own control or are primarily determined by external forces (Rotter, 1966; deCharms, 1968). In spite of these ambiguities, the distinction between internal and external motivation is still useful.

A study reported by Tannehill (1974) illustrates the difference between internal and external influences. The experiment involved a college psychology professor and college sophomores. The professor took two groups of 12 students to the woods. A low hill separated the groups so that, although aware of each other, neither group could see what the other was doing.

To the first group the professor assigned the task of chopping wood. He provided them with axes and agreed to pay them $2.00 per hour to chop, split, and stack logs. He assigned the second the same task at the same rate of pay, except that they could *only* use the *blunt* edge of the ax blade. They were *not* to use the sharp blade of the ax under any circumstances.

The first group, using the sharp edge, enjoyed the work and friendly competition sprang up. They could see the results of their efforts and began to take pride in the growing stacks of chopped and split logs. The second group, required to thump the logs with the blunt edge of the ax, began to mutter about what the professor could do with this experiment. When they threatened to quit, the professor offered them more money to stay.

More money worked at first. However, soon even the increase in money was not enough to hold the second group to the task. The professor then continued to raise the hourly rate until he was paying $12.00 per hour when the last student ended the experiment, threw down his ax in disgust, and stalked off.

This example clearly portrays both internal influences (those related to the performance of work) and external influences (money). A combination of both influences motivates users, but the designer has more control over internal influences.

Even though designers cannot be held totally responsible for user motivation, they should try to identify work, interfaces, and aids that motivate favorably. For instance, designers should want to find out why people work at all, what influences them to work hard, and what influences them to care about and be proud of their performance. The following sections contain discussions of motivation.

Differences in Users

The work population continues to change. New people entering the work force (many of whom work in newly designed systems) are different from those of even 10 or 20 years ago. They tend to be younger, better educated, and less conforming than the previous generation—and many more are women. What might have motivated workers 10 or 20 years ago may not do so now. In general, workers expect their jobs to provide more satisfaction and personal development. Consider the following example:

> L. S., age 58, entered the work force 29 years ago; he speaks English (a second language) only moderately well and has an eighth-grade education. His son, T., is 21, has a college degree, and is actively interested in politics. The same company employs both men. L. S. works a molding machine and T. manages the stockroom.
>
> T. heard from plant gossip that his father's work was secretly being damaged so that the company could claim a tax loss that it badly needed. T. confronted the plant supervisor and was angered to have the rumor confirmed.
>
> He exposed the situation to his father, who was considerably less concerned than his son. L. S. felt that as long as he was paid for his work he had no reason to care what happened to the items he produced. T. disagreed and soon left for a job with another company. L. S. continued to work at his molding machine and to collect his pay every Friday afternoon. Different reasons motivated the two men to perform.
>
> The father seemed to be at least partially motivated by the external influence of pay. He did not care so much about what happened to his work once he finished it. He simply did what was required and collected his pay at the end of the week. On the other hand, the internal influence of the work itself motivated his son—the feeling of pride and satisfaction from a job well done. T. was motivated by a different purpose and toward a different set of objectives than his father.

Instincts, Drives, and Motives

The Greeks, almost 2500 years ago, believed that people did things for two reasons—to seek pleasure or avoid pain. Unfortunately, some•managers still feel that *all* user motivation in systems can be explained in this way: users try to avoid the pain of work, while they attempt to gain maximum pleasure through the highest possible earnings and benefits.

Instincts

The development of the concept of instincts was another attempt to explain motivation. Instincts were thought to be an unconscious, basic part of our human nature. People were felt to have little choice over whether or not they would try to satisfy their instincts. Instincts were considered natural forces that drove people to behave in certain ways. However, one of the outstanding characteristics of the whole idea of instincts was the wide diversity among psychologists in trying to identify a basic set of instincts universal to all people. Sigmund Freud, for example, attempted to explain people's performance in terms of three essential instincts: sex, self-preservation, and death. By 1924, psychologists had identified nearly 6000 separate human activities that supposedly arose from instincts. Shortly thereafter the term became unpopular.

Drives

The idea of instincts was replaced with the concepts of *drives* and *motives*. The drive concept infers that certain impulses within the individual induce very specific behaviors. This concept holds that an attempt to reduce a physiological drive or an aroused bodily condition supposedly motivates people to behave in certain ways. Adherents of this concept believe that a person becomes motivated whenever an imbalance exists. For example, when a person's body needs water, a thirst drive is created and a person's behavior is directed at reducing the thirst by turning on a faucet to draw a glass of water. Drinking restores the balance and thirst diminishes.

A worker's basic physiological need for rest or sleep should not be overlooked.

Many drives have been studied, including hunger, thirst, sleep, and temperature regulation. For our purposes it is sufficient to understand that in some situations motivation is like filling in a gap, a gap created by some specific physiological need. System designers should be aware of the fact that people may be motivated in some situations by basic physiological drives. The design implications here are simple: Design systems so that people's basic physiological needs can be routinely satisfied. For example, users should not be deprived of sleep. Lack of sleep and the upset of certain other body rhythms, such as those that accompany the phenomenon of jet lag, may produce degraded performance. People who are thirsty, hungry, or need to use a rest room may have difficulty concentrating on the performance of a system activity.

Motives

Incentives that are not physiologically based but are learned are called *motives*. Motives are seen as a function of experience, usually some sort of social experience. Consider an athlete who practices and trains for several hours daily. Surely he or she is not motivated to do so because of a physiological imbalance or drive. Similarly, a student who puts in extra effort to get to the top of the class is not attempting to fill a basic physiological need. In both cases, it seems some special desire on the part of the individual to excel motivates the behavior. People frequently do things because they feel good doing them, because these activities represent a positive value for them, or because they simply have learned to enjoy performing the tasks.

Integration of Need Theories

Psychologist Abraham Maslow was perhaps the first psychologist to pull together what was known about drives and motives as motivators. Maslow (1954, 1970) suggested that people have five major *needs* in varying degrees and intensities (see Figure 6-2).

Figure 6-2 Maslow's hierarchy of motivational needs.

Maslow contended that when the lower-order needs, such as the physiological needs and the need for safety, are satisfied, they lose their potency as motivators and new, higher-order needs emerge. The importance of higher-order needs does not decrease, Maslow says, even when they are satisfied. In fact, for self-actualization, increased satisfaction leads to increased need.

Although some specific ideas of Maslow's have received little empirical support, the designer can benefit from his conceptualization of human motives. In particular, a designer should attempt to create a climate in which users can develop their potential. Such a proper climate usually includes provision for greater worker responsibility, autonomy, and variety. In Maslow's terms, these provisions would provide for higher-order need acquisition and satisfaction. The lower-order needs must be accommodated before needs higher up the scale will even be considered. Designers should see to a person's basic physiological and safety needs by providing proper work schedules, sufficient heating and air conditioning, opportunities for recreation, adequate safeguards against dangerous equipment, and so on. A designer should not expect users to seek to satisfy higher-level needs—feel good about themselves and their work—until their physiological and safety requirements are met.

Motivating System Users

History of Work Motivation

Even though considerable research has been done on motivation, until relatively recently there was little systematic thought given to motivating system users. Most of the major approaches for improving user motivation in the past 50 years have had little, if anything, to do with the design of work in a system. Consequently, they were beyond the control of a designer. These approaches included the following:

- Reduced hours and longer vacations
- Increased wages
- Better benefit packages
- More profit sharing
- Better training of supervisors in employee counseling services
- Improved organizational planning.

In the late 1950s, Hertzberg et al. (1959) conducted a review of the literature on job satisfaction studies. Based on the review, he and his co-workers interviewed a large number of employees, asking them to talk about periods at work when they felt exceptionally good or exceptionally bad. For the workers they interviewed, they concluded that feelings of strong job *satisfaction* come principally from *performing the work itself.* The actual work motivators seemed to be

- Actual achievements of the employee
- Recognition received for the achievement
- Increased responsibility because of performance
- Opportunity to grow in knowledge and capability at the task
- Chance for advancement

On the other hand, feelings of *dissatisfaction* are more likely to be attached to the context or surroundings of the jobs, including

- Company policies and administration
- Supervision, whether technical or interpersonal
- Working conditions
- Salaries, wages, benefits, and the like

Hertzberg (1966) found that conditions related most closely to user *satisfaction* are not simply the opposite of those related to *dissatisfaction.* Hertzberg's findings have direct implications for system designers. Among other ideas, he suggested that motivating influences can and should be built into a system. Providing a computer operator with *feedback* of status information during excessive delays or allowing a system designer to leave his or her "mark" on a finished product as a symbol of recognition and responsibility are possibilities for improving motivation. Satisfying work is often seen as work that has value in itself—work that appears *worthwhile* to the user. Satisfying work also seems to be work that is complete in itself so that each user is responsible for a *recognizable product.*

In a series of studies conducted at the American Telephone and Telegraph Company, Ford (1969) incorporated Hertzberg's ideas in the actual redesign of work. In an effort to make some jobs more satisfying, Ford created work modules aimed at increasing task meaningfulness and feedback, as well as employee responsibility, achievement, recognition, and autonomy. Billing clerks, for example, who worked as a team, were responsible for sending out charges on toll bills at staggered dates through the month. The team was in charge of five subsections of a large geographical area. Productivity was low, due dates were missed, and costs were high. The team approach was discarded. Each worker was given responsibility and total control of getting out bills for a given subsection on time. User performance improved when each worker was given the freedom to organize his or her own bills.

Motivation and Basic Design Decisions

Function Allocation

As early in the design process as *allocating functions* to people, the designer must consider user motivation. People generally cannot reliably detect very low levels of visual, auditory, and tactual stimuli. Even so, designers commonly make the mistake of routinely allocating such difficult functions to people regardless of the potential motivational effects. Consider how difficult it would be to come to a job every day and listen for 8 or more hours for barely detectable infrequent sounds or stare at a CRT for hours searching for suspicious shadows that may indicate a gun is hidden in a piece of luggage. Signal detection would most likely improve by allocating the function to a machine. This would be much better than having to deal with the degraded human performance, costly turnover, or absenteeism that often accompany poor user motivation.

Human Performance Requirements

Once we identify the major functions that a human will perform, we must further consider each function in terms of its *human performance requirements.* Setting out human performance requirements requires an understanding of user motivation, since these requirements include statements about errors, manual processing time, training time, and job satisfaction. If these requirements are not realistic, motivation is affected. For example, to require the total training for a new user on a complex system to be completed in just 2 hours may be absurd. Such a requirement, if taken seriously, could have a tremendously negative impact on the motivation of trainees exposed to a training program with little chance for success. A similar effect on motivation could be predicted if the allowable time to complete one's work was far too short or the required accuracy was unrealistically high for the conditions. Designers should take time to study similar systems and to use such historical data when developing realistic human performance requirements.

Designing Work Modules

The next major step in the design of systems is the *analysis and synthesis of the tasks* included in the system and their *integration into work modules.* This process includes determining the system structure, identifying tasks, preparing a task flow chart, and identifying work modules. These activities constitute a significant part of the basic design process that can lead to acceptable levels of human performance. User motivation considerations must also occur in this stage of system design. *Designers should build into the design of work modules those characteristics that will best ensure motivation in users.* It is here that designers can have the greatest impact on motivation.

Motivation and Work Characteristics

Many users will respond favorably to work that has the following characteristics (Porter et al., 1975):

- The user feels responsible for a *meaningful portion of the work.*
- The performer considers outcomes of the performance *worthwhile.*
- Sufficient *feedback* is provided for the user to adequately evaluate his or her performance.

Meaningful Portion of Work

Turner and Lawrence (1965) have suggested the term *autonomy* to describe the degree to which users *feel personally responsible for their work.* Work designed to have high autonomy will provide users with the feeling that they "own" the outcomes of their performance. A clearly identifiable product will result—a product that the user can initial and in which the user can take pride when all goes well or feel discouraged about when it does not work out well. For example, a key operator can take pride in a totally accurate set of input, a telephone operator can take pride in an exceptionally well handled call, or a factory worker can take pride in a well-constructed part. In all these cases, the designer provided users with an identifiable "something" that results from good performance.

On the other hand, work with low autonomy causes users to feel that the same results will be accomplished (good or bad, more or less work) no matter who performs the task. Low-autonomy work generates no feeling of ownership and no identification with a product or contribution. The implication for designers is clear. The performance required by each work module should lead to the development of an identifiable product. The term *product* is used here to indicate any identifiable outcome of performance, but the outcome should be clearly associated with the performer. The product could be a typed letter, the completion of a maintenance routine or computer program, or the building of an entire generator.

McGregor (1960) argued that the more a person becomes involved in his or her work the more meaningful and satisfying the job becomes. McGregor believed that employees are quite capable of making significant decisions in their work and that such autonomous decision making is in the best interests of a company. He felt that management's role was not to manipulate employees to accept orders from above, but to assist workers in developing their abilities, skills, and interests so that they maximally contribute to organizational effectiveness. McGregor's approach suggests some interesting implications for motivating workers through the design of systems. For example, what motivates one person may have little effect on another. Designers should design systems to be as flexible as possible to account for these individual differences. Probably the optimal condition in most systems occurs when the overall structure exists, but the worker has great flexibility to act as an individual within the structure.

Weiner and Kukla (1970) proposed another approach to understanding autonomy in motivation. Weiner found that people usually attribute success and failure to one or more of the following sources: ability, effort, task difficulty, or luck. Ability and effort are called internal factors because they are part of the person. Task difficulty and luck, on the other hand, are referred to as external factors. He suggests that people are more motivated to continue work when they can attribute success on a job to ability and effort, rather than to task difficulty and/or luck. Consider the frustration of a young worker assembling electronic components, who has just put together his first module and cannot figure out how he did it. Surely his motivation to continue is far less than a more experienced co-worker who has just established a new piece-rate record for the plant. In the first case, success was attributed to luck, while in the second it was attributed to ability and effort.

Worthwhile Work

A designer should also provide work that appears *worthwhile* to a user. If a person must perform activities that appear to have little value to anyone, it is unlikely that the performance will be taken very seriously by anyone, including the performer. Porter et al. (1975) proposed two ways a designer can develop work that will be experienced as meaningful by users.

The first concerns providing the user with a *complete* or *whole piece of work*. The work should provide a clear beginning and ending, with the user having *total responsibility* for what takes place in between. And, most importantly, a product's change due to the user's performance should be *highly visible* to both the user and others.

In a telephone company, for example, one person often has total responsibility for producing a telephone directory for smaller towns. The person begins with a collection of names of telephone subscribers and, using a computer, works with a printer and others to produce an error-free directory on schedule. The user has total responsibility from start to finish for production of the directory. An older and alternative approach that did not take into account this "wholeness" principle had different people responsible for various pieces and parts of the total process. In the latter case, several people produced the directory, and no one person had the full feeling of satisfaction for an accurate directory produced on schedule.

When others consider the performance worthwhile, they also consider the performer worthwhile. When this occurs, along with open feedback channels that function correctly, the user feels worthwhile and motivated to continue the high-level performance. This assumes that a product is available to be evaluated. To simply observe the behavior of an individual does not provide much of an insight into a person's performance. Recall how difficult it was for Barton Hogg, who had the students digging in the sand for lead, to judge their performance by observing their behavior. To judge how good a performance is, a product must be produced. A designer should provide for the production of a worthwhile product.

A second way to help ensure that work appears worthwhile to a user is to provide for *variety in performance.* Performing a variety of different activities gives a person the chance to develop and exercise different skills. Many people remain interested in the work performed as long as skill building continues. Once the skills are fairly well developed, the work may not be as interesting. By providing a variety of different experiences, a designer can considerably extend the time for skill building.

As skills develop, performance becomes more and more automatic. When this occurs, a user has more time for thinking, reading, or socializing while performing the work. In one computer-based system, the users' performance became so skilled (automated) that they could read magazine articles while effectively taking reports from customers over the telephone. Obviously, the work developed for these users contained a limited variety of different experiences.

When skills have been developed and the work is routinely, almost automatically, performed, many users seem to gain little satisfaction from the work itself (internal influences). These users may begin to receive more satisfaction from other aspects (external influences) in the work situation, influences external to the actual work performance. In some work situations, thinking about water-skiing after work, reading a novel, watching television, playing a radio, doing crossword puzzles, and visiting with co-workers are all possible while performing the assigned work.

Socializing, in particular, is very common in many systems. In one word-processing unit, the users were all highly skilled at keying and would spend a good deal of time talking with one another and using the telephone. Neither their typing accuracy or productivity was questioned. Even so, it was very interesting to watch what happened when a tape containing dictation needed to be typed. The dictation was always passed to the newest person in the word-processing center. As one typist observed, "It takes too much thinking to type from a dictated tape, you get left out of things that are going on around you." Typing from the dictated tape is potentially more challenging, which requires the building and exercising of new skills. But, in this situation, once a person has given up on internal motivating influences in favor of external ones, the external influences seem to be preferred. In fact, people may view "meaningful work" as an interruption and even an irritant once relevance shifts to external influences.

When considering both wholeness and variety, the designer usually has only partial control over the outcome. To help overcome this problem, designers should develop relatively small work modules, ones that contain few tasks. This enables the management of a new system to combine the smaller work modules in ways that best satisfy the requirements of wholeness and variety. Management usually has one critical additional piece of information that most designers never have; managers know *exactly* what kind of people were hired to perform in a system.

Designers, then, should develop meaningful work modules, each with a definite product, and provide the modules to management, who match the precise work required with the specific people hired to do it. Management may find whole jobs do not motivate some users. In this case, management may more profitably have several people performing smaller work segments. On the other hand, some users may respond very well to the wholeness idea and, in these cases, a combined set of modules can give these people work that is intrinsically satisfying. The same logic can be followed for giving users work with or without variety. From a designer's point of view, the most reasonable approach is to provide management with the flexibility to go in either direction.

Feedback

A designer also has a major responsibility to ensure that users receive adequate *feedback* from their performance. If a key operator inputs large quantities of material into a computer each day and never receives any idea on its acceptability (was the volume great enough and accurate enough?), then the key operator would have little chance to feel worthwhile or to know if the work is important. Clearly, a product exists and somebody most likely considers it important, but if the user does not know this, then it is all meaningless and the potential for motivation is nil.

Feedback can come from the work itself, but only if the designer has taken time to develop a way for this to happen. For example, a computer could be programmed to inform a key operator when certain errors occur or even to give an evaluation of the volume and accuracy at the end of each day, compared against other days or other operators. The designer has total control over this level of feedback. It should be provided for each work module. This means that at a minimum each work module should result in a worthwhile product and provide feedback to users about their performance.

Skinner (1953) contends that improved performance depends on feedback because any improvement results from how we were rewarded for doing the same or similar things in the past. Performing in an acceptable way depends on our past experience and the contingencies of reinforcement (reward or punishment) that we have learned. In designing a computer system, for example, verbal messages praising a naive user at various stages of interaction can have a reinforcing effect. The user will likely be motivated to continue interfacing with computers because he or she found it rewarding to do so in the past.

Effectively using the contingencies of reinforcement is a powerful way to motivate people to achieve an acceptable level of human performance, particularly while learning a new system. Appropriate rewards built into the ongoing operation of a system can help ensure an acceptable level of motivation over a long period of time. Remember, the total work experience should be positively reinforcing to a user. A poor situation cannot be patched up by an occasional, and perhaps not even work related, pat on the back.

A system could be developed to provide feedback on user performance to management, with the hope that users will receive the necessary feedback from management. Unfortunately, management does not always make proper use of this type of feedback. Management may totally ignore the information, forget to tell a user, not report on a regular basis, figure that no benefit can be gained from telling users about their performance, or use the feedback information to punish certain users. Thus, a designer may provide the opportunity for management feedback to take place and later find that it is never used or not used properly.

System feedback often makes delays seem more tolerable to system users.

Responsiveness of Users

Remember, not all users will consider the existence of internally motivating influences as desirable. The design recommendations just discussed, which provide for a clearly defined product, worthwhile work, and the necessary feedback, apply primarily to *users who are still sensitive to internally motivating influences.* Condry (1977) and Lepper and Greene (1979) reported many conditions where external influences appeared to destroy the effect of internally motivating influences.

We have reason to believe that, as users gain experience performing work that does not contain internally motivating influences, over a period of time the users come to rely heavily on externally motivating influences, such as money, time off, or opportunity to talk to others. A study by Kornhauser (1965) of automobile assembly line workers suggests that, when work is designed without internally motivating influences, users will eventually devalue internal influences in favor of external influences (cf. Porter et al., 1975).

Consider this in regard to the example reported earlier of the students chopping versus thumping wood. Assume that no other jobs were available and the "thumpers" had to perform the activity on a regular basis. These students would soon give up expecting any motivating influences from the work itself and come to depend totally on externally motivating influence (money, other benefits, socializing). Furthermore, even if given the opportunity to work with the "choppers" who were performing work that *did* contain motivating influences, the "thumpers" may continue to rely on *externally* motivating influences and ignore the internal influences that were part of the work.

Unfortunately, we have gone through several years (even decades) where people were expected to perform work that was totally devoid of internally motivating influences. This being the case, a designer may frequently inherit people from other systems where designers ignored the idea of building in internally motivating influences. It may take quite some time before performance improvements related to internally motivating influences will result for large numbers of workers. Nevertheless, a designer has a responsibility to develop work activities that take into account what is now known about motivating potential users.

Motivation, Satisfaction, and Performance

For years psychologists and others have been interested in the relationship between motivation, satisfaction, and performance. Research generally shows that the relationships that do exist often vary from individual to individual and from situation to situation. According to some researchers (e.g., Brayfield and Crockett, 1955; Vroom, 1964) little evidence exists to suggest that satisfied users consistently exhibit higher levels of motivation. One can easily envision a user who is quite satisfied with some aspect of his or her job but who is not motivated to perform for any number of reasons; for example, he or she likes the social aspects of the job but does not care about the actual work.

One reason for this dichotomy is the difference between *job satisfaction* and *performance satisfaction.* A person's job includes much more than a person performing an activity in a particular context. A job is frequently considered in terms of (1) the time and difficulty to get to it and back home, (2) the number and types of other workers, (3) opportunities to participate in nonwork activities, (4) length of breaks and cleanliness of the break area, (5) quality of food in the cafeteria, and so on. A designer cannot hope to positively affect all aspects of a job, but he or she can strive to elicit satisfaction from the much narrower consideration of the actual work performed.

It *has* been demonstrated that increased motivation leads to improved job performance (cf., Hertzberg et al., 1959). Users motivated by the work generally perform well. However, designers should not confuse user satisfaction and motivation. Design decisions that lead to increased motivation will not necessarily lead to increased satisfaction. Motivated workers in a system tend to be effective performers (assuming that they have developed the needed skills), while in many cases the most that can be said about satisfied workers is that they are not dissatisfied.

Satisfied workers may or may not be motivated toward a high level of performance. It seems to depend on whether their satisfaction results from motivating influences that come from the work itself or if the satisfaction results from motivating influences that are external to the work. People may be satisfied because their jobs give them the opportunity to be around and visit with other people. The ideal situation from a designer's point of view is depicted in Figure 6-3. In this case the user is motivated by internal influences of the work, performs well, and feel satisfied with his or her performance.

The likelihood that a user will be motivated to perform well depends on the extent to which the activity was found to be rewarding or satisfying to perform in the past. Designers can develop work situations in which individuals gain satisfaction from the work itself. If this satisfaction from work performance is rewarding to the user, the user probably will be motivated to continue performing at about the same level.

Exercise 6: Determining What Makes Games Fun

Purpose: To provide an opportunity for determining what makes an activity fun (intrinsically motivating).

Method: Compare and evaluate at least two different computer-based games (one that is popular and one that is not popular). The games can be those used on a personal computer, with a television set, or in a video arcade. *Play* the games yourself, *observe* others playing

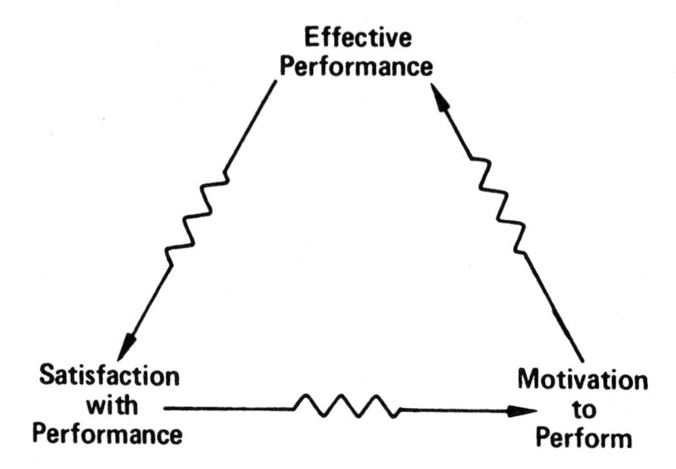

Figure 6-3 Model of the ideal work situation.

the games, and *question* the other players. Determine why some games are more fun than others.

Reporting: Record your findings in a report according to the format presented in Exercise 1A. List five fun-related characteristics of the most popular game. List five ways that these fun-related characteristics could be incorporated into a word-processing program to make the program more satisfying, and possibly more motivating.

Part 3

THE ACTIVITY

The Activity Element

With Chapter 7, we begin an examination of the activity element of the human performance model. As noted earlier, the designer does not usually have much control over the specific people who will be using the new system. Similarly, the designer often has limited control of the physical and social context in which the performance occurs. It is the *activity* element over which designers have the most control.

The topics that comprise the activity element of the human performance model will be discussed in more detail in the next few chapters:

Chapter 7—Iterative Design and Prototyping

Chapter 8—Usability Studies and Usability Testing

Chapter 9—Product Analysis and Definition

Chapter 10—Input and Output Devices

Chapter 11—Task Analysis

Chapter 12—Interaction Issues

Chapter 13—Presentation Issues and User Guidance

Iterative Design, Prototyping, and Usability Testing

In an iterative design approach, interface design decisions are made, tested, and revised. In general, each iteration improves the product (cf. Bailey, 1993). The iterations can be made within each stage of a stage-based system development process or within the prototypes being developed in less structured processes.

Iterative design helps to ensure that the best of all possible design decisions are made. Chapter 7 describes iterative design and discusses creating software prototypes to help in making initial decisions. Chapter 8 covers usability studies and usability testing and how they can be used to help in making initial decisions and to identify product weaknesses. Major usability testing topics include (1) formal evaluations, (2) inspection methods, (3) performance testing, and (4) operational evaluations.

Designing professional, high-quality user interfaces is extremely difficult. Designers cannot expect to regularly outguess their users. Using information from user interface studies is the best way to determine the goodness of design decisions in advance, and usability testing is the best way to verify and identify weaknesses of design decisions once they are made.

A frequently used (and generally more expensive) alternative is to release untested systems to users. Users are then left with the burden of complaining about inadequate user interfaces or unacceptable functionality. Changes may then be made in future revisions of the system. These changes may take years to make. The costs associated with having users deal with poor user interfaces can be horrendous. In most cases these costs are due to having slow user performance and excess errors.

Product Analysis and Definition

Chapter 9 begins a discussion of the three major stages related to user interfaces. The first stage-based activity addresses product analysis and definition. This stage addresses three major user-related activities conducted early in the development of a system: (1) determining and verifying functionality, (2) determining user profiles, and (3) determining computer profiles, including input and output devices.

Chapter 10 covers many of the issues related to the physical interaction carried out with displays and controls. The purpose of this chapter is to discuss the importance of having well-designed displays for receiving (sensing) information and well-designed controls for sending information (responding). These controls can be either device based or screen based. The overall environment, including workplace layout, should be optimized to enhance the usability of displays and controls and to ensure user comfort.

To achieve the best and most acceptable human performance in systems, human factors (ergonomics) technology should be applied early in the system development process. Designers should focus on ensuring that the system is designed with the user in mind from the very beginning. Designers should not wait until screens need to be developed to begin thinking about product users. Many early design decisions can have substantial impact on how users interact with a system.

User Interface Design and Development

An *interface* is the boundary shared by interacting components in a system. The essence of the interaction is communication, or the exchange of information back and forth across the boundary. In systems requiring communication between people, between computers, and between people and computers, there are three possible exchanges:

(1) One person communicating with another person (directly or through the computer)

(2) One computer communicating with another computer

(3) One person communicating with a computer

If any of these exchanges are degraded, the outcome is poorer performance. While all these exchanges are important, the following chapters primarily address human–computer communication issues. The issues that must be optimized in order to ensure efficient communication between people and computers.

Chapters 11, 12, and 13 discuss critical issues about designing the human–computer interface. Chapter 11 addresses the beginning of the actual user interface design process, for which a *task analysis* is performed. Virtually all professional, high-quality user interfaces begin with a task analysis. There are two major task analysis approaches: (1) decomposition oriented and (2) scenario oriented. The first is used in the initial design of traditional user interfaces, while the latter is used to assist in the initial design of object-oriented user interfaces. The first is more process oriented, while the latter is more object oriented.

Chapter 12 discusses major human–computer *interaction* issues, and Chapter 13 covers human–computer *presentation* issues and user guidance.

7

Iterative Design and Prototyping

System Development Methodologies

Considering System Performance and Human Performance

There are hundreds, if not thousands, of different ways to design computer-based products, applications, or products. From the beginning, all the approaches should include a careful consideration of human performance. The human, by far the most complex of all system components, requires considerable attention to achieve acceptable performance.

Not all system design and development methodologies take into account the professional creation of a user interface. Many still focus almost exclusively on getting newly coded software to work on a previously selected computer. It was not until graphical user interfaces were in common use that we started using and reusing common elements. Also, with the use of more and more toolkits to assist in the development of systems, we were finally able to borrow some of the elements of previously designed user interfaces. Until that time, and with most existing character-based systems, each new product has a unique look and feel. At least the graphical user interface allows users to experience some consistency from application to application.

The quality of the user interface is at least partially dependent on the way the overall system is designed. In some cases, the design process does not allow time for developing a professional user interface. In other situations, there is time to do the interface, but the tools do not allow the required iteration (prototyping and usability testing) to make it happen. In still other cases, the user interface has been previously designed by some unknown group of toolkit builders, and we must use what they think is a good interface (or do a lot of expensive work-arounds).

Because the user interface must be designed within a variety of different system development methods, it is important to have a user interface design approach that is independent of how other parts of the system are being created.

Major Design and Development Activities

No matter what design strategy is used, there seems to be agreement that the following activities must be done.

- Identify *user needs* and other information to establish a final set of system functions (i.e., determine *what* the system will do)

- Make numerous *design* decisions to determine the best way to have the work completed (i.e., determine *how* the system will do what needs to be done)
- Implement the design by doing the necessary *programming* and *writing* the required documentation and training materials
- Conduct appropriate *tests* to ensure that the software and hardware work, and work well together

System Design Philosophies

Most system design and development methodologies fall into one of the following categories: (1) a structured design method, (2) an iterative prototyping method, or (3) some combination of the two.

Structured Design Methods

The structured approach is the more traditional one and is still assumed by many to be the best (particularly for very large systems). For example, virtually all integrated computer-assisted software engineering (CASE) tools assume that a structured approach is being used. Those using the structured system design approach, which is also known as a top-down or waterfall approach, believe that it is necessary and possible to complete all activities in one step or stage before moving on to the next.

The stages of a methodology may be few or many, and activities within each stage can also vary considerably. The stages help partition the development process to make it more manageable. Each stage requires the completion of numerous activities before moving to the next stage.

All structured methodologies are future oriented. They require designers to accurately visualize the ultimate outcome and then make decisions that will make it happen. The system may not be completed for 6 months to 2 years or more. The initial decisions obviously become the basis for later decisions, and later decisions are always constrained by early decisions. Designers make all decisions while focusing on the system as it will eventually become.

The structured methodology assumes that it is possible to do it right the first time, that everything should be correctly completed within each stage, and once documented, the initial decisions should not be changed. Once completed, the information in each stage is considered cast in concrete. Because so many decisions are interrelated and based on earlier decisions, making changes either within or between stages is not easy.

The structured strategy is patterned after long-standing, traditional ways of designing and developing houses, bridges, and airplanes. For example, when building a new home, first comes a discussion of the family's needs, then a plan (drawing or blueprint), and finally the actual finished house. Some low-level testing takes place by building inspectors while the house is being built, but most serious user testing takes place after the family has moved into the house.

It is easiest to change the location and size of rooms during the initial discussion of homeowner needs. It is more difficult to change the rooms on the blueprints. It is extremely difficult (and expensive) to move a room or wall after construction is completed. The same is true when using the stage-based strategy for creating computer systems.

Unfortunately, the times when the changes can be most easily made are the times when users have the least information for understanding the system. As users gain experience with a system, they have new ideas as to how to improve the system.

Iterative Prototyping Methods

An alternative system design strategy is to use iterative prototyping. This strategy assumes that the system design and development process is too difficult and complex to do without making and correcting mistakes. For example, even after users and designers agree on a "final" set of functions for the new system, they know that some new functions eventually will be added. The initial set is recognized as only a partial set. The complete set of needs will only become apparent as users have experience with viewing or using a prototype of the proposed system.

The iterative prototyping approach is not future oriented (like the stage-based approach), but is "now" oriented. What you have, as reflected by a prototype, at any given time is what you get. The state of progress is reflected in the development of the prototype and related documentation. The prototype and associated information reflect the sum total of all design decisions.

The iterative prototyping method requires the availability of adequate prototyping and usability testing tools. Also, there should be a strong commitment to make the necessary changes as the usability testing identifies weaknesses.

One of the most effective ways of ensuring acceptable system performance comes from designing a system from the *outside in,* that is, prototyping the user interface first and then adding the appropriate functionality as needed (built around the user interface).

Combining Stage-based and Iterative Prototyping Methods

In many, if not most, current system development efforts, both structured and iterative prototyping strategies are used. Some designers still use either one approach or the other, but most systems are the result of effectively combining the two approaches. The pure stage-based approach and the pure iterative prototyping approach probably lie at two ends of this continuum:

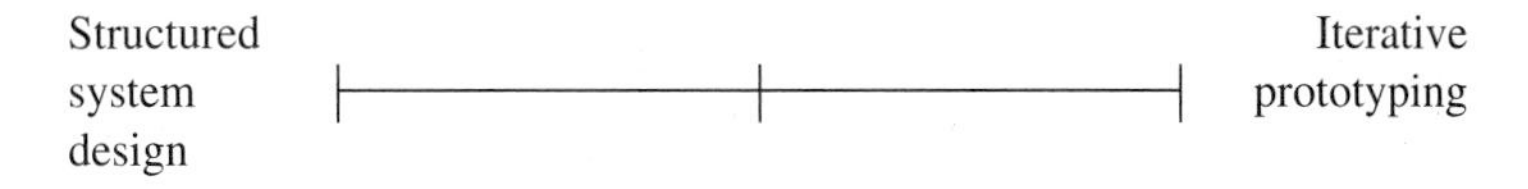

Very large systems, with many designers involved, require the control that comes from using a structured design philosophy. Very small systems are best prototyped directly and can be effectively developed without using a more elaborate structured method. All systems that fall between the two extremes most likely can benefit from using the best ideas from both approaches.

Studies of the Iterative Design Approach

Iterative Design

Gould et al. (1987) developed the Olympic Message System to help validate IBM's usability principles, which included many examples of iterative design. Also, they introduced a "user interface first" concept and were able to demonstrate the effectiveness of using numerous different usability testing methods.

Evidence That Iteration Works

One of the few studies that demonstrate the true value of the iterative design approach was published by Gregg Bailey (1993). He had eight experienced designers (four programmers and four user interface specialists) use a prototyping package (Protoscreens) to design and develop a small system. Each system was then used by three users (one at a time), who attempted to accomplish the activity in the shortest time with the fewest errors. Their performance was videotaped and shown to each designer. The designers then made changes to their systems, and three different users were tested. This process continued until each designer stated that no further improvements could be made. All designers stopped after three to five iterations. A total of 96 usability tests were performed.

The results showed that the test subjects improved an average of 12 percent with each iteration. They completed 19 percent more tasks in the last iteration than they did in the first iteration. Also, the average time to complete each task was 35 percent less in the last iteration.

The total design and development time required by the user interface specialists and the programmers was not reliably different. Even so, the usability skills and knowledge of the user interface specialists enabled them to elicit significantly better initial and overall performance than did the programmers. The user interface specialist's subjects completed 27 percent more tasks and performed in 39 percent less time than the programmer's subjects.

In fact, the average number of activities completed and the average time to complete the activities in the user interface specialists *first* iteration and the programmers *last* iteration were almost the same. This suggests that, even though both groups were able to improve with each iteration, the user interface specialists elicited significantly better performance from their test subjects in the beginning.

Iteration Observations

The iterative design approach seems to offer a definite advantage over delivering the initial design to users. There is little question that the original design (the one usually delivered) almost always can be substantially improved. Also, the better the product design to begin with, the better the results are after iterating. User interface specialists, when compared with programmers that have less user interface design experience, appeared to be able to start with better designed user interfaces. Existing iterative methodologies are probably not capable of finding and fixing all usability problems. Many designers continue to make the comment, "I know what is wrong, I just do not know how to fix it."

Software Prototyping

Modeling User Interfaces

Software prototyping refers to the quick creation of computer systems, applications, and products. Special-purpose prototyping software allows people to rapidly create models of real computer systems. The focus is generally on simulating the user interface and then using the interface to make inferences about underlying functionality.

To potential users, these prototypes can give every impression of really working. Users can select from menus, do data entry, use a mouse to point and click on push buttons in dialog boxes, open and close windows, perform calculations, and enter commands.

Prototyping can be used to help identify user needs, assist in setting system requirements, as an adjunct to written design specifications, to illustrate window design alternatives, and so on. Prototypes can give prospective users the impression of working with real systems. This is like visiting a model home, rather than just inspecting a drawing or blueprint, before constructing a house. This allows early verification of the user interface and system functionality by collecting both human performance and user preference data.

Prototyping encourages iterative design, during which ideas are constantly tried, tested, and changed. Being able to make quick and easy changes allows functions to be systematically added or deleted, and user interfaces refined. Both can be done long *before* programming begins.

There are basically two types of software prototyping tools.

- **Production prototyping.** The most common form of prototyping is usually termed *production prototyping*. Production prototyping software enables designers to design and develop an entire system. These prototyping tools are more like rapid application development tools and serve as a "holder" for design decisions. At some point, the prototype ceases being a prototype and is delivered as the actual system. Unfortunately, many prototypes get delivered to users without having steps taken to make them commercially viable.
- **Rapid prototyping.** Another form of prototyping is usually referred to as a *rapid prototyping*. Rapid prototyping software enables designers to quickly try out various decisions and ideas. A usability test can then be conducted, which enables an evaluation or test of the ideas. Rapid prototyping tools facilitate fast creation and fast changing of prototypes. The create–change–create process continues until a usability goal is achieved.

These prototyping methods will be discussed in more detail in the following sections. Systems developed using prototyping can be very efficient to produce. In addition, if they are combined with effective usability testing, they are much more likely to meet user needs and expectations. Good prototypes, then, help ensure acceptable functionality, speedier performance, fewer errors, less training time, and greater user acceptance.

Concept of Prototyping

Introduction

Software prototypes are usually defined as representations or models of real systems. Prototypes are lifelike representations. A model of an airplane or a car or a ship is simply a representation of a real airplane, car, or ship. Models can be very real without being the real thing. Whenever software prototypes are created, they are representations of some actual system or system component. Prototypes can represent an entire system, a subset of functions, or a specific set of user interface issues.

The basic questions become, then, What are the kinds of things we want the software prototype to do? What must the prototype represent? What kind of functionality do we need to create in order to be able to illustrate the proposed new system?

Prototyping with Existing Software

Prototypes can be created using a wide range of different software tools. They can be created using integrated computer-assisted software engineering (CASE) tools, programming languages, expert system shells, word processors, spreadsheet software, and graphics software.

The functionality necessary to create prototypes obviously exists in most programming languages, such as C, C++, Pascal, BASIC, or COBOL. Using these languages, a skilled programmer can prototype virtually anything. The software is readily available and the fidelity level of the finished prototype can be high. These are advantages of using these languages to prototype.

The applications that we put together using these languages can be very small or very large. In virtually all cases, however, the prototype eventually becomes the system. The main disadvantage of using these languages to build prototypes is that they tend to take a very long time to learn how to use well. People do not become proficient C++ programmers in a matter of weeks. Programming languages are most useful for people who use them constantly.

Prototyping with Prototyping Software

Ideally, prototypes will be developed using an actual prototyping tool, that is, software that was designed to be exclusively used to develop prototypes. The ideal prototyping tool is inexpensive, does absolutely everything we need, and can be used with little or no training.

Unfortunately, many good prototyping tools are expensive, limited in their capabilities, and difficult to learn and use. Almost invariably, software complexity dictates a major resource investment, including considerable training time. Developing complex applications usually means that we must use a complex software tool.

Iterative Design and Prototyping

All software systems should be developed using an iterative design strategy. The main question in most design situations is how to best use prototyping to facilitate the iterative design process. One nice thing about using prototypes is that they assist designers in receiving fast and timely feedback. We actually get the chance to visit the "model home," rather than having to study blueprints. The feedback that comes from actually using a system, like the feedback that comes from walking through a model home, gives designers an opportunity to evaluate how close we are to where we want to be.

Iteration is what makes almost any human-made product outstanding. No one is smart enough to design the perfect product in the first place. In the software world, we are just starting to understand how we can effectively use iteration in the design and development of systems.

The concept of iteration means that we construct a product, evaluate it, and then make changes based on the evaluation. This process continues until we reach our goal. This process allows us to take our original design decisions and refine them by making a series of informed changes. Because it is possible to make changes, we do not need to feel that we always must be right with the first set of design decisions. This approach is obviously different from developing a system based on limited user input and considerable designer imagination.

Iteration is a useful way to go about designing virtually anything. Consider successful Broadway musicals. One reason that they are successful is because of thousands of small changes that get made to the original production. The original is produced in some quiet location, where audience and critic responses are carefully evaluated. Based on responses from the audience, changes are made and then more performances are given. The audience response is again evaluated, and more changes are made. This process can easily take 3 to 5 years before the show is ready for Broadway.

Rapid Application Development

There is considerable confusion in the software development community concerning prototyping. In most other industries, prototypes are used to evaluate ideas. Once designers have learned all that they can from the prototype, they set it aside and develop a commercially viable product. This design strategy focuses more on the new products ultimate usability (in the real world), rather than using prototyping to speed up the overall "producability."

Rapid Application Development Software

Most software that is being sold as "rapid prototyping" software is actually "rapid application development" software. If the goal of using the software is to begin and finish coding the new product as soon as possible, then it is probably a rapid application tool. If the goal is to ensure the best possible usability *before* coding begins, then it is most likely a rapid prototyping tool.

There is nothing wrong with using rapid application developers. In fact, the only way that we will ever be able to create all needed applications is to considerably speed up the current system design and development process by using rapid application development software. True prototyping contributes to speeding up the system development process by reducing the amount of rework, recoding, and redoing that typically goes on. For commercially viable computer systems, issues such as processing efficiency and maintenance (making future changes in response to changing user needs) need to be more directly addressed.

For example, in the area of knowledge-based system development, this process is becoming commonplace. The initial system can be constructed using an expert system shell. Once the system is performing satisfactorily (i.e., the functionality is correct), it is recoded using another language (e.g., C or C++). This helps make the system more efficient when running. In situations such as this, the expert system shell is used primarily to develop and refine the prototype.

Special-purpose Prototyping Software

Using the Best Software for the Purpose

With all the prototyping capability available in existing software, why have special software for rapid prototyping? The best answer is that existing software requires too much time and offers too little in return. Using a third- or fourth-generation language requires a person to prototype much more than is desired in most applications. In addition, because of the built-in power of these languages, they can be difficult to learn and to remember how to use. Also, using word-processor or spreadsheet programs to do prototyping does not allow an easy way for screens, windows, and fields to be linked together. Thus, they provide too little functionality.

A specialized prototyping tool offers reduced functionality and reduced development and learning time. Also, we generally can run a prototype on different computers, moving easily from computer to computer within certain hardware environments.

Frequently, design teams use prototyping to help in developing products, applications, or systems. These teams can consist of systems analysts, programmers, usability (human factors) specialists, writers, and trainers. All join in the common goal of creating a usable system. Using a prototyping tool helps to enable work to be done at its fullest capacity, because of increasing the communication among members of the group. It is particularly useful when all members of the team know how to use the prototyping tool or are at least familiar with its capabilities.

Occasional Use of Prototyping Software

Most programming languages require regular use in order for programmers to remember how to use them efficiently. Rapid prototyping tools, on the other hand, tend to be used infrequently. These prototyping tools should support use by people who use them for a few days or weeks and then do not use them again for several months (i.e., occasional users). Prototyping software may sit on the shelf for long periods of time, and then something comes up and suddenly, we have a need to prototype an application. A good prototyping tool must be powerful enough to do what needs to be done, but simple and easy to use.

Some software tools were specifically built to do prototyping. Others were initially constructed to speed up the development of systems, but were then used to create software prototypes. A good example is Apple's Hypercard. It was introduced as a way to quickly develop simple applications. But many people realized that it could be used to prototype systems that were far more complex.

Advantages of Prototyping

Quick Results

One of the strongest arguments in favor of prototyping is that it generally can be done in less time and for less cost. There is great value in being able to quickly create software models or representations. This allows both designers and users the opportunity to look at the models, use them, evaluate them, and do usability testing.

Good software prototypes give the impression to users that they actually work. Because they seem to work, designers are able to experiment more. They can try different alternatives and ask more "what if" questions. This allows designers to proceed in a systematic, step-by-step manner toward a specific design goal.

Reduced Costs

Probably the most significant advantage of using prototyping is that it can reduce the cost of designing and developing computer systems. The research and experience currently available suggest that the overall development process can be substantially shortened, and the associated costs can be reduced by as much as 40 percent.

Improved Communication

Another important advantage of prototyping is the increased capability of improving communication. Whenever we have two or more people involved in the design and development of a system, there is the potential of having faulty communication. This includes communication among programmers, between analysts and programmers, between users and developers, between different users, and between user interface specialists and designers. With prototyping we have a means for all people associated with a new system to experience the exact same product, as opposed to merely reading about it.

Managing Change

Prototyping encourages numerous changes to be made in a product while it is being designed and developed. The "C" word is one that system developers attempt to avoid as much as possible. Whenever someone mentions "change" to most designers, it immediately translates into increased costs and missed schedules. If too many changes are required, there will not be time to complete all that was originally intended.

Users Change

During the development of a product, many elements can change. Potential users continue to change as they view and use each new iteration. As they experience each new prototype, they have new ideas. They ask, "Why don't you do this?" or "This is what I really want," or "I think this is the way that we should do it." Making these kinds of observations and recommendations is good. It is part of the iterative process.

Software Changes

The software generally must change in response to the changes that come about in users. This occurs, however, only if the developers have the skill, inclination, and flexibility to make the changes. They can change the system functionality, the data or knowledge, the user interface, or facilitators.

Hardware Changes

We can actually, through good prototyping, decide to change the hardware. Assume that we had originally decided to use the traditional keyboard and mouse as the input device and a CRT-based monitor as the output device. After prototyping, we may decide that it is better to use a touchscreen for input. The prototype and usability testing may also suggest that we should augment the touchscreen with more voice output.

Documentation and Training Changes

Effective prototyping can also encourage changes to the documentation and training. Here, again, we do not have to stay with the first thing that was produced. Unfortunately for users, most designers end up delivering the first document that they write.

Traditionally, we have not allowed much iteration with documentation and training because we did not have quick and easy ways to do it. Much of the difficulty had to do with the use of paper as the medium for delivery. As the system design and development become more computer based, we are capable of developing more prototypes and, after usability testing, make easier changes.

Fidelity Model of Prototyping

Introduction

In computer-based systems, at least three readily identifiable components can be prototyped. These include the system's functionality, the system's user interface, and the system's physical characteristics.

Functionality

The functionality of a system includes all the processes the system is expected to carry out. In other words, functions are processes carried out to accomplish the purpose of the system. In a word-processing system, for example, the functions include the ability to enter, change, and delete characters, words, or paragraphs. In a spreadsheet system, the functions include the ability to add, subtract, multiply, and divide rows, columns, or sections of numbers.

User Interface

The user interface, perhaps better stated as the human–computer interface, of a system includes the software and devices necessary to accomplish the system's functionality. It is the means of accessing and interacting with the functionality. Numerous user interface issues must be dealt with in the design and development of systems.

Physical Characteristics

The physical characteristics of a system are generally the hardware features of input and/or output devices. Examples include the layout of buttons and displays on a hand-held or desktop telephone set, alternative layouts of calculator keyboards, and voice input. Rather than creating expensive hardware devices, the main features of certain types of input and/or output devices can be prototyped. This allows developers to evaluate many different configurations without the cost of building the physical devices themselves.

Concept of Fidelity

When a representation of an object or idea closely matches that of another, we have a *high-fidelity* representation. Thus, when a prototype closely matches the real system, we have a high-fidelity prototype. As previously discussed, we can have prototypes that focus on the functionality, user interface, or physical characteristics (or all three). Although it is possible to do, constructing high-fidelity prototypes of all three major components is the same as developing the system itself and can be time consuming and costly.

Generally, we use prototyping to construct a portion of one or more of the components. We may consider only one or two functions, a small section of the user interface, or the layout of numeric push buttons on a large keyboard. When doing this, we frequently have prototypes that are good enough to help in making certain decisions, but do not accurately represent the overall functionality, user interface, or physical characteristics of a system. Thus, they may be considered as *low-fidelity* prototypes.

In most systems, prototyping of the three major components can range from very low fidelity to very high fidelity, depending on the needs of the developers. For example, a prototype may reflect low-fidelity functionality (the illusion of processing), high-fidelity user interface, and low-fidelity physical interface.

Production Prototyping

Introduction

It is possible to differentiate between at least two types of prototyping, *production prototypes* and *rapid prototypes.* This section briefly describes and discusses production prototypes.

Characteristics of Production Prototypes

A *production prototype* is an evolving system. It is a platform for implementing design decisions. The end result is a full-scale working representation of a real system. Once completed, both the system functionality and the user interface are high fidelity because they not only represent the real system, but they *are* the real system.

This is the single, most clear-cut characteristic of a production prototype—when finished, it is an operational system. Developers start with an application development software tool that works on a specific hardware platform. They make initial design decisions that are programmed into the prototype. They continue to make decisions and, while doing so, add to the prototype. In this way the new system gradually grows, develops, and matures. Both developers and potential users are able to look at the evolving new system, comment on it, and evaluate it.

The system takes form rapidly, which sometimes causes confusion between rapid development of a system and being able to rapidly create prototypes. Because the software can be produced quickly, some consider the rapid development of a system using this form of prototyping as doing rapid prototyping. It is more accurate to refer to this system development philosophy as *rapid application development* using production prototyping software.

Throughout the development process the new system is considered a prototype—a production prototype. Eventually, the system is completed, possibly because the due date has arrived, and it is delivered to users. The production prototype is now an operational system. In the minds of many designers, production prototypes have the magical ability to change overnight from a prototype to the real system.

Because they are evolving systems, production prototypes have life spans that can last for many years. Generally, production prototypes are developed using the software and hardware that are intended for the final operational system. Thus, the development environment tends to be the same as the one used for operating a system. Having a relatively high fidelity representation of the system provides the potential for usability testing.

The primary focus of production prototyping is to develop software that can be converted as quickly as possible into a real system. Many have built-in automatic code generators to facilitate the conversion from a prototype to a real system. To effectively use code generators requires a level of definition in the prototype that can make doing prototyping hard to learn and use. Developers are required to make decisions at a relatively low level of detail and to code these decisions into the prototype.

Possible Problems

Because production prototypes begin as systems in embryo and continue to grow and develop as new decisions are made, some possible difficulties are inherent in this process.

Built-in Biases

One of the most common problems occurs when something is built into the prototyping software itself that is not consistent with good design principles, a style guide, or what a specific group of users typically expect in a system. Where these biases exist in the prototyping software, a developer must learn to live with them.

One early software package used for production prototyping, for example, required users to press the ENTER key to move from field to field on the same screen and the END key to move from screen to screen. This caused considerable confusion and frustration for users who were used to pressing the ENTER key to do both.

The developers of the prototyping software make these decisions so that users of their prototyping software will not have to. Developers may not want to do it this way, but there is really no alternative. Developers of production prototyping software have had to make many user interface decisions so that people using the prototyping software will not have to take the time to do so.

Limited and Inefficient Systems

Some systems that result from using production prototyping software tend to be quite narrow in focus (to adapt to the software platform). Also, the programs themselves may not be very efficient in terms of how quickly the code is able to process. For example, we conducted a study to determine the amount of code required for one small application. An automatic code generator created about 1200 lines of Pascal code to do what an experienced programmer was able to do with only 120 lines of Pascal code. Obviously, the more lines of code needing to be processed, the longer the system will take to do the processing. Although they are improving with each new generation, some automatic code generators still tend to produce systems that run relatively slowly.

Lack of Design Documentation

Designers have found that it is difficult to produce good design documentation as production prototypes are being developed. Numerous design decisions are made and changed before anything is recorded.

Production prototyping software is generally included as part of most CASE tools. It is the platform that supports the developing system. The better CASE tools provide ways of developing design documentation as the prototypes are developing. But, even so, many design decisions are not recorded, making it difficult to justify some of the decisions and to maintain the system later.

Difficulty in Making Changes

Production prototyping software is a platform for creating new systems, where the prototype becomes the system. One advantage is that changes can be made more easily than in more traditional system development methodologies. This allows faster and more frequent changes to a developing system. But one disadvantage is that, because it is so difficult to learn how to use and remember what to do and so time consuming to build the initial prototypes, there is still little motivation to make changes to a prototype. Developers still do not like to discard an existing prototype in favor of new ideas. The investment is too great to encourage much experimentation. The commitment to original design decisions is very similar to what we find in traditional system development efforts.

Summary

Production prototypes are generally high-fidelity simulations because they eventually become the actual system itself. They are system development platforms that allow developers to take design ideas from beginning to end. The initial creation of system components is easier than more traditional methods, but there is still a tendency to avoid making changes to existing prototypes. The developing system continues to be called a prototype until it is finished, and then it is referred to as a completed system.

Rapid Prototyping

Definition

Rapid prototyping software provides a fast and easy method for creating directly usable software. This is software that can be produced without designers having to do any coding. In many respects, rapid prototypes are illusions. A skilled designer can create prototypes that effectively create illusions that give the impression that there is much more to a software system than really exists. Rapid prototypes can be models of entire or partial systems.

The goal with rapid prototypes is not necessarily to end up with a real system. The intent is to try out different alternatives until developers and users are satisfied with the model. Because the model actually works and gives the impression of the actual system, it is useful for trying out ideas.

Basic Components

The basic components of a rapid prototyping tool include (1) a way to design and develop windows and screens, (2) a way to link together fields, windows, and screens (i.e., to allow transitions within and between windows and screens, (3) a run-time program, (4) a means of conducting usability tests, (5) a means of capturing or "photoing" existing windows or screens, and (6) a way of printing the results.

Designing Windows and Screens

Perhaps the most common activity performed with rapid prototyping tools is to design and develop windows and screens. With prototyping tools, "what you see is what you get" (WYSIWYG).

Linking Fields, Windows, and Screens

A rapid prototyping tool gives users the ability to quickly and easily link fields to fields, windows to windows, and screens to screens. This component of rapid prototyping software is sometimes referred to as the *process controller* or the *control mechanisms.* Changes from one field, window, or screen to another are prompted by *events.* The actual changes themselves are referred to as *actions.*

Running Prototypes

Once designers have put together some fields, windows, or screens and completed the necessary linking, the prototype should run like an actual system. This is best handled by a special *run-time* component. The best run-time components are stand-alone so that prototypes can be distributed to other members of the development team, including users, for comments. Most run-time programs do not require users to pay a license fee for their use.

Usability Testing

Another integral part of a rapid prototyping tool is having a module that allows designers to develop and conduct usability tests. These tests involve having actual users use the prototype and collecting performance data.

Quick Illustrations

Rapid prototypes are used to try out, evaluate, and communicate design decisions. Generally, rapid prototypes can be thought of as software that allows you to quickly develop models of other software. It can be done in a very short time and usually does not end up being used as the final system. Unlike production prototypes, rapid prototypes are "off to the side" and are used primarily to try out new ideas, to develop and show demonstrations, to record needs information, and the like. The final prototype does not necessarily have to end up in the final system, although it could.

High Fidelity Simulations

Rapid prototypes provide designers with the ability to have software simulations of high enough fidelity that users not only get the "look," but also the "look and feel" of a real system. These prototypes are much more than a "slideshow" series of screens. They allow users to make decisions on when and how to move from field to field and from screen to screen.

Short Life Spans

Rapid prototypes can have very short life spans. Frequently, the resulting prototypes are truly "throw away." There is great value in having the ability to quickly create software and, if it does not do exactly what is desired, to erase it and try another approach. This will only work in situations when the designer's commitment to creating the software is low enough to allow it to be discarded. If little time was spent in designing and developing a prototype, then one does not feel too bad about getting rid of it. If a good deal of time (weeks or months) is spent generating the software, then it can be very difficult to throw away or make major changes.

Few system designers can spend weeks or months developing software to illustrate an idea, show it to others, and then cast it aside. They usually argue that any new ideas should simply build on the software already developed. There is a point when too much time and effort have been invested in a software product to throw it away or even make major modifications. The main question becomes, "How much time and effort is a person willing to put into something before they are unwilling to discard it?" In virtually all production prototypes this threshold is reached very soon. Too much is required from the designer before even a simple working prototype is available. But with rapid prototypes, the investment can be so small that starting over is still a possibility. Thus, one of the big differences between production prototyping and rapid prototyping is the level of investment made to create the prototype in the first place.

Conception to Delivery Prototyping

A rapid prototype can be used from the conception of an idea through to the point where a system is delivered. Rapid prototypes are valuable in all design stages. Consider that we can create a prototype long before we have even decided to construct a new system. We can use it to consider alternative ways of building a system. We can create prototypes as part of a thorough needs analysis and can put together two or more sets of alternatives to help evaluate and define system requirements. We can prototype two or more different design options that represent alternative ways to go about doing the exact same thing. Finally, we can develop prototypes to help optimize the user interface, on-line documentation, and training.

Dealing with Change

We already talked about change as it relates to prototyping in general. With rapid prototyping we focus directly on change. Even when allowing changes to take place, we can still shorten the whole process. We can do more in less time.

We know before we begin that design errors are going to be made. They are made in any product development process. If we know in advance that they are going to be made, the best strategy is to make our best design decisions, evaluate them and find the errors, and then make the necessary corrections. The most important place to catch these errors is early in the design process, while identifying user needs and determining system requirement. This eliminates having to go back and reconsider some of the early decisions that have others stacked on them. Also, it allows many issues to be resolved before one line of code is written.

Rapid prototyping not only allows change, it also encourages change. Rapid prototyping allows change to be a natural, normal part of the development process. It also allows better management of changes. Once we accept the fact that it is okay to make changes, we put ourselves into a position where we can better handle the fact that changes are going to be made. Rapid prototyping has as a goal to bring something up and let somebody look at it with the idea that changes will be made.

Prototyping by Users

Many users who have never designed or programmed anything in their lives are using rapid prototyping software to tell designers, "This is what I want from you." When users show designers what they want, designers who understand the realities (constraints and limitations) of creating computer-based systems suggest changes. When done right, we do not start out with what we want to end up with. We start out with something that allows us to get to where we really ought to be.

Possible Limitations

One of the biggest difficulties we have is that rapid prototyping tools tend to allow us to do prototyping of systems that we cannot deliver. We could create prototypes that could not be developed using traditional programming in the time available. So there is a potential problem of promising more functionality and of having a more sophisticated user interface than could be programmed in any reasonable time.

For example, when running a prototype on a personal computer, we can get less than 1-second time. But when running the exact same system on a mainframe, the response time could be 5 seconds or longer, so the original prototype demonstration that we gave, showing how well the system would perform, could be totally misleading.

We have also found situations when users look at a prototype and say, "Yeah, that's it. That's exactly what I had in mind. I'd like to start using it tomorrow." And then, of course, we must give them the bad news, "We cannot deliver this system for another 3 or 6 months."

Usability Requirements

Setting User Performance Goals

To help designers to know when a prototype reflects a level that can be delivered as an operational system, a set of *usability requirements* is frequently used. These requirements usually relate to the performance of the human and computer working together to achieve a particular goal. They generally include such performance characteristics as the time to perform activities, accuracy levels, the time necessary to develop unique skills, and user satisfaction. A designer must know these requirements to make informed decisions concerning the goodness and completeness of the prototyped user interfaces and facilitators.

Measuring User Performance

When establishing human performance requirements, we should always develop a way of measuring each one. If a requirement relates to accuracy, there must be a meaningful way to measure errors. If the requirement concerns user processing time, there must be a meaningful way to measure user speed. This could mean time per customer contact, number of items produced per hour, or average keystrokes per day. Usually, the problem is deciding what measure (from among many possibilities) gives the best indication of the performance. With a little imagination, any human performance can be meaningfully measured (cf. Gilbert, 1978, p. 29).

As a minimum, human performance requirements should include statements concerning errors, user processing time, the training time necessary to ensure the minimum skills, and job satisfaction. If we do not clearly state usability requirements from the beginning, we cannot expect human performance considerations to be taken seriously while the product is being developed. In fact, in the absence of clearly defined human performance requirements, software and hardware requirements will become the focus of the design team. And when the system is operational, people will be left to perform as best they can without adequate provision for ensuring an acceptable level of human performance.

Telephone Installation Example

For example, a new computer-based system was being designed to provide telephone installers with the information needed to make a residence telephone installation. The computer printout work order used in the old manual system contained many different items of information. If one or more of the information items on the work order were incorrect, the installer had to stop work, call back to the office, and wait for the correct information. The time lost in correcting the error meant that fewer customers would be serviced per day.

If all information on the work orders were always correct, an installer would be able to make every installation without ever checking back with the office. The actual payoff to the telephone company of having accurate information on each service order was the servicing of more customers per day. Obviously, the more customers serviced per day by each installer, the fewer installers, trucks, and other personnel required.

The system was evaluated, and it was determined that the probability of having an obstruction-free installation was about 0.81(81 times out of 100). This accuracy level was not acceptable to potential users of the new system. They wanted the installer to make at least 90 to 95 percent of the installations without having to stop and call back for accurate information. Thus, in the new system, the information had to be made more accurate. It was found that the accuracy level for *each item* contained on the work order would have to average about 99 percent to have an obstruction-free installation 90 percent of the time. Each item would have to average 99.5 percent to have an obstruction-free installation 95 percent of the time.

As the new system was being designed, considerable cost and effort were expended to ensure these high accuracy levels.

Error Correction Example

In another system, users would spend their day updating a billing database. Many errors would be detected and corrected as they were keying. Once they finished a half-day's work, the computer would automatically run a series of more sophisticated error-detection programs and print out other errors. Users would then correct their own errors. The average time to correct these errors was about 12 minutes per error. This was considered an unacceptable amount of time for efficient system operation.

When a replacement system was proposed, management levied a human performance requirement for the new system of only 7 minutes per error, which was achieved. Reducing by 5 minutes the average time to reconcile errors helped to reduce the cost of operating the system during a 1-year period from $144,000 to $96,000.

Other human performance requirements could be associated with training time. For example, a requirement could be that "The total time to train clerical personnel to perform the basic activities should not exceed 3 days." Another human performance requirement could relate to job satisfaction. For example, "After performing an activity for 6 months, employees should respond in a positive way to their work, as measured by the XYZ questionnaire."

Stage-based Iterative Design

Designing User Interfaces in Stages

Most good system development efforts are based on at least two main principles: (1) clearly defining the problem and (2) planning and carrying out a strategy to attain the desired solution. Defining a problem helps to establish the need for a new system. The emphasis then shifts to determining and carrying out a useful development plan or strategy. Nearly everyone is familiar with the necessity for good planning, and many have benefited from their own plans or the planning of others. Building a backyard patio, going on a camping trip, or transferring to a different city all require considerable planning and much attention to detail. As complicated as such activities can be, the development of a system can be considerably more complex. Consequently, it is helpful to have an effective means of planning and executing the design process so that it can be carried out as efficiently as possible.

Partitioning the user interface development process into a series of meaningfully related groups of activities called *stages* makes the process more manageable. Each stage contains a set of activities that usually requires completion before moving on to the next stage. Generally, the work in each stage is finished before the work in the next stage begins.

Major Stages

Only three stages are required when using an iterative approach for the design and development of user interfaces:

Stage 1: Analyze and define the product

Stage 2: Design and develop the user interface

Stage 3: Design and develop facilitators

These three stages help to focus attention on user interface issues long before the first screen, window, or user interface object is designed. This points up one of the most important considerations in the design and development of quality user interfaces. *The earlier we begin making user interface decisions, the better will be the user interface.* If the first systematic interface decisions are made for screens, windows, or report layouts, it may be too late to have a quality user interface. From the very beginning of a new product or system, a set of user interface activities should be performed to help to ensure a quality user interface.

If the application is being developed using a structured methodology, attempt to integrate these three user interface stages into the larger set of stages. If an iterative prototyping approach is being used, simply make sure that these major activities are done at some time while the system is being developed.

The stages can be expanded to show typical activities in each in more detail, including prototyping, usability studies, and testing.

Stage 1: Analyze and Define the Product

- Determine system objectives and performance specifications
  - Develop idea prototyping (as needed)
  - Develop demonstration prototyping (as needed)
- Determine functionality
  - Develop user needs prototyping (as needed)
  - Develop system requirements prototyping (as needed)
- Verify functionality
- Develop user profiles
- Develop a computer environment profile
- Conduct appropriate usability studies and testing

Stage 2: Design and Develop the User Interface

- Conduct a task analysis
  - Develop design prototypes (as needed)
- Develop interaction capabilities
  - Develop design prototypes (as needed)
  - Develop user interface prototypes (as needed)
- Develop presentation capabilities
  - Develop design prototypes (as needed)
  - Develop user interface prototypes (as needed)
- Conduct appropriate usability studies and testing

Stage 3: Design and Develop Facilitators

- Develop user guidance capabilities
  - Develop user interface prototypes (as needed)
- Develop documentation
  - Develop documentation prototypes (as needed)
- Develop training
  - Develop training prototypes (as needed)
- Conduct appropriate usability studies and testing

Freestyle Iteration Example

Iterative Design

One of the best published examples of iterative design being used in the design and development of a successful product is related to Wang's Freestyle (Perkins et al., 1989). Freestyle is a multimedia mail system that allows effortless shuffling of electronic forms from desk to desk over a variety of different networks.

The Freestyle project had as two of its design goals to minimize learning time and to not deliver hardcopy documentation. Its design methodology was centered around an iterative design philosophy by which the designers constructed a prototype and then conducted usability testing. The results of the usability testing were used to make changes to the prototype. This process was used until the product met the initial design goals.

The designers concurrently designed and developed the product software and a computer-based (on-line) tutorial. The focus of usability testing was to identify any nonintuitive aspects of Freestyle and then make changes either to the product or the on-line tutorial.

Test Items

The usability testing focused on conducting a series of performance tests. The performance test items included having test subjects

- Open a Freestyle page
- Scroll to the bottom of a page
- Create a title page
- Add a voice message to a document
- Electronically staple the pages together in a document
- Mail a copy of a document
- Throw away a document
- Exit the system

These same test items were used for all usability tests so that any improvements could be measured.

Prototype, Test, Make Changes

The Freestyle designers developed four different prototypes and conducted five performance tests. Each test involved five different test subjects. In general, the iterative process proceeded as follows:

- Developed the first prototype
- Conducted the first test (five participants)

- Developed the second prototype and constructed an initial tutorial
- Conducted the second test (five new participants)
- Developed the third prototype and revised the tutorial
- Conducted the third test (five new participants)
- Developed the fourth prototype and revised the tutorial
- Conducted the fourth test (five new participants)
- Conducted the fifth test without using the tutorial (five participants)

After finding nonintuitive features, designers could make changes to the product software, the tutorial, or both.

One nonintuitive feature found with performance testings was the way that users tried to access computer-based help. In the first prototype, subjects erroneously tried to select the "info" icon to get help. The "info" icon was originally intended for a totally different purpose. In the second prototype, no changes were made to facilitate users accessing help. However, testers observed that subjects continued to seek help on each icon. Based on the results of the second performance test, designers provided users with improved help in the tutorial with the third prototype. The third performance test clearly illustrated that using the tutorial to provide help on icons did not work well. In the fourth prototype, subjects received help by dragging the "info" icon over and touching any other icon. This worked exceptionally well and was implemented this way in the final product. The ultimate solution became apparent only after numerous iterations.

Integrating Product and Tutorial Design

There were at least two reasons for including the tutorial in with the product. First, it provided an alternative way of solving performance problems that were discovered in the testing. For example, if users were having trouble with some aspect of the product, designers could either try to change the product itself or change the tutorial.

This iterative design approach allowed designers to make trade-offs between the product software and the tutorial software. They reported that, for repeated user errors that occurred due to a mismatch with user expectations, the designers tended to change the product itself. However, for errors that could be eliminated through "one trial learning," they tended to change the tutorial. For example, they found that no person could scroll a page without training, but that no person who was trained had problems scrolling. This was an excellent opportunity to change the tutorial (to teach scrolling), rather than change the product (to make scrolling more intuitive).

A second reason for including the tutorial is that it helped in obtaining a well-designed tutorial. The iterative design process helped to make the tutorial much stronger than it was initially. The improvements to the tutorial enabled an average learn time for the product of 10 minutes (fourth test), whereas the average without the tutorial was 29 minutes (fifth test). This was true even though the final prototype contained far more features than the early prototypes.

Lessons Learned

In this situation, the designers found it helpful to develop prototypes and conduct usability testing while the system was being developed. This helped keep mainstream product development activities from outrunning usability testing. The designers also found it useful to have the freedom to modify either the software or the training. This prevented unnecessary documentation from being developed on the intuitive aspects of the product.

The Wizard of Oz Example of Iterative Design

Designing an Electronic (e-mail) System

Another good example of how an iterative approach was used to develop a user interface is reported in the creation of a character-based electronic mail system (Good et al., 1984). In this system development process, designers first developed the basic architecture of a command-driven electronic mail system. Once they had a prototype that could do the basic set of functions, the first users were tested.

In their tests, only people that had no experience with electronic mail systems were used. The new system did not have a help facility nor did it have any menus. In addition, no documentation or training was provided. Each subject was tested individually and sat alone in a room with a computer and the new system.

The activities (test items) that they were to perform looked like the following:

- Have the computer tell you the time.
- Get rid of any memo that is about morale.
- Look at the contents of each of the memos from Sam Jones.
- On the attached page there is a handwritten memo; using the computer, have Denise Brown receive it.
- See that Jan Smith gets the message about the keyboard study from Lou Delaney.
- It turns out that the Jones memo that you got rid of is needed after all; so go back and get it.

The testing procedure allowed each subject to enter whatever words that came to mind in attempting to solve each activity. If they entered the correct word or words (usually a one-word command), the system processed their instructions (e.g., indicated the time). However, if they entered a word that was not known to the computer, their entry was immediately sent to a person sitting at another computer (in an adjoining room). That person quickly determined what the subject was trying to do and sent the correct command to the computer. The computer carried out the action and then responded to the test subject as though the correct command had been entered originally.

For example, a subject might enter "tell me the time"; the person at the other computer would read the request and enter the correct command "time," at which point the computer then responded to the original user by showing the correct time. The main advantage of this testing technique is that it allows a natural human–computer interaction in which the flow of information is not continually interrupted by error messages. No error messages were ever sent to test subjects, leading them to believe that all their entries were correct.

While the users are interacting with the computer, a list is made of the words that users tried and that the computer was unable to process. After every three subjects, the computer program is changed by adding the most common entries tried by the users. Then three new subjects are individually tested. The updated iteration of the system allows more of the user's entries to be processed directly, without the help of the person on the other computer.

In the development of the user interface for this electronic mail system, a total of 67 subjects was used. The first version of the system was only able to recognize 7 percent of the words tried by naive users, while the final version was able to handle 76 percent of the commands.

After the system was completed, eight people experienced in using electronic mail were tested using the exact same test items. Every one of them was able to complete all the activities correctly. In addition, they all expressed that they liked the system. Thus, this testing approach, which is known as the *Wizard of Oz method,* showed that a system that was developed based on the feedback of novice users was even satisfactory for experienced users.

Designing a Help System

Carroll and Aaronson (1988) demonstrated a creative means for evaluating help facilities by simulating a help facility using a Wizard of Oz technique. Based on initial experience with a system, they developed 73 error messages. As users participated in usability testing and made errors, the testers either sent one of the canned messages or a message that they created "on the fly." Of the 164 messages sent, 55 percent had not been developed in advance. Users did not suspect that the "on the fly" messages were not software based.

Exercise 7: Designing a Bank Query System

Purpose: To become familiar with the iterative design process.

Method: Assume that you have been charged with the responsibility for designing a system for a bank that allows users to interact with the bank on their personal computers. The new system only allows users to view their accounts. The accounts include a checking account and two different savings accounts. Design a paper prototype (simulate screens or windows on white paper) that shows how typical bank users will query these accounts. Have at least three people evaluate your paper prototype. Make the appropriate changes to your prototype based on the evaluators comments.

Reporting: Prepare a report following the directions in Exercise 1A. Include your original and final prototypes as an appendix to your report. What types of changes did your reviewers suggest? Was it worthwhile having others evaluate your ideas?

8

Usability Studies and Usability Testing

Introduction

The human is complex. People performing simple tasks are more complex. Users interacting with a sophisticated computer are even more complex. It is not possible to correctly make all required design decisions without substantial information about potential users, their proposed activities, and the environment in which the tasks take place.

When developing computer-based systems, there are two major types of data collection: studies and tests. *Usability studies* are used to help to make good initial decisions and may be conducted at any time in the design process. These studies are especially important when decisions are being made as to the feasibility of developing a given product, determining what functions should be included, and in helping to make design decisions. *Usability tests* are used to evaluate design decisions after these decisions have been made. They require a product of some kind to evaluate, such as a specification or prototype.

General Data Collection Considerations

Introduction

Systematic collection and analysis of data are important during the system development process. In fact, good data collection serves as important input to the design process. All too often, however, data collection is poorly planned and carried out (or not carried out at all), usually because designers are eager to "get on with the work" or are not familiar with data collection techniques. Not surprisingly, many design decisions in such situations range from disappointing to disastrous. Some general considerations are briefly discussed.

Collecting Data in a Timely Fashion

A need to collect data may arise any time within the system development effort when there is insufficient information to support a necessary decision. When critical information is lacking, steps should be taken to acquire it. The only alternative is for a designer to guess, which is a poor second choice. A wrong decision (or series of wrong decisions) could lead to degraded human performance over the life of the system.

Even though information of various types is needed throughout the entire system development process, it is usually true that the earlier in the system development process that data collection takes place the more valuable the results will be. In other words, data should be collected when needed, without waiting for "a more convenient time." The later a requirement for data is acted on, the less chance there is for the collected data to have a major impact on the overall system.

Cost Justifying Data Collection Decisions

Data collection can be expensive and involved and should be performed only when it will help assure system success. Data should not be collected and/or analyzed for its own sake. The amount of effort put into data collection should be commensurate with the question being asked. In some situations, the alternatives are not of sufficient consequence to warrant data collection. On the other hand, if the best alternative decision is not readily apparent and the selection is vital to having acceptable human performance, data collection activities and associated expenses are probably justified.

Using Others' Data Collection Results

Certain data collection efforts should not take place without a thorough literature search to see if the information of interest has been collected by someone else. Therefore, a logical first order of business in many collection efforts is to attempt to discover if the needed information is already available and has been published elsewhere. Appendix A contains a list of several human performance resources where studies by others are reported.

Sampling

When conducting studies (surveys) or usability tests, there is always an issue concerning how many and what kinds of people to involve. Designers always have the option of including all the people in the group (i.e., include the entire user *population*). Unless it is a very small group, this approach is usually not practical. If they cannot use all people, then they have to select a few users from the much larger group. They must select a portion or *sample* of the total population. Sampling lets us obtain information from a small group of potential users (the sample) and then make inferences about their behavior, performance, or attitudes to a much larger group of users (the population). To make strong inferences, the sample should be truly representative of *all* possible users.

As the size of a user sample increases and more of the population is included in the sample, the sample becomes a better estimate of the population. However, in an industrial situation, this should not be taken as an argument for always using large numbers of people. Ideally, a designer will select a sample just large enough to comfortably represent the population. In general, the fewer people involved, the less the study or usability test will cost, but the more risky are the resulting design decisions.

The two most commonly used types of sampling in studies and usability tests are random sampling and subjective (judgmental) sampling. In selecting a sample, one of the first and most important steps is to decide what kind of sampling procedure should be used. Each study has its particular characteristics that make one type of sampling more appropriate than others. To use random sampling, for example, when subjective sampling is acceptable, may add greatly to the cost of the study.

Random Sampling

Random sampling does not mean haphazard or aimless selection. Random sampling implies selection governed wholly by the laws of chance. This means that the selection of individuals to include in a sample is completely independent of human decision. Each person should have an equal chance of being selected and placed in the sample.

Random sampling should be used when the entire population contains basically the same kind of people. In this procedure, the people included in the sample are drawn at random from the entire population. A common procedure for making this random selection is to first assign a unique number in serial order to each person in the population of interest and then, using a table of random numbers, make a random selection of the serial numbers (the people associated with the numbers) for inclusion in the sample.

More information is included in the Statistics section of the Appendix.

Subjective Sampling

Any sampling procedure that does not follow precise statistical principles can be thought of as a subjective or judgmental sampling procedure. In subjective sampling, subjects may be selected by one or more persons who choose (hand pick) the particular people whom they feel best represent the population. This approach is used, for example, when selecting people for performance testing. Frequently, this kind of selection is necessary because of cost considerations, but it has two limitations that a designer should keep in mind: (1) subjective (judgmental) samples are subject to the personal bias of the individuals responsible for the selection, and (2) the fact that the selection of people is not objective precludes the use of some statistical techniques.

Usability Studies

Most usability studies use a relatively small number of different methods to carry them out. These studies frequently include making direct observations, distributing questionnaires, and conducting interviews.

Making Observations

There is some value in having designers themselves make whatever observations are needed. This helps them get firsthand information. However, when this is not possible, other people with the necessary skills should be found and trained to make these observations.

Data collection through direct observation is used in situations when the information to be gathered is of an operational nature. For example, if a designer wanted to know how many employees come to work late or leave early in an existing system, it is not a good idea to ask the individuals themselves. In such a situation, it would be far more reliable to run a tally of time cards. Or if time cards were not available (or not accurate enough), it might be best to have a trained observer stand at the entrance and make a count.

Data collection by observation requires careful planning. A plan for taking samples must be developed to determine how many situations will be studied and which will be selected. Such considerations as whether the population should be represented by a few people conducting all activities or more people conducting a few selected activities must be taken into account. In addition, preliminary work should be done to isolate the specific operations that are important to the study.

Keep in mind that a major problem in the use of observation is that the known presence of an outside observer can change the actions of the people being observed. Usually, depending on the particular circumstances, workers become used to an observer after a short time, and their usual habits and work patterns take over.

Stop watches and other counters are used when observations must be time or volume related. Typically, such observation is done with special clipboards that hold the collection form and the watch or counter in convenient working proximity. Collection instruments and observation activities should be tested to be sure that the methodology works and the data gathered are usable.

Using Questionnaires

Questionnaires can provide a large amount of data economically and relatively quickly. Questionnaires may be either self-administered or administered by a designer. *Self-administered* questionnaires are usually set up for the respondent to read the instructions, question himself or herself, and write the reply. *Interviewer-administered* questionnaires may be accomplished by face-to-face conversation or by telephone. Each type of questionnaire has its strong points and problems. More specific information on questionnaire development can be found in Appendix B.

Conducting Interviews

Face-to-Face Interviews

When time, budget, and other considerations permit, the face-to-face interview offers the best chance of collecting the most complete and usable information. This is because interviewers can

- Elicit a higher rate of response and encourage more thoughtful answers to the questions
- Explain unclear questions
- Note possible distractions and better control the distractions and the interview

The face-to-face interview, however, is more expensive, so it is important to weigh relative costs and results. Also, telephone interviews remove a measure of interaction and sensitivity between interviewer and respondent. However, they are more quickly conducted and usually cost less.

Telephone Interviews

At least two special requirements apply to telephone interviews. Unfortunately, these constraints restrict the volume and scope of data that can be gathered. In general, the interviews should be short, certainly no longer than 15 minutes. Second, the information requested must be readily available. If the respondent has to search files, contact assistants, or take other special steps to get information, the telephone interview is apt to produce either an incomplete information or mere guesses.

Combining Techniques

Many usability studies use a combination of methods for data collection, including observations, in-person interviews, telephone interviews, and self-administered questionnaires.

Example of a Usability Study

A good example of conducting usability studies to help understand design alternatives and user preferences was reported by Henneman and his colleagues (1990). He was designing a customer-activated terminal for use in restaurants. Just as automatic teller machines (ATMs) allow users to carry out certain banking tasks, the customer-activated terminals (CATs) were being developed to allow customers to order at fast-food restaurants. These systems are different from many other computer-based systems because typical users may have no prior computer experience, no opportunity to adapt to any design limitations, and no opportunity for training.

The iterative methodology used in this development process included an initial data collection process, which consisted of numerous different usability studies. The goal of these studies was to identify interface-related characteristics of the major system components. These studies involved interviews with over 200 restaurant customers, making videotapes and audiotapes of over 150 customer transactions, collecting lists of customer orders, making direct observations and making physical measurements of an existing restaurant, and conducting structured interviews with store personnel.

Based on the information collected in these usability studies, researchers were able to establish numerous important *user* characteristics. The demographic data, for example, indicated that 66 percent of customers were under age 40 (average = 34), 95 percent had graduated from high school, and 53 percent had completed or enrolled in college. Many of the customers already had experience using VCRs (55 percent), personal computers (42 percent), and ATMs (40 percent). After conducting these usability studies, the developers concluded that potential users were not likely to be intimidated by using the new customer-activated terminals.

They also conducted usability studies to determine certain user characteristics having to do with ordering behaviors. For example, they found that 79 percent visited the restaurant at least twice a month (25 percent at least eight times a month), 83 percent reported that they "always" or "frequently" ordered the same items, and 75 percent reported that they "always" or "frequently" know what they are going to order before they arrive. Based on these usability studies, the developers concluded that the new customer-activated terminal should be designed to emphasize customer ordering, not to try to sell certain food or drink items.

Other usability studies were conducted on certain *task* characteristics. They found, for example, that a typical transaction consisted of a customer entering the store, standing in line and reviewing the menu, placing and paying for an order, obtaining a receipt, waiting for the order, and then picking up the order. Other findings include the average order size during lunch time of 2.8 items and at other times of 4.1 items. Based on these results, the developers decided to not use complex scrolling and to not allow any error-correction capabilities. They also found that 46 percent of all orders were one of six lunch specials plus a beverage. From this information, the developers reduced the scope of the initial system to ordering only lunch specials.

A final set of usability studies was conducted on *environmental* characteristics associated with the new product. They found that the restaurant that would first use the new product was located in a busy commercial mall in Boston. The menu included pasta, pizza, and sandwiches. The restaurant had a constant level of lunch and dinner traffic, which kept three point-of-sale terminals busy throughout the lunch period. Also, there was little available floor space in the restaurant where the new customer-activated terminal could be mounted. This suggested that the new product probably would be mounted on a counter top.

Once the usability studies were completed, the developers used the information to guide them in developing a prototype. After a prototype was in place, they began using usability testing.

Usability Testing

Introduction

Many, if not most, system test programs place primary emphasis on the evaluation of the software and/or hardware components of a system. Any information gained about user performance is usually a by-product of these tests and tends to be of little value for ensuring an acceptable level of human performance. A good usability test program, on the other hand, will consist of a series of tests *developed specifically to evaluate human performance and user preferences.*

Usability testing enables designers to identify many (if not most) user interface problems long before a product is delivered. When combined with effective prototyping, this iterative design process is sometimes referred to as *usability engineering* (Good et al., 1986). The method usually requires designers to set acceptable usability levels (goals), conduct usability testing and make required changes, and then repeat this process (iterate) until the initial usability goals are met.

Finding the Ideal Solution

High-quality usability testing helps to identify the *ideal solutions* to design problems. In most systems there is an ideal user interface design. When the design is suboptimal, or less than the best it could be, users will demonstrate degraded performance. There is evidence that, no matter how much the system is used, users will never be able to achieve the performance level that they could have with a better user interface.

Dutta and Proctor (1993) reported a study in which stimulus–response compatibility effects were shown to persist even after extended practice. They had participants perform as follows:

Group 1 pressed a left key for a left stimulus and a right key for a right stimulus (direct mapping).

Group 2 pressed a left key for a right stimulus and a right key for a left stimulus (indirect mapping).

After 2400 trials, group 1 still showed an advantage in both speed and accuracy over group 2 ($p < 0.01$). The authors concluded that the failure to eliminate compatibility effects was related to fundamental human information processing factors. Extensive training and experience were not sufficient to overcome difficulties from incompatible assignments.

Early usability testing will provide designers with insights into design areas that require changes. Early test and evaluation enable detection of design deficiencies before they become cast in concrete. In general, as the development of a system progresses, it becomes more and more expensive to make major design changes.

Measuring Usability: Performance versus Preference

Measurable usability parameters fall into two broad categories: objective *performance measures,* which indicate how capable the users are at using the system, and subjective *user preference measures,* which assess how much the users like the system. Performance and preference measures do not always match.

Most people (71 percent) either perform well and like a system or perform poorly and dislike a system. A large percentage of people (29 percent), however, perform well and *dislike* a system or perform poorly and *like* a system (Nielsen and Levy, 1994). In fact, the correlation between performance and preference is a relatively low 0.44, suggesting that the experiences surrounding a person's performance can explain only about 19 percent (0.44^2) of the observed variance in the users' preferences. For highly *experienced* users, the correlation is even lower at 0.36, which means that performance can explain only about 13 percent of the observed variance in the users' preferences. Keep in mind that this study considered only user preferences as expressed after they had actually *used* a product, application, or system.

Another study (Bailey, 1993) had designers attempt to predict the performance level of alternative user interfaces. Eighty-one computer professionals were shown a demonstration of a small system and then selected the one that they thought would elicit the best performance. They then used prototypes of the interfaces and were given an opportunity to change their original selection. There was no significant difference between their first and second preferences. The same people then performed using each of the four different interfaces. As is clearly shown in Table 8-1, 95 percent of the designers selected an interface method that did not elicit the fastest performance.

Table 8-1 Preferences and Performance Scores for Alternative User Interfaces

	Designer Preferences (%)	Average Time per List (seconds)
Typed entry	5 percent	9.9
One-level menu	18 percent	11.1
Two-level menus	40 percent	12.2
Multiple menus	37 percent	14.7

A follow-on study reported similar results (Bailey, 1993). A study was conducted to determine the best screen-based control (widget) for a new application. Forty-six computer professionals practiced using four different methods and then selected the one that they thought would elicit the fastest performance. The same people then participated in a performance test. The results are shown in Table 8-2 (there was no reliable difference between the open selection lists and the radio buttons). Again, a relatively large percentage of the designers preferred interface methods that did not elicit the fastest performance.

Table 8-2 Preferences and Performance Scores for Alternative Widgets

	Percent Preferring Each Widget	Average Time to Perform (seconds)
Open selection lists	76 percent	58
Drop-down selection lists	13 percent	90
Radio buttons	7 percent	53
Typed entry	4 percent	72

Tullis (1993) reports a study in which he examined whether experienced designers could pick the best user interface for a particular task. He had designers select the best interface method from among seven alternatives. He then had actual users perform an activity using each alternative. These included using two different drag and drop methods, push buttons, radio buttons, drop-down selection lists, and two different entry field methods. Twenty-eight programmers were shown screen images of the seven user interfaces and were requested to rank order them from best to worst based on their expectations for user performance. The correlation between their rankings and the users' performance data was a very low –0.07.

We can conclude from these studies that the preferences of system designers do not always lead to the best performance-related design decisions. Most designers seem to believe that designing to meet their preferences will ultimately provide acceptable levels of performance for users. Because there is little evidence that this is true, there is a strong argument in favor of performing extensive usability testing during the development of new products, applications, and systems.

It is interesting that people can have strong preferences, even in situations when they are unable to reliably discriminate between the options. Givon and Goldman (1987) reported on a study in which their participants showed clear preferences without being able to reliably discriminate among the items being evaluated. For example,

Designer: Is window A different from window B?

User: I can't tell.

Designer: Which do you prefer?

User: Window B.

Types of Usability Tests

There are four major types of usability tests (Bailey, 1992):

Formal evaluations

Inspection evaluations

Performance tests

Operational evaluations

Formal Evaluations

GOMS

Formal evaluation methods are used to evaluate the adequacy of user interfaces while sitting in your office (generally with the help of a computer). Probably the best known approach is GOMS (Card et al., 1986). GOMS stands for goals, operators, methods and selection rules. Using the GOMS methodology, designers can predict performance times for a set of accurately performed activities. These performance times can then be compared for two or more alternative ways of performing the same activities, enabling designers to find the most efficient approach.

Gong and Kieras (1994) applied a formal GOMS model approach to the design and evaluation of a user interface. Use of the method helped them to identify many usability problems. They report that a redesign resulted in (1) a 46 percent reduction in learning time and (2) a 39 percent reduction in execution time. However, the task-level GOMS predictions tended to be less than half the actual times. There are ongoing attempts to automate GOMS (cf. Byrne et al., 1994).

Other Methods

Other formal evaluation methods have been proposed that enable designers to predict both performance times and training times. These include the cognitive complexity theory (Kieras and Polson, 1985), programmable user models (Young et al., 1989), and cognitive walkthroughs (Polson and Lewis, 1992).

The formal evaluation methods that are currently available focus primarily on routine perceptual and cognitive skills for expert users (erroneous responses are usually ignored). After evaluating currently available methods, Gugerty (1993) concluded that the work involved in using formal evaluation methods is considerable when compared to the benefits gained.

Inspection Evaluations

Inspection evaluations should begin as soon as there are products to evaluate. These products can include a list of commands, proposed error messages, initial screen or window designs, proposed user interface objects, outlines of work procedures, or a rough draft of written instructions. Obviously, items do not have to be in final form to be inspected and to make changes.

Inspection evaluation methods are low-level evaluation approaches. They include

Walkthroughs

Observational evaluations

Questionnaires

Guideline-based reviews

Heuristic evaluations

Scenario-based reviews

Walkthroughs

Walkthroughs involve designers, programmers, potential users and others working together to evaluate a product. Most important user involvements are evaluated by this group as they proceed step by step through typical human–computer interactions. In some multimedia-based systems, this process is also referred to as *storyboarding*.

Observational Evaluations

Observational evaluations usually involve testers watching subjects perform using a prototype of a new product, application, or system. Testers take notes on user comments, unsuccessful attempts to perform certain tasks, number of times that a document is used, number of help requests, and the like.

Questionnaires

Questionnaires can be administered to potential users to help determine their attitudes concerning a new product, application, or system. These questionnaires can be given to subjects prior, during, or after any contact that they may have with a new system.

Guideline-based Reviews

These reviews are conducted by comparing design decisions against published guidelines and standards. The guidelines and standards may have been developed within an organization or externally. External standards for character-based systems can include documents such as the "Design Guidelines for User–System Interface Software" (Smith and Mosier, 1984). Available standards for graphical user interfaces include style guides such as Common User Access (CUA), Windows, and Motif. The most ambitious set of user interface standards is available from the International Standards Organization as ISO-9241.

Heuristic Evaluations

Heuristic evaluations are a method for finding usability problems in a user interface by having a small set of evaluators examine an interface and judge its compliance with usability principles (heuristics). The resulting observations represent an evaluator's opinion about what needs to be improved in a user interface (Nielsen, 1992). The evaluation can be conducted on a specification, a paper prototype, or a full working prototype. When conducting a heuristic evaluation, evaluators simply inspect the user interface and look for deficiencies. When more than one person is used to conduct a heuristic evaluation, it is best to have each evaluator do an independent evaluation and then combine their results.

When a heuristic evaluation is conducted, evaluators inspect an interface with respect to a set of usability principles (i.e., heuristics). Nielsen (1994) attempted to identify the heuristics that best explain actual usability problems. After evaluating seven sets of published usability heuristics (totaling 101 unique items), he identified 53 heuristics that accounted for about 90 percent of the variability in a set of usability problems. The top seven heuristics only accounted for 30 percent of the variability. He concluded that it is not possible to account for a large part of the variability with a small, manageable set of usability heuristics.

When doing heuristic evaluations, better evaluators can find up to twice as many problems as those with limited usability experience. Novice evaluators, those with little or no usability experience, tend to detect between 29 percent (Nielsen, 1992) and 37 percent (Virzi, 1990, 1992) of existing problems. Lewis (1994a) found that novice test participants only found 16 percent of *known problems.* Highly experienced usability specialists can detect from 46 percent to 61 percent of existing problems (Nielsen and Molich, 1990). However, good evaluators can miss easy problems, and poor evaluators can find some hard problems.

When conducting a heuristic evaluation, it does not seem to matter much whether the evaluator inspects a paper specification or an actual prototype. About the same percentage of problems is detected in both cases (Nielsen, 1992).

Scenario-based Reviews

These reviews are similar to heuristic evaluations, except that the evaluators are usually representative subjects, performing typical tasks, using a working prototype. While performing the activities, the subjects "think aloud," or vocalize their thoughts about what they think is wrong with the system. Their responses are recorded by a tester and later combined with those of other subjects to help in understanding the deficiencies in a system or product. This approach may also be referred to as using *think-aloud evaluations.*

Issues surrounding the impact of the think aloud method on performance have been studied. The research shows that subjects who talk (think aloud) while performing committed *fewer* errors and took *less* time to complete the tasks. These investigators concluded that verbalizations directly available in short-term memory do not appear to change task performance (Wright and Converse, 1992).

In another study, the effect of thinking aloud on the value of user verbalizations was examined. The authors evaluated three conditions:

- Think aloud while performing
- Think aloud while viewing a video right after the study
- Think aloud while viewing a video 24 hours after the evaluation session

They found no significant differences among the three conditions (Ohnemus and Biers, 1993). This suggests that having test subjects convey their observations and attitudes up to one day after having performed still allows them to indicate their feelings and intentions.

Estimating the Number of Usability Problems in a Product

Nielsen and Landauer (1993) proposed a method for estimating the number of problems that actually exist in a new system. They suggest that if a heuristic evaluation or a scenario-based review finds a certain number of problems, and we have an idea of the relative success of these methods, then we can predict how many problems still exist in the new product. This is calculated using the following formula:

$$N = \frac{E}{P}$$

where

N = estimated number of problems in the interface

E = problems found in a single evaluation

P = probability of finding the average usability problem

For example, if eight problems were identified using a scenario-based review, and we assume that the probability of finding a problem is 0.40, then the actual number of usability problems is estimated to be 20 (8/0.40 = 20).

Number of Evaluators

An estimate of the number of evaluators required to conduct an effective heuristic evaluation or scenario-based review can be calculated using the formula shown in Figure 8-1 (Virzi, 1990). This formula is used to determine the cumulative binomial probability for the likelihood that a problem of a certain probability will occur at least once:

$$1 - (1 - p)^n$$

where

p = mean probability of detecting a problem

n = number of heuristic evaluators

Figure 8-1 Formula for calculating the number of the required number of heuristic evaluators.

Several studies have shown that the average number of problems detected in a heuristic evaluation is about 40 percent (Virzi, 1992). This means that any one evaluator will find (on average) about 40 percent of the problems found by *all* evaluators. This does not mean that all important problems will be identified by the evaluators or that those that are identified will actually have an impact on user performance.

By using this probability with the formula, we can predict how many evaluators will be needed to find a certain percentage of "all" problems. For example, if two people independently conduct a heuristic evaluation, together they should be able to identify about 64 percent of the problems.

$$1 - (1 - p)^n = 1 - (1 - 0.40)^2 = 1 - (0.60)^2 = 1 - 0.36 = 0.64$$

Carrying the example further, if three evaluators are used, they should be able to detect about 78 percent of the problems.

$$1 - (1 - p)^n = 1 - (1 - 0.40)^3 = 1 - (0.60)^3 = 1 - 0.22 = 0.78$$

Finally, if nine evaluators are used and their observations are combined, we should expect that about 99 percent of the problems could be identified.

$$1 - (1 - p)^n = 1 - (1 - 0.40)^9 = 1 - (0.60)^9 = 1 - 0.01 = 0.99$$

If a designer desires to discover 90 percent of all problems from a set with an average probability of occurrence of 0.40, the heuristic evaluation would require at least five participants. Obviously, if the evaluators were only able to detect an average of 25 percent, it would require more people to identify 90 percent of the problems. On the other hand, if they were able to detect an average of 55 percent, they would need fewer evaluators.

Thus, as the number of heuristic evaluators increases, so does the probability of finding problems in the user interface. To obtain the most information for the least cost, designers try to use the fewest number of heuristic evaluators.

Evaluation Teams

Scenario-based reviews can be conducted with individuals working alone or people working in two-person teams. One study reported that two-person teams were able to identify an average of 29 percent of the problems, with a range from 15 percent to 42 percent (Wright and Monk, 1991). The use of teams does not appear to increase the number of problems identified (Hackman and Biers, 1992).

Reliability of Evaluations

Designers should *not* stop testing once a series of Inspection evaluations has been completed. Regardless of how good the commands, messages, windows, screen-based controls (widgets), or user guides may appear, experience has shown that we cannot merely ask someone to look at something and expect a reliable determination of its effectiveness. For many years, we have known that such evaluations are unreliable, even when evaluators are highly skilled human performance specialists (Bailey, 1974; Rothkopf, 1963).

Performance Testing

This part of an overall test and evaluation program usually consists of conducting *performance tests* of individual components and groups of components as they are being developed. The primary purpose of these tests is to determine whether the product or the process elicits the necessary level of human performance to meet the requirements established for it. When deficiencies or weaknesses are discovered, there is an opportunity for redesign and for retesting of the altered components.

Performance testing in computer systems originated during the early development of large computer-based systems at Bell Laboratories in the early 1970s (Bailey and Kulp, 1971; Bailey, 1972). Since its inception, this type of performance testing has been successfully used in the development of numerous computer-based systems (cf. Martin, 1974, 1975, 1979). These test programs have led to substantial improvements in human performance in computer products, applications, and systems.

In one test program, for example, 130 tests were conducted over a 20-month period (Martin, 1975). The test program used 448 subjects. The original tests elicited an average error rate of 13.1 percent. Based on recommendations in test reports, designers made extensive changes to the user interface. Retests showed that errors were reduced by 35 percent, which resulted in substantial cost savings once the systems were installed.

Performance tests consist of well-controlled simulations of people performing activities using product, application, or system prototypes. Sometimes these tests are conducted on complete systems, but they are usually carried out on pieces or parts of an application. These tests require people that are typical of potential users. The assumption is that a person's performance on this test will be very similar to his or her performance in a real working situation.

The user profile is used to select test subjects. After potential users have been trained using only test-related training materials, a performance test session is conducted in which the users perform using a sample of the actual tasks in the new system. These performance tests should be designed to represent the actual on-the-job environment as much as possible. Human performance data are collected during the test sessions and provided to designers in detailed test reports.

Performance tests usually focus on collecting speed, error, and training time data. However, there are numerous other performance-related variables that can be evaluated to help in understanding performance deficiencies in a product, application, or system. Some designers even make physical and/or physiological measurements of subjects to try to understand how products can be improved. These studies have measured the finger pressure on keys, blood pressure levels as people interact with a system, heart rate changes, and even electroencephalograms (EEG). One physical measurement that is becoming more popular is evaluating eye movements (cf. Benel et al., 1991). Eye tracking can be used to supplement other measures of performance. The gaze parameters of most interest include the time spent fixating on each region of a screen and the delays associated with moving from region to region on a screen.

Preparing for a Performance Test

A performance test requires much preparation. In general, there are four major activities: planning, conducting the test, evaluating the outcome, and preparing and conducting retests. Each of these major activities contains numerous other activities.

- Planning
  - Define the purpose of the test
  - Identify a set of tasks (test items) that will adequately exercise the system
  - Plan how information will be collected from the tests, and determine how information will be recorded
  - Decide questions for use as a debriefing interview once the test is completed
  - Prepare the test facilities
  - Find a set of users representative of the potential user population
- Conduct the test while monitoring the test participants doing the tasks
- Evaluate the outcome
  - Analyze and summarize the performance data
  - Collate observations, comments, and questionnaire results
  - Identify potential problems with the product as it is currently developed
  - Determine possible fixes to the product
- Make changes
  - Agree with other designers and/or testers on changes and implement solutions
  - Test the system again to see the effect of the changes
  - Make changes and retest (if needed)

Some of the most critical activities will be briefly discussed.

Planning

Because people are so different and the activities that they are expected to perform vary so widely, human performance tests tend to be much more complicated than most engineering or physical tests. People are constantly changing. They learn, they become bored, and they are influenced by what happens to them while performing. To get dependable results in the face of incessant change, the designer has to use special techniques. Designers should be aware that certain precautions are necessary in order to have a meaningful test.

Selecting Test Subjects

The designer should decide who will serve as test subjects. Two kinds of decisions are involved: the *number* of people to be tested and the *kinds* of people to be tested.

You should use enough test subjects to get *reliable* (repeatable) results, but not so many as to unreasonably increase the cost of the test. If reliable results can be obtained with five test subjects, it is wasteful to test fifteen. However, there is no easy way of deciding in advance how many test subjects are needed for a particular test. It is not uncommon to conduct performance tests with four or six test subjects.

Correct Number of Subjects

The appropriate number of subjects to use in a heuristic evaluation or performance test is the number that accomplishes the goals of the study as efficiently as possible. If the number of evaluators or test subjects is larger than necessary, it will increase the cost of testing and possibly even the time taken to develop the system. If the number of evaluators or test subjects is too small, the usability testing process may fail to detect problems that could reduce the usability of a product.

A slight variation in the binomial probability formula previously discussed (see Figure 8-1) can be used to help to estimate the appropriate number of evaluators or test subjects (Lewis, 1993). The cumulative binomial probability for the likelihood that a problem of a certain probability will occur at least once is shown in Figure 8-2.

$$1 - (1 - p)^n$$

where

p = probability of the event occurring

n = number of test subjects

Figure 8-2 Formula for calculating the minimum number of test subjects.

Unclear Icons Example

Consider the following example. Assume that the icons designed for use in a window will on average be confusing to 50 percent of potential users. In this situation, the likelihood that the icons will confuse any one user is 0.5, and in a well-designed performance test, the likelihood that the icons will confuse any one test subject is also 0.5. This means that the chance of identifying the potentially confusing icons, using only one test subject, is only about 50–50.

However, if two representative test subjects are used, the probability of the confusing icons being identified rises to 0.75. This probability is calculated using the binomial formula just presented. The following calculations assume that the probability of the event occurring is 0.5 and the number of test subjects is 2.

$$1 - (1 - p)^n = 1 - (1 - 0.5)^2 = 1 - (0.5)^2 = 1 - 0.25 = 0.75$$

Carrying the example further, if three test subjects are used, the likelihood that at least one of them will have difficulty when using the confusing icons is 0.87.

$$1 - (1 - p)^n = 1 - (1 - 0.5)^3 = 1 - (0.5)^3 = 1 - 0.13 = 0.87$$

Thus, as the number of test subjects is increased, so is the probability of finding problems in the user interface. In this example, if seven users participate, the likelihood that at least one of them will be confused by the icons is 0.99.

$$1 - (1 - p)^n = 1 - (1 - 0.5)^7 = 1 - (0.5)^7 = 1 - 0.01 = 0.99$$

Using seven test subjects obviously makes the chances of detecting the confusing icons during testing very high. Table 8-3 shows the probability of detecting usability problems, based on the problem probability.

Table 8-3 Likelihood of Performance Test Subjects Having Problems

Problem	**Number of Test Subjects**									
Probability	**1**	**2**	**3**	**4**	**5**	**6**	**7**	**8**	**9**	**10**
0.05	0.05	0.10	0.14	0.19	0.23	0.26	0.31	0.34	0.37	0.41
0.10	0.10	0.19	0.27	0.34	0.41	0.47	0.53	0.57	0.61	0.65
0.15	0.15	0.28	0.39	0.48	0.56	0.62	0.68	0.73	0.77	0.80
0.25	0.25	0.44	0.58	0.68	0.76	0.82	0.87	0.90	0.92	0.94
0.50	0.50	0.75	0.87	0.94	0.97	0.98	0.99			
0.75	0.75	0.94	0.98	0.99						
0.90	0.90	0.99								

Note that the *problem probability* in the table can be considered as a rough estimate of problem severity. The more severe the problem (e.g., 0.90), the more likely that someone will have trouble with that portion of the user interface. The less severe the problem (e.g., 0.05), the more likely that most users will have little difficulty, which means that the need to identify it and fix it is less. Obviously, this is not always true. There may be some serious potential problems lurking in the interface of a system that will only be uncovered by users with very special skill sets. These may occur infrequently, but when they do, the results could be so serious that the user cannot use the system at all. Nevertheless, identifying items that are problems for 50 to 75 percent of potential users, when three or four test subjects are used, seems to be a reasonable and cost-effective way to proceed.

Using Representative Subjects

The second decision is an easier one. People used as subjects should represent as closely as possible the eventual users in the real world. A designer should make every effort to find people that are truly representative of future users.

To a large extent the goodness of performance tests depends on how well the designer has been able to select test subjects that have the characteristics of the people who will ultimately work in the system. Some of the human characteristics that are most important include the following:

- Age
- Sensory characteristics: visual acuity, auditory acuity, color perception
- Responder characteristics: body dimensions, strength, handedness
- Cognitive characteristics: general intelligence, problem-solving ability
- Motivational characteristics: cooperativeness, initiative, persistence
- Training and experience: level of general education, amount of specialized training, and specialized experience

For example, if test subjects have an average age of 23 and most have college educations, while the user population has an average age of 54 and most have high school educations, the performance of the younger and more educated test group may show little resemblance to that of the actual user population.

Determining Test Items

A careful selection of test items helps to ensure that *all important conditions are tested.* If properly selected, the test items can measure the ability of design, interfaces, and facilitators to achieve an acceptable level of human performance. A detailed set of human performance requirements, such as maximum acceptable error rate, processing time, training, and user satisfaction, should accompany each task.

It is usually not practical to include one or more examples of every system input, output, and performance characteristic in a test. Therefore, an important part of developing a human performance test is deciding which features are most important and should be included in the test. The *validity* of any test is determined to a considerable extent by how successfully the designer has been able to identify and include representative test items.

When conducting usability tests to measure the adequacy of certain components that are used infrequently, make sure to simulate the proper conditions. Wright et al. (1994) point out that it is similar to providing a hiker with a walking stick; the stick may not be used much until the going gets tough.

Simulating the Context

A good test should take into account the context or environment in which users will be performing once the system is operational. The designer may need to include a full range of conditions, considering both the physical and social environments. Telephone messages that can be easily heard in a quiet setting may be difficult to hear in a busy office. Trucks that can be easily driven in temperate climates sometimes cannot be used in the arctic region where drivers are hampered by several layers of protective clothing, thick insulated shoes, and bulky gloves.

If a large range of contextual or environmental conditions exists, the designer must select a representative sample to include in the test. At the very least, test conditions should include values near each extreme and one in between. For example, if a system is designed to operate in illumination levels ranging from bright sunlight to semidarkness, tests should be conducted in both bright sunlight and semidarkness, as well as in an illumination level in between. If a representative subject performs satisfactorily at both extremes of an operating condition and at the point in between, it is generally safe to conclude that the same level of performance can be expected in real-world environmental conditions. If human performance is degraded at one extreme, however, it may then be worthwhile to make tests at other values to discover at what point the performance begins to deteriorate.

Every test has environmental conditions that may distract a person from adequately performing an activity. Frequently, these conditions are so common that designers ignore them. For example, having several children and a dog in an automobile may affect a driver using a car telephone, yet a designer may overlook these distracting conditions during testing.

The designer should make every attempt to ensure that the test subjects, test items, and test environment are all representative of the real world.

Reporting Test Results

All tests should result in some form of summary and analysis. If there is statistical treatment of results, it should be relatively simple, involving no more than summary tables containing averages and other easy to understand descriptive statistics. As an example, consider collecting and reporting results associated with errors in a computer system. Errors can be very costly. In fact, the costs to detect and correct clerical errors can be as much as $10 per error or even higher (Bailey, 1973). Simulation testing (particularly work module testing) can help to identify the design deficiencies that may lead to excess errors.

Long-distance Testing

Much usability testing is being done long distance using modems. Also, usability testing frequently is done at customer's premise because of the availability of test subjects. Rowley (1994) recommends that when testing is done "on the road" we do the following:

Schedule the software development with field testing in mind.

Try out usability testing locally before going on the road.

Select portions of the software and task scenarios for field testing that are adequately complex, yet feasible.

Get field coordinators involved early.

Keep the test setup minimal.

Video camera

Signal splitter for second monitor

Small lapel microphone

Be prepared to make last minute schedule changes.

Be particularly sensitive to culturally influenced terminology.

Operational Evaluation Methods

Operational evaluation methods are those used *after* a product, application, or system is delivered or made operational. There is no question that it is valuable to watch people using a new product, application, or system and to talk with them. However, usually few substantial design changes can be made at this point. Indeed, most designers apply most of the information gained from such an evaluation in the *next* system he or she develops. From a human performance point of view, it is usually the performance testing, not the operational evaluation, that is a designer's *final opportunity* to make meaningful changes.

When conducting an operational evaluation, the very process of making observations may distort what is being observed. Users may not be as spontaneous or natural when they are being observed. However, the major difficulty with using these methods is that it is much too late to have a major impact on the product being evaluated. Any significant changes must be made in future releases of the product. This means that major changes may not be made for 1 or 2 years.

Many methods can be used, including

- On-site observations and interviews
- Surveys (questionnaires)
- Diaries
- Electronic monitoring

On-site Observations and Interviews

These methods are probably the most common ways of collecting information about an existing system. These methods are usually conducted while actual users are performing real work. This enables evaluators to see how users are interacting with the new system, to ask questions, and to identify any potential usability problems.

Surveys (Questionnaires)

Surveys or questionnaires are used to collect information from actual users concerning their attitudes about a new system. This information can be compared with the attitudes of users of other products as measured by the same instrument (questionnaire). Also, the attitudes can be measured after 6 weeks and then again after 1 year to see if there are improvements. Some designers like to collect satisfaction information in a before-and-after picture. They administer a satisfaction questionnaire to users on the old, existing system (before) and then the exact same questionnaire to users after they are experienced on the new system (after). A standardized questionnaire that measures user satisfaction (Questionnaire for User Interaction Satisfaction) is published by the Human Computer Interaction Laboratory at the University of Maryland.

Diaries

Diaries are used to regularly and systematically collect user's attitudes and concerns about a new product. This diary study technique is helpful in pinpointing critical issues while a product, application, or system is used in the workplace. Typically, users are requested to record daily activities in log form, either in a book or in a computer file. They stop whatever they are doing and make a record every 30 to 45 minutes. Diaries can be kept for 1 or 2 weeks or longer. In one study, designers used this method to determine what sources people used when learning a new system. They found that 42 percent of new users tried things until something worked, 41 percent read the manual, and about 17 percent learned by asking someone else (Rieman, 1993).

Electronic Monitoring

Electronic monitoring refers to having certain user interaction events automatically captured and stored by the computer. The software can be designed to capture the number of times each error message appears, the help messages most frequently used, the commands used, the menu items used, which primary windows and dialog boxes are accessed, and so on. The main advantage of collecting these individual usage statistics in this way is that it is totally unobtrusive (transparent). The users do not know that their performance is being monitored. Also, designers can collect data on large numbers of users for a long time, which provides a much better picture of user behaviors as they gain experience (Teubner and Vaske, 1988).

Predicting Human Performance

The principal purpose of each of the usability testing approaches just discussed is to act as a predictor, a way of forecasting what will happen when the system is actually operational. When we conduct a usability test or evaluation, we expect that the outcome will enable us to make valid statements about future user performance and preferences. However, the usability testing techniques that we have discussed differ greatly in their predictive power. Some are relatively good; others give results that must be interpreted with great caution. The variable that largely determines the predictive power of a usability test is the closeness or *fidelity* with which the test conditions reflect those in the real world.

Test Fidelity

The most direct way of finding out about the performance of people in the real world is to observe them working in the real world. However, when developing totally new systems, this is not possible until a system is operational. Performance testing has less fidelity than observations made once a system is operational. In fact, performance tests can cover a wide range on the fidelity scale. Some are highly realistic, while others may be so abstract that they bear little resemblance to the real world that they are supposed to represent. The lower the fidelity, the more difficult it is to make accurate predictions.

Most inspection evaluations have little fidelity. Rather, their intent is to review materials for completeness, accuracy, and the like. Therefore, it is very difficult to predict user performance in the real world from pretest evaluation results.

Comparing Heuristic Evaluation versus Performance Testing

A study that compared the effectiveness of heuristic evaluations with performance testing (Karat et al., 1992) found that performance tests located significantly more usability problems. They also reported that the cost per problem found was much less when using performance tests than when using heuristic evaluations.

Another study (Virzi et al., 1993) used three different usability testing methods to evaluate a high-fidelity prototype of a voice mail system. The usability methods used included

- A heuristic evaluation with usability experts as evaluators
- Scenario-based reviews (think-aloud testing) in which naive subjects commented on the system as they used it
- Performance testing, during which task performance times and errors were collected from test subjects.

A total of 26 unique problems was identified by the three usability testing methods. All three methods were roughly equivalent in their ability to detect a core set of nine usability problems (about 37 percent of all problems). The performance testing found three problems (13 percent) not identified by either of the other methods. Finally, the two nonperformance testing methods identified 12 problems (the remaining 50 percent of all problems) that were not found by performance testing. This points to one of the major problems with inspection evaluations, which is that a "problem" identified by these methods may or may not be a problem that will ultimately affect user performance (these problems are sometimes referred to as false positives or simply "fluff").

Thus, many problems identified by an inspection evaluation that are suspected to be serious performance-related problems may not be problems at all. One study (Bailey et al., 1992) reported that of the 29 potential problems identified in a heuristic evaluation only 2 eventually had a negative impact on either user performance or user preferences. The same findings held for both character-based and graphical user interfaces.

Testing Cost and Ease of Making Changes

The fidelity of test conditions is, unfortunately, directly related to cost and inversely related to the ease with which meaningful changes can be made to a system. The more we find out about human performance in a system and the less uncertainty we have concerning this performance, the harder it is to make substantial changes to the system that could improve the performance. For example, the least expensive method is conducting a formal evaluation or possibly an inspection evaluation. But they also provide the least information about changes that can be made to improve ultimate human performance. On the other hand, observational evaluations give fairly good indications of performance, but most meaningful changes will be very costly. A highly realistic performance test can be more costly than the early evaluation methods, but it also improves on the ability to predict future human performance and to identify deficiencies early enough that less costly changes can be made.

Accuracy Objectives

To determine if a system that is being developed will meet overall accuracy objectives, each field of data should be assigned an *accuracy objective* based on the data's criticality to system success. The determination of these accuracy objectives should be made *before* work module testing begins. For example, the telephone number data file may need to be 99.8 percent accurate, whereas the terminal location data file may be required to be only 96.5 percent accurate. To increase development and operational costs by trying to make the terminal location file more accurate is unnecessary. However, it may be of vital concern and very cost effective to do whatever is necessary to maintain the telephone number file at a 99.8 percent accuracy level.

Test results can then be reported in terms of the impact that human performance has on the accuracy of each specific data item file. For example, a test may show that requiring an operator to collect a customer's telephone number and to enter the telephone number on a form introduces about 2 percent error. Another test may show that a second handling of the data, transcribing the telephone number from form A to form B, adds another 1 percent error, and that a third handling, keying the data, adds still another 0.5 percent error. For every 100 telephone numbers being added, then, an average of three or four are incorrect (even after computer edits and validations). Thus, with a continual stream of new additions to the telephone number file, the accuracy of the file could be expected to drop from a required 99.8 percent to 97.0 percent or lower. Faulty human performance adding 3 to 4 percent error means that the existing combination of basic design, interfaces, and facilitators does *not* elicit the necessary level of human performance to meet accuracy objectives.

A good test report should show not only that the accuracy objective is not being met, but also suggest reasons why it is not being met. By knowing that the accuracy objective for any data file is not being met, the designer can make the necessary changes to prevent errors from occurring.

Cost-effective Performance

System performance is frequently measured in terms of cost effectiveness. As far as errors are concerned, the system is cost effective only if it produces information at an *accuracy level* at which users can perform their functions with an acceptable level of interruption and delay. The acceptable accuracy level may be extremely high (such as 99.8 percent for telephone number data) or much lower. The overall cost to purify the data, however, must be compared to the costs associated with using data containing a certain amount of error. The latter costs include those incurred because of work completion being delayed (e.g., a telephone installer searching for a nonexistent address), customer irritation (e.g., a customer refusing to pay a bill because calls are listed that were not made), and the impact of errors on other data (e.g., an incorrect telephone number associated with a correct name and address may result in a customer not being billed for calls). It may be, for example, that a telephone installer must complete 95 out of every 100 installations, rather than 81 out of 100, in order to complete the installation process with the least overall cost to the telephone company. If so, then significant system changes may be required to improve the accuracy of the data. This usually means that human performance must be improved in some way.

User Acceptance

System performance can also be measured in terms of user acceptance. Experience has shown, for example, that users have great difficulty accepting a new system that produces less accurate information than that available in the system it replaced. For example, if the accuracy level of the data in the new system is not high enough to carry out telephone installations at the same rate as with the old system, some substantial changes must be made to the new system so that more accurate information is available to the user. Conducting meaningful tests helps a designer to determine fairly early in the development process if the desired accuracy level will be obtained once the system is operational.

Example of a Performance Test

The following example illustrates in more detail the kinds of decisions a designer must make when planning and conducting performance tests. The new system was developed for the job of a repair service attendant (RSA). It consists of software, CRT screens, a user guide, training materials, and a performance aid. An RSA receives calls from telephone company customers when telephones or telephone lines need repairs. He or she records the customer's explanation of the problem and provides the trouble-report information to the individuals who make the repairs.

When contacted by a customer, the RSA first enters the customer's telephone number into a computer using a QWERTY keyboard, and the computer responds with a copy of the customer's file on the CRT screen. After verifying name and address with the customer, the RSA enters a coded description of the customer's trouble, determines call-back information and information concerning access to the telephone, and gives the customer a date when the trouble will be repaired. The RSA report is then electronically transmitted to the appropriate repair bureau.

Test materials were prepared and eight representative test subjects selected. After completion of a 4-day training class, the test subjects were given the 5-hour performance test. The distribution of each type of trouble call used for the test corresponded closely to the actual distribution of calls received in the real world. A special computerized data base was used for the test. This test database contained fictitious information for 94 customers, as well as appropriate CRT screens.

Simulated customer calls were made to each test subject by trained testers. Testers, as customers, used the same prepared script for each call. Each tester made a predetermined number of calls and attempted to exactly duplicate the form and content of each call for each subject. Testers and RSAs were in separate rooms and communicated using telephones.

In general, the simulation proceeded as follows:

1. The customer would indicate a call attempt by pressing a button, which made a clicking sound in the earphones of the test subject (RSA).
2. The RSA responded by stating, "Telephone repair service, may I have the telephone number you are reporting, please?"
3. The customer gave the telephone number.
4. The RSA entered the telephone number, waited for the customer's record to appear on the CRT screen, and then verified the customer's name or address.
5. The RSA requested a description of the trouble and asked necessary questions.
6. The customer made only those responses contained in the script; otherwise, the response was "I don't know."
7. The RSA then asked for call-back and access information and gave a commitment (date and time for repairing the telephone).
8. The customer, who acted either cordial or irate during the call, then made an appropriate closing comment and the call was terminated.

Subjects were tested in groups of two, three, or four, depending on the number of subjects available for each test. There was one tester for each subject. The testers rotated through the subjects. For example, customer 1 would talk with RSA 2, RSA 1, RSA 3 and then return to RSA 3 or RSA 2 with another call. This procedure helped to reduce the possibility of subjects and testers anticipating each other's remarks or behavior. It also required RSAs to wait for calls on occasion or to get several calls in rapid succession. Both situations are typical of the real world.

Information on 10 different human performance measures was collected for each subject during the customer–RSA interaction. This required each tester to determine when errors occurred (usually when RSAs neglected to ask for or provide certain information) and to keep a record of the time spent by each RSA on each call. Information on other measures was gathered after the test session by comparing what the RSAs actually entered into the computer with what they should have entered. Based on these results, designers made changes to the system that improved human performance.

Validity of Performance Tests

How valid is the performance test technique? That is, does this test technique produce a measure of human performance that reliably predicts future job performance?

Recall that a performance test is used because we assume that it will reflect what would happen in the real world if the system were released immediately, without further revisions. Validity is concerned with how closely the test scores represent actual job performance.

Comparisons of human performance speed and accuracy scores for the performance test just discussed are shown in Table 8-4 (Bailey and Koch, 1976). Note that the test provided fairly accurate predictions of on-the-job performance. With error data, the predictions were best for the first week or two of work. For the speed data, the predictions were best after 3 or 4 weeks on the job.

Table 8-4 Summary of Work Module Test and On-the-Job Monitoring ($N = 8$)

Performance Measure	Performance Test	On-the-Job Performance			
		Week 1	Week 2	Week 3	Week 4
Average time per customer (seconds)	109.5	143.6	126.5	123.5	114.2
Overall error rate	0.17	0.15	0.17	0.18	0.22

As discussed earlier, the closer a performance test represents real-world conditions, the better will be the predictive value of the results. This assumes that the test contains and exercises a reasonably good sample of the actual activities to be performed on the job. We cannot overemphasize that the greater the fidelity of the test, the better will be the estimate of performance in the real world and the more useful will be the results.

Reducing Liability Using Usability Testing

Usability testing should begin as soon as feasible and continue even after the system is completed. One reason for this is to help reduce any liability designers may have assumed with their design decisions.

MacPherson versus Buick Motor Company in 1916 was a landmark case in determining liability suits. The plaintiff had been injured when a wheel collapsed on the auto he was driving. He had purchased the car from a dealer who had bought it for resale from the Buick Company. Suit was brought against Buick. Buick based its defense on the fact that, while it had sold the car to the dealer, it had no contract with the injured party and thus was not liable. The dealer in turn claimed that it was not liable because it had not manufactured the car. The judge ruled that the Buick Motor Company was indeed responsible. He stated that the manufacturer had a duty to inspect its products for defects and that failure to do so constituted negligence (Hammer, 1972).

A system designer has the same responsibility to his or her customers. Releasing a system that elicits degraded human performance constitutes negligence on the part of the designer.

The liability analogy can be taken even further. Over the years, the courts have greatly reduced the need for an injured person to show negligence on the part of a manufacturer, assembler, or retailer. The courts have held that, since an injured person generally does not have the technical capabilities necessary to prove that negligence existed, the burden of proof is on the manufacturer to show that negligence did *not* exist. The injured person can recover damages if he or she can reasonably show that a product had a defect in design or manufacture when it left the manufacturer. The manufacturer then has to prove that it was not negligent or prove that the user was negligent; that is, the product was misused. The burden of proof has clearly shifted from the user to the designer.

In liability cases, negligence is defined as the failure to exercise a reasonable amount of care so that injury or property damage does not occur. It is possible that in the foreseeable future the definition could be expanded to include not only injury or property damage, but also situations when a designer's failure to exercise a reasonable amount of care leads to substantially degraded human performance and unreasonable operating costs for users.

Developing a new system, like almost any new product, is a dynamic and complex process. This type of development activity involves elements of invention and creativity. The designer rarely duplicates even portions of systems already in operation. Systems under development are systems that have never been built before. To some extent, there is even some uncertainty about whether they *can* be built. As evidenced by the large number of negligence cases that crowd our courts, the risk of making incorrect design decisions is high. Because designer errors can lead to user errors, it is critical that adequate testing is done throughout the development process.

Exercise 8A: Conducting a Study

Purpose: To provide an opportunity to develop and conduct a study by developing and administering a questionnaire.

Method: Develop a questionnaire that contains at least five items (see Appendix C for assistance). The purpose of your questionnaire is to determine what people believe "user friendly" means. Administer your questionnaire to at least six people (individually) and then summarize the results.

Reporting: Summarize your findings in a report using the format described in Exercise 1A.

Exercise 8B: Conducting a Usability Test

Purpose: To provide an opportunity to develop and conduct a usability test.

Method: The goal of this usability test is to (1) determine how much slower participants are when reading from the computer monitor versus reading from a paper document, and (2) determine what changes can be made to the computer-based information to speed up the subject's reading speed.

To prepare for this usability test, you must do three things:

1. Using any word processing software that you have available, type the textual information found on the first four pages in Chapter 1 of this book (ignore the material in the table).
2. Make a photocopy of the same pages (do not include the table).

3 Develop five multiple-choice questions from the three pages. Try to create questions that can only be answered correctly by reading and understanding the material.

After completing these preparations, you should have a computer-based copy and a paper-based copy of the same text. In addition, you should have a five-item test that can be used to evaluate the reader's comprehension of the material.

Select four test subjects. Have the subjects read out loud the *paper-based* test material (pages of text). They should be told to "read as quickly and as accurately as possible." Record the time for each person. After they finish reading, administer the comprehension test. Calculate the average time required for all subjects to read the paper-based material.

Use the average time required to read from the paper-based material as the goal for someone reading from the computer-based material. Have a new test subject read the same material from the computer monitor. Time how long it takes and then administer the comprehension test. Make whatever changes you think necessary to speed up the reading (and the comprehension) of the computer-based material. After making the first changes, test another new person on the computer-based material. Make more changes and then test another person. Continue this process until your test subjects are consistently and reliably reading as fast from the computer monitor as the original subjects did from the paper-based material.

Reporting: Prepare a report on your activities using the outline shown in Exercise 1A.

9

Product Analysis and Definition

User Interface Model

Previously, we discussed the user interface model and focused on issues concerning potential users and the iterative design process. Chapters 9 through 15 address the 10 user interface processes shown around the outside of the model in Figure 9-1. These 10 user interface processes will be briefly discussed prior to focusing in this chapter on the first two: (1) verifying functionality and (2) developing user profiles. The third interface process, developing the computer profile, will be dealt with in Chapter 10. Chapters 9 and 10 help establish *what* will be done, *who* will perform the activities, and *what* computer-related devices will be used. These all help to establish the *constraints* imposed on designers of the user interface.

System Functionality

Correct Functionality

Functionality refers to the features that a computer product performs. The final set of functions is generally identified by system users working with system designers. The functions of a system should reflect the true needs of actual system users. Identified functions eventually become system requirements. A correct set of functions is required in order for a computer-based product to realize its true potential. Determining the correct set of functions usually requires designers to observe users working with an existing system. If there is no existing system, designers must rely on their observations of how users do their work without a system and on users' descriptions of their expectations. Often functions can be identified through a formalized series of meetings between users and system designers (e.g., joint application design or JAD sessions).

It is not possible to have a truly usable system, no matter how well designed the user interface if the correct set of functions is not included, that is, if the system does not do what users require. Many existing systems have poor interfaces but are still used because they include the correct set of functions. But you cannot have a well-designed user interface sitting on the wrong set of functions. If the system does not do what users need to have done, the system will not be used (nor will the user interface).

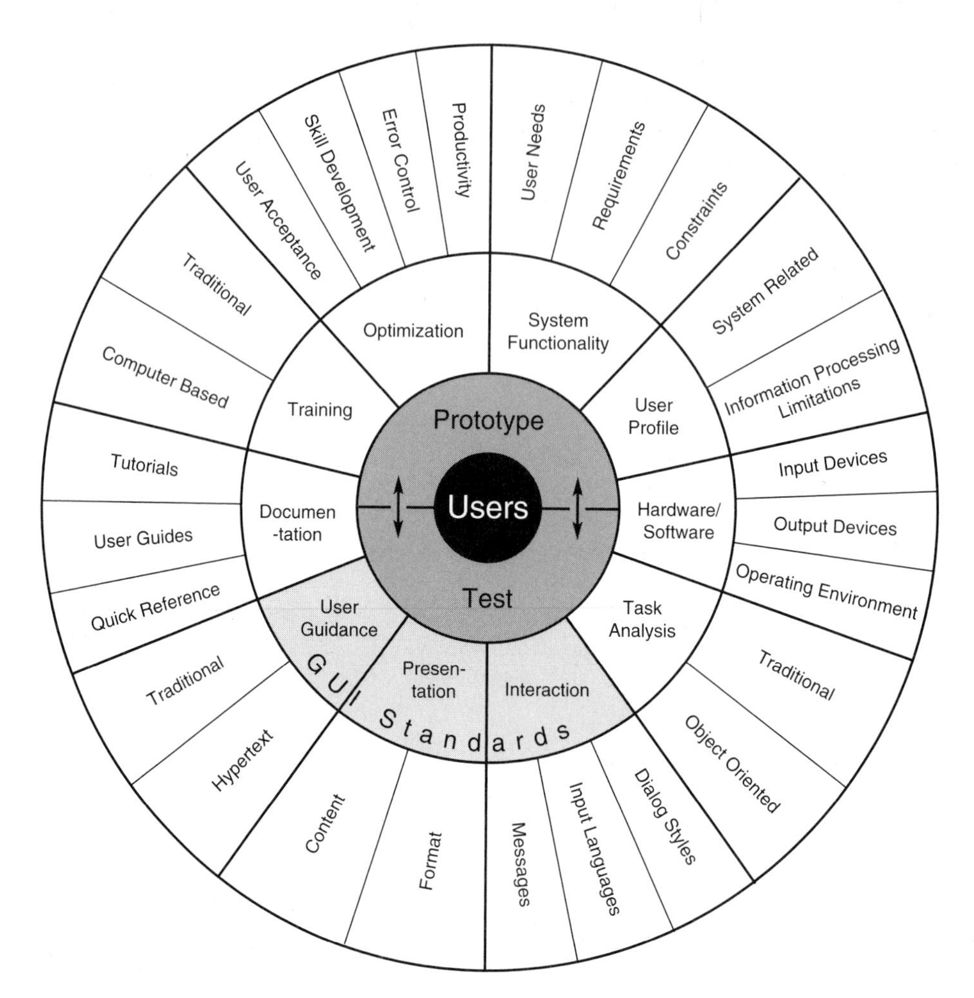

Figure 9-1 User interface model.

User interface designers may or may not be involved in helping to determine the functions for a system. However, they should take steps to *verify* that the selected functions do what users need to have done, to verify that the proposed system functionality truly matches the needs of potential users and that nothing important has been overlooked.

Minimal Functionality

Another major activity of user interface designers is to ensure that the number of functions included in a system is not too large. The more functions a system has, the more complex the user interface must be, making it almost impossible to have a system that is even marginally usable. In general, the fewer the functions, the more likely the interface can be made easy to use.

User Profiles

The user profile consists of brief descriptions of typical system users. Experienced system designers recognize that they cannot effectively develop generic systems, that is, a system with one user interface that anybody and everybody can use. User interface designers have a much better chance of having a successful user interface if the system is designed for specific users with a set of clearly defined system–related characteristics.

Most user profiles contain information that identifies the following:

- How much computer experience typical users have. This includes both general computing experience, as well as experience with specific systems.
- How much business knowledge users have and how this knowledge is applied in the new system.
- The amount of time users will spend with the new system. *Frequency of use* is usually divided into three categories: regular use (daily), occasional use (weekly, monthly) and infrequent use (seldom). Regular users can remember what to do and how to do it much better than occasional and infrequent users. Infrequent users use documentation much more than the other two categories.

Designers should also specify the *usability requirements* associated with each class of users. Usability requirements represent the performance goals once a system is operational. For example, the user interface could have a usability requirement of being able to ensure that service representatives could easily handle 25 customer calls per hour. If performance testing demonstrated that this requirement could be met, then the user interface has been met by this particular performance goal.

Computer Profile

The input and output devices, plus the operating environments and windowing systems selected for an application, can have a large impact on the quality of the user interface. User interface designers should participate in selecting the best input devices, such as the keyboard, mouse, pen-based controls, and touchscreens. In addition, they should be involved in selecting the best output devices, including display screens (e.g., CRTs) for visual output and speakers for auditory output.

Many older systems continue to have user interfaces that are constrained by using older technologies, for example, keyboards without function keys or mouse availability, monochrome and character-based display monitors, and exclusive use of command-line entries. User interface designers should also evaluate the strengths and limitations of the available operating and/or windowing systems (e.g., VMS, DOS, UNIX, Microsoft Windows, and X Windows).

Being able to design, develop, and deliver quality user interfaces depends considerably on how well the input and output devices and operating environments are selected. These choices, for example, will dictate whether designers can use color, different sizes and styles of type fonts, sound (including voice), graphics, and so on. Frequently, the market rather than the designer helps to dictate a lot of these choices. If the current market demands Microsoft Windows and voice input, a DOS system without voice input just will not be acceptable. This is true no matter how well suited and easy to use the product is.

Task Analysis

After the user and computer profiles are prepared, designers are ready to begin designing the user interface. The first step is to conduct a *task analysis.* The task analysis is conducted on activities that are either totally performed by users or completed using a human–computer interaction. These tasks are evaluated and recorded either in *flow charts* or *scenarios.*

Interaction

Human–Computer Interaction

We previously defined the user interface as consisting of input devices, output devices, the information input by users, and the information output by the computer. Each time users use an input device (control) to convey information *to* a computer and the computer responds an *interaction* has taken place. This aspect of human–computer communication (the interaction) is the most difficult to design and probably the most critical part of any user interface.

In traditional character-based user interfaces, many of these interactions take place by having users enter commands. The system then responds with information and/or messages. In some cases, particularly with regular users who quickly become power users, a well-designed command language may provide the best interaction procedure. Interaction considerations can be divided into those concerned with dialog type itself and those concerned with the language used to carry out the dialog.

In graphical user interfaces, most interactions take place when users select screen-based controls (widgets) contained on the desktop or in windows. These controls can include menu items, push buttons, check boxes, drop-down lists, icons, or others.

System Messages

Most information passed *from* the computer to users is contained in messages: status messages, error messages, notifications, prompts, progress indicators, and the like. Users are expected to read these messages and, frequently, to take appropriate action. Well-designed messages allow users to know where they are and what is expected in a system and allow human–computer interactions to proceed in a smooth, uninterrupted fashion.

Different types of problems require different types of error messages. Many systems rely on messages generated as a result of an error being detected by the system, to provide sufficient information to quickly correct the errors. Well-designed error messages provide information on what the error is, why it was detected as an error, and what the user should do to correct the error condition.

The status of a system relies on appropriate messages. This is particularly true when a computer process is started that may take longer than a few seconds. If the processing will take up to 10 seconds, the cursor may change shape (e.g., become an hourglass). If the processing takes longer than 10 seconds but less than 1 minute, a progress or percent-done indicator may show.

Presentation (Display) Features

Once optimal interaction capabilities are determined for a new system, the next step is usually to determine the best set of presentation or display features. These features include ensuring proper content, selecting appropriate colors, optimizing the layout, positioning important information, and others.

Traditional character-based interfaces usually involve users receiving text and numbers. Graphical user interfaces can extend the presentation of computer responses to color graphics, animation, photographs, sound, and full-motion video. Also, the richer presentation media provide opportunities for designers to provide high-level visualization capabilities.

User Guidance (Help)

Most modern computer systems provide on-line assistance to enable users to quickly resolve their problems. Well-designed help facilities are usually context sensitive. This means that the computer knows exactly where the user is positioned in the system when the help request is made. The resulting help message can respond to a specific problem, rather than providing general information.

User guidance can be as complex as having error messages connected to a computer-based user manual, which is connected to computer-based training. Under these conditions, users can pursue information until the question is answered. Most user guidance systems today use hypertext concepts for linking and organizing user guidance information.

Documentation

Many systems require documentation to be fully used. This is because the existence of some activities that the system can perform is not obvious. Another reason is that some activities are too complex to be intuitive, that is, to be used without assistance. User documents generally fall into two major categories: user manuals or quick reference guides. These documents can be either paper based or computer based. Many system designers automatically provide a document with their systems. Others try to establish a clear-cut need for documentation before taking the time to develop it.

Some experienced system developers consider all computer-related text as part of the documentation for the system, including messages, help facilities, and traditional documents.

Most computer-based documentation is delivered in the form of hypertext. Hypertext consists of text modules (usually less than one screen or window of text), links to and from the modules, and buttons for accessing the modules.

Training

Most new computer systems require users to learn how to perform new activities. In many systems, users simply discover the best way to proceed by trial and error. In other systems, particularly systems with many functions, very complex functions, or functions that are used infrequently, the learning process may be more structured. This is usually referred to as *training*.

Training can be as simple as a few "over-the-shoulder" instructions, to providing a brief tutorial, to a several day lecture-based, hands-on training course. The amount of training provided with each application should be related to the needs of system users. Two of the most important decisions have to do with providing either self-paced learning or lecture, and if self-paced learning is selected, providing either a discovery or tutorial strategy.

Optimization

Once all user interface components have been developed, they should be integrated. This is done (1) to demonstrate that they work well together and (2) to determine if they are adequate to ensure easy and consistent access to system functionality.

In software systems, the user interface is generally evaluated in terms of its ability to ensure a high level of productivity (speedy performance), with few (if any) errors and requiring minimal (if any) training. In addition, user interfaces should be designed to ensure a high level of user acceptance. In many systems, each of these four criteria are quantified (converted to numbers) and become the target benchmark for the user interface.

For example, a real estate system may have a user performance benchmark of allowing one sales assistant to adequately support two sales agents, because of being able to do the job faster and with fewer errors or oversights. Another performance benchmark could be that this performance level could be achieved with 12 hours or less of formal training and 1 week of practice. Finally, a preference benchmark could be that users will score 90 or higher on a user satisfaction questionnaire. When these benchmarks are met, the system is optimized.

Stage 1: Analyze and Define the Product

System Objectives and Performance Specifications

Major Activities

From a usability point of view, during this stage designers are primarily concerned with

- Verifying that the functionality truly does allow users to do what they need to do
- Identifying the intended users of the new product, application, or system
- Selecting the appropriate computers, input–output devices, operating systems, network systems, and the like

What is done in this stage and how well it is done will affect not only the next design stage, but the entire development process. In this stage, designers will make decisions that could have considerable influence on human performance and user acceptance once the system is operational.

Definitions

System objectives and performance specifications should be clearly identified before the design of a system begins. *System objectives* usually are stated in general terms to avoid their revision in cases when performance requirements may be altered in response to technological, budget, or other constraints. Examples of system objectives are to prepare a White Pages telephone directory, develop a new automobile, or create a space vehicle that will go to and return from Mars. The system's *performance specifications* should state what the system must do to meet its objective. Generally, these requirements are derived from a careful study of user needs, typically involving interviews, questionnaires, site visits, and work studies.

The system performance specifications usually include a set of requirements and a set of constraints. *Requirements* help to identify the purpose of the system as a whole and the operational characteristics or performance requirements that detail specific objectives. System *constraints,* on the other hand, are limits within which accomplishment must take place.

System Objectives

A system may be developed to achieve a variety of objectives. Achieving these objectives requires that system users be provided with information that is necessary, as well as accurate, timely, and complete. These *needs* of users should be clearly spelled out in this stage. The focus here is to clearly understand *what* the new system will be able to accomplish. It will be more costly, in terms of both time and money, to make changes to the basic functionality once designers have begun to make design decisions. Designers have a major responsibility for ensuring that the requirements of users are clearly and comprehensively defined. In the next stages, these requirements will be further refined and considered in relation to *how* they will be performed.

Potential users include all people who may be involved with the operation of the new system. It is important to identify the specific needs of different types of existing and potential users. If these needs are well understood, the new system can be designed to support the users as fully as possible. One way to start this process is to consult the person or persons who first identified the need for a new system. Additional information on user needs may be collected through observation, interviews, and questionnaires.

System Requirements

The development of system requirements is an iterative process that extends throughout this stage. Since this activity occurs long before the assignment of functions to be performed by people or a computer, the designer must be content with ensuring that all important user needs and expectations have been identified and are reflected in the stated system objectives.

If a system is developed that does not meet user needs and expectations, the users can reject the new system in a variety of different ways, ranging from acts of sabotage to the quiet use of an alternative (usually manual) system. Other forms of rejection are to work slowly, purposely make excessive errors, never quite "catch on" to what is required, and constantly complain of how dissatisfying it is to work in the system.

Examples of performance specifications include that the White Pages directory must be produced in 3 months, the automobile must get 100 miles to the gallon and comfortably carry eight people, and the space vehicle must carry a crew of three and a payload of 10,000 pounds.

Determine Alternatives

We should point out that many designers tend to figure out how to develop a system and *then* proceed to identify objectives and performance specifications. This approach creates two problems. The most obvious is that system objectives are defined to fit a preconceived design. It is best to identify all objectives and performance specifications and then work in the design characteristics.

The second problem is that a premature design discourages the development of design alternatives. *It is usually best to develop several alternative designs before selecting one to use.* Crossley (1980) reports on a study in which four experienced designers were asked to independently design a simple bell-crank lever. Not surprisingly, the four people produced significantly different designs. What interests us even more is that none of the designers produced what turned out to be the best design. This and other studies suggest that the first design solution to come to mind is rarely the best. In fact, each iteration can improve the performance level from 12 percent (Bailey, 1993) to as much as 30 percent (Nielsen, 1994).

Determine System Performance Specifications

Once a clear and accurate set of system requirements is identified, system performance specifications can be determined. The purpose of this activity is to develop a quantitative description of

- What is to be done
- How well it is to be done
- How to measure what is done and how well it is done

The kinds of system performance criteria related to human performance that we are considering tend to be quite general, such as cost and time savings, decreases in accidents, reductions in required people, improved customer service, and better employee work opportunities.

The designer should be especially careful when evaluating the potential impact of the new system on human performance. Keep in mind that the functions to be done have not yet been allocated to users or the computer, making it difficult to make any precise estimates. Even so, the designer must be careful not to establish system performance specifications that will hinder future decisions concerning achieving acceptable human performance.

When considering system performance specifications, the designer should always take into account the context in which the system will operate. From a human performance point of view, this could include the nature of the labor market, existing labor agreements, skills in the existing work force, and the existing organizational structure. Characteristics of a system's context can greatly affect human performance. A designer would not want to develop a system, for example, that required large numbers of skilled people if the system is to be used primarily in a sparsely populated rural area.

Designers must determine if the same kinds of people employed in an existing system will be employed in the new system. To evaluate this decision, the designer would have to have a good understanding of the existing user population.

Another possible decision may be that the new system will reduce by 50 percent the number of people required to perform the work. Such a decision suggests that the computer will perform many more functions than in the past. Designers should be aware of the major human performance implications of this decision. Some (perhaps many) of the existing people will lose their jobs once the system is ready. This could cause considerable hostility toward the new system while it is being developed and when it is operational.

Bank Example

As an example of the types of activities that can take place while defining system functionality, consider the situation when a bank is having an extremely difficult time because of large numbers of customer overdrafts. In most cases these overdrafts are not due to fraud. In fact, the bank conducted a study and found that the vast majority of people who had overdrafts simply did not know how much money was left in their bank account when they wrote checks. The bank's management decides that it needs a way of helping people to keep better track of their bank balances. They decided to develop a new system that would reduce the number of overdrafts. The new product would help customers to know how much money they had in their checking accounts at any given time. After much discussion, the bank's management agreed on the following objective for their new system.

System Objective

Provide a means for assisting customers to be more aware of their checking account balance. This objective is general enough to allow designers considerable freedom in finding a good solution, but specific enough to convey to the designers exactly what the system ought to do. The designers also are provided with the following set of requirements and constraints:

Requirements:

Should be convenient for most customers to use.

Most customers should be able to accurately use the new approach.

Most customers should be interested in using the new approach.

The system should be easy to use by most customers.

Constraints:

Customers' understanding of what is required by a bank concerning their account varies widely.

Customers have a wide range of math skills.

Customers have a wide range of attitudes toward maintaining an accurate balance (from having no interest to having personal accountants).

The system designers took the objective and performance specification (and all other information that they could collect) and came up with several alternative ways to design a new system. Some of the alternatives are listed next.

1. Customers fined $100 for each overdraft.
2. Customers rewarded financially for having no overdrafts.
3. Provide special software for customer's personal computers.
4. Customer telephones the bank and updates all transactions once a day and receives balance at that time (the telephone call is required or the bank will not honor any checks written that day).
5. Bank mails a statement to customers each time that they write a check.
6. Bank sends balance and all transactions to the customer's home computer on a daily basis (the bank buys a large number of home computers and makes them available as rentals for people who do not own a computer).
7. Bank provides extra money that is then borrowed by the customer to cover overdrafts (this allows overdrafts to continue, but with less penalty to the customer).
8. Customer is required to telephone for clearance before each check is cashed. Each check must contain the clearance number to be honored. Balance is given when call is made.
9. Customers are allowed to cash checks only at places of business that have a direct hookup to the bank so that the balance can be adjusted as the check (more like a credit card) is cashed.

Function Analysis

Earlier in this chapter, we showed several activities that should be performed at appropriate times in the development of a new system to help to ensure an acceptable level of human performance. The first major process is *function analysis.* This involves analyzing and describing all functions that the new product, application, or system will perform. Once the need for a new system has been demonstrated, it is necessary to identify the major functions for the product. A *function is a statement of work,* and if all functions identified are successfully completed, then the system should meet its objectives. Functions usually reflect the most general, yet differentiable, means whereby the system requirements are to be met. There was a time when people performed *all* functions in a system. Changes came about when animals and machines began to be used to supplement human muscle power. Thus, the earliest function allocation came about when people began dividing work among people, animals, and machines.

System Functionality versus User Interface

Computer systems can be divided into issues associated with the basic *functionality* versus the *user interface*. System functionality refers to the work actually done by the system, that is, the functions performed by the system. The work can be done by a person alone, the computer alone, or the human and computer working together in some interactive way. The user interface refers to all issues associated with accessing and using the system functionality.

It is possible to have a system with good functionality and a poor user interface. Over the past several years, many such systems have been developed. Goodwin (1987) provides several examples of how improving the user interface, without changing the system functionality, resulted in improved system performance. Perhaps the most interesting was a study in which system A had full functionality and a poor user interface, and system B had limited functionality (far fewer functions) and a good user interface. The results showed that participants actually used significantly more functions using system B (the system with the fewest functions).

A well-designed system will include both the required set of functions and easy access to the functions. To put it simply: The user interface provides access to system functionality. The access can be awkward, slow, and error prone or efficient, natural, and error free. The access can be hard to learn and remember or transparent and easy to remember. The access can be frustrating, tedious, and boring or fun and satisfying (like many computer-based games). Another observation from these and other studies is that fewer functions generally reflect less complexity, which can lead to better user interfaces.

Determine Functions

Definition

Once system objectives are set and system performance specifications are determined, the factual basis of the system can be established. At this point the basic requirements for the product have been decided. This means (we hope) that several possible alternatives have been evaluated and the best selected. As stated earlier, a function is simply a statement of work that must be accomplished for the system to meet its objectives. Designers usually identify these functions and may prepare system-level flow charts to display the functions. Depending on how much time is available, detailed narrative descriptions can be prepared for each function. It is vitally important that human performance considerations begin as soon as functions begin to be identified.

Avoid Allocating Functions

A designer should be aware of the tendency in this stage to go beyond a simple description of the work to be performed and to make decisions about *who* will perform each function (people or the computer). Some designers prepare functions that describe not only *what* will be done, but also *how.* It is not uncommon to find that, when functions are assigned to people or the computer during this early stage, wholesale assignments are made to machines, leaving the leftovers for people to do. *These assignment decisions are premature* and could have a considerable negative effect on human performance once the system is operational.

Functions can be more systematically analyzed, with allocations more appropriately made, later in the developmental process. The best approach is for designers to concentrate on establishing the factual basis of the new system (i.e., ensuring that all required functions are identified).

In the bank example, the alternative selected was to *design, program, and issue a special interaction device.* The device changes a user's account balance as deposits are made and checks are written. (To simplify the example, problems relating to a monthly reconciliation of the checking account will not be addressed.) The major human performance-related activity during this stage is to identify potential system functions. The major functions are

- Determine the account balance
- Add money to the account when a deposit is made
- Deduct money from the account when a check is written
- Store the account balance

In this stage, it is most important to ensure that the list of functions be fairly exhaustive. We should emphasize *what* must be done to satisfy the system objectives, taking into account the system requirements and constraints.

Verifying Functionality

In some product or system development efforts, user interface specialists may not be involved in identifying the initial set of functions. Even so, it is critical that the proposed set of functions be verified for correctness. It is not possible to have a well-designed user interface associated with the wrong set of functions. If the functions do not allow users to do exactly what they need to do, the user interface, which gives them access to these functions, is useless.

Function Analysis Techniques

Two major techniques are used to identify system functions or, in other words, *what* a new system will do. These two techniques are joint application development and design and participatory design. Both are used by designers to help to identify, evaluate, and document system functions.

Joint Application Development/Design

Joint application development and design (JAD) is a methodology used to develop computer systems. The basic approach is to conduct an intensive 2- to 4-day workshop.

This workshop is most useful for determining basic system functionality and eliciting system requirements. The workshop is usually attended by designers, potential users, user management, and a JAD facilitator. In fact, the key to having a successful JAD workshop is to have an experienced facilitator who is capable of helping the group to move through various steps toward developing a solution.

A full set of system requirements can be defined and some early design decisions made and recorded. The workshop results in a fully documented application solution that can include a working model (software prototype), tentative screen and/or window designs, report formats, data elements, processing routines, operational flows, and so on.

Probably the most important advantage of this methodology is the possibility to accelerate the design and development process. A well-done JAD session can consolidate the activities of problem definition, requirements definition, and early application design. For small- to moderate-sized systems, for which these activities may have taken from 3 to 6 months, they can be completed in less than a month.

Spending less time on front-end activities helps to reduce the required resources and time to produce a system. JAD advocates report that the elapsed time of projects can be reduced by up to 40 percent by using a workshop to accelerate the definition of requirements and making early design decisions. This process lays a good foundation for making better user interface decisions later.

Participatory Design

Participatory design (PD) originated in Scandinavia and assigns responsibility for system development to users (Carmel et al., 1993). Thus, users are continually involved throughout the design process, and their managers are excluded. Its main goals include accentuating the social context of the workplace and promoting workers' control over their work and lives (i.e., to empower users). The primary assumptions of participatory design include

- Giving workers better tools instead of having their work automated
- Recognizing that users are best qualified to determine how to improve their work
- Users are not involved just during meetings, but continually. Designers are essentially working for users.

Contextual Inquiry

One of the most widely used forms of participatory design is a method referred to as *contextual inquiry* (Holzblatt and Beyer, 1993). The goal is to more effectively involve customers in the design and development of new computer products. Contextual inquiry attempts to bring customer data into design through a well-defined sequence of activities.

It is basically an interviewing and analysis process and provides techniques to get data from users *in context*. Information is secured while potential users are working at real tasks in their current workplace. Several members of a design team interview several customers at the same site simultaneously. The information is then compiled and placed around the walls of a special room, referred to as the "think tank." All members of the team help in organizing the information into useful affinity diagrams, work model diagrams, and prototypes.

Functional Thinking

One of the most important advantages of the stage approach to system development is that it allows designers to concentrate on the most important decisions one at a time as the process proceeds. For example, when thinking in terms of functions, designers are encouraged to separate work from the means used to achieve the work.

Singleton (1974) suggested two advantages to functional thinking in this stage of the design process. First, it produces a *common language* for all specialists. For example, power generation may prompt a chemist to think of a chemical reaction, such as the burning of gasoline as a source of power; an electrical engineer to think of an electric motor that gets its energy from a wall socket; a mechanical engineer to think of the wind and windmills; and a solid-state physicist to look to solar cells. Each specialist has his or her own idea of what power generation entails, but each clearly understands the abstract concept.

Second, Singleton sees functional thinking as the means to an integrated design. He views many current designs as primarily done in one physical discipline, with other disciplines tagged on as an afterthought or as a quick fix for an unforeseen problem. An example might be the catalytic converter included on many automobiles. This is a quick, chemical solution to the old problem of clean, efficient personal transportation. An alternative approach would be to attack the problem from a more fundamental position, which may lead to better solutions, such as fuel injection or electric cars. If a functional approach had been taken from the beginning and if all the necessary specialists were thinking in functional terms and working toward an objective together, a more integrated design might have resulted.

John (1980) suggested another advantage of functional thinking. She observed that many theories of creativity and innovation regard the definition of the problem in the broadest terms possible as the first step in achieving true novelty in a solution. Thus, if a system is expressed in terms of its functions, its problems are expressed in their most abstract form and the best chance for innovation exists from the beginning.

Function Analysis

Before an allocation of functions can be made, each function must be well defined and reduced to a level at which the allocation process is meaningful. Thus, one of the first steps in dealing with functions is to identify, analyze, and describe each function to be performed in a system.

User Profiles

Introduction

Once functions are identified, two major activities can be carried out to help to ensure an acceptable level of human performance. One of these is to identify user profiles, and the other is to select or identify the computer environment, including input and output devices. User profile issues will be discussed next, while the computer environment issues will be discussed in Chapter 10.

Documenting User Capabilities

One of the most important activities a designer can do is to clearly identify the computer-related characteristics of potential users. When this information is agreed upon by the design team and written down, it represents a *user profile.* As the user interface activities continue to develop, each designer should have these user characteristics in mind. In fact, each designer should be able to clearly envision the characteristics of typical users who will be interacting with the computer.

The user profile is a detailed description of minimum acceptable qualifications in terms of the skills and knowledge required to efficiently and effectively perform the computer activities. It provides a way of screening people in order to have the exact kinds of people in the new system that the designer had in mind when he or she completed the task analysis and designed the computer product.

Estimating the Kinds and Numbers of People

If all individual user profiles were combined into a single document, it would provide an accurate description of the types of people needed to perform in the new system. The number of each type of person can then be estimated from human performance test results. By having a good idea of the types and number of people required to perform in a new system, personnel information can be made available to user organizations long before the system is operational.

Types of User Information

Two kinds of information that designers may wish to collect include (1) the availability of human resources (numbers and types) and (2) the basic characteristics of people in the potential user population, including information on sensing, cognitive, and responding capabilities.

For example, suppose that the policy has been adopted that the new system will be designed so that it can be operated by existing employees. In such a case the designer will want to make sure that good information is available about the existing work force. The designer will want to ensure that information about these human capabilities and limitations is available for use in the next stage. In the bank example, the potential users include all bank customers. Obviously, there will be a broad range of skills and knowledge.

Types of Users

The longer computers are with us, the more varied are the types of users. Users were once a select group of mathematicians sharing a common interest in computer technology. Today this group includes scientists, engineers, managers, clerks, children, educators, salespeople, and students.

Systems should be designed taking into account certain characteristics of potential users. No system can be designed with the assumption that the user will be virtually anyone who happens to step forward and use the system. All good systems have a clearly defined user group. Thus, the designer constantly looks toward the ultimate end user of the system and makes design decisions to best benefit that user. The designer recognizes that a new system is designed for a specific type of person and cannot elicit an acceptable level of human performance from all people in the world. Therefore, the designer must understand in advance the design-related characteristics of potential users.

Several considerations concerning users need to be addressed with new systems. One of the most important has to do with identifying the characteristics of typical users and then designing specifically for them. This usually includes identifying any unique strengths or limitations that these typical users may have.

Managers versus Clerical Users

Management Users

Consider, for example, the differences between two typical user groups: managers and clerks. These two user groups are different enough to require unique design decisions. Managers tend to have little understanding, appreciation, or even interest in directly using many computer systems. Most managers will use the system only occasionally and will not waste time on a system without realizing some relatively straightforward, direct benefits. Because they are in a position to reject a system, managers require systems that are highly usable from the very beginning. In addition, managers have the ability to have others use the computer for them; consequently, if the computer is difficult to use, they have the option of having someone else use the computer to generate the needed information.

Clerical Workers

Clerical workers, on the other hand, may have to use a system or lose their jobs (e.g., a word processor). Clerical users tend to be regular users and usually recognize the value of using the system in terms of getting more done in shorter periods of time. Probably more than any other major user group, clerks require protection from having repetitive, boring, and routine tasks.

Expert and Power Users versus Novice Users

In many cases one must decide if the system is being designed for experts or novices. Experts tend to be more knowledgeable about their domain and know how to apply that knowledge in the use of a new system (Kolodner, 1983). Novices tend to have little understanding of either the domain or the new system. It should be noted that there is a difference between expert users and power users. Expert users have expertise about the domain in which they work, whereas power users have expertise about the computer product that they are using. Ideally, typical users will be both expert users (with the domain) and power users (with the computer product). The design should reflect these differences.

Users That Change

Any computer-based system will have users with characteristics that change as they gain experience. Ideally, as people change the system could also change. One design objective should be to build human–computer interfaces with critical elements that change as the users become more experienced. These systems may be either adaptable or adaptive (cf. Oppermann, 1994). Adaptable systems allow users to make changes as they gain experience. Adaptive systems automatically change as users gain experience. There is some evidence that adaptive systems can lead to better user performance (Trumbly, 1994).

A person learning to pound a nail can learn the task by using exactly the same tool he or she will use after mastering the skill. This is not the case with computers. When learning to use the computer, people may require a tool (particularly software, but sometimes hardware) much different from the one that they will use after gaining proficiency. As people change, their computer-related needs change.

Mental Models

People who have had little direct exposure to computers tend to have vague mental models. Their mental model of what is going on inside a computer tends to be very unclear. As with most computer users, the mental model that people have when driving their cars can be very unclear. For example, drivers do not have a very clear understanding of what goes on inside an internal combustion engine. Nevertheless, they can be extremely good drivers and can reliably drive a car from one point to another even over long distances. Their driving capabilities have little to do with understanding the difference between a carburetor and an alternator, until the engine begins acting in a peculiar way.

The same is true for users of most computer systems. They should not be required to understand the difference between an 8-bit and a 16-bit microprocessor or the implications of having 640K of memory versus having 64MB of memory. However, a fairly good mental model may assist a person in understanding why the computer takes longer under some conditions than under other conditions. Having a clear mental model, for example, may also help a person to understand why certain information is automatically stored, while other information is not.

Of course, there is really no need for many users to understand exactly how computer processing takes place. The required clarity of the mental model depends on the type of system and how it is to be used. In general, if users tend to routinely perform the same types of actions on a regular basis, then an unclear mental model may suffice. However, if the application requires users to understand and apply a wide range of different computer actions in unplanned and spontaneous ways, then a better mental model of the system would be helpful. Usually, the more experimentation with the system that is required, the clearer should be the mental model.

Satisficing

Computer products should be designed to accommodate the skills, knowledges, and needs of typical users. Many users do not take full advantage of all the procedures provided for them. When people use only the minimum set of functions, and typically use these functions in inefficient ways, this is referred to as *satisficing.*

Research seems to suggest that many users do not learn to efficiently use all functions contained in many systems. A study reported by Huuhtanen et al. (1993) suggests that about 15 percent of users are *power users,* about 80 percent are satisficers, and about 5 percent are novices.

The satisficing strategy is defined as follows:

Having reached the performance level where routine tasks are done fairly efficiently, they seem content that they know enough,

Satisficers rarely expend the effort to find more efficient methods.

An example would be deleting a paragraph character by character rather than identifying a block and then pressing the delete key (Santhanam and Wiedenbeck, 1993).

Having extensive experience with a product does not seem to guarantee that users will learn even the basic functionality (Nilsen et al., 1993). Most users seem to learn a subset of features. They learn just enough to get their work done, and not necessarily in the most efficient way.

One investigation of a software product showed that the speed of performance improved markedly (doubled) over the 16 months of the study. Even so, the most skilled users were still almost half as fast as experts. For example,

Originally: 46 minutes

End of study (16 months later): 25 minutes

Experts: 13 minutes to do the same tasks (these experts have learned more efficient strategies)

Acceptance Factors

Past Experience

Experience has shown at least four user characteristics that seem to have a relatively strong influence on the acceptance of new systems. First is past experience with computer systems in general and the specific type of system being developed. For example, there seems to be a greater likelihood of acceptance by users that have had experience on other computer systems. Also, as users gain experience with a new system, the probability of acceptance of that system increases. Thus, user acceptance appears to be closely linked with both general and specific system experience.

Type of Work

A second consideration has to do with the type of work a person is doing and how the computer is supposed to assist in doing the work. Traditionally, managers and salespeople have not readily accepted systems. A main reason for this is that managers are paid for managing and salespeople are paid for selling. If the computer allows them to manage better or sell more, then it is accepted. If not, the new system will find great difficulty in being accepted.

Thus, user acceptance appears to be closely linked with the type of work a person performs. If the new system clearly helps them to perform faster, with fewer errors or greater satisfaction, then it will be used. If the new system merely holds the promise of making things better or is viewed as an interruption or nuisance, then there is less chance of it being used.

Level of Enthusiasm

A third consideration is the enthusiasm level of potential users. How well have they been prepared for the new system? How well managed have been user expectations for the new system? Even if the new system has the potential of substantially speeding up whatever work is being done, if the new users have not been well prepared, they could refuse to use the system or take steps to make the new system look bad. In either case, acceptance could be very slow.

Age of Users

Finally, there seems to be a relationship between user acceptance and age. In general, the younger the users are the better. There seems to be more resistance to using computers by older workers. In some cases, there are fears associated with not being able to learn or remember how to do certain things. Others are more concerned about doing something wrong that may break the computer and then not knowing how to "fix it" (cf. Gomez et al., 1986).

Interrelated Factors

All the factors just discussed are interrelated. Probably the worst case is to have a person that has had little or no experience with computers, who works in sales (on commission), who perceives that taking time to learn the new computer system is a waste of time, and who is nearing retirement (early sixties). A group of users in this category may find that the typical "plain vanilla" user interface is not good enough to ensure use and acceptance. Even when a system has all the functionality that is needed for a given task, the system may not be accepted if the interface is not easy to use (Goodwin, 1987).

User Satisfaction

Measuring user satisfaction helps designers to know if they have been successful in these endeavors. There has been a recent interest in developing an independent measure of user satisfaction (Chin et al., 1988). This instrument is not yet perfected, but it does provide hope that in the future we will be able to reliably measure user satisfaction as systems are being developed.

Number of User Interfaces

Designers should make sure that *all* user interfaces are considered when designing a system (Malde, 1986). There is a tendency to focus on the ultimate end user and to ignore other users who also must interact with the system. For example, a bank commissioned an ergonomics evaluation of a new computer-based automatic teller machine (ATM). These machines are in common use for making withdrawals, deposits, and other transactions without the need of a bank employee.

The bank was particularly interested in an in-depth review of the user interface. As the review progressed, we identified *five* user interfaces:

1. Customers
2. People who load the money and proof the machine two or more times daily
3. People sitting at VDTs in centralized customer service organizations that receive notifications about problems with ATMs and make arrangements for fixes
4. People who do both software and hardware repairs to the machines
5. People who use microcomputers both before and after the system is installed to format the computer screens and develop efficient interactive procedures

The bank was originally interested only in the customer interface, but soon found that there was more payoff by improving the user interface for the other user groups as well.

This is typical of the multitude of user interfaces for many systems. Here is another example. Many expert systems must consider the user interfaces for ultimate end users, the clerical people who keep the knowledge base up to date, the knowledge engineer who is making continual updates to the system, and others.

Selecting Users

Plato proposed a series of tests for the selection of the guardian of his ideal republic. From this, we can assume that he desired some control over who was selected to help to ensure that the guardian would perform adequately. Similarly, for the same type of control, designers should propose guidance for the selection of people to work in their new systems. There is probably nothing quite as discouraging as developing a system for college-educated, experienced users, only to find that the people that are typical users of the new product are high-school graduates without experience.

Preparing User Profiles

A designer should be especially concerned about two aspects of the selection process for typical users. First, an explicit and clear profile for each type of potential user should be prepared. The profile should consist of several detailed statements concerning the basic skills and knowledge a potential user should possess in order to use or to begin learning how to use the new system. It is always important to specify the skills that the designer had in mind for each user *before* the user is exposed to additional training in the new system. In systems that require little or no training, these basic skills alone may be all that is needed to perform adequately.

Selecting Qualified Users

Second, the designer should provide ideas and suggestions on how to efficiently and effectively select people who meet a particular user profile. Decisions concerning this aspect of selection can be critical to the ultimate level of human performance obtained in the operation of a new system. The designer should specify the skills, knowledge, and attitudes required of users and, where possible, the best tests for measuring these characteristics. This process is illustrated in Figure 9-2. Designers should know best who they had in mind as potential users and should take time to specify, in some detail, the types of people that they assumed as users. When these suggestions are made available, management can better select the people who will most likely perform adequately in the new product, application, or system.

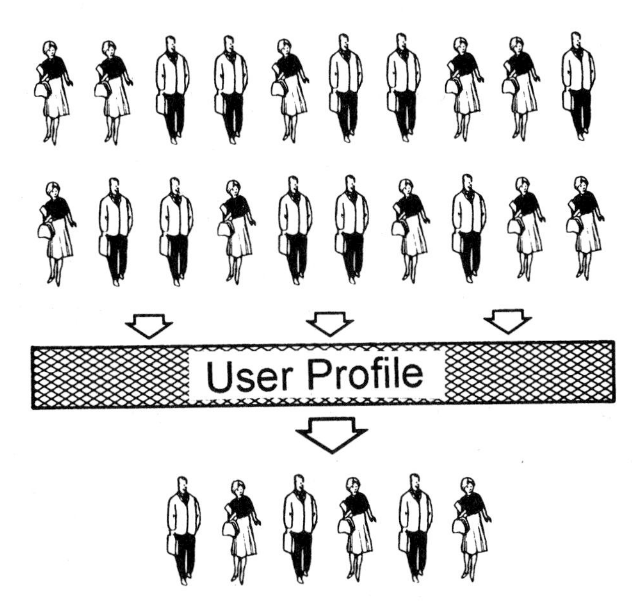

Figure 9-2 How a user profile can help in the screening and selection of potential users.

People differ in many ways. A designer needs to keep this in mind when making basic design decisions, as well as decisions about interfaces and facilitators. This particularly needs to be kept in mind when *specifying* the types of people needed to perform the system. The differences that exist from one person to another (in such things as education, experience, and attitude) provide a foundation for identifying and selecting those people who *best match* the requirements for the activity.

We observe that people differ in such obvious physical traits as height, weight, or hair color. We are also well aware of differences in less obvious traits, such as intelligence and personality. As we meet people we recognize these differences: some are dark, others light; some are heavy, others thin; some are bright, others dull; some are excited about their work, others are bored. Such simplified descriptions are usually of little value. They imply that each trait can be described adequately in terms of only two categories (for instance, tall–short, pleasant–disagreeable, superior–inferior). Such a classification may be convenient, but it is not adequate for selecting people who match the requirements of a job. We need more discriminating classifications of people. Within any trait, the differences can best be discriminated by adding more and more categories. Well-constructed tests are used to make these finer discriminations possible.

If designers can establish that the existence of a certain trait will help achieve an acceptable level of performance, then management should be so informed. It can then select people who possess that particular trait. For example, intelligence, as measured by an intelligence test, usually is related to performance in almost any kind of training activity. Generally, those with higher intelligence learn faster. Thus, the trait of intelligence is associated with performance in a training course, and an intelligence test may prove useful in identifying and selecting those people with a high probability of doing well.

Long History of Testing

The use of tests to assist in the systematic selection of people to fill various work assignments began in the United States about 1918 and has expanded rapidly ever since. Thus, for a period of at least 70 years, psychologists have produced an abundance of psychological tests that measure traits related to human performance. Many of these should be of interest to designers. From the assembly-line worker to filing clerk to top-level management, there is scarcely a type of work for which some kind of test has not proved helpful. Unfortunately, few designers are ever exposed to available tests that may help in selecting the best person for a particular job.

If a trait is well understood, there are usually many equally satisfactory alternatives for measuring it. The problem becomes one of determining the right test for each selection question and taking the necessary steps to ensure that the test does indeed help in selecting people who are potentially the best performers. There are published lists of available tests (cf. Buros, 1974; Buros et al., 1961), as well as lists of critical reviews for many of the tests (cf. Buros, 1978). Many designers will not be able to spend a great deal of time reviewing and evaluating potential tests. Even so, they should be aware that these tests exist and take steps, usually with the help of a testing specialist (industrial psychologist), to help to ensure that they are used.

Validating a Test

A word of caution is in order. Selecting a useful test is not easy. Even testing experts can make a wrong decision. A designer should be very much concerned with finding out *for sure* whether a test will actually help to improve performance. This is called *validating* the test.

Test validity is concerned with how closely test scores are related to the performance that is being predicted. A test should be put to use only when there is clear evidence that it is valid. Unfortunately, some people pick tests on "looks" or using some other vague criteria and do not take time to ensure that the test is measuring what they think it is measuring. Many such tests turn out to have little or no relationship to performance. Once you have selected a test, it may be necessary to acquire the help of a test specialist to ensure that the test is valid.

Using Existing People

Frequently, new systems make use of people who are left over from the old system. In some cases few, if any, new people are hired until after the new system has been operational for quite some time. In the interim, selection criteria should be used to match the skills, knowledge, and other traits of people from the *old* system with the activities to be performed.

Providing User Profiles

If a designer is unable or unwilling to specify detailed selection criteria (including proposed tests) that may be helpful in selecting people, at the very least he or she should provide a user profile with the new system. Ideally, the entire selection process should begin with detailed statements of relevant characteristics of potential users that a designer had in mind when making design decisions.

We need to make one final point. Selection is concerned with choosing one person from among many to perform a given activity. However, where people are already available, the emphasis should shift to choosing the one *activity* best suited to the strengths and weaknesses of a particular person. The primary question then is, "What activities will best use the individual's strengths?" or "On what activity will each individual perform best and feel the greatest satisfaction?"

User Profile Questionnaire

When attempting to identify the essential characteristics of potential users to include in user profiles, the following questionnaire may be of value. Designers should feel free to modify the questions or to add and delete questions to help to tailor the questionnaire to their needs.

1. How many years have you used computers?
2. How many years have you used a graphical user interface (GUI) such as Microsoft Windows, the Apple Macintosh, or UNIX workstations?
3. Currently, how often do you use a computer?
    - a Once a month or less
    - b Several times a month
    - c Several times a week
    - d Several times a day
4. How do you rate your skill and knowledge with computers?
    - a Little
    - b Some

c Average

d High

5 How do you feel about using computers?

a They make me very anxious

b They make me a little anxious

c Somewhat confident

d Very confident

6 Where is your copy of the system located?

a On my own personal computer

b On a network

c Both on my computer and on a network

7 How much do you use the keyboard (versus the mouse) in the system?

a As little as possible

b Some of the time

c Most of the time

d All the time (if possible, I never use the mouse)

8 How much do you use the mouse (versus the keyboard) in the system?

a As little as possible

b Some of the time

c Most of the time

d All of the time (if possible, I never use the keyboard)

9 Do you consider yourself a touch typist when keying characters using the entire keyboard?

10 Do you consider yourself a touch typist when keying *numbers* using the numeric keypad?

11 How do you characterize your use of the system?

a Mandatory (I must use it)

b Discretionary (I could select another product if I desired)

12 Who had the most influence in deciding whether to secure the system for your organization?

a Someone else did

b I did

13 How much pressure are you usually under to complete your work while using the system?

a Extreme (almost unbearable) emphasis on getting the work done fast b Much emphasis on getting the work done fast

c Some emphasis on getting the work done fast

d Little or no emphasis on getting the work done fast

14 How much accuracy is required when you are using the system?

a Extreme emphasis on having perfect accuracy; no errors are tolerated

b High emphasis on having high accuracy; very few errors are tolerated

c Some emphasis on accuracy; a few errors are expected and tolerated

d Little or no emphasis on accuracy; most errors are ignored

15 On average, which of the following best describes the number of interruptions you experience when working the system?

a Numerous interruptions (about one every 5 minutes)

b Many interruptions (about one every 15 minutes)

c Some interruptions (about one every 30 minutes)

d Few interruptions (about one per hour or less)

16 How would you rate the size of the text characters that you must read while using the system?

a Some are too small to be read accurately

b A little small

c About right

17 If English is not your native language, how many years have you been able to read English?

18 What is your highest degree?

a No degree

b High school

c Associates

d Bachelors

e Masters

f Doctorate

19 What is your age?

20 What is your sex?

a Female

b Male

21 How do you rate your typing speed?

a Less than 30 words per minute

b 30 to 60 words per minute

c 60 to 90 words per minute

d More than 90 words per minute

22 How do you rate your typing accuracy?

a Many errors

b Moderate errors

c Few errors

d Very few errors

23 Which is your preferred hand when eating and writing?

a Right hand

b Left hand

24 Do you have difficulty discerning or discriminating among certain colors?

a Yes

b A little

c No

25 What disabilities do you have that make using the system more difficult?

a None

b Inability to use one or more fingers

c Inability to use one or both hands

d Severe vision problems

e Severe hearing problems

f Severe motor disability

g Other ______________

26 How often do you use the system's on-line help?

a Frequently

b Occasionally

c Seldom

d Never

27 How often do you use the system's user manual?

a One or more times a day

b A few times a week

c A few times a month

d About once every 3 months or less

e Never

28 Where do you keep your copy of the user manual?

a On my desk

b On a shelf or in a cabinet near my desk

c On a shelf or in a cabinet outside my work area

d Do not have one

29 Which of the following do you most prefer when first learning to use a computer system?

a Reading a paper document

b Working with an on-line tutorial

c Asking questions of experienced users

30 Which of the following strategies do you prefer when learning new software?

a Taking a formal training course (having others show me how to use the product)

b Using a step-by-step set of procedures that are contained in an on-line tutorial or paper document

c Using the product without formal training or a paper document and discovering (through trial and error) how it works

31 If you attended a formal training course on this product, how long did it last?

a Did not attend a formal training course

b One day or less

c Two days

d Three days

e More than 3 days

32 Do you learn better by listening or reading?

a Listening

b Reading

Exercise 9A: Estimating User Interface Characteristics

Purpose: Applying critical front-end factors to determine the potential differences in user interfaces.

Method: Describe the user interfaces for each of the following profiles.

Profile 1: The new system will be used daily by highly skilled users. The system will run on a mainframe, using character-based screens, and supports only command-line entries. Available programmers are highly skilled. The system is not being designed to improve user productivity, but to improve record keeping. The primary usability requirement is that all users be satisfied with the system.

Profile 2: The new system will be used three to five times a month by users that have a fair understanding of computers. The system will run under Microsoft Windows on a 80486, 66-MHz personal computer with 8 Mbytes of memory. The programming team is moderately experienced with Visual Basic (the programming language), the Windows style guide, and available toolkits. The using organization is expecting to moderately improve their productivity with the new system.

Profile 3: The new system will be used four times a year in the central office by people who are monitoring their own and their competitors sales performance. This is about the only time most of these users will ever touch a computer. The system will run on a powerful UNIX workstation with the capability of high-level graphics, direct manipulation (drag and drop), and document-centered processing. The programming team has limited object-oriented analysis and programming skills. The using organization is expecting to substantially improve their productivity with the new product.

Exercise 9B: Collecting User Profile Information

Purpose: To use the user profile questionnaire to collect and synthesize information concerning one class of potential users.

Method: Make photocopies of the user interface questionnaire. Find six people who are using the same software package (e.g., the same word processor) and have them fill out the questionnaire. Summarize their results into a one-page user profile.

Exercise 9C: Preparing a User Profile from Making Observations

Purpose: To provide an opportunity to prepare a user profile by observing people actually performing the work.

Method: Go to a fast-food restaurant (e.g., McDonald's, Burger King, Wendy's) and observe the person who is taking orders, taking the money, and entering the cost of items into the register. If the same person is not performing all three of these activities, then choose a person to observe who is entering the cost of items and making change for customers. After determining exactly what the person does, prepare a one-page user profile for the work being performed. What assumptions do you think designers made concerning the people who would be using the cash register?

Record the skills, knowledge, and attitudes required of the person doing the work. Determining this information may require an interview with the person being observed or a member of management. Remember, be as specific as you can. Try to prepare the user profile so that *another person* could use it to select the exact person now working in the job.

Reporting: Submit a one- or two-page user profile.

10

Input and Output Devices

Introduction

This chapter focuses primarily on the human–machine interface that typically exists in the design of commonly used equipment (e.g., automobiles, airplanes, control rooms, and computers). Sometimes issues surrounding these interfaces are considered the domain of the "traditional" human factors or ergonomics specialist.

Displays and Controls

An advertisement in a major magazine uses human performance considerations as a principal reason why people should buy the BMW automobile manufactured in Germany. Part of the advertisement reads:

> All seats are orthopedically molded; all individual seats are infinitely adjustable. Controls are within easy reach and all displays are instantly visible in an innovative three-zone control panel that curves out toward the driver in the manner of an airplane cockpit. So thorough is the integration of human and machine that the driver literally functions as one of the car's working parts.

As this advertisement correctly states, three of the most human performance considerations in a successful system are controls and the relationship of the displays and controls to each other and to the operator.

At times, the distinction between displays and controls becomes somewhat blurred. A light switch, for example, is a control, but the position of the switch conveys information about the status of the system (the electricity is on or off); hence it also functions as a display. Another instance is when automobile drivers derive critical information from controls, such as a "soft" brake pedal or a vibrating steering wheel. These controls are obviously displaying important information.

Flight 007

Many do not realize how important effective displays and controls can be to the success of a system. The following description of a terrible and shocking disaster helps to illustrate the tremendous importance of the human–machine interface. Among other things, this story is an example of misreading displays and misusing controls.

On September 1, 1983, a commercial airliner (Korean Airlines flight 007) was shot down by a Soviet military aircraft, killing all 269 people on board. Unknown to its flight crew during this middle-of-the-night flight, the plane had wandered off course. When hit by two air-to-air missiles, flight 007 was flying over Soviet air space. Consider some major events in the last 2 hours of the flight before it crashed into the Sea of Japan. The times begin at 1:30 A.M.

1:30 Flight 007 enters Soviet airspace for the first time (over Kamchatka Peninsula).

1:37 Four Soviet interceptors take off and try to locate flight 007; they never do and eventually return to base.

1:58 Flight 007 flies back into international airspace (over the Sea of Okhotsk) but on a direct course for Soviet bases on Sakhalin Island.

2:44 Flight 007 is picked up by Sakhalin radar.

2:56 The first Soviet interceptor from Sakhalin is airborne (at least six interceptors, three SU-15s and three MIG-23s, were eventually involved).

3:12 One Soviet SU-15 pilot reports that he can see the target visually (the aircraft's running lights?) and is ready to fire if so ordered (2½ hours after first being seen on Soviet radar).

3:13 The SU-15 pilot was told to try to electronically identify the intruder (using IFF) to at least make sure it was not a Soviet aircraft. (Radio contact was not possible because Soviet interceptor aircraft are equipped only with UHF radios, while commercial airliners use HF or VHF radios.)

3:15 The SU-15 pilot reports no electronic response from flight 007 and that his weapons are switched on and ready to fire.

3:16 Flight 007 again enters Soviet airspace (over Sakhalin); the copilot of flight 007 asks Tokyo air traffic control for permission to climb from 33,000 to 35,000 feet. The SU-15 pilot is told to try to signal the intruder and force him to land.

3:17 The SU-15 pilot is asked by his ground control, "Do you see the target or not?" He responds that he can see the flashing navigation lights.

3:20 The SU-15 pilot is told to make one final attempt to signal the intruder with cannon fire and, if unsuccessful, to shoot down the aircraft. The Tokyo air traffic controller gives permission to flight 007 to climb to 35,000 feet.

3:21 The SU-15 pilot, flying about 1 mile behind and 3000 feet below flight 007, tries to attract its attention by firing four bursts from his cannon (about 120 shells). At the distances involved, the shells would not have been bright enough to be seen by the flight 007 crew.

3:22 Flight 007 slows as it climbs to 35,000 feet, causing the SU-15 interceptor to almost fly past. The Soviet pilot angrily complains to ground control that he has missed an opportunity to fire.

3:23 Flight 007 reports to Tokyo air traffic controller that he is at 35,000 feet. The SU-15 continues to drop back in preparation for a strike.

3:25 The SU-15 pilot reports that he is once again locked on the target.

3:26 Flight 007 is only about 1 minute away from being back in international airspace (over the Sea of Japan). The Soviet SU-15 pilot fires two air-to-air missiles and then reports, "The target is destroyed."

3:27 The flight 007 copilot tries to radio Tokyo, but only a few words are intelligible: "rapid decompression . . . descending to one zero thousand. . . . " There is no indication that they knew they had been hit by a missile. Flight 007 makes a semicontrolled descent for about 8 minutes, down to 16,000 feet, and then all control is lost as it crashes into the sea.

The cause of this disaster can be tracked back, at least partially, to a data-entry error made several hours earlier, an error made before flight 007 ever left the ground. Prior to taking off, the first crew member of flight 007 to enter the cockpit was the flight engineer. One of his responsibilities was to turn on the aircraft's inertial navigation system (INS) and to enter the aircraft's present position (both latitude and longitude).

There are three INS units in the aircraft, a main unit for the pilot and two backup units for the copilot and flight engineer. Normally, only the pilot's INS unit controls the plane; the other units are used for monitoring purposes. The flight engineer needed to enter the exact same information separately into each unit.

He began with the pilot's INS unit and apparently entered the current position as W139 degrees longitude instead of W149 degrees. This placed the plane about 300 miles east of its current location. No error was detected by the system; the computer could detect entry errors of latitude, but not longitude.

After making all the necessary entries, the pilot's INS unit was switched to another mode and the latitude and longitude coordinates were no longer displayed. The engineer then began putting the correct coordinates into the copilot's INS unit, causing an amber warning light to come on. The computer had detected a mismatch between the erroneous W139 entered in the pilot's unit and the correct W149 entered in the copilot's unit. Because the warning came on while entering numbers into the copilot's unit and because he could see that the numbers he had entered were correct, he assumed that the system was "acting up" again, and *turned off the warning light* (i.e., deleted the error message).

The INS unit's extreme sensitivity to changes in latitude, sometimes caused by changes to the aircraft's position while being serviced or changing ramps for passenger loading, occasionally makes the computer reject correct coordinates. A shift of the aircraft by only a few feet can cause the amber warning light to turn on. In such cases, the flight engineer can either turn off the warning light or take the additional time to find out exactly what is wrong. The latter course may end up requiring him to reenter *all* the numbers again.

After finishing the copilot's unit, the flight engineer proceeded to enter the numbers into the third (his own) INS unit. When he entered the W149 and the computer again found a mismatch with the original W139, the system did *not* send another error message by turning on the amber light. If the error message is turned off once, the system is designed to not react to any further discrepancies among the INS units.

Now all three INS units seemed to agree. However, the pilot's unit, which was the only one actually used when flying the aircraft, was different from the other two. The first steps had been taken that would cause flight 007 to fly a different course than the one intended. Consider some of the user interface issues:

1. The flight engineer had made a data-entry error by substituting a 3 for a 4 (was it correct on the original flight plan?).
2. The data-entry device was constructed so that this type of error could occur (was it a preparation or keying error?).
3. The user did not self-detect the error (was feedback adequate?).
4. The computer did not detect the error (is it possible to do so?).
5. Once another entry was made (to the copilot's INS unit) so that the computer could match the two entries, the system detected the error and sent a warning or error message (the amber light).
6. The warning display (i.e., the error message) was too general; it did not indicate exactly what was wrong and what should be done. In the absence of precise information, the flight engineer mistakenly guessed at the source of the problem and then turned off the warning (should it be possible to turn off a warning without fixing the cause?).
7. The system was designed to *not report* other errors once a user had turned off the amber light (should a system stop reporting future errors after a user has neglected to respond to one?).
8. Shortly after takeoff, the pilot made a navigational error when making a slight change to the flight plan and did not detect or correct the original error.
9. Neither of the critical errors was detected, at least partially because
    a. The pilot left the cockpit within a few minutes after taking off, leaving all monitoring to his flight crew.
    b. The feedback being received by the flight crew seemed to be consistent with information on their INS units (they were never aware that their backup units did not match the pilot's main unit, the unit that was flying the airplane).

In flights like this, there is little to do after takeoff and before landing except to periodically monitor the instruments. Because the flight crew expected the airplane to be accurately programmed and most information that was being received confirmed this expectation (at least on their displays), all indications by the computer that the plane was off course were either not seen, not understood, or ignored.

The original data-entry error, when combined with system design errors (e.g., a poor error message, the ability to turn off all subsequent warnings), a subsequent navigational error, and a low alertness level by the flight crew combined to help produce a disaster that shook the world.

Visual Displays

This section covers some major human factors considerations in the design of visual displays. The requirements for displays are developed by combining characteristics of human information processing with information gathered during the task analysis. In fact, determining the need for a display, as well as the types of information each display should present, is an integral part of the task analysis process. The most common visual displays are CRTs, dials, indicator lights, digital readouts, computer printouts, and paper forms.

Task-related Considerations

To design good displays, a designer should know as much about the tasks to be performed as possible. Designers should consider carefully the way each item on the display will be used, its importance, when it will be used, and specific characteristics of users, such as visual acuity or skill level. Different uses of the same display during normal operation versus an emergency situation must also be considered. It is usually a good idea to obtain a sample group of potential users to test prototypes of the displays. The tests should be conducted in a context that reflects as close as possible the real-world uses of the displays.

Selecting the correct type of display for a given application is one of the most important decisions a designer can make. First, the designer decides what sense to use. If, for example, the eyes are already heavily occupied (as when driving an automobile in heavy traffic), then it is probably best to select a display that uses hearing or touch. If, however, the eyes are generally free and the task requirements tend to favor visual information, then visual displays may be advantageous. The designer should not make this decision unless he or she has a good overall idea of the required activity.

Favor visual presentation of information if

- The auditory sense of the user is overburdened.
- The message is complex and/or long.
- The message deals with a specific location on a panel.
- The message must be referred to later.
- The user works primarily in one location.
- The receiving location is so noisy that some auditory messages may be missed or misheard.

Other Major Considerations

Three of the most important variables we should consider when designing or selecting visual displays are the following:

1. Type of display (based on what the user needs to know)
2. Information content and format (the information should be directly useful)

3 Physical characteristics (character design and size, background brightness, density or clutter, etc.).

Electronic Displays

There has been an expanded use of electronic displays for the presentation of a wide variety of information. Cathode-ray tubes (CRT), light-emitting diodes (LED), liquid-crystal displays (LCD), gas-discharge panels, and other types of light-emitting displays are being used for information displays varying from single-character displays to those with 40 or more lines of data, to maps and pictures.

Probably the most popular is the CRT. Research and development have provided a broad base of application for the CRT. The primary advantages of using CRTs are high writing speed, high resolution, simple addressing, full color capabilities, full range of gray scales, storage capability, and a large range of screen sizes.

Disadvantages of CRT use include bulkiness of the equipment, curvature of the screen, high voltage required, relatively delicate equipment (vacuum tubes), and limitations of maximum screen size. The development of CRT flat panels is helping to alleviate some of the problems.

Many display technologies are available to designers. Eight of these technologies will be briefly described, and then a procedure is presented to assist designers in selecting the most appropriate technology. The designer can make similar types of trade-offs for display technologies not discussed here.

Cathode-ray Tube (CRT)

As indicated, the CRT is the workhorse of displays and has been adapted or modified to meet many requirements. It is fairly reliable with an acceptably long life and is produced quite inexpensively. Among knowledgeable system designers, however, the CRT is chosen for numerous applications because of its tremendous flexibility. CRTs are available in a variety of sizes and shapes, provide gray scale and color, can have reasonably good resolution, can provide storage capability, and can be addressed with both raster and stroke patterns.

Flat-panel CRT

The conventional CRT has great flexibility in information display. However, a major disadvantage in some applications is its depth; as the displayed image size is increased, so is the length of the tube. Flat-panel CRTs have been developed that are as thin as 5 centimeters.

Light-emitting Diode (LED)

The LED has been used successfully in calculators, wristwatches, instrumentation readouts, and discrete miniature lamps. Its popularity is based on the combination of good luminance, low cost, low power, high reliability, and good compatibility with integrated-circuit technology. Larger one- and two-dimensional arrays of LEDs have been developed for message readout and graphics displays.

Electroluminescence (EL) Panels

The EL (more properly field-excited electroluminescence) display is potentially compatible with requirements for alphanumeric readout and graphics. EL panels can produce a reasonable gray scale and dynamic range, have a wide acceptance angle for viewing, can be fabricated in sizes ranging from a few centimeters to greater than a meter, and are potentially capable of very high element density.

Plasma Displays

The plasma display panel offers a good alternative to the CRT in some applications, possibly for largescreen picture-on-the-wall television. Plasma displays have been used for alphanumeric readouts in single rows, full screens, and matrix panels for graphics and video applications.

Liquid-crystal Displays (LCD)

LCD technology is one of the most popular and most developed of the flat-panel display types. Rather than emitting light energy, the LCD controls or modifies the passage of externally generated light. For many applications, the LCD is a good choice for a single- or multiple-character alphanumeric readout. The characters can be made any size, the contrast is typically adequate, costs are very low, and voltage and power requirements are compatible with battery sources. Graphic and video displays, some using full color, are also being used.

Mechanical Displays

Digital Displays

Displays frequently are needed to present quantitative information. For reading precise, static, or slowly changing numerical values or making exact numerical settings, a digital display, such as that shown in Figure 10-1, is usually preferred. Digital displays

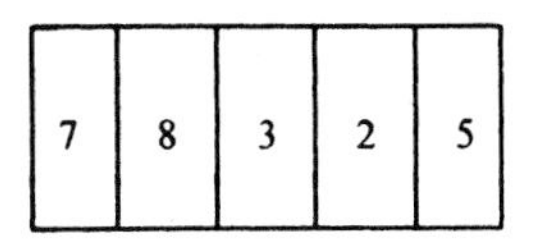

Figure 10-1 Digital display.

require little space, and the amount of accuracy available is usually only limited by the number of digits displayed. However, there are some disadvantages:

- Determining the rate of change is difficult.
- Reading rapidly changing display values is difficult.
- Interpolating is difficult when two numbers are partially visible in a window.
- Gauging distance to a boundary (control limit, danger zone) is difficult.

Fixed-scale, Moving-pointer Displays

Where interpolation between numbers is required or rate or trend information is important, a fixed scale with a moving pointer is usually preferred. The fixed-scale, moving-pointer display is particularly good for making check readings, and it usually has a simple and direct relation between pointer motion and motion of a setting knob. This type of display comes in various forms, as shown in Figure 10-2.

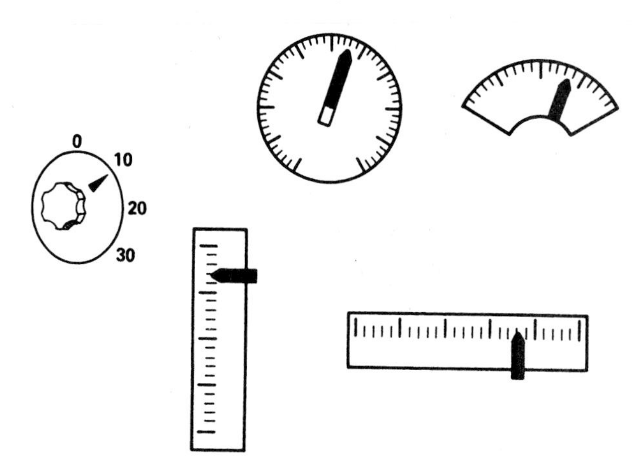

Figure 10-2 Examples of fixed-scale, moving-pointer displays.

A moving pointer against a fixed scale is also good for qualitative check reading. If several dials are to be scanned rapidly, orient pointers so that the normal position of the pointer is at 9 o'clock (Figure 10-3). Remember to orient dial scales so that the increase in range will be read from left to right or from bottom to top.

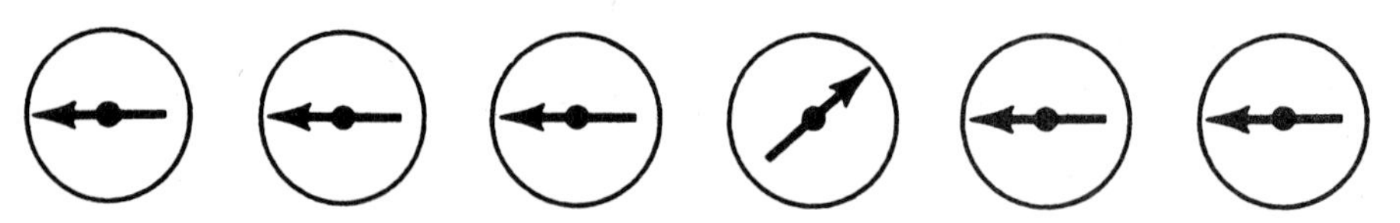

Figure 10-3 Preferred layout for check reading several displays.

Two disadvantages of fixed-scale, moving-pointer displays are that they frequently require more panel space than some other types of displays, and the scale length is fixed and thus limited.

Information Coding

The information format of a display is determined by how the information needs of a user can be best presented. Does the information have to be an exact replica of the source, or can it be coded? If coding is necessary, there are many options.

Color Coding

Frequently, displays require a certain amount of coding. Color coding is widely used, with colors representing various categories of information.

Colored lights are generally used in situations that change and where the environment is not fixed, such as changes in the condition or status of various pieces of equipment. Lights can also flash on and off as an attention-getting device. Brightness and rate of flash are only moderately useful when coding information, so it is best to think of lights as having only two major properties that can be varied: color and on–off status.

Some standard practices dictate the use of certain colors to code information. Generally, red represents a warning. Amber indicates caution and green indicates that a system is operating normally or is in a ready or available status. Therefore, these colors should *not* be used to indicate conditions other than those that they typically represent.

The introduction of color CRT has led to more frequent uses of color coding. Designers of color displays should consider the interaction of a label's color with the color associations of words in the label. *Grass,* for example, is often associated with *green* (Warren, 1974). Scheibe et al. (1967) found that a mismatch between the color in which a word is presented and the color with which it is commonly associated tends to slow down identification of its color. This sort of mismatch also slows down reading, may impede the performance of concurrent activities, and can contribute to misidentification of a word or message. To ensure rapid and accurate user response to displays, it is important to avoid such color mismatches.

To illustrate the extent of today's use of inconsistent color coding schemes, a study of color coding of displays in nuclear power plant control rooms found that red denoted "on" or "flow" and green denoted "off" or "no-flow" (Parsons et al., 1978). In addition, the investigators found that this use of red–green coding had been mingled with military and other color coding schemes to the point where almost every designer used his or her own preferred scheme. To avoid this kind of confusion, designers should rely on standard color coding schemes.

Warren (1980) conducted a study to determine the degree of association between color and words in alarm and status labels. With few exceptions the words had one or more colors strongly associated with them. These strong color associations should be considered when displays are developed, both to help user performance and to avoid user identification errors. Table 10-1 recommends the color to use when presenting certain words. For example, CRITICAL, ALARM, and POWER should always appear in red; ON, RUN, and ACTIVE in green; and STANDBY, MINOR, and AUXILIARY in yellow.

Table 10-1 Recommended Colors for Alarm and Status Words

Word	Color
active	green
alarm	red
clear	white
critical	red
disable	red
emergency	red
enable	green
failure	red
major	red
minor	yellow
normal	green
off	black
on	green
on-line	green
power	red
run	green
standby	yellow
stop	red

Adapted from Warren, 1980.

Color Associations for Users

Courtney (1986) had a group of people from the United States indicate associations between certain colors and a limited number of common concepts. The results were as follows:

Strong associations (percent agreement)

- Red for stop (100 percent)
- Green for go (99 percent)
- White for cold (96 percent)
- Red for hot (95 percent)
- Red for danger (90 percent)
- Yellow for caution (81 percent)

Weak associations

- Green for safe (61 percent)

Red for on (50 percent)

Blue for off (32 percent)

There are two cautions to be observed when using color coding in displays. First, a relatively large percentage of users (particularly men) will be color blind or color weak. Second, the color associations in different cultures may differ from those shown in Table 10-1. For example, a study of color associations in China (Courtney, 1986) showed a preference for red to indicate "on," blue to indicate "off," and white to indicate "cold." Also, the study showed that, even where there was general agreement between the two cultures, the *strength* of the preferences could differ. For example, nearly 100 percent of a United States population preferred the color red for "stop." In the Chinese population, only 49 percent agreed on the color red for "stop" (the other major contenders were black, yellow, and white).

Finally, designers should not use color as the only means of coding. Color should always be used with other forms of coding (refer to the example in Figure 10-4). To be effective, color coding requires an adequate amount of light, preferably white light, to show the true hues. If the visual displays are not lights themselves (e.g., signal lights), be sure that adequate lighting will always be available when the displays are used. For example, if people need to read maps or charts in an aircraft cockpit, automobile, or truck cab under subdued lighting or even red lighting (to help preserve night vision), the color coding will not convey information well.

Size Coding

Size coding is another way to display various information categories. A small square, for example, might be used to represent one data category and a large square might represent another. Dot coding is useful in representing multiples of the same category. One dot on a map might represent a population of 3000, two dots might represent 6000, and so forth.

Geometric Shape Coding

Geometric shapes can also be used to code various categories of data. Circles and triangles provide good discrimination; good road maps make use of these symbols. Pictorial representations provide effective category distinctions. For example, the number of telephones installed in a given period could be shown by a telephone silhouette and a number associated with it. Pictorial representations such as icons are used effectively on CRT displays in some applications.

Numbers, Letters, and Words as Codes

Much of the information on visual displays is in the form of words, numbers, or some combination of letters and numbers. A problem in some visual display design situations is keeping the number of words to a minimum to help reduce the display size or to make the most effective use of available space. Codes including abbreviations can reduce display size. However, if codes for items must be used, these codes should be clear and meaningful to users. The less meaningful the code, the more training required to learn that code, and the greater the number of errors that can be expected when it is used.

Redundant Coding

Differences in meaning can be emphasized by using *redundant coding* methods, such as position, color, or labels. Figure 10-4 shows a familiar display using redundant coding. Redundancy with the stoplight, for example, includes color (red), position (top), and the word STOP.

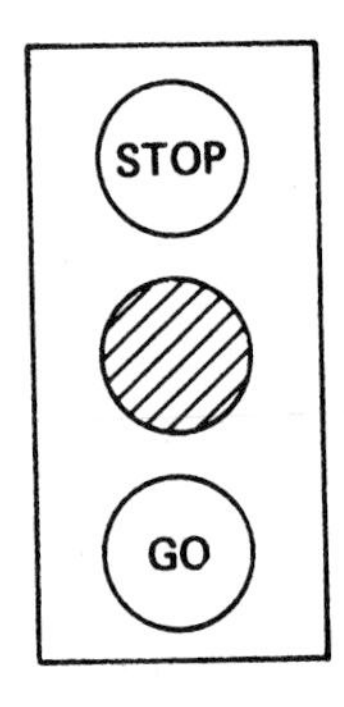

Figure 10-4 Traffic light showing redundancy.

Limit Coded Information

Designers should use warning lights sparingly to avoid the Christmas tree effect. Too many lights on a panel make it difficult to determine what any one of the lights means. During the 1979 Three Mile Island nuclear power plant accident, over 100 alarms, most of them visual, went off with no way of suppressing the unimportant ones and identifying the important ones (Kemeny, 1979). Therefore, it is usually desirable to use lights only in situations where a preferable way to give the user information is impossible.

If both information and warning lights are used on a panel, the warning lights should be a different color or three to five times brighter than the information lights.

In addition, warning lights should be well within the user's range of vision. We can learn a lesson from the Three Mile Island accident. Some of the key visual displays that would have helped early diagnosis of the problem were actually located on the back of the control panel (Kemeny, 1979). Important warning displays should also be located within the primary visual area, the area where the user is looking most often.

The effects of having too much coded information are pronounced when work load is high and exposure time is short. It is important not to overburden a user with cluttered displays. When considerable data are presented in a visual display, the display should be formatted so that a user can easily locate and read relevant information.

Selecting a Coding Technique

Several factors are involved in choosing the appropriate coding technique in display design. The first is the kind of information to be displayed. A lot of information usually requires an alphanumeric code. On the other hand, the status (on–off) or the presence or absence of trouble usually requires a light.

A second factor is the amount of information to be displayed. Some codes are better than others in yielding a maximum number of items that can be reliably represented. By assessing both the kind and amount of information to be displayed, a designer can make an initial estimate of the coding possibilities and eliminate those that appear unacceptable.

A third factor, the space requirement for the code, may be critical. To use an extreme example, a designer could indicate danger or a malfunction by a red warning light or a message in large letters spread across a CRT screen. Obviously, the red warning light will occupy much less space and would be the signal to use if space is at a premium.

Another factor to consider is ease and accuracy of understanding. The coded information should be understood promptly and accurately. This means simplifying display coding. The greater the amount of data on a display, the more difficult and time consuming it becomes to search out relevant codes. When a great deal of data is displayed, the user may fail to detect some of them or may misinterpret irrelevant data as relevant. Errors of this type are likely to increase as the amount of data increases, available search time decreases, and the user's work load increases.

A final consideration in the selection of coding techniques for displays is the interaction among displays at any given time. Designers should consider the possibility of other displays causing distraction. Warning lights should quickly alert users to something wrong—this is a useful distraction. However, a status light continually presenting "power on" should not interfere with the reading of other displays.

The designer must consider code compatibility and code discriminability. Code *compatibility* is good correspondence between the data to be coded and how they are coded. For example, exact quantitative data, such as the number of telephones installed, should be displayed by numbers. On the other hand, a qualitative item, such as the functioning mode of a system, should be represented by a coding technique that has appropriate qualitative connotations, such as red for danger.

Code *discriminability* permits the observer to distinguish one coded value from another. This requires recognition of the word, character, or symbol used for coding. The design, familiarity of the character or symbol, and the number of distinct categories used all affect recognition.

Physical Characteristics

Many nondigital status displays have similar physical characteristics. To best understand these characteristics, a designer should be familiar with the following terms (see Figure 10-5):

Scale range: the numerical difference between the highest and lowest value on a scale.

Numbered-interval value: the numerical difference between adjacent numbers on a scale.

Graduation-interval value: the numerical difference represented by adjacent graduation marks.

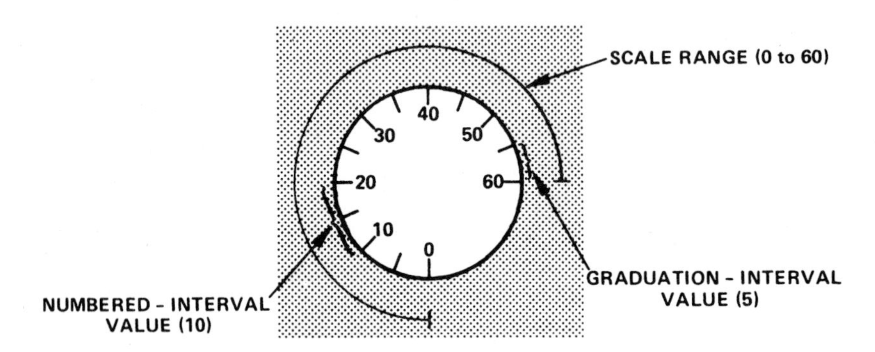

Figure 10-5 Important scale characteristics.

Scale Selection

Before selecting a scale for an indicator, a designer should decide on the appropriate scale range and should estimate the reading precision required. Figure 10-6 gives examples of different levels of scale precision.

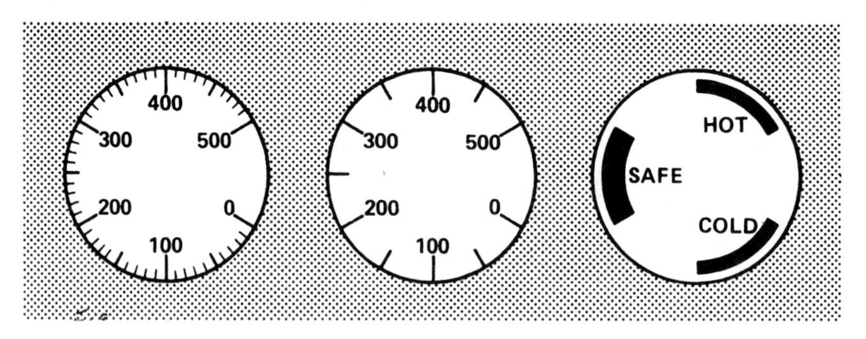

Figure 10-6 Examples of scale precision (adapted from Van Cott and Kinkade, 1972).

If possible, all displays should indicate values in an immediately usable form so that users need not perform a mental conversion. Transformed scale values can be found in jet aircraft engine tachometers that have been calibrated in percent rpm rather than actual rpm. For the pilot, this has several advantages. Maximum rpm differs for different engine models and types. Transforming the scale values into percent rpm relieves the pilot of the necessity of remembering operating rpm values for different engines. In addition, the range from 0 to 100 percent is more easily interpreted than a range of true values, such as 0 to 8000 rpm, and the smaller numbers on the dial make a more readable scale. In Figure 10-7 the two tachometers illustrate these advantages. The dial on the left can be read more easily and precisely than the one on the right.

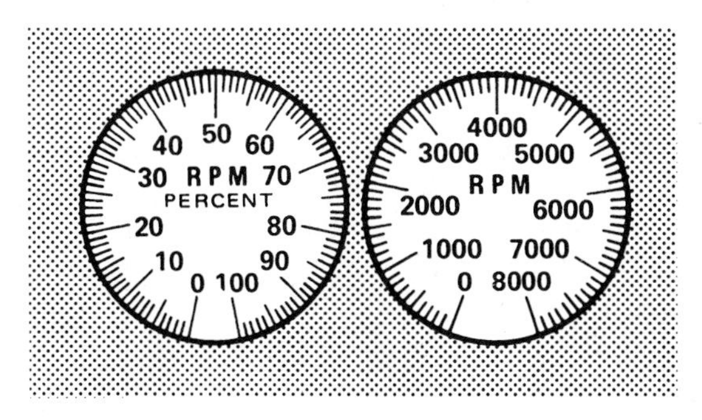

Figure 10-7 Sample tachometer dials.

Scale Design

There must be enough separation between scale indexes to make reading easy. In addition, cues should be provided for determining differences between major and minor graduation marks. More specific recommendations for scale dimensions depend on the illumination level at the dial face.

The following recommendations apply to indicators that are reasonably well illuminated. Assuming high contrast between the graduation marks and dial face, adequate illumination levels on the dial face, and reading distances of 13 to 28 inches, the following recommendations should be observed (see Figure 10-8):

- The minimum width of a major graduation mark should be 0.0105 inch.
- Although graduation marks may be spaced as close as 0.035 inch, the distance should not be less than twice the stroke width for white marks on black dial faces or less than one stroke width for black marks on white dial faces.
- The minimum distance between major graduation marks should be 0.5 inch.
- The height of major, intermediate, and minor graduation marks should not be less than 0.22, 0.16, and 0.09 inch, respectively.

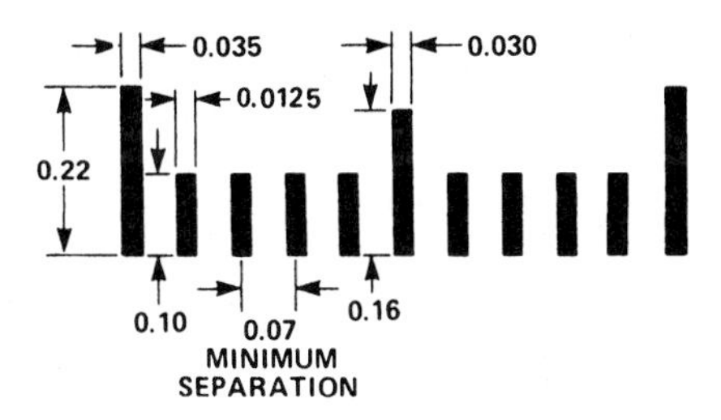

Figure 10-8 Recommended scale dimensions for high and low illumination (adapted from Van Cott and Kinkade, 1972).

When indicator scales must be read in lower than normal illumination, the additional aid of varied stroke widths of major and minor graduation marks becomes important. The recommended minimum dimensions shown in the bottom illustration in Figure 10-8 apply to scale design for low illumination levels. These dimensions should not be considered as fixed values in the sense that other factors, such as scale size, number of graduation marks, and importance of indication, are not given equal consideration, but should be considered as models for relative dimensions. For instance, we assume a reading distance of 28 inches for the dimensions in Figure 10-9; for other reading distances, assume a proportional increase or decrease in the recommended scale dimensions (see Table 10-2).

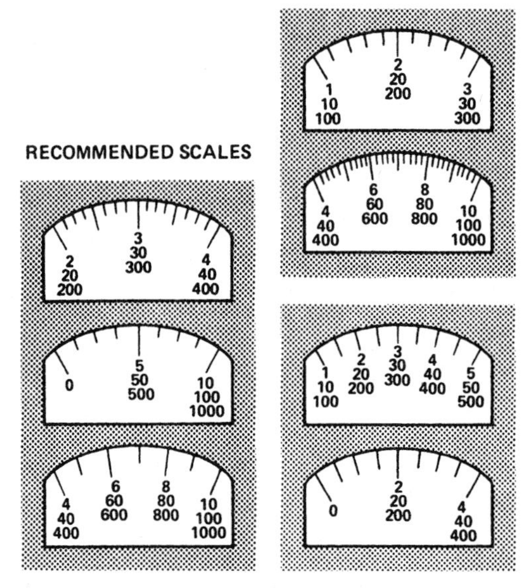

Figure 10-9 Examples of displays with acceptable graduation interval values and scale-numbering systems (adapted from Van Cott and Kinkade, 1972).

Interval Values

Some combinations of graduation-interval values and scale-numbering systems are more satisfactory than others. The following recommendations will assist the designer in selecting the most readable scale (see Figure 10-9).

- The graduation-interval values should be one, two, five, or decimal multiples thereof. Graduation interval values of two are less desirable than values of one or five.
- There should be no more than nine graduation marks between numbered graduation intervals.
- Normally, scales numbered by intervals of 1, 10, 100, and so on, and subdivided by 10 graduation intervals are superior to other acceptable scales.

- Ordinarily, scales should be designed so that interpolation between graduation marks is not necessary; but when space is limited, it is better to require interpolated readings than to clutter the dial with crowded graduation marks.

With this information in mind, the designer can select the most suitable interval values.

Table 10-2 Scale Numbering Recommendations

	Height (in.)	
Nature of Markings	**Low Luminance**	**High Luminance**
Critical markings, position variable (numerals on counters and settable or moving scales)	0.20–0.30	0.12–0.20
Critical markings, position fixed (numerals on fixed scales, control and switch markings, emergency instructions)	0.15–0.30	0.10–0.20
Noncritical markings (identification labels, routine instructions, any markings required only for familiarization)	0.05–0.20	0.05–0.20
For 28-inch viewing distance. For other viewing distances, increase or decrease values proportionately.		

Adapted from Van Cott and Kinkade, 1972.

Scale Interpolation

Quantitative scales should be designed for reading to the nearest graduation mark. For instance, a scale with a range of 50 read to the nearest unit should be numbered by 10s with a graduation mark for each unit, as shown in Figure 10-10(A). If less space was available for this scale, it would appear as in Figure 10-10(B). But the graduation marks on this scale may be too crowded for accurate and rapid reading, particularly under low illumination. Such situations call for a scale that may require interpolation as, for example, in Figure 10-10(C). This scale had a graduation-mark spacing that is more acceptable for low illumination. Also, this scale requires only a simple interpolation of one unit between graduation marks. Limited space might require interpolation in fifths or even tenths of a unit, but such interpolation could increase reading errors.

Number and Letter Size and Style

Designers should ensure that numbers and letters on indicator dials, panels, and consoles are as clear as possible, taking into account space restrictions and range of illumination. Recommendations in Table 10-2 and in the following paragraph apply to displays viewed under ordinary ambient lighting.

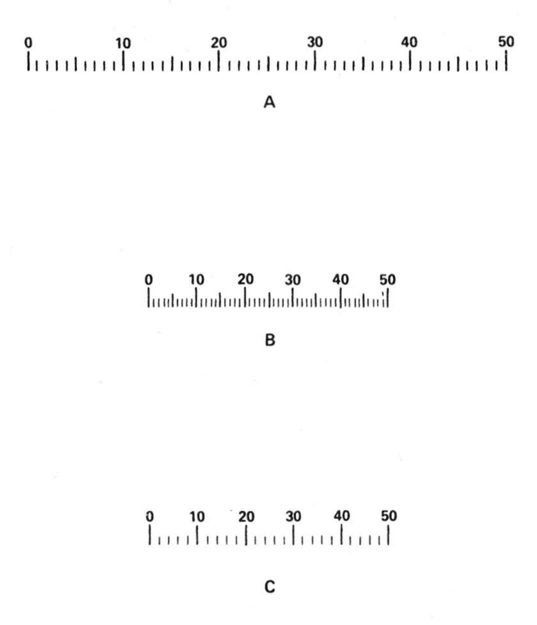

Figure 10-10 Sample quantitative scales.

The width of numbers should be three-fifths of the height, except for the 4, which should be one stroke width wider than the others, and the 1, which should be one stroke width wide. The width of uppercase letters should be three-fifths of the height, except for the *l,* which should be one stroke width, and the *m* and *w,* which should be about one-fifth wider than the other letters. For both numbers and letters, the stroke width should be from one-sixth to one-eighth of the character height.

Scale Layout

Numbers should increase in a clockwise direction on circular and curved scales, from bottom to top on vertical straight scales and from left to right on horizontal straight scales (see A in Figure 10-11). Except on multirevolution indicators, such as clocks, there should be a scale break between the two ends of a circular scale. When the scale has a break in it, the zero or starting value should be located at the bottom of the scale (B), except when pointer alignment is desired for check reading. In this case, the zero or starting value should be positioned so that the desired value is located at the nine o'clock position (C). The zero or starting value on multirevolution indicators should be at the top of the scale (D).

In general, on circular scales it is better to place numbers inside of the graduation marks to avoid constricting the scale. However, if ample space exists, place the numbers outside the marks so that they are not covered by the pointer (E). On vertical and horizontal straight scales, place the numbers on the side of the graduation marks opposite the pointer, and align the graduation marks on the side nearest the pointer and step them on

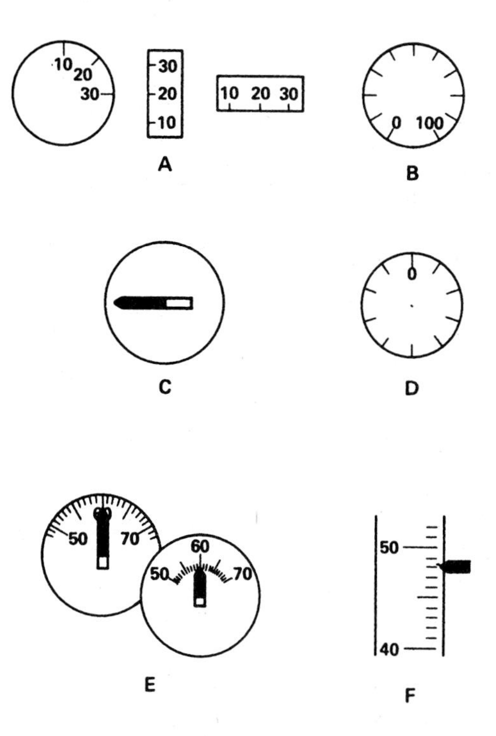

Figure 10-11 Examples of scale layouts (adapted from Van Cott and Kinkade, 1972).

the side nearest the numbers (F). The pointer should be to the right of vertical scales and underneath horizontal scales.

Zone Marking

Zone markings (Figure 10-12) indicate various operating conditions on many indicators, such as operating range (upper, lower), danger limits, or caution. These zone markings may be color coded.

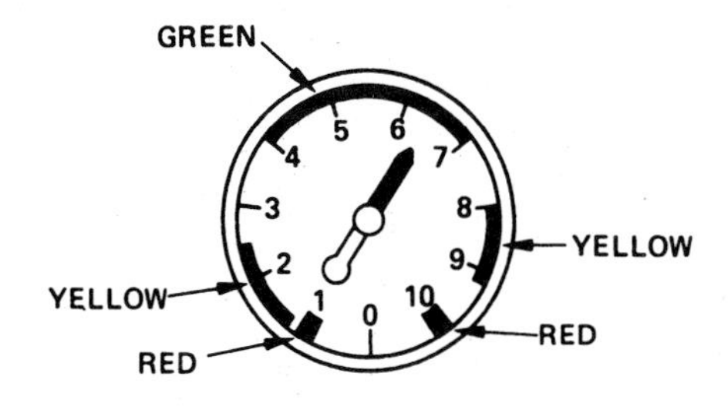

Figure 10-12 Example of zone markings (adapted from Van Cott and Kinkade, 1972).

Auditory Displays

In addition to visual displays, systems frequently use many auditory displays. Auditory display of information is preferred when

- The message is relatively short (yes or no, time of day).
- Response time to the message is important (get out fast!)
- The message need not be referred to later.
- The vision of a user is already overburdened (pilots, surgeons).
- The receiving location is not suitable for the reception of visual information (e.g., too light or too dark).
- The user's job requires considerable movement (visual messages require that the receiver must be looking where the message will be displayed).

A common use of auditory displays is for alarms and warnings. When designing this type of auditory message, a designer should consider the following alternatives:

- If the user will ever be a considerable distance from the auditory signal while performing other tasks, the signal should be loud, but low frequency (i.e., low pitched).
- If the user goes into another room or behind partitions, the signal should be low frequency.
- If there is substantial background noise, the signal should be of a readily distinguishable frequency.
- If an auditory signal must attract attention and the preceding design features are not adequate, a designer might consider modulating the signal to make it even more noticeable.
- The alarm should cease *only* after the user responds appropriately to the cause of the alarm.

Auditory displays can be used for continuous monitoring or tracking. Examples include sonar and various navigation technologies. Auditory displays are finding important use in biofeedback systems, in which changes in muscle tension level or skin temperature are displayed by a corresponding change in the pitch of the feedback signal.

Also, it should be noted that the sound made by any piece of equipment is to some extent an auditory display. Equipment users frequently depend on these sounds to understand the state of the system. Consider, for example, the sounds made by a disk drive when reading or writing a disk.

Controls

People use controls such as keyboards, mice, steering wheels, knobs, levers, push buttons, and toggle switches to interface with systems. Controls usually enable a user to make a change in the system and often are used with displays. A designer can determine the best type of control only after he or she knows exactly what is required of the user.

Most controls serve four kinds of functions:

1. *Activation:* an on or off switch or some other binary action, such as pressing keys on a keyboard.
2. *Discrete setting:* a control set to a position representing any of three or more discrete system responses. Automobile gear positions such as park, neutral, reverse, and drive fall into this category.
3. *Quantitative settings:* individual settings of a control device that vary along some continuous quantitative dimension. The volume adjustment on a radio is usually a quantitative setting.
4. *Continuous control:* constant control of equipment (e.g., steering an automobile, maintaining a constant water pressure).

Selecting Controls

Since the type of control used will vary with task requirements, designers must know how to select controls for particular situations. This selection can be done best only after a thorough task analysis.

When selecting an appropriate control, consider four major areas:

1. Function of the control
    a. What is its purpose?
    b. What does it affect?
    c. What must be controlled?
    d. How important is the control?
    e. Is it critical to the successful operation of the system?
    f. Does it provide only a minor adjustment?
2. Task requirements
    a. What degree of precision is required?
    b. How fast must the setting be made or the control activated?
    c. Is it an emergency control that must be reached immediately after a particular condition is noticed?
3. User information requirements
    a. What must be done to help the user locate the control?
    b. Is it a single control on a small panel, or is it grouped with many other controls on a large panel?
    c. What are the user's needs for determining the control setting or the effects of changed settings?

d How quickly must the user be able to determine its existing setting or enter a new setting?

4 Work layout

a Where should the control be located?

b How much space is available?

c How important is the control?

d How does its positioning affect operator efficiency?

Designers should distribute controls so that no one limb is overburdened. When accuracy and speed of control positioning are important and when it is not necessary to apply moderate-to-large forces, use hand controls instead of foot controls. And for precision and speed, use one-hand controls instead of those operated with both hands. Radio tuning, for example, is a fine adjustment task best suited to finger-operated rotary knobs. Even though it is best to operate steering wheels with two hands, two-handed controls most often work best when large forces are required.

Consider foot controls when

- The application of moderate-to-large forces (greater than about 20 to 30 pounds), whether intermittent or continuous, is necessary.
- The hands are overburdened with controlling other tasks.

Assign controls requiring large or continuous forward applications of force to the feet. Although a considerable number and variety of controls can be assigned to the hands, each foot should not have more than two controls assigned to it, and these should require only fore-and-aft or ankle flexion movement.

Coding Controls

After the designer selects the appropriate control, he or she must decide how a user will know what function it controls. Making controls easy to identify decreases the use of a wrong control and reduces the time required to find the correct control. Proper control coding not only improves user performance, but also reduces training time. There are a variety of ways to do this.

The five most common methods for coding controls are labeling, color, shape, size, and location. Several methods of coding should be combined to achieve maximum differentiation and identification. An emergency control labeled EMERGENCY that is larger than surrounding controls and is red would be more readily identified than if only one of these coding methods was applied.

The choice of coding method depends on the following six factors:

1 Total demands on the user when the control must be identified

2 Extent and methods of coding already in use

3 Illumination of the user's workplace

4 Speed and accuracy with which controls must be identified
5 Space available for the location of controls
6 Number of controls to be coded

Labeling

The simplest way to indicate a control function is to label it. Well-designed labels aid initial learning and subsequent performance by allowing the user to immediately identify the control being used to carry out a particular function. Essentially, such information tells the user that if he or she manipulates a control in a certain way it will produce a specific effect.

Observe the following general recommendations for labeling:

1. Locate labels systematically in relation to controls (usually above the controls).
2. Design labels to tell what is being controlled, for example, gear position, brightness level, function key, or number.
3. Make labels brief. If space permits, do not use abbreviations (e.g., CR for the return key) or icons (e.g., a bent arrow for the return key). If abbreviations must be used, use only common abbreviations.
4. Only employ unusual technical terms when absolutely necessary and only when they are familiar to all operators.
5. Do not use abstract symbols (squares, stars, etc.) when they require special training.
6. Use a letter and number style that is standard and easily readable under all conditions.
7. Locate labels so that they can be easily seen.

Labeling may allow rapid visual identification of the control. But it could require a large amount of panel space to avoid a cluttered appearance. When a user operates a control or positions other devices on the panel, he or she may obscure a particular control. Nevertheless, in many applications the advantages of labeling usually far outweigh the disadvantages.

As a rule, label any control that appears on a panel. A written description in the form of a label should tell the function and describe the setting of the control (if appropriate). Follow consistent labeling practices. Labels should be clear and unambiguous and present only essential information.

Labels should be on or very near the items they identify, and they should be placed horizontally to facilitate reading. When space is limited, place labels vertically only if they are not critical to safety or performance. Location of labels should be consistent—do not place a label above one display, but below another. Label every console or rack, panel, functional group, control or display, and control position with labels graduated in size, increasing approximately 25 percent from smallest to largest in the following order: (1) control position, (2) control or display, (3) functional groups, (4) panel, (5) equipment console or rack.

Place nonfunctional labels (nameplate, manufacturer, or part number) inconspicuously so that they will not be confused with operating labels.

Identify groups of related controls by enclosing them within a borderline and labeling the groups by common function. If there is a set of controls for tuning some device, these controls should be grouped together and enclosed by a thin line. At the top of the enclosing line, the label TUNING should also appear.

The designer must also consider letter height, width, spacing and style, and contrast between letters and background. If labeling alone is not enough to distinguish one control from another, varying the color, shape, or size of the control may prove useful as an additional coding technique. Vary the color of the panel surface to indicate that the controls located in that portion of the panel have a particular characteristic or that they control a particular portion of the system.

Color Coding Controls

Color coding is most effective when a specific meaning can be attached to the color, for example, red for emergency controls. As pointed out earlier, the use of color coding depends on ambient or internal illumination. Keep in mind that color should not normally be used as the *only* or even the primary method for coding controls.

The color coding scheme should be standardized for the equipment or family of related equipment. System-wide standardization is recommended so that users moving from one job to another can readily learn to do the new job. A function represented by a blue control on one job should also be represented by a blue control on another job in the same system. The same color code should be applied for all functionally related codes. Certain colors have associations that should be reinforced. Red, for example, should be reserved for emergency controls. Green is frequently used to indicate safety equipment of various sorts. Green may also be used for important or frequently used controls if consistent with other system controls. The connotation of green as a *go* condition or a *safe* condition must be kept in mind when applying green as a color code for controls.

Shape, Size, and Location

In addition to labeling and color coding, controls can be coded by means of shape, size, and location. Using different shapes and sizes is especially useful when controls must be identified without the use of vision. Using different shapes provides for tactile identification of controls and aids in identifying them visually. When feasible, select shapes that suggest the purpose of the control. In addition, it is important to use shapes that are easily distinguished from one another when touched.

Some keyboards, for example, make the F and J keys more concave than the other keys to help users know when they are properly located on the home row. Also, some numeric keypads put a small nipple on the 5 key to help users maintain the proper location.

Controls can be coded by size, but if the operator must rely on touch alone, usable sizes are quite limited. However, the ability to discriminate size by touch is relatively independent of shape discrimination; hence, a limited number of different sizes can be used with different shapes. Keep in mind that different sizes and shapes are less effective if the user is wearing gloves. When the user cannot feel all the controls before selecting the proper one, only two or at most three different sizes of controls should be used (small, medium, and large).

Designing Controls

Follow these general principles when designing controls.

- Critical and frequently used controls should be located within easy reach of a user.
- The force, speed, accuracy, and range of body movement required to operate a control should never exceed the capability limits of the *least capable* user. In fact, these performance requirements should be considerably less than the abilities of the least capable operator.
- The total number of controls should be kept to a minimum.
- Control movements should be as simple, easy, and natural as possible (i.e., they should conform to user expectations whenever possible).
- Control movements should also be as short as possible and exhibit the necessary "feel" to allow the operator to make accurate settings.
- When a control needs to be powered (provided with a mechanical advantage, perhaps hydraulically or electrically), artificial resistance may be required.
- Control actions should result in a positive indication to the user.
- Control surfaces should be designed to prevent the finger, hand, or foot from slipping.
- Controls should be designed and located to prevent or reduce the probability of accidental operation. When accidental operation would result in a critical situation such as personal injury, these controls should be provided with guards.

Special Controls

Hands and feet traditionally have been the only parts of the body used for operating controls. Theoretically, at least, any of a large number of body responses—movements of the knees, hips, elbows, shoulders, head, or eyes—could be used to actuate control devices. In fact, knee levers have been used for years as a standard control on sewing machines, and head movement is now used to aim guns in some aircraft.

Designers should be alert for situations where a user may not be able to use hands or feet. Some users may lack the use of one or more limbs either because they are multiple amputees or are paralyzed. Designers should take into account the potential for using less conventional control methods. A nod of the head, for example, could be used to type a character, start a motor, summon an elevator, open a refrigerator door, or turn on a television set.

Arrangement of Controls and Displays

Introduction

Even if all displays and controls are properly designed, unless they are organized in a logical manner, a user is likely to read the wrong displays or lose valuable time hunting for the proper control.

User Expectations

Probably the designer's most important consideration when arranging displays and controls is to make decisions that are consistent with what the user expects. With controls, for example, some direction-of-motion expectancies seem almost natural—pushing a throttle forward to increase forward speed or turning a wheel clockwise to turn right. Designers should avoid relationships between controls and displays, as well as between controls and vehicle motion, that yield an unexpected direction-of-motion relationship. When the user expectations are not obvious, a designer should conduct a study to determine what the expectations are.

Grouping

When using a large number of controls and displays, their grouping should aid in determining

- Which control affects which display
- Which control affects which equipment component
- Which equipment component is described by each display

Controls and displays are most commonly grouped by function and by sequence of use. In functional grouping, controls and displays related to one function are put together in one panel area. In addition, *sets* of controls and displays within functional groups can also be put together.

Grouping of controls and displays by sequence of use helps to provide efficient user movements and reduces the requirement of retracing or skipping around the panel. This type of grouping should provide for movements from left to right and from top to bottom.

Population Stereotypes

Ensure that the movement of a control leads to the most expected movement of a display or vehicle. The recommended direction-of-movement relationships are usually intended to ensure natural relationships. Natural relationships refer to control-movement habit patterns that are consistent from person to person without special training or instructions; they are responses that individuals make most often and are sometimes referred to as *population stereotypes.*

The following general direction-of-movement rules are applicable:

1. The direction of movement of a control should be considered in relation to (a) the location and orientation of the user relative to the control, (b) the position of the display relative to the control and the nature and direction of the display's response, and (c) the change resulting from the control movement, either in terms of motion of moving components (landing gear, automobile wheels, etc.) or in terms of some dimensional quantity (volume of a radio receiver or brightness of a CRT screen).
2. The preferred direction of movement for most hand controls is horizontal, rather than vertical, and is fore and aft, rather than lateral.
3. All equipment that the same person uses should have the same control–display motion relationship.
4. Control-movement relationships are particularly important when they result in vehicle movement. A movement of a control to the right should result in a movement to the right, a right turn, or right bank of the vehicle.

Placement

Displays and controls should be arranged so that a user can see or use them from a normal working position, without excessive shifting of the head or body. In addition, they should be arranged to elicit efficient patterns of movement. Displays and controls should be adequately identified so that they can be located quickly and without error. Their arrangement should provide expected direction-of-movement relationships. Finally, controls should be spaced far enough apart and away from adjacent structures to permit adequate grasp and manipulation through an entire motion range.

Computer-specific Controls and Displays

Introduction

As briefly discussed in Chapter 9, the quality of a computer's user interface depends (at least partially) on the characteristics of the computer, the operating system, the windowing system, and the networking system. Other computer-related issues include the primary and secondary input devices (e.g., keyboard, mouse, touchscreen) and the primary and secondary output devices (e.g., monitor, printers, speakers). All computer-related issues that limit the designers ability to create a truly optimal interface should be described in the *computer profile*. Note that the computer profile, like the user profile, imparts *constraints* on the product.

Most computer products offer a wide variety of different possible configurations. Some possibilities include the following:

- Sun SPARCstation, mouse, full keyboard, large color monitor, running X Windows and Motif 2.0
- Macintosh, mouse, full keyboard, color monitor, running System 8

- Pentium/120, full keyboard, mouse and SVGA color monitor, running Microsoft's Windows95
- 486/66, full keyboard, mouse and VGA color monitor, running OS/2 Warp, NetWare, and Common User Access (CUA)
- 286/20, full keyboard, character-based color monitor, running DOS

Because the computer environment in which typical users are working is usually less than the development environment, designers must be careful to limit their decisions to what most users have available to them.

Input and Output Devices

Two of the most important computer-related decisions include selecting the appropriate input and output devices. Many systems currently use a keyboard and mouse as the primary input devices and a monitor and printer as the primary output devices. But these devices may change depending on the functions to be performed and the types of users that will typically be working with the new product.

Common Input Methods

Keyboards

The familiar standard typewriter keyboard has been around for over 100 years and will continue to be used for many years to come. The arrangement of keys on the standard keyboard, also known as the QWERTY keyboard (for the first six letters on the top row of keys), is very similar to one introduced as early as 1872 and used on the first commercial typewriter. The origin of the QWERTY keyboard layout is not clear. However, a story that has become traditional is that the keyboard was originally designed to minimize the possibility of striking conflicting keybars. This was accomplished by separating those keys most often used on consecutive strokes. This means that the keyboard may have been designed to *slow down* keying.

The original keyboard seems to have survived largely because the particular manufacturer continued to hold sales leadership in the new industry. Other keyboards were introduced, but none of the manufacturers were successful enough to have their machines and keyboard arrangements accepted.

From time to time, various "improved" keyboards have been proposed, but none has gained widespread acceptance. Figure 10-13 shows the QWERTY layout plus two examples of alternative keyboard layouts. The QWERTY keyboard arrangement elicits an acceptable level of human performance in most situations. Several generations have learned to use this keyboard with reasonably good speed and accuracy.

Designers usually do not have the option of designing the basic keyboard in new systems. However, it is conceivable that in the future there will be a requirement to totally optimize the human–computer interface in some systems. At that time, it may be required to change the keyboard to match what we now know about the human performance of keying.

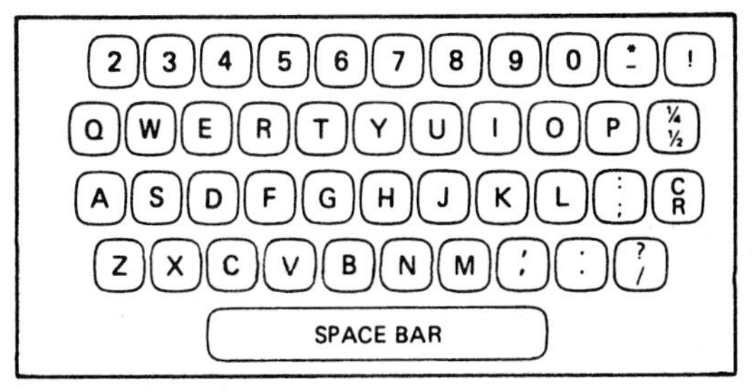

The Standard Qwerty Keyboard Layout

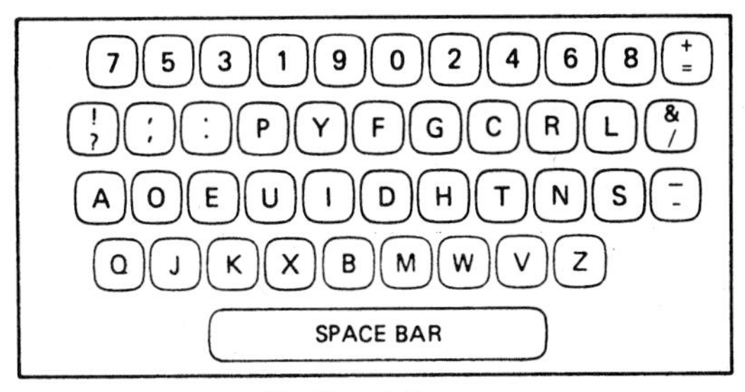

The Dvorak Layout

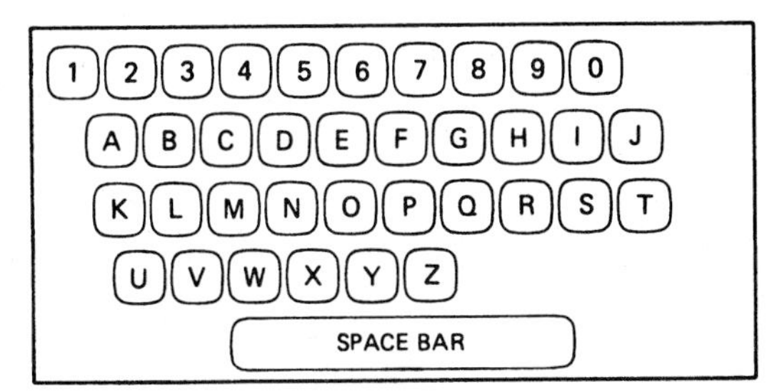

The Alphabet Layout

Figure 10-13 Examples of keyboard layouts.

Griffith (1949) conducted a study on the QWERTY keyboard and found several major areas where improvements to the keyboard would result in better human performance. He reported that on the QWERTY keyboard 48 percent of the motions to reposition the fingers laterally between consecutive strokes are one-hand motions rather than the faster two-hand motions. Making fast one-hand motions is more difficult than making two-handed motions in the same way that it is more difficult to beat a drum rapidly with one stick than with two. He reported that one-hand motions consume about 75 percent more time than two-hand motions.

The second problem Griffith identified is that the QWERTY keyboard requires reaching from the home row to another row for 68 percent of the key strokes. He felt that a well-designed keyboard would substantially reduce this reaching. When keying English text, it is possible to have over 70 percent of the strokes made from the home row. Another problem is that the QWERTY keyboard overloads the left hand with 56 percent of the total strokes. A well-designed keyboard probably should have the left hand carrying less than half the stroking load.

Probably the most well-known attempt to systematically design a keyboard that takes into account these concerns is the simplified or Dvorak typewriter (see Figure 10-13). This typewriter was designed about 1932 by August Dvorak and William Dealey. Several studies have suggested that the Dvorak keyboard may improve human performance (both speed and accuracy). Some studies, particularly a study sponsored by the General Services Administration of the U.S. Government (Strong, 1956), failed to show any advantage for the Dvorak keyboard.

With so many new computer systems being built and so few trained typists, a new question about keyboard layout has arisen. For these inexperienced people, it seems likely that a keyboard arranged in alphabetical order would be preferable to the standard QWERTY keyboard. It has been shown, however, that an alphabetic layout of the keys is no more advantageous than the standard QWERTY layout (Michaels, 1971). Performance on the two keyboards was essentially equal for *unskilled typists.* Hirsch (1970) also studied the typing performance of nontypists on the standard QWERTY keyboard and on an alphabetically arranged keyboard. His results suggested that an unskilled person actually could enter correct data faster on the QWERTY keyboard than on the alphabetic keyboard.

We will make one final note on keyboard design. In 1926 a German named Klockenberg published a book dealing with the design and operation of the typewriter. He described how the keyboard layout required the typist to assume postures of the trunk, head, shoulders, arms, and hands that were unnatural, uncomfortable, and fatiguing. Klockenberg suggested a number of improvements to the keyboard layout, some of which are still valid. Probably his most interesting idea was to separate the keyboard sections allotted to the left and right hands to alleviate tension in the typist's shoulders and arms (see Figure 10-14). Many of the newer split keyboards incorporate many of the same ideas. Kroemer (1972) published a study showing that this layout produced better performance.

Keyboard Guidelines

The following guidelines should be followed to obtain the best possible performance for full-size keyboards.

- *Key size:* The key tops should be square or slightly rounded and have a diameter of about 13 millimeters (mm) [0.5 inch (in.)].
- *Key separation:* The adjacent edges of keys should be about 6 mm (0.25 in.).
- *Key surface:* The key tops should be concave and treated to minimize glare.
- *Key displacement:* Key displacement should be at least 3 mm (0.125 in.).
- *Key resistance:* Resistance should be in the range from 2 to 4 ounces.
- *Key labeling:* Key symbols should be etched to resist wear and colored with high-contrast lettering.
- *Key mounting:* Keys should be securely mounted and firmly fixed in place to minimize horizontal movement.

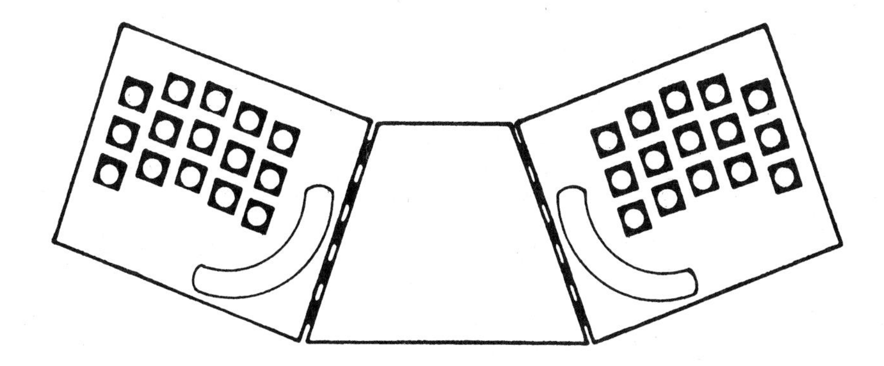

Klockenberg keyboard (1926)

Figure 10-14 The Klockenberg keyboard introduced in 1926, and the Maltron keyboard introduced in 1978.

- *Keyboard angle:* The keyboard should be capable of being adjusted up to an angle of 15 degrees from the horizontal.
- *Keyboard height:* The home row (ASDFGHJKL) of the keyboard should be located below the user's elbow height.
- *Highlighting:* Functional highlighting of the various key groups should be accomplished through the use of color coding techniques.
- *Palm rest:* A palm rest should be provided.

Telephone Keypad

Researchers (Marics, 1990; Blanchard et al., 1993) attempted to determine where the Q and Z should be included on current telephone keyboards. The studies found no reliable difference in user errors or speed among the alternatives. However, the majority of users preferred having Q on the 7 key, and Z on the 9 key. The Q and Z were assigned to these keys based primarily on preference (rather than performance) data.

Other Input Devices

Many other input devices are used with computer-based systems. Designers should select the input devices that will elicit the best performance. This is not always easy. For example, the most popular input device, other than the keyboard, is the mouse. One study (Karat et al., 1986) found that in some cases touchscreens provide much better performance than the mouse. A number of improvements have been made to the accuracy and quality of touchscreens (Potter et al., 1988).

In addition to the more familiar input devices, many others should be considered in the design and development of new systems. These range from foot-operated cursor-positioning devices (Pearson and Weiser, 1988) to speech recognition systems (Petajan et al., 1988). In some cases, even pictures can be used to navigate through information spaces (Egido and Patterson, 1988). There are many options, and designers need to evaluate the needs of users and then select the most appropriate device or devices.

Preferences for Input Devices

Mack and Lang (1989) compared input devices for nonentry tasks: mouse, stylus, touchscreen, and keyboard. Users performed tasks that included viewing, comparing, merging, copying, verifying, and modifying. They found no reliable speed differences between devices. However, except when comparing the touchscreen and keyboard (50–50 split), users had some clear-cut preferences:

88 percent preferred the mouse over the touchscreen

75 percent preferred the mouse over the keyboard

63 percent preferred the stylus over the mouse

63 percent preferred the stylus over the keyboard

Evaluating the Mouse

Hill et al. (1991) conducted an extensive evaluation of the mouse uses. They concluded that individual mouse buttons 1 and 2 were easy to use and were perceived as easy to use. Mouse button 3 was hard to use and was judged as difficult (even pressing mouse buttons 1 and 2 together was perceived as easier than pressing mouse button 3).

Mousing with the Preferred versus Nonpreferred Hand

Kabbash et al. (1993) studied users performing with the mouse, stylus, and trackball. The participants used either their preferred or nonpreferred hand to perform a pointing and a dragging task. Users were significantly faster (about 20 percent) with the preferred hand only when using the mouse and stylus. There was no difference in errors.

Touchscreen Interfaces

Touchscreens have received considerable research (cf. Plaisant and Sears, 1992; Sears et al., 1993). They found that typing using a touchscreen keyboard was much (25 wpm) slower than using a traditional keyboard (58 wpm). Also, users tend to touch below the target (Leahy and Hix, 1990).

Comparing Two-finger Typing and Handwriting Speeds

Brown (1988) evaluated two-finger typists. The main question was whether two-finger typists could write or type faster. The results showed the following speed and accuracy levels:

Memorized text

Writing: 31 wpm, 0.1 error

Typing: 37 wpm, 1.6 errors

Copying text

Writing: 22 wpm, 0.9 error

Typing: 27 wpm, 5.2 errors

Computer-based Writing Recognition Accuracy

Wolf et al. (1991) assessed writing recognition accuracy by a computer. She reported that discrete writing had a recognition accuracy of 92.6 percent. System recognition accuracy was reliably better for numbers (98 percent) and uppercase characters (96 percent), than for the full character vocabulary (90 percent). Adequate training for the system was achieved by having users only enter three samples per character.

If the computer attempts to train itself to recognize the handwriting of individual users, the accuracy of the automatic training is as follows (Santon et al., 1992):

	Percent Recognized
Novices (no experience)	57
10 minutes experience	91
3 hours experience	97

Gestural (Pen-based) Input

Wolf (1988) used a spreadsheet program to determine if participants were faster with gestural interfaces or a traditional keyboard interface. She found that gestures took only about 72 percent of the time taken with the keyboard. In a follow-on study, Wolf (1992) compared gestural, keyboard, and mouse interfaces. Users performed several Lotus 1-2-3 problems. She found that gestural interfaces were from 24 to 58 percent faster and more preferred.

Optical Character Recognition (OCR) Devices

Many organizations are using scanners to input textual information. These devices can considerably increase the speed and accuracy of input under certain circumstances (Cushman et al., 1990).

Overall OCR text entry is faster than manual entry only if 94 percent of the characters are correctly recognized.

To obtain finished documents with no more residual errors than typed documents requires 98 percent correct recognition.

Data Glove

Johnsgard (1994) had participants move from a start push button to a rectangle using both the mouse and a data glove. He found that the mouse was faster ($p < 0.001$) and had lower error rates ($p < 0.001$).

Foot-operated Mouse Pointer

Anderson et al. (1993) evaluated a foot-operated input device ("mole") for use with computers. The mouse emulated the functionality of a mouse. The mole was accurate enough to select individual characters, but was approximately 5 to 10 percent slower than the mouse.

Speech as Input

Martin (1989) observed that speech input should be used in situations that have short transactions and high interaction. Speech is less efficient in tasks with long think times. For example, dictating a letter is only about 50 percent faster than keying it, even though people can speak three to five times faster than they can key.

Rudnicky et al. (1994) point out that there are numerous uses of speech as input in systems. These include the following:

Speech as a shortcut: rather than opening a file by traversing many levels of hierarchy, such as saying "open budget."

Hands busy, eyes busy: such as changing the font style while a user is typing or changing a drawing tool while the user is drawing.

Information retrieval: for example, "Find all documents from John received after March 1."

Portable applications: As computers shrink in size from desktop to notebook to subnotebooks, the keyboards will be more difficult to use or even nonexistent. In this case, speech can be used as an alternative to small keys.

Discrete Speech

Biermann et al. (1985) report that the average speech rates is about 47 words per minute. The number of unique spoken words required to complete activities is very domain dependent and ranges from only 237 to 2550. This compares with using a written vocabulary of about 16,500 unique words.

Accuracy of Automatic Speech Recognition

Word-level accuracy rates using speech as input are as follows (Rudnicky et al., 1994):

Isolated words	Percent Recognized
Digits	99.6
Alphabet	96.0
Words (5000 words)	98.0
Continuous speech	
Query (1000 words)	97.0
Dictation (5000 words)	95.0
Dictation (20,000 words)	87.0

The accuracy of speech-recognition devices tends to decline during an extended period of continuous use (Frankish et al., 1991). Template retraining was conducted in mid-session. The retraining restored the accuracy level, which remained for the remainder of the activity.

Natural Language Interactions

Guindon (1988) studied how users interact with *on-line advisory systems* using natural language. Users asked questions (requested advice) about a task by typing the question into the computer. The results showed that users conversed differently with the computer than with a human listener. Their requests resembled spoken discourse more than written discourse. The participants used

- Short, simple utterances
- Few passive sentences
- Many fragmentary utterances
- Few pronouns (it, they, them)

Gaze as Input

It is possible to use gaze to make menu selections. Consider the following example (Jacob, 1990):

User: Eye focuses on the "File" item on the menubar (1 second).

Computer: The "File" pulldown menu appears.

User: Eye focuses briefly on each menu item (150 milliseconds).

Computer: Each menu item is highlighted.

User: Eye focuses longer on the "Open . . ." item (1 second).

Computer: Dismisses the menu and shows the "Open File" dialog box.

User: Eye scans the "Open File" dialog box for available files (150 milliseconds per item); then the eye focuses on the "Cancel" button (1 second).

Computer: Dismisses the dialog box.

Fitts's Law as a Tool in User Interface Design

Fitts's law suggests that big targets at close range are acquired faster than small targets at a distance. The formula has provided good performance predictions for the mouse, joystick, and others. However, it remains difficult for designers to use the formula to help in the selection of appropriate input devices when developing new systems (MacKenzie, 1992).

Computer Output Devices

Users should select the output device with characteristics most suitable to a particular application. Because numerous output devices are now available for use in computer systems, designers need to carefully evaluate user needs and then select the output device that is most appropriate.

Although cathode-ray tube (CRT) displays are still the most popular choices, numerous other possibilities exist. As discussed earlier, these include liquid-crystal displays, tactile displays, auditory displays (including automated speech generation), graphical displays (including three-dimensional graphics), and others.

A computer display should be designed to suit the particular conditions of its use. For example, the designer should consider the viewing distance. The maximum viewing distance will influence the size of the character heights and graphic heights shown on the display. Usually, designers assume about a 400- to 700-millimeter (16 to 28 inches) viewing distance from a user's eye to a computer's visual display.

CRT Displays

Designers should keep in mind that each condition involved in the design of CRTs presented next interacts with all other conditions. Statements that we can make regarding a particular condition are limited since the effect on the user of a given level of one condition depends on the levels of other important considerations.

- ***Luminance and brightness:*** Luminance is the radiant intensity of a surface measure in millilamberts (mL). The luminance level of the pages in this book in a well-lighted room is about 50 mL; however, any luminance above 25 mL or so is usually acceptable, assuming adequate contrast. This level may be used as an estimate of the recommended luminance for symbols on CRT displays. Brightness is the subjective impression of luminance and typically is not uniform across a CRT screen. All displays should provide the user with the advantage of controlling luminance (brightness) by adjusting appropriate display controls. CRT displays are usually hindered rather than helped by general lighting in the work area. Using general illumination with CRT displays requires careful design to minimize glare on the cover glass.
- ***Contrast:*** Contrast is the ratio of background luminance minus symbol luminance to background luminance plus symbol luminance. Gould (1988) recommends a contrast ratio of 30 (contrast of 94 percent) as preferred and 15 (contrast of 88 percent) as acceptable.
- ***Flicker:*** If a CRT is not regenerating fast enough, it appears to flicker. In general, a regeneration rate of 60 cycles per second (hertz or Hz), which is the rate of home television receivers, is probably sufficient to prevent the perception of disturbing flicker on monochromatic screens. The refresh rate should be higher than 70 Hz for color monitors.
- ***Resolution:*** The resolution required for a CRT display depends primarily on the viewing distance of the user and the application of the display. A designer should ensure that image quality is adequate by providing the level of resolution needed for accurate viewing. The angle subtended on the retina should be at least 15 minutes of arc. This usually means that the minimum acceptable character size is usually 8 points in pixel-based systems or has a dot matrix height of 7 (minimum) or 9 (preferred) in character-based systems (Buckler, 1977; Stewart, 1976).
- ***Character style:*** A mix of upper- and lowercase is best for readability. Generally, the use of only uppercase or only lowercase characters should be avoided. Each character-generation method has its own set of constraints and distorting properties that need to be taken into account.
- ***Character height:*** The character height for comfortable reading should not be less than 15 minutes or larger than 24 minutes of arc. The preferred height for most reading tasks is about 20 minutes of arc. Character height should never be less than 10 minutes or larger than 45 minutes of arc (Human Factors Society, 1988).

- ***Character width:*** Generally, character width should be between 70 and 85 percent of character height.
- ***Character spacing:*** Spacing between adjacent characters should be at least 18 percent of the character height.
- ***Interline spacing:*** Distance between lines should be at least 50 percent of character height. A distance of 66 percent is preferred.

Are CRTs a Health Hazard?

One of the most frequently asked questions is what is the relationship of CRT use to the health complaints that seem to be related to their use. Frequently, the problems tend to interact with each other. Improper eyeglasses, for example, may be the source of some problems of neck pain. Consider people wearing bifocals and needing to read CRT characters from the lower lens. They will usually need to raise their heads to look out of the bottom of the glasses. If this happens for a long period of time or frequently throughout the day, there may be added stress on the neck muscles and neck pain may be the result.

Although many studies indicate a relationship between frequent and continuous use of visual display units (including CRTs) and certain health-related complaints, the cause of the problems does not appear to be the CRT itself. In other words, there is no evidence that harmful radiation is emitted by properly functioning CRTs. Most problems are caused by the way the work or workplace is designed.

Users with Disabilities

Many research studies show that users with disabilities should have both input and output devices that take into account the type and severity of the disability. For example, Casali and Chase (1993) collected time and error data with subjects using the mouse, trackball, graphics tablet, cursor control keys, and a joystick. They found that for users with impaired hand and arm function the best input device was the trackball.

Other researchers have dealt with issues having to do with output devices. For example, Mynatt and Weber (1994) present and discuss the issues and solutions of having a graphical user interface for blind users.

Work Area

The general area where work is done should be optimized as much as possible. Some of the most important considerations are as follows:

- ***Workspace:*** The work area (workspace) used with keyboards should be at least 750 millimeters (30 inches) wide and about 500 millimeters (24 inches) deep. This should accommodate most input documents. Obviously, if source records require more space, then a larger workspace should be provided.
- ***Work surface height:*** The work surface top that is used for source documents and writing should be about 750 millimeters (29 to 30 inches) above the floor.

- ***Leg room:*** Unobstructed leg room should be provided.
- ***Noise generation:*** Noise generated by the video display terminal should be less than the ambient noise level of the environment before the device is installed (about 50 to 60 decibels). Other factors, such as continual high-pitched whir or hiss also should be eliminated from the equipment.
- ***Heat generation:*** Heat generated by the equipment should be dissipated in a way that does not expose the user to excess heat either directly or through an uncomfortable increase in ambient room temperature.
- ***Ambient light level:*** Provision should be made for the satisfactory viewing of the CRT in a room having an ambient light level from 70 to 100 footcandles.
- ***Illumination:*** The illumination should be sufficiently bright, uniform, and free of flicker. The required luminance depends on the visual task to be performed. In workplaces with computers, an illuminance in the range of 200 to 500 lux, measured on the work area of the work surface, is normally sufficient. Fewer than 200 lux may be desirable for computer tasks not involving hard copy, while more than 500 lux may be desirable for tasks involving the use of poor quality paper documents.
- ***Glare:*** *Direct glare* is produced by light sources within the field of view (including windows). *Reflected glare* is produced by reflections from glossy surfaces. In general, glare increases with the brightness and proximity of a glare source. Glare reduces contrast at the retina and thereby reduces visual performance. Some sources of glare may cause discomfort without degrading performance. There are several techniques for minimizing glare in the workplace.

  Light source (laminar) glare should be controlled by using shields (shades) or indirect lighting.

  Equipment should be located or the room arranged so that bright sources (e.g., windows) are not in the visual field while viewing a screen.

  Window light should be controlled by using reduced transmission glass, louvers, drapes, or shades.

  Computer displays should be positioned to avoid glare. This may require a device that allows users to tilt or swivel the display.

  Glare may be reduced on a display by employing antiglare treatments for the display screen. These include diffusing surfaces, antireflection coatings, or faceplate filters.

Locating Computer Monitors

- ***Location:*** The monitor should be located so that it may be easily read by users in their normal working position.
- ***Orientation:*** The screen should be perpendicular to the user's normal line of sight.
- ***Reflectance:*** Displays should be arranged to minimize or eliminate reflectance of the ambient illumination from the glass or plastic cover.

- ***Screen contrast:*** Contrast on the face of a monitor should be 3 : 1; that is, the luminance of the background should be at least three times the luminance of characters.
- ***Operator controls:*** Brightness and contrast controls should be provided for the user.
- ***Adjacent surfaces:*** Surfaces adjacent to the screen should have a dull matte finish. The brightness range of surfaces immediately adjacent to the monitor should be between 10 and 100 percent of screen brightness.
- ***Document holder:*** If needed, a document holder should be available that can be freely adjusted by the user and has a surface area larger than the documents being transcribed.

Exercise 10: Evaluating Display-control Compatibility

Purpose: To demonstrate the wide range of variability (lack of agreement) in human responses.

Method: You will need six subjects for this exercise. Make several photocopies of Figure Ex. 10-1. Give a copy of the page to a subject (you can test individually or in a group). Ask the subjects to indicate for each of the six examples the correct motion (clockwise or counterclockwise).

On items 2 through 6, the black dot is a knob, and the dial to be moved is like the dial used to designate stations on older car radios.

Indicate whether the following knobs should be turned clockwise (C) or counter-clockwise (CC)

1. Make a darker photocopy

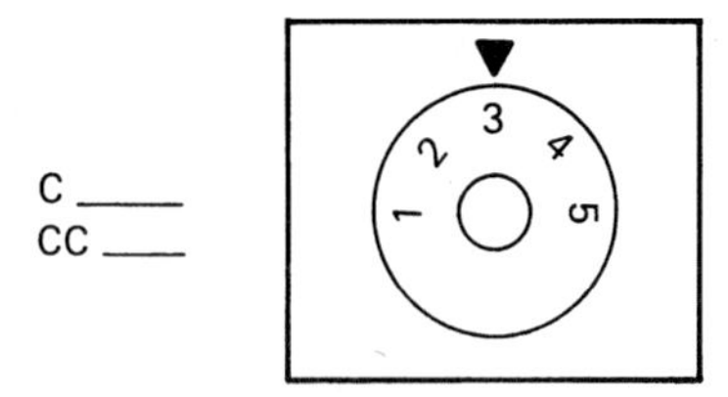

2. Move the dial down

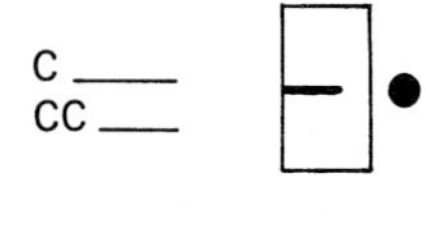

3. Move the dial up

4. Move the dial left

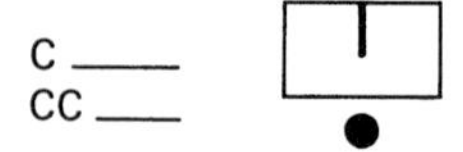

5. Move the dial right

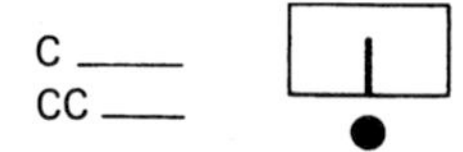

6. Increase the number from 1 to 2

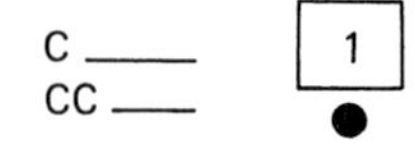

Figure Ex. 10-1

After collecting several sets of responses, summarize the data. Then compare your results on items 2 through 6 with those shown in the following tables.

Direction-of-Movement Stereotypes in Different Cultural Groups (Chapanis, 1975)

	Flemish Workers		Algerian Workers		Moroccan Workers		Moroccan Boys		Moroccan Girls	
Test Item	**CC**	**C**	**CC**	**C**	**CC**	**C**	**CC**	**C**	**CC**	**C**
2	**21**	9	14	5	14	11	8	6	6	9
3	19	11	12	7	15	10	7	7	7	8
4	*27*	*3*	11	8	17	8	11	3	11	4
5	*5*	*25*	5	14	8	17	6	8	**3**	**12**
6			8	11	12	13	7	7	8	7

	African Boys: Luabo and Kamina		African Boys: Kangu		African Men: Kamina		African Girls: Luabo And Kamina		African Women: Kamina	
Test Item	**CC**	**C**	**CC**	**C**	**CC**	**C**	**CC**	**C**	**CC**	**C**
2	*46*	*20*	32	20	19	11	**34**	**16**	14	16
3	*46*	*20*	23	29	18	12	22	28	15	15
4	*47*	*19*	*44*	*8*	20	10	*36*	*14*	18	12
5	*11*	*55*	*9*	*43*	13	17	19	31	**8**	**22**
6	34	32	28	24	12	18	*13*	*37*	13	17

Reporting: Prepare a report that describes what you did and what you found, using the format discussed in Exercise 1A. Include a discussion of how your results compared with those from individuals in other countries.

11

Task Analysis

The Task Analysis Process

Task Analysis Defined

Up until this time in the development of a new product, application, or system, the emphasis has been on *what* functions would be included, *what* people would be typical users, and *what* computer-related input devices, output devices, operating systems, network systems, and the like, would be used. The first step in determining *how* users will interact with a product is to conduct a *task analysis.*

Task analysis is intended to identify the work activities to be performed by users and to match the work to be done with the kinds of people who will do it. A *task* is a term that describes the activities that people perform, and (when used correctly) it is always defined in terms of potential users.

Task Analysis Types

The task analysis process has three main parts. Currently, there are two distinct approaches to conducting a design-oriented task analysis. The *traditional* approach begins with systematically decomposing tasks and ends with identifying work modules. This way of performing a task analysis works well when designing most character-based systems and traditional graphical user interfaces. This approach assumes that the final product will be based on a definite, clearly defined process that users will be performing. The user interface is based on how people will typically perform their work or use the product. It requires that designers know exactly how the system will be used, when it will be used, and how often it will be used.

The task analysis approach used for designing and developing *object-oriented* systems does not make the same assumptions concerning how the systems will be used. The focus is on identifying the objects of most interest to typical users while performing with the new product. There is less emphasis on identifying the final work process. The emphasis is on providing the user interface objects necessary to carry out whatever activities users require and in whatever order they desire. The two approaches will be referred to as traditional and object-oriented.

Identifying Tasks

The first part of the process, *identifying tasks,* (1) either breaks down the original functions into smaller and smaller pieces until arriving at the task level or (2) attempts to directly identify certain activities at the task level.

Describing Tasks

The second part of the process is to describe the results of the identification process. For the tasks that are decomposed into smaller pieces, the second step consists of describing each task and possibly organizing all tasks into a *flow chart* that will accommodate the variety of different transactions that a new system must accomplish.

For the tasks that are directly assigned, the second step consists of ensuring that the assigned *task statements* are written at a level, using the proper vocabulary, that readily communicates to typical users of the task.

Identifying Work Modules or Scenarios

For tasks that are decomposed, the third step has to do with identifying *work modules* by synthesizing the tasks previously identified into manageable modules of work. Having systematically derived work modules assists in the design of interfaces and the preparation of facilitator materials, such as instructions, performance aids, and training. For tasks that are directly assigned, the third step consists of developing *task scenarios* that can be used to evaluate alternative ways of interacting with the computer.

Traditional Task Analysis

Historical Background

In the early 1900s, Frederick Taylor (1911) proposed the general structure of this approach for developing work. He called this new way of dealing with work *scientific management.* The design of jobs was central to his notion of scientific management, as illustrated in the following statement:

> Perhaps the most prominent single element is the task idea. The work of every workman is fully planned out by the management at least one day in advance, and each man receives in most cases complete written instructions, describing in detail the task which he is to accomplish. . . . This task specifies not only what is to be done but how it is to be done and the exact time allowed for doing it. These tasks are carefully planned, so that both good and careful work are called for in their performance. (p. 59)

Basically, this approach suggested that the work should be carefully analyzed and then assigned to appropriate workers with specific instructions on how the work could be accomplished most efficiently. Once the work is designed and people selected to perform the work, the new workers are to be adequately trained to ensure that they perform the work exactly as planned.

Designers attempt to answer the following questions in a typical task analysis.

What tasks are typically performed?

What are the most critical tasks?

What important steps are performed within the tasks?

How frequently are the tasks performed?

What information is needed to complete the tasks?

What output should result from task completion?

What performance standards are assumed for completing the tasks?

How do the tasks typically (generally) flow to complete an activity?

Concept of Decomposition

Designers can break down activities to a task level at which they can all be performed by "off-the-street" new hires or a level at which they can be performed by people who have acquired many high-level skills. This process is usually referred to as *decomposition.* The designer has considerable flexibility in determining which functions will be broken down into low-level tasks and which into high-level tasks. Higher-level tasks require little training if a person has acquired the necessary skills elsewhere.

Identifying Tasks

Two primary considerations are associated with identifying tasks.

1. Determine the user's knowledge and skills and derive skill-level categories
2. Derive lower-level activities until reaching the task level (the decomposition process)

The process may be repeated any number of times until each activity is assigned a single skill level. Secondary considerations in the identification of tasks include meeting a *full-advantage objective.* Designers should attempt to develop tasks that will ultimately take full advantage of the user work force. This is difficult to quantify, but during the analysis process most designers gain a feel for what is meant by taking full advantage of the skills available in their user population, and this should be reflected as the tasks are being identified.

Determine User's Knowledge and Skills

For a designer to determine existing user knowledge and skills, he or she must become very familiar with the potential users. Much of this information will be contained in the user profiles. In some cases, additional data collection may be required. Originally, a designer can use a rough description of existing user knowledge and skills to help to derive *skill-level categories,* which he or she then can use to break down tasks. In some cases, the extent of domain-related knowledge and skill can be estimated based on the time that the person has spent with a company (see Figure 11-1).

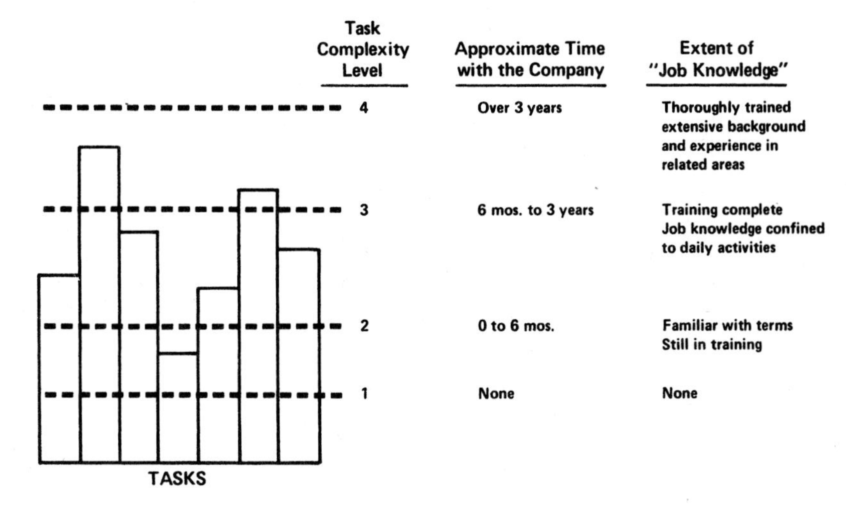

Task Complexity Level
Approximate Time with the Company
Extent of "Job Knowledge"
4
Over 3 years
Thoroughly trained extensive background and experience in related areas
3
6 mos. to 3 years
Training complete Job knowledge confined to daily activities
2
0 to 6 mos.
Familiar with terms Still in training
1
None
None
TASKS

Figure 11-1 Simplified example of skill-level categories.

Identify Tasks

Once a set of skill-level categories is established, it is possible to start identifying tasks. Having skill-level categories is critical to this process because they can be used as a *stopping rule.* The process of breaking down functions into subfunctions and then breaking down each subfunction into lower-level subfunctions makes it difficult to know when to stop the breaking down process. The best stopping point is usually *where a work description can be read and understood adequately by a user with a minimum of training.* If the level is too high, users will not be able to effectively deal with it, and if the level is too low, the user may feel it is too simple, perhaps even demeaning. The value of this task analysis approach is that it generally can be done very quickly.

An example of function decomposition is given in Figure 11-2. In this case, the output is 6 and the input is the scores of 5, 6, 7, 8, 4, and 6. The function to be performed is to *determine the average.* The difficult part comes in identifying all the subfunctions required to convert the input to the output. In this example, three things must be done: (1) the scores must be added, (2) the number of scores must be counted, and (3) the sum of the scores must be divided by the total number of scores. The three subfunctions are each lower-level activities; only when they are *all* completed will the function itself or the higher-level activity have been accomplished. This is the basic process for conducting a task breakdown.

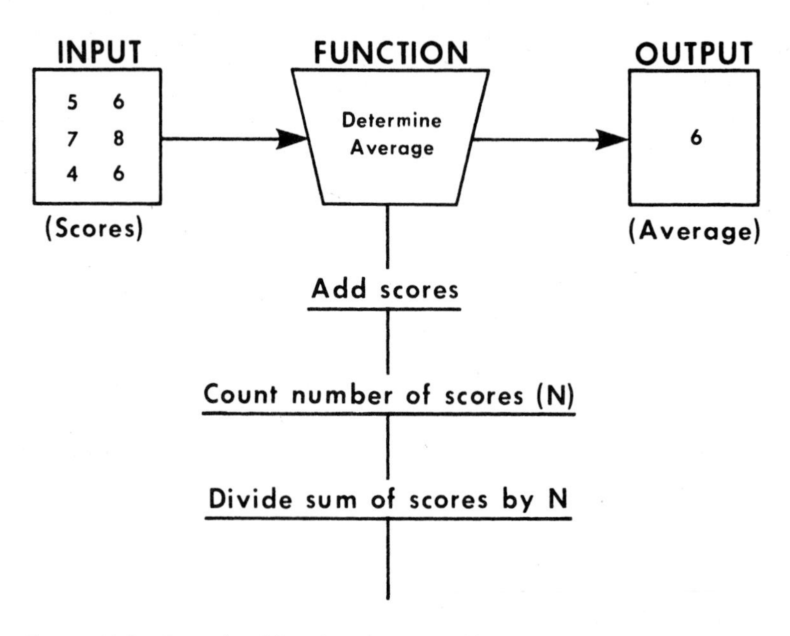

Figure 11-2 Example of function decomposition.

Mutually Exclusive and Exhaustive

There are two rules for the subfunctions. First, each subfunction should be independent (mutually exclusive) of the others. Second, the resulting subfunctions must be exhaustive. This means that everything necessary to accomplish the main function must be included in the subfunctions. The function will be accomplished only if each subfunction is performed.

Example: Prepare and Serve a Salad

As another example of the relationships of functions and subfunctions, consider the following. Suppose that the original function is to *prepare and serve a salad.* In this case, the designer derived three subfunctions necessary to accomplish this effort (shown in Figure 11-3). First, note that in our example the three subfunctions are all at about the same

Figure 11-3 Example of breaking down a function into subfunctions.

level of detail. Notice also that these activities are mutually exclusive; that is, they are separate and distinct from one another with no overlapping activities. In addition, they are exhaustive; on this level they represent all the activities required to achieve the goal.

Assume that the designer focuses on the "prepare salad" function first. To break down the function further or determine more detailed activities (or subfunctions), he or she must first make a design decision. What kind of salad is wanted: lettuce, jello, Waldorf? This information should be available in the system objectives. If the salad is to be a Waldorf, for example, the designer can move to a more detailed level of functioning, as shown in Figure 11-4, as well as into a new region of definition.

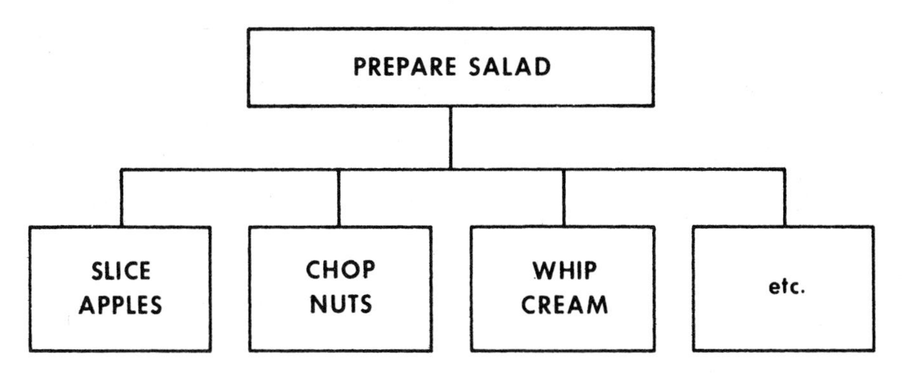

Figure 11-4 A more detailed breakdown.

Figures 11-3 and 11-4 are examples of moving to different levels of functional detail. Those functions at a more detailed level are an explanation of the whole from which they are derived. For example, "prepare salad" is related to "slice apples," "chop nuts," and "whip cream," activities inherent in preparing Waldorf salads. Notice that these functions are not exhaustive, as implied by the box labeled "etc.". All the subfunctions are mutually exclusive and at about the same level of detail.

The activity of breaking down functions into subfunctions continues until we have identified all tasks. *A task is always defined in terms of the potential user.* For highly skilled users (e.g., a master chef), the task level would be much higher than for inexperienced users (e.g., a person learning to cook). Thus, we continue to break down a function into subfunctions until we reach a level where we can match each task against the desired skill levels, or a precise set of user characteristics, or possibly even a particular user. When the typical user can read the task statement and understand exactly what to do, we are at the *task* level. This is very similar to breaking down computer activities until we reach the point of writing code that a computer can understand.

Describing Tasks

Once the tasks have been defined, each should be described. Some designers like to develop a flow chart that shows the interdependencies of all tasks as they relate to various uses of the new system. Flow charts can help in identifying duplicate or omitted tasks and may result in adding, dropping, or combining of tasks. This approach to task analysis tends to be very forgiving, and duplications and omissions can be taken care of very efficiently. Once all tasks are accounted for in a task-level flow chart, the process of synthesizing tasks into work modules can begin.

Work Module Design

Once tasks are identified, adequately described, and flow-charted, the next logical step toward ensuring the desired level of human performance is to identify *work modules.* A *work module* is a set of tasks that a user accomplishes as a part of, or all of, his or her job (Soth, 1976). It is a basic unit of work. Usually, one or more work modules are combined to form a job.

Work modules can be made up of tasks derived from several *different* functions (see Figure 11-5). In this figure, work module A consists of five tasks from function 1. However, work module C consists of two tasks from function 1, two tasks from function 2, and one task from function 3.

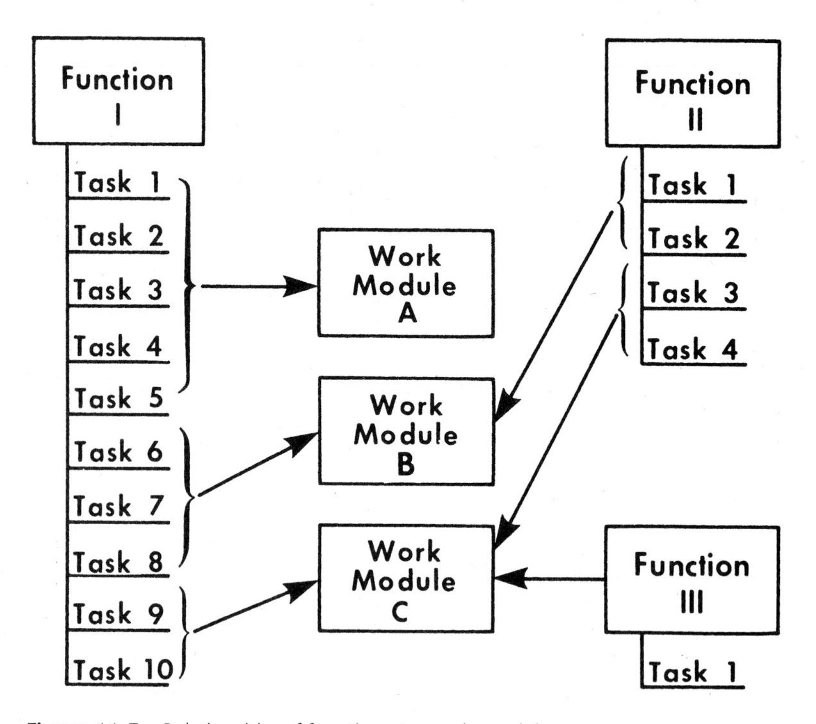

Figure 11-5 Relationship of functions to work modules.

Recall that a function is nothing more than a statement of required work. A work module, however, is a description of work to be performed by a specific type of person. The main objective of the designer is to effectively match the work with potential users.

Consider the following when grouping tasks together into a work module.

- Data relationships
- Skill level for a task
- Task relationships and sequence
- Time dependencies
- Special human–computer interaction considerations

Tasks that have the same or closely related data items used together are candidates for the same work module. For instance, task 1 might require the use of the billing name of the customer account and task 4 require the use of the billed amount. These tasks use related data and should be considered for inclusion in the same module, providing the other four considerations are met.

Tasks of the same or almost the same skill level are also candidates for combination. If tasks 1, 2, and 4 require a high skill level and task 3 requires moderate skill level, these can probably be combined, since a high skill level would be required for the module. However, it is usually not practical or economical to combine high-skill-level tasks with low-skill-level tasks.

One reason for combining tasks with similar skill levels is that it makes it easier to prepare written instructions, performance aids, and training materials. Another reason is that users are given an opportunity to perform activities at about the same difficulty level. For example, a work module in a system with severely degraded human performance was analyzed. The work module was found to contain a total of 38 tasks (most work modules contain 6 or fewer tasks). Even worse, the tasks ranged from being very simple to extremely complex, as shown in Figure 11-6. The manual processing time was excessive and the training time was considerably greater than expected. In fact, when some students finally learned how to perform certain tasks in the work module, they had already forgotten how to perform others. The tasks were analyzed again by usability specialists. As a result, some tasks were combined, and three new work modules were developed from the original, poorly designed work module. These three new work modules are illustrated in Figure 11-7.

It was not possible to have the tasks in the redesigned work module at exactly the same level, but the differences were minor when compared with the original work module. There are fewer tasks in each of the three new modules, which greatly assisted training for each of the modules. As people gained experience on the first work module, some received the opportunity to perform on the second module also. As the new system matured, some of the people who had begun on work module 1 and progressed to work module 2 eventually progressed to perform the work contained in work module 3.

Tasks that are related, whether by data or nature of the work, or that have a sequence dependency should be considered for inclusion in the same work module. This indicates that the normal flow of the work must be anticipated and kept in mind when forming human work modules. (Keep in mind that work modules that are combined usually do not include tasks for transmitting information from one to another.)

These criteria can be used to construct a cohesive and logical set of related tasks for assignment to one work module. This is a convenient way to design work since these basic units can be assembled and built into a variety of jobs either by designers or by users.

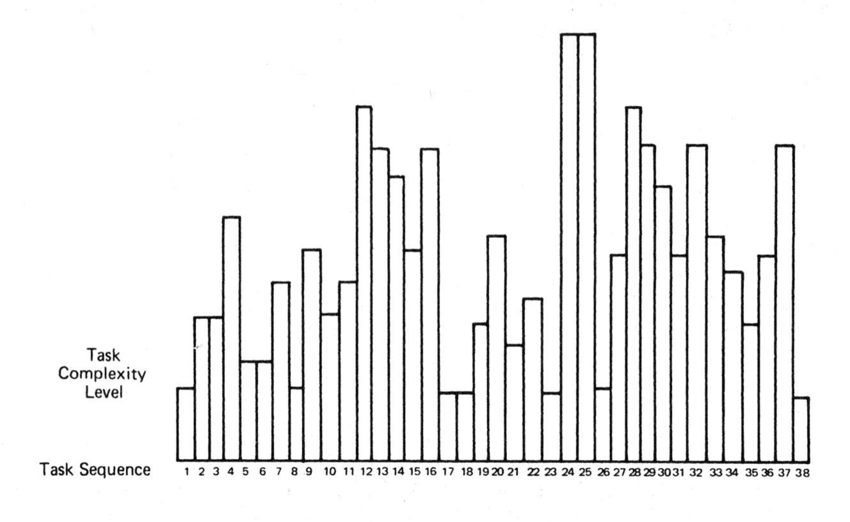

Figure 11-6 Example of an exceptionally large work module in a system with degraded human performance.

The number of tasks in most work modules ranges from four to nine. Probably the ideal size is about six tasks. If the volume of input to a work module is great, more than one person can perform the same human work module. In fact, under conditions of heavy volume, one work module may become the job for several people.

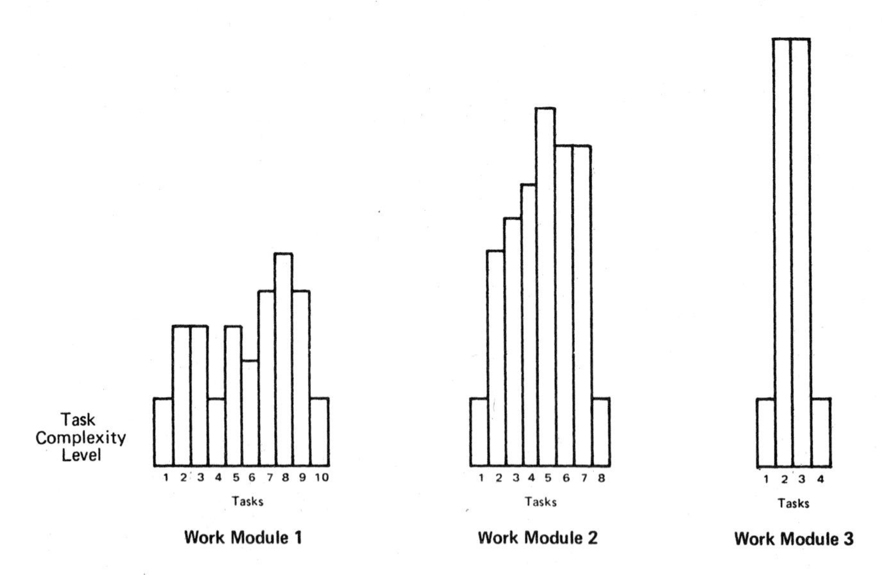

Figure 11-7 Example of three new work modules made from one large work module.

Function Allocation

Allocation Strategies

Singleton (1974) provided an interesting survey of function allocation approaches. In the early 1950s there were numerous attempts at cold, logical comparisons of people and machines. The problem was seen as a simple one of determining the relative performance of people and machines in a variety of different activities. This idea for allocating functions resulted in several lists comparing human capabilities with machine capabilities. The original developers of these lists envisioned a designer considering a system's functions one at a time. Each function would be carefully analyzed and then compared with established human and machine performance criteria and allocated accordingly. This *comparison* allocation approach has proved to be of limited value.

Leftover Allocation

Probably the most popular allocation strategy is *leftover* allocation. With this approach, as many functions as possible are allocated to a computer, and the functions left over are done by people. The erroneous assumption is that people can do and are willing to do any type of work assigned to them. Thus, they are allocated the leftovers.

Economic Allocation

In some systems, the primary criterion for allocating functions is *economic.* In its simplest form, the question is, "For each function, is it less expensive to select, train, and pay a person to do the task or design (or acquire) and maintain a computer to do the same task?" This *economic* allocation approach is still used in many systems.

The economic allocation approach produces a major flaw in many systems. The designer should consider the overall cost of a system when making allocations. This includes the cost of *design* and *operation.* Money saved by shortening the design process may be lost many times over the life of a system due to expenses caused by inappropriate allocation of functions. For example, to keep costs down, the time for the design process of one system was cut from 24 to 20 months. To save time, certain allocations that should have been made to computers were made to people. Over the life of the system, these inappropriate allocations could cost many times the amount saved by shortening the design process.

Telephone company studies of several computer-based systems have shown that it costs at least four times more for people to do the work than it does for computers. In many well-designed systems, it costs about the same to develop computer programs as it does to develop user material. When considering costs associated with the allocation of functions, a designer should carefully consider both developmental and operational costs.

Humanized Allocation

Some systems have used the *humanized task* approach. This concept essentially means that the ultimate concern is to design a job that *justifies* using a person, rather than a job that merely can be done by a human. With this approach, functions are allocated and the resulting tasks designed to make full use of human skills and to compensate for human limitations. The nature of the work selected for people should lend itself to internal motivational influences. The leftover functions are allocated to computers.

Flexible Allocation

Some function allocation problems have been effectively dealt with using a *flexible allocation* approach. With flexible allocation of functions, users can vary the degree of their participation in an activity. In other words, *users* allocate functions in the system based on their values, needs, and interests. This concept is particularly relevant in appropriately designed computer-based systems in which a person can allocate his or her own functions by making the necessary adjustments to the software. This is now becoming more common.

In some cases, a machine can be designed to take over those things that a user finds difficult, as well as those things that the user simply does not feel like doing at any specific time. A good example in common use is the autopilot in a commercial aircraft. Once the aircraft is airborne and headed in the right direction, the pilot can switch to an autopilot (computer) that carries out routine in-flight functions. This is probably the most elegant solution to the function allocation problem: to design a system so that certain functions can be *allocated by users*.

Duration of Function Allocation

The function allocation process actually takes place throughout the design of a system. Usually, designers make the most substantial allocations early in the development of a system and make minor changes later to finely tune the system. The allocation process is iterative. A designer originally allocates each function either to a human, machine, or human–machine interaction. This allocation is then used as a working model until more is learned about system characteristics, when some changes to the original allocation should be made.

The allocation process is very forgiving, particularly when dealing with software or computer-based systems, which usually can be changed more readily than solely hardware based systems. A designer should review the allocations critically several different times until the system is operational. It is particularly important to carry out a detailed evaluation during usability testing to determine whether an early function allocation decision is degrading human performance and hampering the overall operation of the system.

Indirect Allocation

Often, by the time a designer can systematically allocate some functions, others already could have been allocated either directly or indirectly. Indirect allocations may take place in at least three different ways. The first deals with certain allocations that a designer's management makes. For example, management may decide that a certain set of activities must be computerized. Designers may find themselves in a situation in which certain functions have been allocated long before the designers became involved.

A second form of allocation is determined by the requirements levied against the system. If, for example, a certain function must be reliably performed in less than 100 milliseconds, then it certainly will not be assigned to a human. The requirement itself establishes that it will be done by a computer. Thus, another group of functions may be allocated indirectly by setting system-level requirements that are so stringent that a designer no longer has an option.

A third situation is the traditional function allocation that designers are supposed to perform. Ideally, *all* functions should be systematically allocated by designers. In many systems, however, functions are allocated by all the sources just discussed: management decision, system requirements, and, finally, designers.

An Example of Alternative Allocations

In considering various alternative function allocations, it is sometimes helpful to construct a grid such as that shown in Figure 11-8. This particular grid was used to evaluate different ways that computer-detected errors could be handled in a system. In the first alternative, the approach is primarily manual; a clerk receives the error message from the computer, reads it, analyzes it, decides who can best make the correction, fills out a form with comments concerning the errors, makes a photocopy, and then either corrects the error or mails the materials to the appropriate person for correction.

	Alternative 1 (clerk)	Alternative 2 (clerk/computer interaction)	Alternative 3 (computer)
CLERICAL ACTIVITIES	• Reads error notification • Analyzes notification • Decides who will correct • Determines error made • Fills out form • Photo Copies form • Sends materials to error-maker	• Analyzes error • Decides who will correct	
COMPUTER ACTIVITIES	• Prepares and prints notification	• Displays error message to clerk • Prepares notification • Stores copy in memory • Determines error-maker • Routes to error-maker	• Prepares notification • Stores copy in memory • Determines error-maker • Routes to error-maker

Figure 11-8 Alternative ways to accomplish a function for reconciling errors.

The second alternative splits the tasks between the person and the computer. The person is assigned the analytical tasks associated with the error and decides who can best correct it. After displaying the error message to the clerk, the computer prepares a notification, stores a copy, informs the clerk who the error maker was, and automatically routes the error to the error maker or anyone else, all at the direction of the person, in this case the clerk, who does the analysis of the error. Here the human makes the decisions and the computer performs the clerical work.

The third alternative has the computer handle the reconciliation exclusively. The computer automatically prepares a notification, stores a copy, determines the error maker by a code on the original input form, and automatically routes each error to the error maker. In this case, the error maker always corrects the error.

Alternative 1 saves a great deal of programming, whereas alternative 3 requires additional programming, as well as a considerable number of advance decisions. Alternative 1 relies heavily on people making good decisions and doing much clerical work. Alternative 3 takes the human out of the loop and has the computer perform both the analytical and clerical tasks. Alternative 3 tends to be less flexible in that all errors are dealt with in the same way. Alternative 2 probably offers the highest degree of flexibility and efficiency because people do the analytical and decision-making tasks, whereas the computer automatically cares for clerical tasks after people make decisions. Even so, any one of the alternatives may be an acceptable way to accomplish the function.

Designers may prefer alternative 1 in a system where fewer than ten errors occur in a day and where people that are *regularly* assigned to other tasks can easily handle these errors. Alternative 2 is probably best in situations where numerous errors occur daily, from those errors that are easily and quickly corrected to those that require considerable correction activity, usually by the error maker. Alternative 3 will also probably work best in a system where many errors occur daily—the kinds of errors that are to be corrected by the error maker.

Functions Are Always Allocated

Frequently, the time needed for carefully and systematically analyzing functions is not available. In many cases the requirement to meet a schedule simply does not allow sufficient time for all required development processes to be done adequately. When this occurs, the function allocation process may be greatly abbreviated. But keep in mind that, systematically or not, functions are *always allocated.*

As the design development proceeds and uncertainty about each function lessens, the allocation of some functions may change. As mentioned earlier, a good designer's prerogative, duty, and responsibility are to make changes to the allocation where these changes will enhance human (or system) performance.

If done well, a function allocation will lead to a smooth integration of human and computer components. Ideally, we would always allocate activities to people that they do well and enjoy doing. We would try to avoid allocating activities to people that would be boring or confusing to them and that may lead to a general feeling of dissatisfaction. It is curious that we have the power to create enjoyable jobs for people, yet frequently we end up with systems in which many jobs are menial and tedious and in some cases could be performed by well-trained chimpanzees.

Revising Original Allocations

At the end of the systematic function allocation process, it is a good idea for designers to document the major reasons for each function allocation. Frequently, after the system has been under development for even a few months, it is difficult to remember why a function was originally assigned to the computer as opposed to a person. If allocation changes seem warranted as development proceeds, it is helpful to go back to the original allocation to determine its rationale.

Object-oriented Task Analysis

Compared with Traditional Task Analysis

A task analysis will proceed differently if the goal is to identify user interface objects. The major difference between the two approaches discussed in this book is that one focuses on decomposing functions down to a task level, while the other focuses on directly identifying user interface objects. In the latter case, the traditional decomposition process actually tends to make finding user interface objects more difficult. The traditional task analysis approach tends to be *process* oriented. The object-oriented task analysis approach tends to be more *product* oriented. Another difference is that rather than identifying tasks and converting them into work modules, designers identify tasks and organize them into scenarios.

Tasks and Scenarios

The major task analysis processes include (1) directly identifying potential user *tasks,* (2) organizing the tasks into alternative *scenarios,* and (3) evaluating and selecting the final set of task-level scenarios.

Object-oriented task analysis assumes that the product consists of a collection of objects that cooperates to achieve some desired functionality. The focus is not on decomposing the tasks, but in breaking the tasks into meaningful parts. Each part then represents some object (or class of objects) from the problem domain.

Traditional task analysis does not work well with the principles of object-oriented design. Using traditional task analysis as a front end to object-oriented design may lead to failure. This is because the traditional task analysis is based on a model that is biased toward traditional decomposition. Traditional task analysis does not lead to objects because it focuses on processes (e.g., operations or actions). This approach helps to discover smaller and smaller processes and may actually cause designers to avoid focusing on objects.

It is difficult to focus first on process and then effectively determine meaningful objects. It is usually better to identify user interface objects as soon as possible and then determine the underlying processes. Too much process orientation can lead to weak object orientation.

Object-oriented Technology

Object-oriented technology is an approach to software construction that shows considerable promise for solving some of the classic problems of development. The core concept behind object technology is that all software should be constructed out of standard, reusable components wherever possible.

The key mechanisms underlying object technology include objects, messages, and classes. Objects are software packages that contain related data and procedures. In many cases, objects can correspond to real-world business objects. Messages are the means by which objects communicate. In essence, objects request services from one another, working together to carry out essential business operations. Classes are templates for defining kinds of objects. They help to organize information about business objects in a natural, intuitive manner.

Object technology offers significant benefits. Computer systems can model the actual elements and behavior of an organization. Because object systems are easily modified, this model can rapidly adjust to changing business conditions and strategies. Companies that adopt object technology will be able to more readily adapt to changing markets, react to new opportunities, and adjust their operations to better meet their customers' needs. Instead of being obstacles to change, their information systems will become enablers of change.

Abstractions

Object technology has been effectively used to reduce the complexity of user interfaces. Designers attempt to identify the right set of tasks (abstractions) for a given product. This process leads to simplified descriptions of products and systems and is an exceptionally powerful technique for dealing with complexity. Designers attempt to identify the most important aspects of a domain by focusing on important aspects and ignoring nonessential details. The domain emphasizes certain details or properties while suppressing others. It highlights details that are important to users, while suppressing details that are not important to the user interface.

High-level abstractions are more generalized and contain less information. For example, a "living thing" class is at a higher level of abstraction than the plant class upon which it builds. A plant class is at a higher level of abstraction than a flower class. The lower (deeper) the abstraction is the more detailed the information. Eventually, the abstraction becomes the actual product, application, or system.

When designers are preparing abstractions, they should be careful to include the words that form the vocabulary of the problem domain. Describing and naming things properly is often treated too lightly by designers. The tasks should be stated at the right level of abstraction—neither too high nor too low. For example,

Do this: "Look at the cat sitting on the porch."

Not this: "Observe the quadruped reclining on the veranda."

A *task* should be at the abstraction level at which typical users can read and readily understand their part in a human–computer interaction activity. The correct words will be meaningful to intended users.

Developing Scenarios

Introduction

A *scenario* is a means of documenting task statements that includes activities of both the user and the computer. Task scenarios are used to illustrate typical use of a product. It is a good idea to construct scenarios for all task-level activities that are fundamental to the product's operation.

Developing scenarios helps designers to better understand and refine task-level performance. The user's involvement is combined with the computer's involvement in these scenarios. This helps to illustrate how the two components can work together to accomplish a system goal. Scenarios should reflect an understanding by the designer of the strengths and limitations contained in both the user profiles and the computer-related profiles.

Scenarios describe how the human and computer, working together, will accomplish the most important activities. Developing scenarios can be the most critical part of the user interface design process. All subsequent activities depend on how designers decide that the interaction will take place.

The typical format for a scenario is as follows:

User: Does this

Computer: Does this

User: Does this

Computer: Does this

The computer actually makes two responses. The first is quick, appropriate visual or auditory feedback, and the second is an information response to the user's action, for example, a new window shows. It is the second response that is of most interest when completing a task analysis.

The designer should decide whether the computer or the user will begin (initiate) each transaction, and this should be reflected in the scenario.

ATM Example

A good example of a scenario is the one given next of a person using an automatic teller machine (ATM) to withdraw $100 from her savings account.

Computer:	Shows initial screen requesting the customer to insert a bank card
User:	Inserts a bank card
Computer:	Requests a customer's personal identification number (PIN)
User:	Enters the PIN
Computer:	Shows banking options (withdrawals, deposits, view-only, etc.)
User:	Selects "Withdrawal"
Computer:	Shows account options (checking, savings)
User:	Selects "Savings"
Computer:	Requests the amount of the withdrawal
User:	Enters "100"
Computer:	Confirms the previous entries—"You wish to withdraw $100 from your savings account?" (Yes or No)
User:	Selects "Yes"

Computer: Delivers the money and a receipt

User: Retrieves the money and the receipt

Computer: Delivers the customer's card

When developing scenarios, designers should address only the most simple, straightforward, basic set of activities. They should focus on the few tasks associated with the *critical path* and should not include error checking or any *exceptions* processing.

Interaction (Dialog) Styles

Designers must decide which interaction method is best for a specific product. Typical interaction styles include

Conversational with speech input and output

Direct manipulation with icons and high-level graphics

Menus and lists (fewer graphics)

Form-filling with typed input and printed output

Commands

Designers must determine the best interaction (dialog) style for achieving acceptable performance in the new product. If the answer is not clear, then the scenarios could reflect two or more different interaction styles. In some situations, it is a good idea to even include two or more task-related variations of *each* interaction style.

Using Iteration with Scenarios

When developing scenarios, designers should begin simply and then eventually expand the initial scenarios. Many designers use an iterative approach: they iterate a little, then design a little. The adequacy of scenarios can be evaluated using walkthroughs, storyboards or paper prototypes, or computer-based prototypes.

Scenario Wording

The level at which the scenarios are written is important. Write the scenarios so that a typical user can read and understand them. The vocabulary used in the scenarios will be eventually used to help to identify important user interface elements.

Keep the wording of the scenario general so that detailed design decisions can be made at a later time when we have more information.

Do this User: Opens the XYZ window

Not this User: Goes to the menu bar and clicks on the File option, then clicks on the Open menu item, then watches the XYZ window open

Do this User: Dismisses the window

Not this User: Uses the mouse to move the mouse pointer to the Close pushbutton and then clicks the left mouse button

Grocery Shopping System Example

Assume that a product is to be developed that will allow people to purchase groceries using their computer, which is connected to a store's computer. The first step is to identify the major activities required for a shop-at-home customer to make an order. These activities should be abstracted at the task level.

The process of identifying tasks requires making many design decisions. This means that the development process has moved from determining *what* is to be done to determining *how* the work will be accomplished. We have moved from *analysis* to *design.*

The only activities that will be dealt with are those that require the knowledge and skills of potential *users.* Computer activities will be assumed, but not analyzed. The computer processing is included only as an assumed part of the user activities.

The activities in the new grocery shopping product that require human processing are as follows: These activities will be identified as *tasks.*

Accessing available grocery items (i.e., entering the "store")

Selecting the desired grocery items

Paying for the grocery order

Indicating a method of obtaining or receiving the order (i.e., shop, go to a pickup window, have the order delivered)

After identifying potential tasks, designers then must determine the interaction (dialog) style that is most appropriate for the new product. Once the interaction styles are identified, a series of scenarios can be developed. Examples of scenarios for different interaction styles are shown in Figures 11-9 through 11-12.

Access available grocery items

U: Obtains a booklet with all grocery items listed, along with a code for each item.

U: Reads and marks the desired grocery items in the booklet.

Select the grocery items

U: Types the "grocery" command and presses the ENTER key.

C: Shows the "grocery system" prompt.

U: Types the "grocery list" command.

C: Shows the "grocery list" prompt.

(continued)

Figure 11-9 Grocery shopping system example using a *command* style.

(Figure 11-9 continued)

U: Types the code for one grocery item and presses the ENTER key (repeats until all items are entered).

C: Shows the "grocery system" prompt.

U: Types the "list items" command and presses the ENTER key.

C: Shows a list of all items, including the entered codes and a brief (one line) text description for each.

Pay for the grocery order

C: Shows the "grocery system" prompt.

U: Types the "amount" command and presses the ENTER key.

C: Shows the total cost of the grocery items.

U: Types the "pay cash" command and presses the ENTER key.

Indicate method of receiving the order (i.e., shop, go to a pickup window, have delivered)

C: Shows the "grocery system" prompt.

U: Types the "pickup" command and presses the ENTER key.

C: Shows the time when the order will be ready for pickup and the address of the pickup location.

Access available grocery items

C: Displays a menu of available computer-based home applications.

U: Uses the cursor keys to move a menu bar to the "grocery system" choice and presses the ENTER key.

C: Displays a menu of available options in the "grocery system."

U: Uses the cursor keys to move a menu bar to the "list groceries" choice and presses the ENTER key.

C: Displays the beginning screen for the grocery system showing several function key options.

Figure 11-10 Grocery shopping system example using a *form-filling* style.

(Figure 11-10 continued)

Select the grocery items

U: Obtains a printout showing available grocery items and related codes; then selects and circles the desired grocery item codes.

U: Presses the F2 key to access an entry form.

C: Displays an entry form.

U: Types the code for a grocery item and presses the ENTER key (repeats this process until all items are entered).

C: Displays the list of items on the entry form.

U: Presses the F3 key to access a summary list of items.

C: Displays a list of all items, including the entered codes and a text description.

U: Reviews and confirms the list of items.

Pay for the grocery order

U: Presses the F4 key to calculate the cost of the items.

C: Displays an "amount owed" field showing the cost.

U: Inspects the amount owed and, if acceptable, presses the F5 key to access a list of payment methods.

C: Shows a menu of payment methods

Cash

Check

Credit card

U: Uses the cursor keys to move a menu bar to the desired option and presses the ENTER key.

Indicate method of receiving the order (i.e., shop, go to a pickup window, have delivered)

C: Shows a menu of obtaining/receiving methods

Shop (from a list)

Pickup at a window

Delivery to home or office

U: Uses the cursor keys to move a menu bar to the desired option and presses the ENTER key.

C: If "shop" is selected, prints a list; if "pickup" is selected, enters a time when ready and location for pickup; if "delivery" is selected, enters a time when ready to be delivered and requests an address.

Access available grocery items

C: Displays a window showing a typical store front and several icons (one icon is labeled "Grocery Items").

U: Selects and clicks on the grocery items icon.

C: Displays a window showing available grocery items.

Select the grocery items

C: On the left side of the window, shows a list of grocery items sorted by major category, then alphabetized within category. On the right side of the window, shows an empty list.

U1: Points at a grocery item in the left list and drags it to the right list (repeats this process until done).

C1: Shows grocery items accumulating in the right list and the accumulating cost.

or

U2: Points at a grocery item in the left list, and clicks.

C2: Moves the item from the left list to the right list and shows the accumulating cost.

or

U3: Points and clicks at several items in the left list (each highlights) and then presses the "Select" push button.

C3: Moves all highlighted items to the right list and shows the cost of the selected groceries.

Pay for the grocery order (for pickup or delivery)

U: When done, selects the "Pay" menu item from the menu bar on the original Grocery Store window.

C: Displays a "payment" window showing Cash, Check, or Credit Card.

U: Selects the preferred payment option and then presses the "OK" pushbutton.

C: Closes the payment window.

Indicate method of obtaining or receiving order (shopping, pickup, or delivery)

U: Selects the "Receive Groceries" menu item from the menu bar on the original Grocery Store window.

C: Displays an "obtain/receive" window showing shopping list, drive-in pickup, or home delivery.

U: Selects the preferred option and presses the "OK" pushbutton.

C: Closes the window, and either (1) prints a shopping list, or (2) displays the time and address for pickup, or (3) displays the time and requests an address for delivery.

Figure 11-11 Grocery shopping system example using a *menu and lists* style.

Access available grocery items

C: Shows all store aisles and shelves (grocery categories) and a small shopping cart.

U: Points at the shopping cart and drags it to the desired aisle and shelf (grocery category) of the store.

Select the grocery items

C: Shows grocery items on the shelves.

U: Points at a grocery item and drags it to the shopping cart (repeats "aisle selection" and "item selection" until done).

C: Shows grocery items accumulating in the cart and a dollar and cent amount accumulating in a window.

U: When done, points at the shopping cart and drags it to a cash register icon.

Pay for the grocery order (for pickup or delivery)

C: Shows a window displaying payment options as icons: Cash, Check, or Credit Card.

U: Points at the shopping cart and drags it to the desired payment-type icon.

Indicate method of obtaining or receiving order (shopping, pickup, or delivery)

C: Shows a window displaying completion options as icons: shopping list, drive-in pickup, or a house (for at-home delivery).

U: Points at the shopping cart and drags it to the desired completion option.

Figure 11-12 Grocery shopping system example using a *direct manipulation and graphics* style.

Exercise 11A: Conducting a Process-oriented Task Analysis

Purpose: To provide experience with conducting a process-oriented task analysis.

Method: Begin with a *function* called "Prepare a Quiche Lorraine." The activities to be performed in this function are given next. Assume that all the activities are allocated to a person to perform, in some cases with the help of a stove.

Prepare a Quiche Lorraine

Set the oven at 450 degrees.
Cook until crisp
    ¼ pound bacon.
    Crumble and set aside.
In about 1 tablespoon of the fat, cook slowly until golden
    ½ cup finely chopped onion.

Line a 9-inch pie pan with ready-made dough.
Bake 5 minutes.
Remove from the oven. Sprinkle into pie shell the bacon, the onion, and
½ pound Swiss cheese, cut small.

Mix
3 eggs, slightly beaten
2 cups cream or milk
½ teaspoon salt
Cayenne and nutmeg.
Strain into the pie shell.

Bake 10 minutes, then reduce the heat to 325 degrees and bake until firm.
Cut in 6 wedges and serve warm, but not piping hot.

For your new Quiche Lorraine system, assume that the user is a 12-year-old with *no* cooking experience and that all the preparation and cooking takes place in the kitchen of a restaurant. Also, assume the following human performance requirements:

1. The user will take no more than 20 minutes per Quiche Lorraine (not including cooking time).
2. The user should properly perform all steps and use the exact ingredients specified for each Quiche Lorraine (no errors).
3. The user will receive no training and must rely totally on written instructions.
4. The user will receive great satisfaction from making Quiche Lorraine using this system.

Reporting: Prepare a *list* of tasks, with a brief *description* for each task to be performed in the system. Next, prepare *a flow chart* showing the order in which tasks are to be performed and other important relationships. Group the tasks into at least three *work modules.*

Exercise 11B: Conducting an Object-oriented Task Analysis

Purpose: To provide experience with conducting an object-oriented task analysis.

Method: Use the information given next to make up a set of potential tasks for each intended user of the system. After identifying a set of tasks for each user, develop a set of task-based scenarios for each user that shows how the new system will typically work.

Reporting: Submit four different task lists (one for each user class) and the related scenarios for each list (make sure to keep a photocopy for yourself).

The Employment System

Background

The goal of this new system is to assist in the hiring of new employees. The product will be used by all those involved in the hiring of prospective employees, including the job candidates themselves, the employment office professional employment specialists and supporting clerical staff, and management-level people who have job openings.

The main function of the new system is to improve the efficiency of the hiring process in an organization with 3500 employees. The organization is growing at a moderate rate and also must replace employees because of normal turnover. They hire about 300 people per year.

Currently, the entire hiring process is conducted without the use of computers. The organization's employment office has six people: one manager, three employment specialists, and two clerks. In general, it is believed that the new computer system should allow the capturing, storing, and processing of virtually all employment-related information. It is not clear what users will do in the system.

Proposed Major System Functions

Store, show, and print available job openings.

Store, show, and print candidate qualifications.

Match job openings and candidate qualifications.

Store and show interview summaries.

Store and show the results of background checks.

User Profiles

Job candidates: Must assume limited or no computing experience because many of the jobs for which they are applying do not require computing skills.

Company managers: Moderate to limited computing skills. Usually familiar with a word processor and possibly a spreadsheet program. Use of this product will be relatively infrequent (e.g., once every month or two).

Employment specialists: Moderate computing skills. Usually familiar with a word-processing program. They fully understand the employment process, including evaluating applications, using the paper-based filing system, and conducting effective interviews.

Employment clerks: High to moderate computing skills, primarily with a word-processing system that they use daily. Good typing and filing skills. Moderate knowledge of the overall employment system, mainly because they tend to stay in their jobs usually for about 1 year before being promoted.

12

Interaction Issues

Introduction

Probably the most significant invention of the twentieth century is the computer. It pervades virtually all aspects of life. In all developed countries and many developing countries it is a rare individual who is not affected in some way by computers.

Designing for an acceptable level of human performance in computer-based products, applications, or systems continues to be difficult. Not giving serious consideration to the human–computer interface frequently results in slow human performance, excessive errors, extra training time, and user frustration.

Unfortunately, many user interface designers have not yet internalized a complete set of general design principles that, if followed, would guarantee the desired level of human performance. When confronted with human–computer interface problems, many designers simply do the best they can, which includes a considerable amount of guessing.

Even so, the collective knowledge about human–computer interaction is growing rapidly. Much of the available information is in the form of standards (e.g., ISO 9421 see Stewart, 1992) and guidelines [e.g., Smith and Mosier, 1983) and style guides (e.g., Motif Style Guide, see Open Software Foundation, 1990, 1994)]. These guidelines are being used more to make initial design decisions. Once a series of decisions is made, they are validated using usability testing.

Beginning in the early 1980s, a number of user-oriented approaches for developing systems have been proposed (Akscyn et al., 1988; Card et al., 1983; Carroll and Thomas, 1988; Dumas, 1986; Gould and Lewis, 1985; Malone, 1984; Morland, 1983; Norman, 1984; Saja, 1985; Shneiderman, 1987; Smith, 1986; Tullis, 1983).

Interaction Concepts

In Chapter 10, we discussed how many of the earliest design decisions are made. In Chapter 11, we discussed issues related to input and output devices. This chapter focuses on the actual interaction, the exchange of information, between a human and a computer. This is probably the most important of all user interface design activities.

User Interface Procedures

Transactions, Procedures, and Systems

A *transaction* is a single exchange of information between the human and the computer. For example, a person may designate a certain menu item (e.g., Print) by pointing at it, and the computer then responds by beginning to print. This is a single transaction. Figure 12-1 shows another example.

> Computer: Displays an error notification.
> User: Moves the mouse pointer to the "OK" pushbutton and presses the "select" button.

Figure 12-1 Example of a complete transaction.

A series of transactions is a *procedure,* and a series of procedures constitutes a *product, application,* or *system.* Conversely, all systems are made up of procedures and all procedures are made up of transactions. The relationship of systems, procedures, and transactions is shown in Figure 12-2.

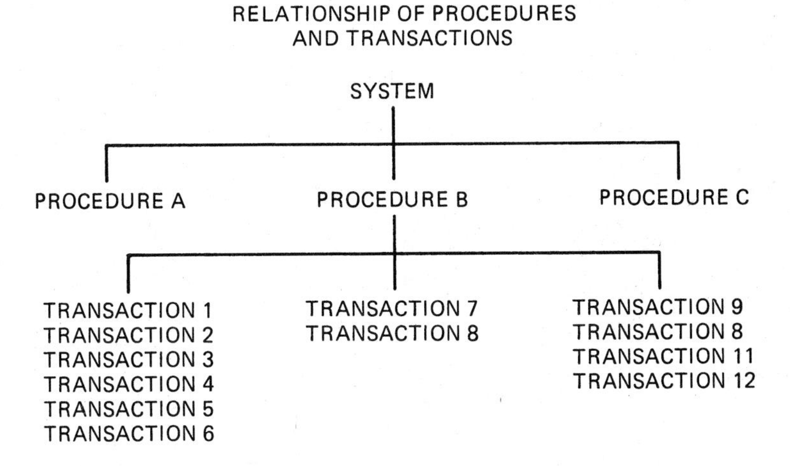

Figure 12-2 Relationship of systems, procedures, and transactions.

Physicians versus Computer Users

Procedures relate to a series of transactions that is carried out to accomplish some objective. A comparison can be made with physicians who are learning to be surgeons. A surgeon in training learns a series of procedures. The first learned procedure is usually relatively simple. As surgeons continue in training, they build up a repertoire of procedures such that, when they are ready to graduate, they know numerous procedures, many quite simple and many very complex. The same can be said for users of computer systems. In the beginning they learn relatively simple, straightforward ways of accomplishing certain actions. As they gain more experience, they learn more procedures. The sum total of procedures that a person has learned constitutes his or her understanding of the user interface.

Natural and Efficient Procedures

Natural and efficient procedures can be achieved by ensuring that the interaction reflects the way the user views the task to be accomplished. It is important for procedures to be as natural and efficient as possible. A natural and efficient procedure is one that fits neatly into a user's expectations of what they think ought to happen. Procedures should minimize the time it takes to carry them out.

It is important to gain an appreciation of how transactions lead to procedures that lead to a product or system. The role of users interacting with computers can best be understood by looking at the protocol of user interactions (i.e., a series of transactions). This means looking at a series of transactions to see how efficiently they go about completing a procedure. Conversely, a procedure can be evaluated in terms of how slow and awkward it is to carry out.

Efficiency Examples

Reducing Inefficient Data Entry Procedures

Numerous procedural decisions need to be made in the design of systems. Take, for example, data entry. Some major procedure-related considerations include the following: (1) the user should not enter the same data twice unless these data are going to be used for error detection; (2) users should not enter data the computer already has or that the computer can generate; (3) users should not reenter a series of data items simply to correct the information in a single data item. Finally, the order of entering data items should be totally consistent with the order found on the source document or input form.

Reducing Inefficient Access of Menus and Lists

Users should be able to bypass a series of transactions if they elect to do so. This is frequently done in situations where users access a menu bar to select an item, go to a pull-down menu, then go to a cascaded menu to find the item for which they are looking. If users know in advance where they would like to end up, they should simply enter the appropriate shortcut keys. This obviously takes much less of the user's time than having to go through each menu separately. It also encourages transition from computer-initiated dialog to user-initiated dialog.

Once a new way of working is learned, the menus can be bypassed. When bypassing some transactions in favor of others, users are learning more efficient ways of completing a procedure. This means that they are using a different set of transactions.

Initiative

Every transaction has an initiator, the user or the computer. In procedures where the *computer* has the initiative, users respond to the computer. In these cases, the computer asks a question and the user responds, or the computer presents a menu and the user selects a response, or the computer presents a screen and the user makes several entries. Computer-initiated procedures tend to be easier for new or occasional users.

In systems where the *user* has the initiative, the computer responds to user inputs. A common example is when the user inputs a command and the computer carries out an action.

In some database management systems, users ask questions to which the computer attempts to respond. In many of these systems, users can make requests in natural language and the computer attempts to understand the words and make an appropriate response.

Another user-initiated procedure, which is a variation from entering commands, is for the user to press function keys or special-purpose keys. A user can press a key such as "delete," and the computer then responds by deleting a character, word, or paragraph. User-initiated procedures are preferred by experienced and regular users. Examples are shown in Figures 12-3 and 12-4.

User activates an input device (e.g., keyboard)
  Is this the best input device?
User uses the input device to convey information to the computer ("print filename")
  Is this the best interaction method (dialog style and language)?
Computer receives, processes, and prepares, to send a response
  Is the processing taking place in a timely manner?
Computer activates an output device (e.g., CRT display)
  Is this the best output device?
Computer uses the output device to convey information to the user
  Is this the best presentation method (optimized format, content, colors)?

Figure 12-3 Usability issues in a *user-initiated* interaction.

Computer prepares to send information
    Is the appropriate processing completed?
Computer activates an output device (e.g., CRT display)
    Is this the best output device?
Computer uses the output device to convey information to the user
    Is this the best presentation method (optimized format, content, colors)?
User activates an input device (keyboard)
    Is this the best input device?
User uses the input device to convey information to the computer
    Is this the best interaction method (style and language)?

Figure 12-4 Usability issues in a *computer-initiated* interaction.

In some direct manipulation systems, users initiate the procedures, but with considerable assistance from the computer. In fact, in some of these systems the initiative shifts back and forth within the procedure. Many of the procedures associated with most pen-based systems are difficult to separate as being user or computer initiated.

Interactions

Human–computer interactions occur anytime users are involved in the operation of an interactive computer product. To be effective, each interaction must be completed (have both a human and computer component). One component initiates an event, which is always followed by the other component making a response. Consistent with our previous discussion, there are usually only two possibilities: (1) the user initiates and the computer responds, or (2) the computer initiates and the user responds.

Whenever the user does something and the computer responds, or the computer does something and the user responds, this is referred as one *interaction*. These interactions need to be as fast and natural as possible. Well-designed procedures tend to elicit good interactions. Graphical user interfaces (GUIs) provide good ways to interact with a system for some users. In other cases, particularly with regular users who quickly become power users, a well-designed command language may provide the best interaction procedure.

Interaction Examples

The concept of interaction is valid for all types of human–computer exchanges of information. Figure 12-5 provides an example of a character-based system, Figure 12-6 provides an example of a traditional graphical user interface, and Figure 12-7 provides an example of an object-oriented graphical user interface. In all three examples, the same goal is achieved, but in different ways.

User: Types "ed" and presses the ENTER key
Computer: Displays the "editor" prompt
User: Types "print filename" and then presses the ENTER key
Computer: Starts the printer and makes the keyboard inactive

Figure 12-5 Example of a *character-based* interaction.

User: Uses the mouse pointer to select a word processing *application* from icons in the primary window
Computer: Displays the word processor's primary window
User: Uses the mouse pointer to open a list of documents
Computer: Displays a list of documents
User: Uses the mouse pointer to select a document
Computer: Displays the document
User: Moves the mouse pointer to the "file" menu item on the menu bar
Computer: Displays the pull-down menu
User: Moves the mouse pointer to the "print" menu item and presses (clicks) the "select" mouse button
Computer: Displays a dialog box requesting information, entire document, number of copies, etc.
User: Moves the mouse pointer to the "Print" pushbutton and presses (clicks) the "select" mouse button
Computer: Starts the printer

Figure 12-6 Example of a *traditional* graphical user interface.

User: Uses the mouse pointer to select a *document folder* from icons on the workplace (background screen)
Computer: Displays several icons, each representing a different document
User: Uses the mouse pointer to select a document and drag it over to touch the printer icon
Computer: Starts the printer

Figure 12-7 Example of an *object-oriented* graphical user interface.

Event-based Interactions

Each time the human conveys information to the computer and the computer returns information, or each time the computer conveys information to the human and the human makes an appropriate response, an interaction (or transaction) has taken place.

Each interaction begins with either the human or the computer creating an *event* that becomes the stimulus for the other component to make an appropriate response. More detailed examples are shown in Figure 12-8 for user initiated and in Figure 12-9 for computer initiated.

Typical user-initiated events
- Typing using a keyboard (making key presses)
  - Entering characters, including spaces
  - Deleting characters
  - Delete/backspace
  - Changing modes
    - Uppercase/lowercase
    - Typeover/insert
    - Keypad: numbers/other
  - Moving location on the screen
    - Page up/page down
    - Home/end
  - Moving the cursor
  - Entering preset functions
    - Function keys
  - Print
  - Pause/break
  - Termination/entry (ENTER)
- Moving the mouse pointer and making clicks
- Handprinting
- Touching the screen
- Speaking characters, words, commands

Possible computer responses (the computer may or may not do additional underlying processing; the focus here is on the interface)
- Displays individual characters
- Displays cursor and/or mouse pointer position
- Displays a written text message
  - In a dialog box
  - In a screen's text area
- Displays an auditory (spoken) message
- Displays graphics and animation
- Displays a photograph
- Displays full-motion video in a window
- Starts and controls an internal device (CD player, tape drive, disk drive, modem)
- Starts and controls a peripheral device (printer, plotter, FAX, lights/temperature)
- Conducts a database search

Figure 12-8 Examples of *user-initiated* events and possible computer responses.

Computer-initiated events (focusing only on the interface)
- Displays a text/data entry field
- Displays questions
- Displays a blank window area (multiline field) with a blinking cursor for character entries
- Displays a blank window area for graphics entry
- Displays menu items, push buttons, selection lists (all widgets?)

Possible user responses
- Typing using a keyboard (making key presses)
  - Entering characters, including spaces
  - Deleting characters
    - Delete/backspace
  - Changing modes
    - Uppercase/lowercase
    - Typeover/insert
    - Keypad: numbers/other
    - Shift/Alt/Ctrl
  - Moving location on the screen
    - Page up/page down
    - Home/end
    - Moving the cursor
  - Entering preset functions
    - Function keys
    - Print
    - Pause/break
  - Pressing the ENTER key
- Moving the mouse pointer and clicking buttons
- Handprinting (pen based)
- Touching the screen
- Speaking characters, words, characters

Figure 12-9 Examples of *computer-initiated* events and possible user responses.

Dialog Style and Input Languages

Many interaction considerations can be divided into those concerned with *dialog style* and those concerned with the *input language* used to carry out the dialog. Two somewhat independent decisions need to be made. The first decision has to do with how the human and computer will interact (dialog style). The second has to do with determining the best language for communication (input language).

For example, when correcting a spelling error using a command-driven word-processing system, the first decision has to do with how this error correction will take place. Will the user backspace? Will users move the cursor to the word that they have misspelled? Will users ignore the spelling error and hope to catch it later? The second decision has to do with the commands that will be allowed. Should the command be delete, substitute, or replace?

How the human and computer combine to go about correcting errors can be very different. In many systems, the actual correction procedure has little to do with the word or words available for issuing commands. Thus, there are two separate decisions to be made, one having to do with procedure and the other having to do with language.

Dialog Styles

A *dialog* is used here to indicate a series of transactions or exchanges of information between a human and computer. The concept of dialog merely suggests that one component somehow sends information and the other component, based on the nature of the information, responds.

Dialogs that take place outside computer systems include two birds exchanging information as they jointly build a nest, a rattlesnake signaling to a rabbit as it prepares to strike and the rabbit's response of moving away, a little girl stroking her purring Siamese cat, or two people talking over lunch. When the exchange of information is no longer two way, the dialog has become a monolog, which is more typical of batch systems.

The dialog styles listed next are intended to be representative of a much larger collection. This concept was originally proposed and developed by Martin (1973) and Ramsey and Atwood (1979). The following dialog styles should give an idea of the variety of human–computer dialogs that can take place within systems.

Commands

Form-filling

Menus and lists

Direct manipulation

Conversational

Dialog styles represent an attempt to provide a structure for better understanding human–computer interactions. Dialog styles build on the previous discussion concerning whether the human or computer initiates the dialog.

In *user-initiated* dialogs, the impetus for a response seems almost trivial when compared with the relative complexity of preparing for a response. Cognitive processing requirements generally become progressively more demanding as we move from computer-initiated to user-initiated dialogs. Because more demand is placed on the user with user-initiated dialogs, both a general understanding of computing and the specific understanding of the systems (i.e., mental models) become more important for effective use.

In an interaction, one component initiates (creates) the stimulus event and the other makes an appropriate (preplanned, programmed) response to the event. When users initiate events or make an event response, they do it to help accomplish work. Each user has a *procedure-completion* goal in mind. Prior to initiating an event or making an event response, users may engage in considerable cognitive (mental) processing. When initiating an event, the cognitive processing includes setting one or more goals.

User-initiated events almost always stimulate a computer response. The response can be as simple as displaying a character or as complex as starting a full-motion video sequence. On the other hand, computer-initiated events may or may not stimulate an observable user response.

Direct Manipulation

Direct manipulation was one of the most significant developments in the early 1980s. Direct manipulation refers to an action-based mode of interaction between people and computers. Direct manipulation theoretically places less load on the human cognitive system and is generally preferred by users over using commands. However, the use of direct manipulation (when compared with menu-based interfaces) did *not* reduce errors and only led to faster task performance for new or occasional users. The difference becomes progressively less as users gained experience (Benbasat and Todd, 1993).

The ultimate extension of direct manipulation (i.e., the *model world*) is the development of *virtual world* systems. These systems create the illusion of being immersed inside the model world. Users are psychologically projected into an *artificial reality* that they can affect through their own actions.

Icons

Usability of Icons

Blankenberger and Hahn (1991) report that words are slower than icons for users to decode, though generally more accurate. Benbasat and Todd (1993) found that the use of icons (versus words without pictures) did not increase speed or reduce errors. However, not all icons are created equal. A bad icon is far worse than a good word.

When assessing the usability of icons in user interfaces, Kacmar and Carey (1991) reported much poorer performance for icons without labels than those having labels (58 versus 86 percent recognition). Pictures without descriptive text (labels) may not be sufficient for adequate user performance when users must form associations. Nevertheless, most studies find that icons continue to be preferred by users.

Designing Icons

When designing icons, the design goals are to (1) ensure that the images suggest the function that they represent, (2) ensure icon discriminability, and (3) demonstrate icon recognition. Effective icon development requires an iterative process in which designers first create numerous candidates, then test alternative designs with representative users, and finally make whatever changes as needed. Keep in mind that the design of an icon is not very important when icons remain in the same location and are regularly used.

Direct Manipulation versus Conversational Interactions

For many (if not most) uses of computer-based systems, a more *conversational* form of interaction will most likely elicit superior performance. This interface uses expressions at the interface that are like utterances in a conversation. The interaction is language based. Changes are made to the system by speaking or keying words. In some cases, changes are made by controlling the behavior of some *virtual partner,* which then carries out the requested actions. This approach employs linguistic utterances rather than physical actions as the primary means of communication.

Is it better to work directly within some model world or work indirectly through some intermediary? Most systems have tasks that are difficult or inefficient to perform directly. These products can be much better accessed using a conversational approach, possibly by delegating to an *intelligent agent.* Many computer-based products are introducing intelligent agents to facilitate the indirect performance of routine activities.

The best guidance is (1) to use direct manipulation with those few tasks that are naturally visual and (2) to use conversational interactions with tasks that can be accomplished more readily verbally.

Conversational Interfaces

Intelligent Agents

Intelligent agents are computer programs that engage and help users. They simplify the use of computers by allowing users to move away from the complexity of command languages and the tedium of direct manipulation. Users are able to move toward more intelligent interaction. Intelligent agents are also known as

Intelligent interfaces

Adaptive interface

Autonomous agents

Knowbots, knobots, softbots, userbots, taskbots

Personal agents

Personal assistants

Well-designed agents offer friendly assistance so smoothly that users are not even aware of their existence.

Some of the most important questions about agents include

Will users want agents?

How might agents feel about people?

Should agents have emotions? (cf. Bates, 1994)

How should users and agents communicate with each other?

How will agents work with other agents?

Should agents have a common corpus of shared knowledge?

Will we have many specialized agents or a few (very smart) humanlike agents?

How will agents obtain new knowledge (learn)?

What types of tasks will agents perform?

How can agents improve human performance?

Personal Assistants

One of the most common intelligent agent metaphors used is of a *personal assistant* who is collaborating with the user. The assistant becomes gradually more effective as it learns the user's interests, habits, and preferences.

Personal assistants can assist users by (1) hiding the complexity of difficult tasks, (2) performing tasks on the user's behalf, (3) training or teaching users, (4) helping different users collaborate, and (5) monitoring events and procedures (Maes, 1994).

Personal assistants are being used to

Filter information

Retrieve information

Manage electronic mail

Schedule meetings

Select books, movies, music, and the like.

Neural Networks

Many intelligent agents and personal assistants "learn" through the use of neural networks. Neural networks represent brain-style computation in which the brain is used as a model of a parallel computational device (Rumelhart et al., 1994). The most important advantage of neural networks is their adaptivity, by which self-optimization characteristics allow the neural network to "design" itself. Large classes of problems appear to be more amenable to solution by neural networks than by other available techniques (Widrow et al., 1994). For example,

Credit card fraud detection

Handprinted character recognition

Loan approvals

Real estate analysis

Speech recognition

EEG waveform recognition

Computer Emotion

One quality that animators feel is important is appropriately timed and clearly expressed emotion. One Disney animator observed, "From the earliest days, it has been the portrayal of emotions that has given the Disney characters the illusion of life (i.e., make them believable)." The emotionless character is lifeless and appears as a machine.

When creating intelligent agents with emotions, clearly define the emotional state of the agent, and then accentuate (exaggerate) the emotion. The emotion should show in the agent's thoughts and actions.

Considering Computers as "People"

People demonstrate a wide range of social behaviors. Users attribute many of these social behaviors to their computers (Nass et al., 1994). This is done even though users know that computers are machines and that they do not actually possess feelings, attitudes, or human motivations. People seem to apply social rules to their use of computers, even though they know the rules are inappropriate.

Responding to Computer Faces

Walker, J. H., et al. (1994) investigated subjects' responses to a synthesized talking face displayed on a computer screen. When compared with answering text-based questions on the same monitor, those participants responding to a talking face made fewer mistakes and wrote more comments. When the face was made more stern in appearance, users made even fewer mistakes, wrote more comments, and spent more time. Users indicated that they liked the neutral face better than the stern face.

Takeuchi and Nagao (1993) evaluated the value of facial displays in a series of studies. They found that the human face communicates conversational signals and emotions. When comparing a system that had facial displays with one that did not, they found that the system featuring facial displays enabled better performance.

Adapting Systems

Applications are designed for classes of users (not individuals) and sets of tasks. Individual users change and tasks continue to vary and evolve. Good user interfaces should change to meet the specific needs of users. Once a system is operational, there are two ways to enable the system to adapt to users: make them *adaptable* and/or make them *adaptive.*

Adaptable systems provide users with tools that make it possible for them to customize certain system characteristics. Users have direct control over all changes. *Adaptive systems* are capable of automatically changing their own characteristics.

The research does not show a clear superiority of one solution over the other (Oppermann, 1994). Studies have shown that users have difficulty using adaptable features in systems and frequently do not use them. Self-adapting systems have inherent problems of their own.

Adaptivity is based on the idea that "computers should be more user literate" in dealing with user intentions. Adaptive systems can support users in learning to use a new system. In addition, they can draw attention to (1) functions that are not being used and (2) more efficient ways to carry out specific operations.

Adaptive systems attempt to match particular user behaviors with particular usage profiles. User errors and usage patterns are used most for making automatic adaptions.

The major steps in an adaptive system are as follows:

The user is observed by the system.

Actions are recorded.

Information about the user is inferred by the system.

Changes are automatically made by the system to the system.

Trumbly (1994) examined the impact of user's computer knowledge level and an adaptive interface on performance. The use of an adaptive interface did significantly improve performance ($p < 0.001$).

Input Languages

Users must convey information to computers concerning actions that they want taken. Probably, the most important consideration in this regard is the *input language* used for this communication.

The input language can be made up of (1) numbers, (2) alphanumeric codes, (3) full words or abbreviations, (4) function keys, and (5) pick and click. User languages have evolved over the years from allowing only the entry of numbers to making verbal statements using full words. In any given system, the users' language can be as simple as pointing to an area of a window and clicking the word. On the other hand, the users' language can be as complex as remembering and entering alphanumeric codes.

Types of Languages

There are two major approaches to languages. The first has to do with those that are related to *direct manipulation* of objects in windows or on the workplace. These include pick and click, drag and drop, and others associated with users directly manipulating objects. These take full advantage of creating model worlds on the workplace, and their natural extension is into virtual worlds (or virtual reality).

A second major approach are the *conversational* languages. Conversational languages can be put into two general categories: commands and natural languages. *Command languages* are special languages developed for a particular system. Usually, one word is assigned to one action. Typically, commands are used to transmit instructions from the user to the computer. Most systems that use commands have a relatively unique command language. Sometimes frequently used commands are assigned to specific keys (e.g., insert, delete, and function keys).

Another category is *natural language.* True natural language suggests that users input to the computer those words that come to mind to carry out an action. These would be the same words used in person-to-person communication. Thus, with natural language input there is no requirement for learning a special set of words in order to communicate with the computer.

Command Languages

Command Structure

Most command languages have a similar command structure. Typically, this is an action word followed by a modifier, which is followed by an object. For example, the action word may be "print," the modifier may be "single space," and the object could be "document A." This particular structure for commands is based on considerable practice and experience in this area.

Systems that use a consistent command structure tend to be easier to learn and easier to use. Unfortunately, some systems mix the command structure by having some commands that have a verb–object format and other commands that are object–verb. This tends to be very difficult for users to learn because of the inconsistency of the command structure.

Command Abbreviations

Commands, in general, should be as short as possible, consistent with the experience level of the user. Truncation seems to be the best way to shorten commands when they are being entered (Ehrenreich, 1985). Inexperienced users may require the command to be a full word in order to feel confident that they have entered the correct command. For example, an inexperienced user may perform better with a command such as "PRINT," whereas an experienced user may prefer to enter a "P."

More experienced users tend to perform better by entering the fewest number of keystrokes. Many systems provide abbreviation alternatives for experienced users by accepting a full word or one or more characters from that word based on what the user decides to enter. For example, with the "PRINT" command, the computer might accept "PRINT," "PRNT," "PRT," "PR," or "P." Frequently, function keys are used to help reduce the number of keystrokes.

The best guideline seems to be to provide users with commands that can be entered with the fewest number of keystrokes consistent with their ability. This may range from entering one character to entering several characters. Obviously, the shorter the command, the faster the command can be entered, and the easier it is for users to prevent keying errors from occurring and to detect those that do.

Command Ambiguity

Commands should be unambiguous. This means that they should be both semantically and perceptually dissimilar. It is not a good idea, for example, to use the command "PRINT" and "WRITE" in the same system when they have two different meanings. These two words are too similar in meaning in everyday use. It is also important to remember that commands should not be visually or auditorily similar. That is, they should not look alike even though they have totally different meanings, nor should they sound alike even though they have totally different meanings. Thus, designers should ensure that commands selected for a system are not easily confused, which includes taking steps to determine whether the commands selected for their systems are already being used by their potential users while interacting with other systems.

Command Guessing

It is not a good idea to overload users with the need to memorize a large number of commands. Regularly used commands are not difficult to remember. The commands that are most difficult to remember are those used occasionally. When a command is not remembered, users either can look it up or guess. If a person attempts to guess the correct command, the probability of being right depends on how the system has been designed (Janosky et al., 1986).

In systems in which each computer action has only one name assigned to it (e.g., "backup" to have the system make a copy of a file), the probability of guessing the correct command is usually less than 20 percent, that is, if the commands are assigned by a designer using what seems to be "most logical," without conducting a survey of users. If a designer surveys potential users and finds out what they would call the computer actions and uses the most common response, then the chances of users guessing correctly are usually between 20 and 40 percent. For example, see the responses in Table 12-1 for the proposed commands used to leave ("quit") an interactive session. In systems in which the computer accepts either a primary command or one of up to five synonyms, the chances of users guessing correctly are usually between 40 and 60 percent.

Table 12-1 Proposed Commands for Ending an Interactive Session

Command Name	Percent of Total
end	35
quit	19
exit	13
finished	10
cancel	3
clear	3
complete	3
done	3
home	3
leave	3
stop	3
all others	2.
Total	100%

From Carter, 1986.

Finally, it is possible to develop command-driven systems in which users can guess the command correctly as much as 95 percent of the time (Furnas et al., 1984). This is done by having the system maintain a table of guesses by users and the action with which they were eventually satisfied. For example, assume that a user wants to make a copy of a file to use as a backup. The user enters the command "backup" and the system responds with the three most likely actions (based on past experience):

- Move back one screen?
- Make a backup copy of the current file?
- Restore the last deleted file?

The user then selects the second choice ("make a backup copy") and the system executes the command. Using this method, the three most likely alternatives can change based on experience gained with users. Also, the option of having alternatives presented at all can be turned on or off based on the experience level of users. Table 12-2 summarizes the information just covered.

Table 12-2 Comparison of Hit Rates (Making Correct Choices) for Command Guessing

Options	Hit Rate (%)
Designer determines command names without conducting a study	Up to 20
Designer determines command names by conducting a survey of potential users	20 to 40
Designer determines command names after conducting a survey and also allows up to five synonyms	40 to 60
Designer allows users to select one of three alternatives presented by the computer	Up to 95

From Furnas, et al., 1984.

Command Categories

In many command-driven systems, there are at least three categories of commands. The first are those that are used regularly and should be memorized by users. In this case, users memorize both the actions and the commands that are associated with implementing these actions.

The second category includes a series of command–action combinations; users learn the actions that they want, but do not try to memorize the specific command that is needed to implement the action. These command–action combinations are usually procedures that are used occasionally, and the commands are not easily remembered. With the commands in this category, it is necessary to provide users with assistance (either computer based or paper based) that helps them to find the appropriate command for initiating the desired action. It should not be necessary for users to memorize these commands. They should focus on memorizing how they can quickly access information related to finding the command.

Finally, a small group of command–action combinations is used so infrequently that users may not even know that they are available. This could also include new procedures that are added to a system to enhance its capabilities. These system capabilities should be provided for users in a document so that they can review what is available when they have special problems. In this case, users attempt (1) to find out whether a procedure exists and (2) to find the command associated with carrying out the procedure. It is seldom a good idea to intermix all three categories into one large user manual.

Natural Language

Many future users of systems will expect the human–computer exchange of information to be similar to what they have read about in books such as Clarke's *2001* (1968). Consider the following excerpt from that book.

C: We have a problem.

U: What is it?

C: I am having difficulty maintaining contact with Earth. The trouble is in the AE-35 unit. My fault prediction center reports that it may fail within 72 hours.

U: We will take care of it. Let's see the optical alignment.

C: Here it is, Dave, it's still OK at the moment. (A screen display is given)

U: What procedure do you suggest?

C: The best thing would be to replace the unit with a spare so that we can check it over.

U: OK. Let us have the hard copy.

Natural language input can be used in many cases for which it is not being used today. For example, natural language is always a candidate for situations when the activity itself limits the words that a person may use to carry out the activity.

Another consideration when trying to decide whether to use natural language is when specific commands or instructions are being given versus using a conversational form of the language. The biggest single problem with using natural language input is that the computer has a difficult time dealing with the context of a situation. This makes many messages ambiguous (Biermann et al., 1985). The computer has trouble understanding different meanings of the same words under different situations. For example, the sentence, "You would not recognize little Jackie, he has grown another foot!" is very difficult for computers to understand. In fact, a computer might interpret this as Jackie now has three feet as opposed to Jackie being 12 inches taller.

Language Trade-offs

There is always a trade-off between the resources spent in the design of systems and the effort required for users to learn and use a new system. For example, natural language requires little training but substantially more design work. On the other hand, the use of a command language is generally easier to design but requires considerably more training by users.

Computer Messages

Computers convey information to users in the form of *messages*. The most useful messages transfer the desired content accurately and concisely. It is essential that messages be complete, and they should *not* require accessing a document to find out what a message means.

A message can be a prompt for more information, it can be a notification generated by an error condition, or it can be purely informational, such as one that identifies the user's location in a program. No matter what its purpose, the message should be concise but clearly understandable to the user. Meaningless words, abbreviations, and codes appearing in messages are of little value and can provide considerable frustration for many users (particularly those who are inexperienced). Keep in mind, however, that a more experienced user may prefer simple (abbreviated) messages, as long as more detailed messages are available when requested. Users should be able to control the amount, format, and complexity of information displayed by a computer.

Messages should reflect the user's point of view, not the computer's, and should be strictly factual and informative. In addition, the computer should present information to the user in directly usable form. A user should not have to search through reference information (either written or computer based) to translate messages received from a computer (that includes error messages). Avoid requirements for transposing, computing, interpolating, or performing other mental gymnastics. For example, the presentation of numerical data that must be scanned and compared with other items probably could be presented best in graphic form. However, if data are presented in graphic form, users should also be able to look at the raw data as an option.

Computer Response Time

Even after years of research, the relationship between computer response time and user performance and satisfaction is not totally clear. About all we can safely say is that slow computer response time (1) reduces the amount of work that users can do, (2) probably has little practical effect on errors, and (3) can be frustrating. One reason that acceptable response times are so difficult to determine is that people's expectations for acceptable response times differ from situation to situation.

Computer response time refers to the time that the user must wait for a computer response following a user input. Obviously, if the time interval is shortened, the human and computer are able to exchange more information in a given period of time. This opportunity to have more information exchanges can improve human performance in many systems.

Too long or erratic response times may have an impact on the user's attitudes toward the system, the user's work habits, the type of work that the computer is to perform, and even the circumstances in which a user will use a computer. User attitudes can affect frequency of use of the computer and also the types of problems for which the computer will be used.

In all but the simplest applications, the question is not merely one of an acceptable response time, but one of a set of acceptable response *times*. Users seem willing to wait varying amounts of time for different types of requests. The amount of time a user is willing to wait appears to be a function of the perceived complexity of the request and the time when the request is made; people will wait longer for "hard" requests or those made at the end of a series of commands, or *closure point,* than during the interaction.

In addition, designers should consider the trade-offs between having optimal response times and the costs required to achieve them. Martin and Corl (1986) report response times in a problem-solving task as long as 5 seconds without performance decrements.

Many characteristics make human-to-human communication effective. Of these, continuity, a give and take of information without long pauses (1 second or less), appears to be one of the most essential. Conversational-type human–computer exchanges should be designed to ensure the perception of continuity.

Meyer et al. (1990) evaluated several different "wait" messages to determine which were best in helping users to feel like less time had passed (while waiting) than really had passed. They found no effect when using static messages (e.g., "Please Wait"). Two dynamic messages elicited the desired effect: (a) gradually increasing line of X's and (b) clocklike graphic filling. To achieve the desired outcome, however, the rate of change should be slower, for example, a change every 2/3 versus 1/3 second.

Developing Object-oriented Interactions

Conducting a Noun Analysis

In Chapter 11, we presented the tasks and related scenarios for a grocery shopping system. We described how designers should conduct a task analysis by developing a set of tasks and then converting the major tasks into sets of task scenarios. The next step in developing a quality user interface is to identify the *objects* that are of greatest importance to typical system users. This is best done by first identifying the *nouns* that were used to describe the work in the task analysis and scenarios; that is, we conduct a *noun analysis* (Abbott, 1983).

To conduct a noun analysis, designers first underline all the nouns in the task statements and then those in the scenarios. A noun is a word that is the name of a subject of discourse. Nouns include the names of persons, places, things, animals, substances, qualities, ideas, actions, states, and so on. The underlined nouns represent candidate objects (verbs can represent candidate operations on the objects).

Noun analysis is quick and an easy way to *begin* identifying user interface components and objects and to place them in their appropriate user interface classes. One major advantage of the technique is that it forces designers to work in the vocabulary of the problem, that is, to use the words that potential users use.

An example of the first occurrence of underlined nouns in the *direct manipulation* alternative for the grocery shopping system is shown in Figure 12-10. All underlined nouns are potential user interface objects. The nouns (without duplicates) from evaluating the scenario are shown in Figure 12-11.

Studies have shown that about 25 percent of useful nouns come from task statements and about 50 percent come from the scenarios. The remaining 25 percent generally show up in both the task statements and the scenarios. Thus, it is important to use nouns from both sources in the initial evaluation of nouns.

Concept of Classification

Classification is part of all good science and helps to facilitate human comprehension of complex issues, including the design and development of computer products. People develop skills in classifying objects beginning at about age 1. Later, children learn to

Access available grocery items
C: Shows all store aisles and shelves (grocery categories) and a small shopping cart
U: Points at the shopping cart and drags it to the desired aisle and shelf (grocery category) of the store

Select the grocery items
C: Shows grocery items on the shelves
U: Points at a grocery item and drags it to the shopping cart (repeats "aisle selection" and "item selection" until done)
C: Shows grocery items accumulating in the cart and a dollar and cent amount accumulating in a window
U: When done, points at the shopping cart and drags it to a cash register icon

Pay for the grocery order
C: Shows a window displaying payment options as icons: Cash, Check or Credit Card
U: Points at the shopping cart and drags it to the desired payment icon

Indicate method of receiving (pickup or delivery)
C: Shows a window displaying completion options as icons: Shopping list, Drive-in window (for customer pickup) or a House (for home delivery)
U: Points at the shopping cart and drags it to the desired completion option

Figure 12-10 Example of the grocery shopping system *direct manipulation* alternative with nouns underlined.

aisles	customer pickup	money amount
cash register	drive-in window	payment options
cash	grocery categories	shelves
check	grocery items	shopping cart
completion options	home delivery	shopping list
credit card	house	store

Figure 12-11 A list of nouns, with duplicates excluded, from the grocery shopping system *direct manipulation* example.

generalize (dogs are animals) and to discriminate (some dogs are beagles). The further development of this skill over the next several years is critical to designing quality user interfaces.

The concept of classification has had tremendous positive impact on the development of knowledge. For example, until the 1700s all living organisms were organized from the simple to the most complex (using an outside view). In the mid-1700s, a Swedish botanist (Linnaeus) suggested a more detailed taxonomy using species. Currently, estimates of the number of species ranges from 5 million to 50 million, with fewer than 2 million having been classified. In the mid-1800s, Darwin used the previous classifications to help him to justify his proposal that natural selection was the mechanism of evolution. Today, the classification of species continues, but the research tends to focus on a more inside view using DNA.

Designers must learn to *discover* whatever organization exists in the work as it is currently being performed. In addition, designers must *invent* a structure for any new product or processes that they develop.

Classifying User Interface Components

Separating into User Interface Classes

Once a set of product-related nouns is identified, designers should attempt to put each noun into one of the major user interface classes. This generally requires an incremental and iterative process.

The user interface classes of most interest include the following:

Task class

Display class

Controls class

Devices class

Objects in the *task class* generally represent containers. These containers hold display objects, control objects, or device objects. The most common objects in the task class are windows and the workplace. Many objects in the task class become either primary windows, secondary windows, or dialog boxes. Objects in the *display class* include all visual and auditory displays. Objects in the *control class* include all the screen-based controls. Objects in the *device class* consist of device-based icons, such as the printer, trashcan, and the like.

User Interface Classes in the Grocery Shopping System

Some of the nouns may be potential instances of the task class for the grocery shopping system. The nouns that represent prospective task objects (windows) are shown in Figure 12-12.

Completion (obtain groceries) options
Grocery categories
Grocery items
Grocery store
Payment (checkout) options

Figure 12-12 Nouns that appear to fit into the task class of user interface classes.

Some of the nouns may be potential instances of the display class. These nouns represent prospective visual displays and are shown in Figure 12-13. Display objects represent information that users will see or hear while using the new product to perform their work.

Aisles/shelves
Amount (cost)
Cash
Check
Credit card
Drive-in window (pickup window)
House (home delivery)
Individual grocery items
Shopping cart
Shopping list

Figure 12-13 Nouns that appear to fit into the user interface display class.

Some of the nouns may be potential instances of the device class. These nouns represent prospective device objects. Device objects usually are represented as icons that can be moved (dragged) around on the workplace or in a window. Device objects can receive messages from display objects. These messages are transferred by touching (slightly overlapping) the display objects and device objects. For example, to print a document, users could drag a specific "document" icon (display object) over and touch the "printer," which is a device object. Other examples of how display and device objects can potentially interact are shown in Figure 12-14.

Touching a *grocery item* (display object) to the *shopping cart* (device object) to add items to be purchased
Touching a *grocery item* (display object) to the *cash register* (device object) to show the cost of an item
Touching the *shopping list* (display object) to the *shopping cart* (device object) to place all items on the list into the shopping cart
Touching the *shopping list* (display object) to the *cash register* (device object) to show the cost of all items on the list
Touching a filled *shopping cart* (display object) to the *cash register* (device object) to show the cost of all items in the cart

Figure 12-14 Examples of the prospective interaction between display objects and device objects when using a direct manipulation dialog style.

Developing Generic Interactions

Once the initial set of user interface objects has been identified and classified, designers should identify the primary windows. Each task object can be evaluated to determine which could become primary windows. The primary windows will contain user interface objects from both the display and control classes.

Background Screen or Workplace

The background screen, which is also known as the workplace, workspace, or desktop, is used in virtually all systems that have a visual interface. This area fills the entire screen and is used as an overall container for visual user interface objects. Users can organize objects in the workplace according to the work activities (tasks) to be performed.

Windows

Windows are areas with visible boundaries that are used to display user interface components or views of object. Windows also can be used to present messages and to prompt for information. There are two major types of windows. The *primary* window is one in which users carry out their primary interactions with computers. The *secondary* window, which may also be known as a dialog box or transient window, is always associated with a primary window. Secondary windows contain information that depends on information in the primary window. Some secondary windows are predefined (reusable) across products (e.g., Open, Save As, etc.).

Windows contain *parts* that are used to assist in human–computer interactions. These include the following:

A *border* that defines a window and allows users to size windows.

A *title bar* at the top of a window that includes the window name, the name of the object, or a short description of the contents. The title bar also contains a window menu button, a minimize (or hide) button, or a maximize button.

A *menu bar* that appears across the top of a window just below the title bar. The menu bar contains a list of choices called *menu items.* By single clicking on a menu, a pull-down menu appears.

A *client area* that contains the objects and messages of greatest interest to users. The organizing elements can include group boxes, separators, and panes (panes can show parts of the same object).

Primary Windows

Primary windows are areas with visible boundaries that are used to show views of an object. Primary windows are the main window in which users interact with objects, while secondary windows (also known as dialog boxes or transient windows) can be associated with primary windows.

The criteria for determining whether a window should be a primary or secondary window include (1) the relative importance of the task, (2) the frequency of use, and (3) the primacy versus assisting role of the window. An example of a primary window is shown in Figure 12-15.

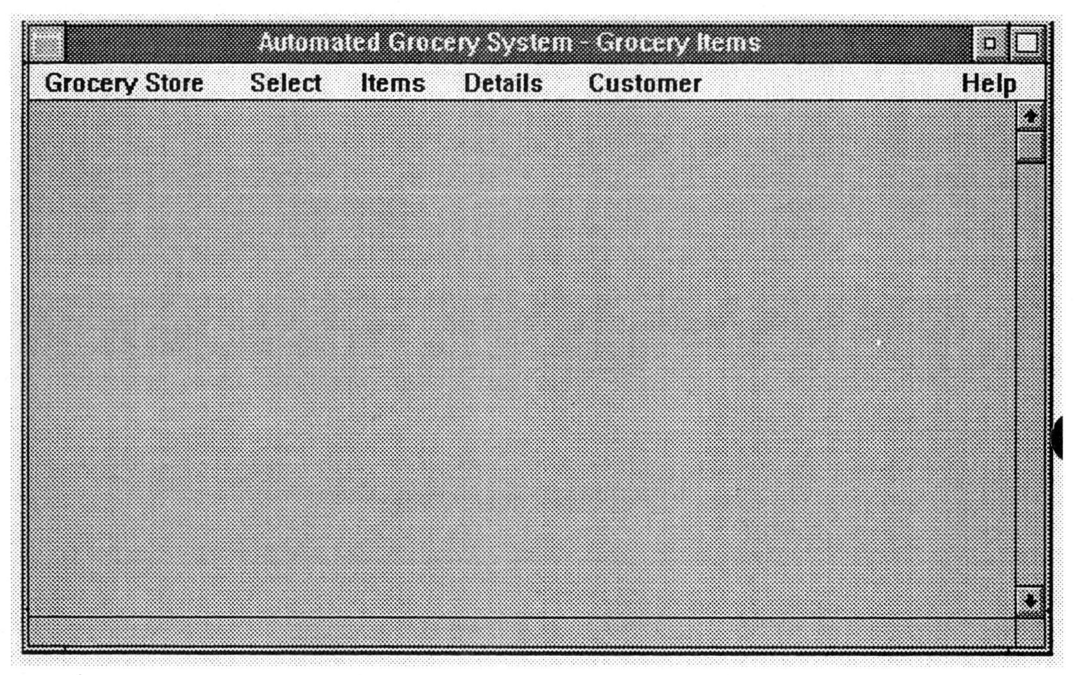

Figure 12-15 Example of primary window.

Views

Each primary window will have one or more *views*. Views represent details about an object and provide alternative ways of looking at an object. Designers should determine the views that will be of most use to users. Users, then, will select which views to use and when to use each view. The views will be made up of displays (usually rough sketches), screen-based controls (e.g., pushbuttons), and devices (initial icon drawings).

Views represent details about an object, with each view showing the object from a different perspective. Multiple views can be displayed by users at the same time (different ways of looking at the same object) in different windows.

Grocery Shopping System Example: Windows and Views

Primary Windows

The potential primary windows and associated views for the grocery store system include (1) *Grocery Store* with four views, (2) *Grocery Items* with eight views, and (3) *Grocery Item Details* with six views. These are shown in more detail in Figures 12-16, 12-17, and 12-18.

View 1 - External store - Picture of typical store front that is accessed by moving the mouse pointer to the front door and clicking.
View 2 - Internal store
Icons/graphics for telephone to connect to customer/store
Access to grocery items
Shopping cart
Checkout line/cashier
Shopping list
Pickup/delivery options
Customer profile, etc.
View 3 - Descriptive information
Address for the store
Telephone numbers for the store
Address for the pickup window
Hours of operation, etc.
View 4 - Help

Figure 12-16 Examples of views for the *Grocery Store* primary window.

View 1 - List of all grocery items available at the store (alphabetized within major category)
View 2 - List of all grocery items previously selected and purchased by this customer
View 3 - "Short list" of the grocery items regularly selected and purchased (daily or weekly)
View 4 - "Long list" of the grocery items typically selected and purchased (weekly or monthly)
View 5 - List of grocery items that are on sale (coupons, discounts, etc.)
View 6 - List of grocery items that are temporarily sold out or no longer available to the store
View 7 - Map of store layout showing aisles of major grocery categories
View 8 - Help

Figure 12-17 Examples of views for the *Grocery Items* primary window.

Secondary Windows

The task objects also suggest the existence of secondary windows. These secondary windows are related to activities occurring within a primary window. Secondary windows are generally used when it is necessary (or useful) to extend the interaction with a primary window.

The possible secondary windows include (1) payment options, (2) completion (obtain groceries) options, and (3) customer preferences. Examples of secondary windows in

Option 1
View 1 - Color photo of package front/back
View 2 - Label text (ingredients, product description, cooking instructions, nutrition information, servings per package)
Single-unit price
Quantity prices (2, 3, case, etc.)
Available sizes
View 3 - Available suppliers, companies
View 4 - Similar products from other companies
View 5 - Coupons/discounts
View 6 - Help

Option 2
View 1 - Color photo of package front/back, showing sizes and variations, e.g., low-salt versions
View 2 - Label text (ingredients, product description, cooking instructions, nutrition information, servings per package)
View 3 - Prices: single-unit price, quantity prices (2, 3, case, etc.), competing prices (other brands), sales or discounts, "best deal" (comparing price per unit)
View 4 - Competing brands: available suppliers and companies; or similar products from other companies
View 5 - Location of item in the store (map with shelf area blinking)
View 6 - Help

Figure 12-18 Examples of views for the *Grocery Item Details* primary window.

the grocery store system are shown in Figures 12-19, 12-20, and 12-21. All three secondary windows are associated with the *Grocery Item* primary window.

Cash
Store charge
Bank
Checking account
Credit card
Direct transfer

Figure 12-19 Example of a secondary window showing *payment options* in the grocery shopping system.

Shop at store (shopping lists)
Drive-in (pickup)
Delivery (and related information)

Figure 12-20 Example of a secondary window showing *completion options* in the grocery shopping system.

Preferences
- Initially set up with a question/answer
- Automatically updated with experience
- Changed directly by users

Contents
- Items usually purchased in a single order
- Brands usually purchased
- Sizes usually purchased per item
- Amount usually purchased
- Preferred method of payment
- Credit history with store
- Preferred method of obtaining groceries (shop, pickup, or delivered)
- Use of coupons or other discounts

Figure 12-21 Example of a secondary window showing *customer preferences* in the grocery shopping system.

Developing Style-related Interactions

The design activities that have been discussed do not require the use of a specific style guide. All style guides support the use of primary (and related views) and secondary windows. But once the windows have been identified, a specific style guide is required to ensure compliance.

The four most popular style guides are as follows:

Windows (Microsoft)

Motif (Open Software Foundation)

Common User Access (IBM)

Macintosh (Apple)

Menus

Sorting-based Menu Categories

Hayhoe (1990) observed that maximum information with the least cognitive effort is achieved if menu categories closely map the *perceived world structure* of typical users. Users should define the categories menu or list items. The categories constructed from group knowledge are best for any individual. This group knowledge can be determined by using pairwise comparisons (Pathfinder software package) or sorting cards into categories. Designers (user interface specialists) and users should work together when assigning names to the categories.

Menu Breadth and Depth

When a menu system is being designed, the optimal breadth and depth of the menu structure must be considered. For many applications the optimal breadth seems to be between four and eight (MacGregor and Lee, 1987). Paap and Roske-Hofstrand (1986) found that the optimal breadth may be much larger for menu structures that have large numbers of options. This is particularly true if the final goal is not easy to determine from the menu bar or the system has slow response time. Bishu and Zhan (1992) found that broad menus elicited faster performance than deep menus and that one-level menus were faster and more accurate than two- or three-level menus.

Organization of Menus and Lists

The way items on a menu are arranged can affect the efficiency of performance. Even though arrangement by functionality provides advantages to people who understand the items well, and an alphabetic list is helpful when the selection names are known, all differences disappear after extended use (Somberg, 1987).

Halgren and Cooke (1993) report that participants were faster with alphabetical and categorical organizations of menus and lists than random ones ($p < 0.05$). They found no significant difference between alphabetical and categorical. Problems with explicit targets were solved faster than those that were implicit, and simple problems were solved faster than complex problems. The effects of problem complexity were more pronounced for the random organization ($p < 0.05$) and implicit target conditions ($p < 0.05$). They found that overlapping categories were detrimental to alphabetical and random organizations, but were not detrimental to categorical organizations. In fact, there was a tendency for overlapping categories to be associated with *fewer* errors when the menu organization was categorical.

Smelcer and Walker (1990) reported that an *alphabetic* organization is best when users know the exact names of desired items, and a *functional* organization is best when users know only the definition of items. Somberg (1987) found that *alphabetic* and *frequency-of-use* arrangements elicited the fastest performance.

Mehlenbacher et al. (1989) reported that *functional* organization is most effective for novices (low frequency of use) and that menu organization makes little difference for experts (slight advantage for alphabetical).

Harpster (1987) observed that performance on *ordered* menus (versus random) was faster. This was true for menus having even two items. The more items, the greater the advantage is.

Menu Selection Methods

Shinar and Stern (1987) compared (1) keying the option number (then ENTER), (2) keying the first letter of the option (then ENTER), and (3) using the cursor keys to move the cursor bar (then ENTER). The fastest was using the first letter and using the cursor bar with five or fewer options. The slowest was using the cursor bar with six or more options.

Other Menu Issues

- **Vertical versus horizontal menus:** Backs et al. (1987) found that search time was shorter for vertical menus than for horizontal menus. Most users (88 percent) preferred the vertical menus.

- **Graying out versus deleting inactive items:** Francik and Kane (1987) reported that selection speed from a menu was fastest when inactive items were deleted rather than grayed out.
- **Spacing in menus:** Williams (1988) observed that double spacing yielded shorter search times when accessing menus. Double spacing was preferred by 85 percent of the participants.
- **Improving pop-up menus:** Walker et al. (1991) reported that pull-down menus elicited 23 percent shorter movement times than pop-up menus to the correct second-level (cascaded) menu. To improve access to pop-up menus (Walker et al., 1991), (1) put entry areas in the middle of the menus (not at the top), (2) make each choice area progressively larger, and (3) make the entire menu wider.

Menu Bars and Pull-down Menus

The first decision usually made using the style guide is related to the use of the menu bar and associated pull-down menus. There are numerous guidelines for developing useful menu bar items. Several of these are shown in Figure 12-22.

Where appropriate, include predefined menu items
Put predefined items in the correct order
Add menu items that will make the product more usable
Include a full set of items on the menu bar (to provide a memory aid)
Identify and use the appropriate mnemonics and accelerators
Use the proper spelling and capitalization rules (e.g., capitalize the first letter of a choice)
When a choice consists of more than one word
    Capitalize the first letter of all words
    Except articles, prepositions, and "to" in an infinitive
Clearly define defaults
Appropriately use "..." and ">>"
Ensure that the menu bar items are assigned meaningful names (determined by designers)
Ensure that the pull-down menu items are located where users expect (determined by users)
Allow users to choose the format in which the menu bar items will appear (text, graphics, or both)

Figure 12-22 Recommendations for developing menu bars.

Examples of menu bars are shown in Figures 12-23, 12-24, and 12-25.

Strengths of Menu Bars

One advantage of menus bars is their ability to serve as a memory aid. This provides the opportunity for users to glance at the menu bar and be reminded of available options.

File
New or New...
Open...
Save
Save As
Print or Print...
Print Setup
Exit

Figure 12-23 Traditional menu bar defaults for the *File* menu item.

[Class name]
    Open as—[View names]
    Print
View
    [View names]
    Sort...
    Include...
    Refresh—[On, Off]
Selected
    Open as—[View names]
    Print

Figure 12-24 An example of object-oriented menu bar defaults for the *Class name, View,* and *Selected* menu items.

Designers should continually strive for shallow menus (menu breadth) by filling the menu bar and providing as many items as needed in the pull-downs. Designers should try to have as few cascaded menus as possible.

Difficulties with Menu Bars

Some potential problems with developing useful menu bars and their related pull-down menus include

- Designers not knowing how the menu will be used and trying to allow for various approaches
- Nouns and their related task objects help to define windows, but there are no similar methods for determining menu items
- Designers must decide if items should be on the menu bar, in the pull-down, or on a pushbutton

Selected
- Open As ->>
- Aisles/shelves . . .
- Available items . . .
- All items previously ordered . . .
- Items typically purchased ->>
- Short list . . .
- Long list . . .
- Sale items . . .
- Unavailable items . . .
- Print
- Exit

Edit
- Remove last item
- Put back last item
- Empty the cart
- Find . . .

Details (for the highlighted item)
- Color photo of package (sizes and variations). . .
- Label text ->>
- Product description . . .
- Ingredients . . .
- Cooking instructions . . .
- Nutrition information . . .
- Pricing . . .
- Competing brands . . .
- Location in the store . . .

Windows
- Cascade
- Tiled
- Most recently used (MRU)

Help

Figure 12-25 Menu bar and pull-down menus for the *grocery items* primary window.

Choices and Emphasis

There are two additional considerations when dealing with menu items and push buttons. These include the appropriate use of *choice types* and *emphasis.*

Choices

Choices are text or graphics that users can choose to modify or manipulate an indicated object. Screen-based controls are the means through which users make choices (e.g., menus, pushbuttons, or selection lists).

Types of choices include

Action choices: an immediate action takes place (e.g., Cancel)

Dialog choices: displays a secondary window with additional choices (. . .)

Cascading choices: Displays a menu with additional choices (>>)

Emphasis

Emphasis is a graphical cue that helps users to distinguish the state of an object. Types of emphasis include the following:

- In-use: shows as diagonal stripes behind an icon, indicating that the window is open on an object
- Selected: shows by changing the foreground and background colors when an object or data item is selected
- Unavailable: shows by dimming the choice that the choice cannot be activated (an audible cue sounds when users attempt to use)
- Source: shows by reducing the contrast (changing every other pixel in the background) of a source object when being dragged using direct manipulation
- Target: shows as a solid thin line around an icon (e.g., around a trashcan icon) when it is acting as a receiver during direct manipulation

Screen-based Controls

Relatively few *screen-based controls* (they are also known as widgets or window elements) are used in most graphical user interfaces. This class of objects enables users to interact with the computer. The most common screen-based controls are shown in Figure 12-26. Each control is briefly discussed, including recommendations for appropriate use, in the following pages.

Menu Items

Definition

Menu items are the controls contained in menus. Types of menus are menu bars, related pull-downs, and pop-ups. Types of choices include action, dialog (. . .), and cascading (>>). An example is shown in Figure 12-27.

Menu items
Pushbuttons
Open selection lists
Drop-down selection lists
Combination drop-down selection lists
Combination open lists
Text entry fields
Check boxes
Radio buttons
Spin lists
Combo spin lists
Sliders
Notebook
Palette
Value set

Figure 12-26 The most commonly used screen-based controls in graphical user interfaces.

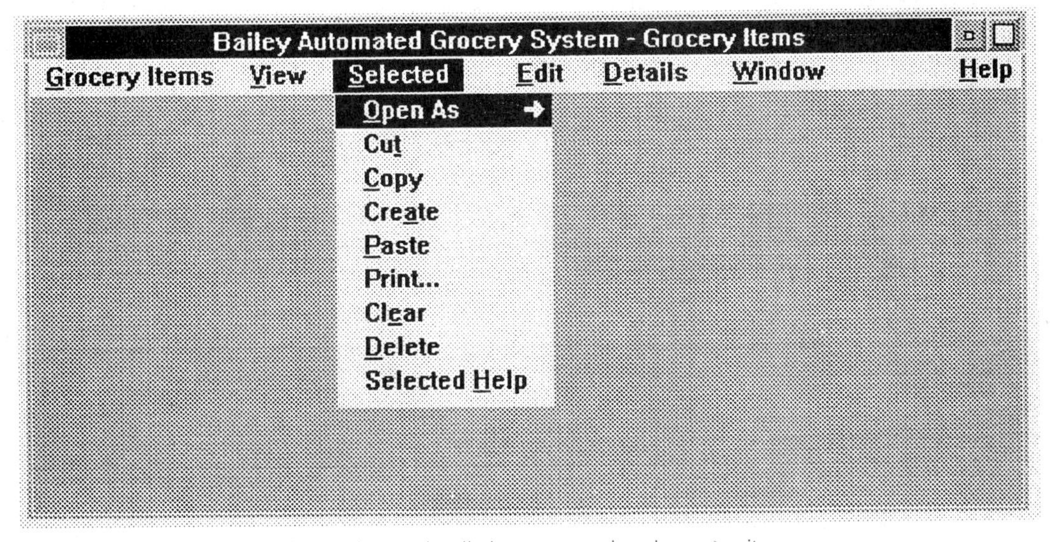

Figure 12-27 Example of a menu bar and pull-down menu showing menu items.

Typical Use

When users are required to make occasional or infrequent choices while performing tasks

Guidelines

1. Provide familiar, fully spelled out descriptions of available choices.
2. Choices should be a single word (if possible).
3. Capitalize the first letter of only the first word of a list item (unless the item contains another word that normally would be capitalized).
4. Place only valid choices in menus (action, dialog, cascading).
5. Put the most frequently used choices at the left or at the top.
6. Use an appropriate and consistent ordering method (frequency of use, categorical, alphabetical).
7. Place at least two choices in a menu.
8. Place as many choices as possible on the menu bar.
9. Place as many choices as needed in each pull-down menu (try to avoid cascading).
10. Place related choices together.
11. Use separators to distinguish groups of related choices.
12. Keep the relative order of identical choices the same across different menus.
13. When appropriate, include predefined choices using the correct terms and proper order.
14. Place product-specific choices either following a group of related predefined choices or at the bottom of a pull-down menu.
15. Provide mnemonics for each textual choice in a menu.
16. Provide shortcut key combinations for frequently used menu items.
17. Provide the ability to define graphic buttons for frequently used menu items.
18. For menu items that cannot be activated in the current context, display unavailable emphasis (dimming).
19. Provide pushbuttons, radio buttons, or check boxes in addition to menu choices when the menus are used frequently for short periods of time.
20. If users are allowed to add and delete menu items, provide a means for restoring the original set of items.

Pushbuttons

Definition

A control containing text or graphics (or both) that represents an action or dialog choice. The control is activated when "pressed" by users (i.e., a single mouse click or finger touch). Examples are shown in Figures 12-28 and 12-29.

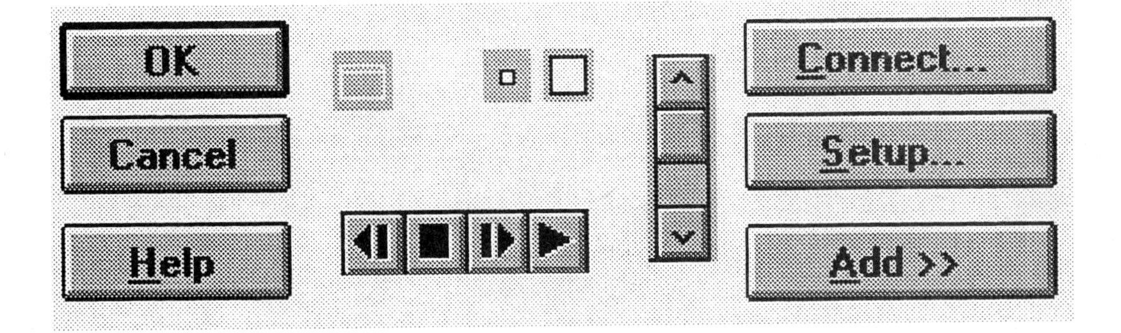

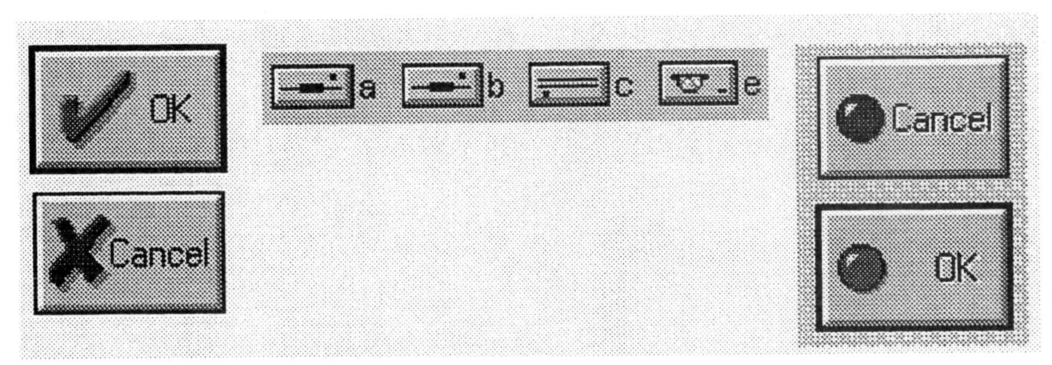

Figure 12-28 Examples of pushbuttons.

Typical Uses

When there is a need to provide convenient access to frequently used action or dialog choices and/or when a menu bar is not provided

Guidelines

1. Capitalize the first letter of only the first word of a choice appearing on a pushbutton (unless the choice contains another word that normally would be capitalized).
2. The label of an action choice pushbutton should clearly indicate the action that will be applied (e.g., Cancel).
3. The label of a dialog choice pushbutton should indicate the window name.
4. The label of a dialog choice pushbutton should include an ellipsis (. . .) following the choice text.
5. If one pushbutton in a group of pushbuttons is used more frequently than the others, make that pushbutton the default.

6 If two pushbutton choices are mutually exclusive (e.g., Undo and Redo), use two pushbuttons and display unavailable emphasis on whichever one is unavailable given the current state of the object.

7 For pushbuttons that provide choices that can be used repeatedly (e.g., arrow buttons on the scroll bar), repeat the action as long as users press and hold the Select button.

8 Place pushbuttons that affect the entire window horizontally at the bottom of the window and justified from the left edge of the window.

9 When a window contains both pushbuttons and a menu bar, duplicate the pushbutton in the pull-down menus.

10 Combine pull-down menu choices to create new pushbutton choices for frequently used combinations of choices.

11 Avoid using group boxes around groups of pushbuttons.

12 Assign a unique mnemonic to each pushbutton choice that has a text label.

13 Use predefined pushbutton labels when possible. Examples include the following:

Apply: Applies changes made to properties choices

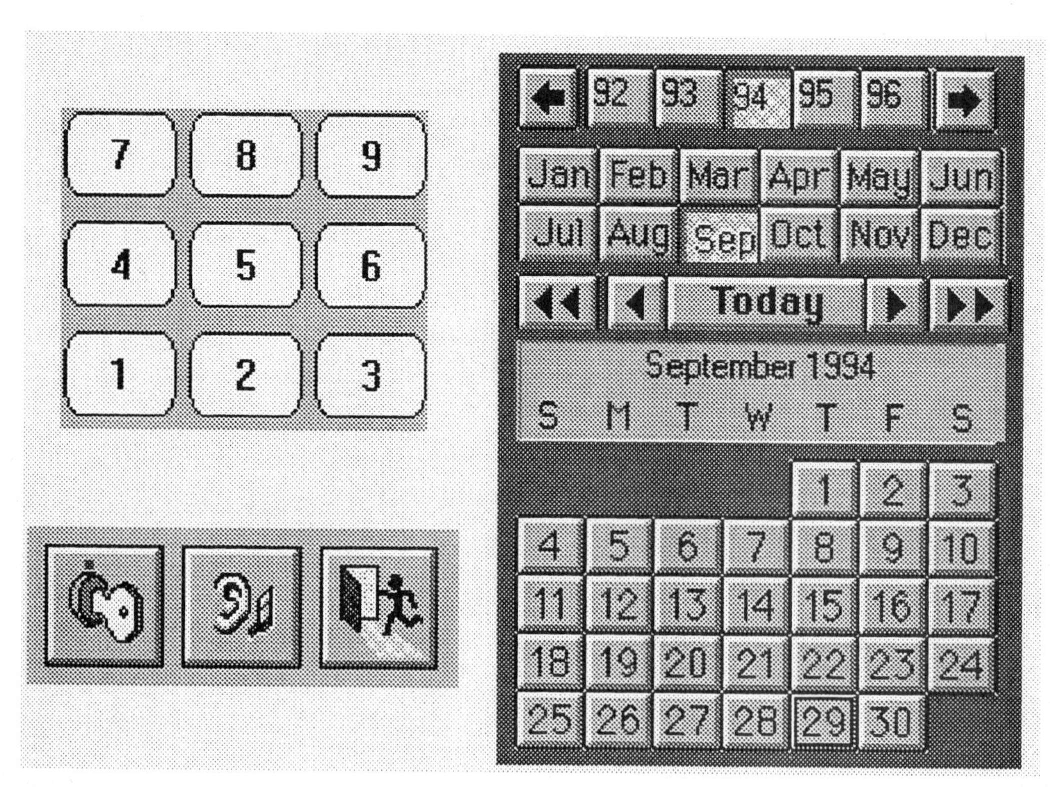

Figure 12-29 More examples of pushbuttons.

Cancel: Removes a window where changes could be made, without applying any changes

Close: Removes a window where changes could not be made

Continue: Resumes a process that has been interrupted

Help: Displays a window containing potential help information

OK: Removes the window and accepts any changes made

Pause: Temporarily suspends a process

Reset: Returns the settings of changed property choices to their last saved state

Resume: Continues a process that was paused

Retry: Tries a process again that has been interrupted

Stop: Ends a process and removes the window

14 If Cancel and Help are used, place them to the right of all other pushbuttons.

15 Avoid using the Close and Cancel pushbuttons on the same window.

16 Whenever an Apply pushbutton is used, provide a Reset pushbutton.

Open Selection List

Definition

A control that contains an open list of items from which users can select. An example is shown in Figure 12-30.

Typical Uses

When decision-making alternatives are required to complete a task

When the window contains sufficient room for an open list

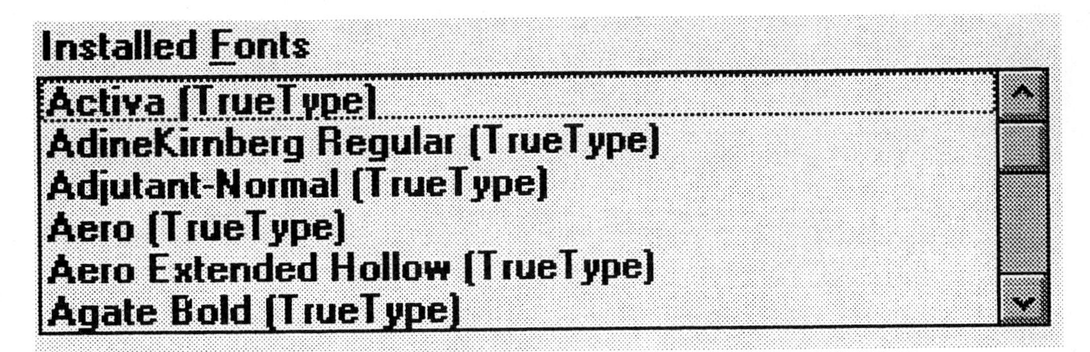

Figure 12-30 Example of an open selection list.

Guidelines

1. Avoid putting product-related controls in a list (e.g., File, Open, and Cancel).
2. Display the elements in an order that is meaningful to users (e.g., alphabetic or chronological).
3. Capitalize the first letter of only the first word of a list item (unless the item contains another word that normally would be capitalized).

Drop-down Selection List

Definition

A control in which a read-only item shows and a selection list is hidden until users take an action to display other list items. The drop-down list usually only allows a single selection. An example is shown in Figure 12-31.

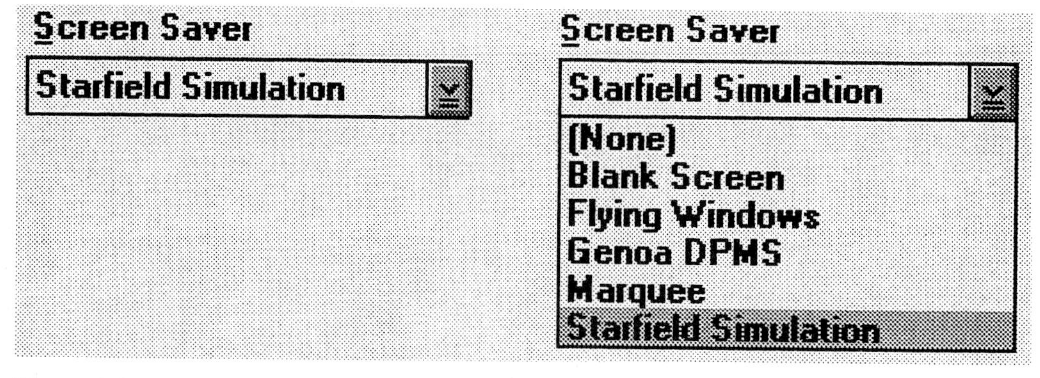

Figure 12-31 Example of a drop-down selection list.

Typical Uses

When decision-making alternatives are required to complete a task

When the list of items does not need to be changed by users

When window space is so limited that there is not enough space to display an open selection list

Guidelines

1. Display an initial value from the list (i.e., the last item selected) as read-only text.
2. Display the list of items in an order that is meaningful to users (e.g., alphabetic or chronological).
3. Capitalize the first letter of only the first word of a list item (unless the item contains another word that would normally be capitalized).

Combination (Combo) Open Selection List

Definition

A control that combines the functions of a text entry field and an open selection list. The combo open selection list contains a text entry field and a list of elements through which users can scroll. Users can either type an entry or make a selection from the list. An example is shown in Figure 12-32.

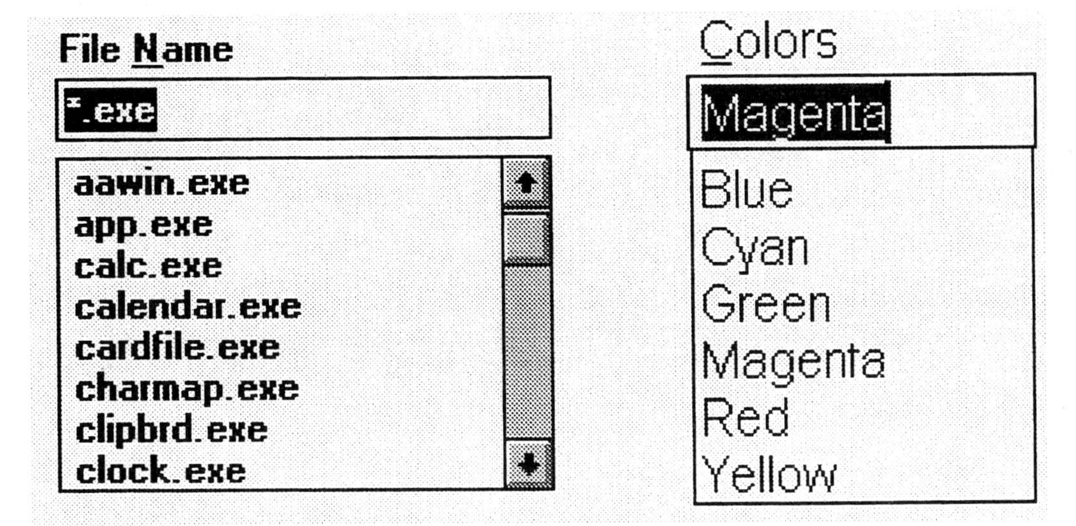

Figure 12-32 Example of a combo open selection list.

Typical Uses

When decision-making alternatives are required to complete a task

When users might have to type values that are not provided on the list

When a set of commonly used alphanumeric characters can be provided to assist users in completing the text entry field

Guidelines

1. Display an initial value from the list in the text entry field.
2. Display the initial value with selected emphasis so that the value is replaced by the first typed character.
3. Display the list items in an order that is meaningful to users (e.g., alphabetic or chronological).
4. Capitalize the first letter of only the first word of a list item (unless the item contains another word that normally would be capitalized).

Combination (Combo) Drop-down Selection List

Definition

A combination list in which one item shows, but the list is hidden until users take an action to make it visible. The combo drop-down selection list contains a text entry field and a list of elements through which users can scroll. Users can either type an entry or, once the list is opened, make a selection from the list. An example is shown in Figure 12-33.

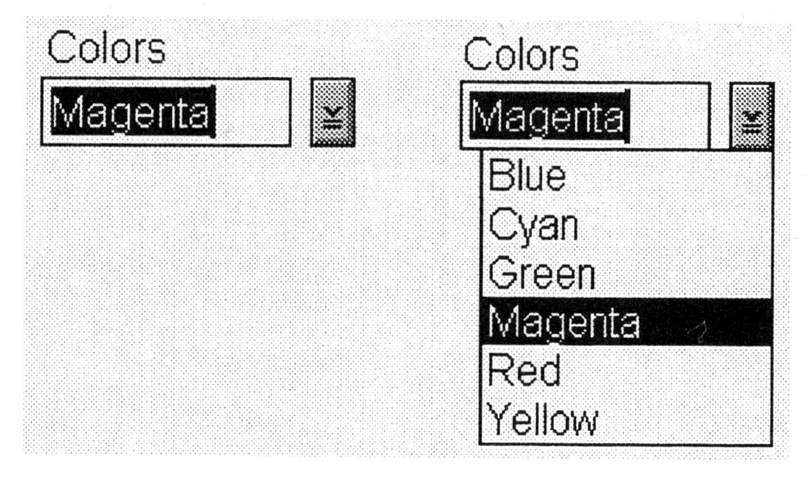

Figure 12-33 Example of a combo drop-down selection list.

Typical Uses

When decision-making alternatives are required to complete a task

When a window does not have enough space to use a combo open selection list

Guidelines

1. Display an initial value from the list in the text entry field.
2. Display the initial value with selected emphasis so that the value is replaced by the first typed character.
3. Display the list items in an order that is meaningful to users (e.g., alphabetic or chronological).
4. Capitalize the first letter of only the first word of a list item (unless the item contains another word that normally would be capitalized).

Check Box

Definition

A control used to display a choice that has two clearly distinguishable states, such as "on" or "off." Check boxes can be used individually or in a group to provide users with multiple choices. An example is shown in Figure 12-34.

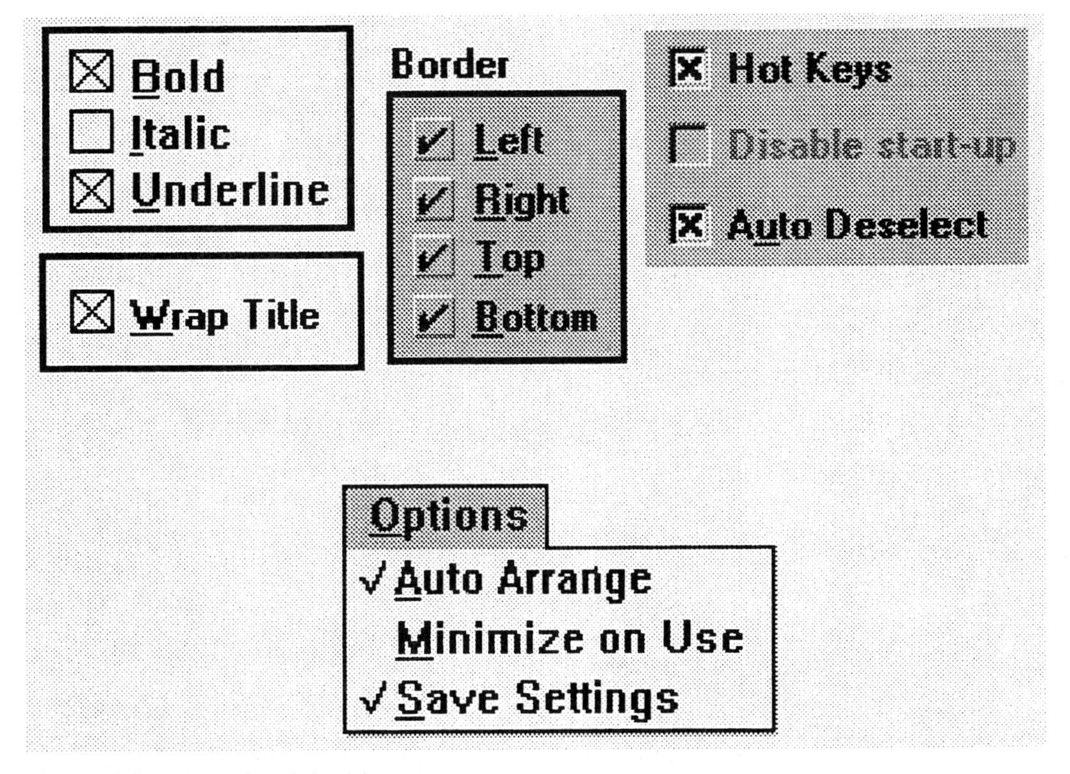

Figure 12-34 Examples of check boxes.

Typical Uses

Use individually, when choices only need to be set to "on" or "off," "yes" or "no," and so on.

Use as a set (group) when choices are not mutually exclusive.

Guidelines

1 Assign a mnemonic to each check box choice.

2 Capitalize the first letter of only the first word of a label (unless the item contains another word that would normally be capitalized).

3. Capitalize the first letter of all major words in the choices.
4. Avoid using check boxes (check marks) in menus.
5. Use instead of two radio buttons if the choice can only be set to "on" or "off."
6. If all the objects or data items have that property turned on, display a mark in the check box.
7. If some, but not all, of the selected objects or data items have that property turned on, fill the box with shading.

Entry Field

Definition

A control into which users type one or more text characters. The entry field can contain one or more lines. Also known as the text entry field, text field, text region, or text box. An example is shown in Figure 12-35.

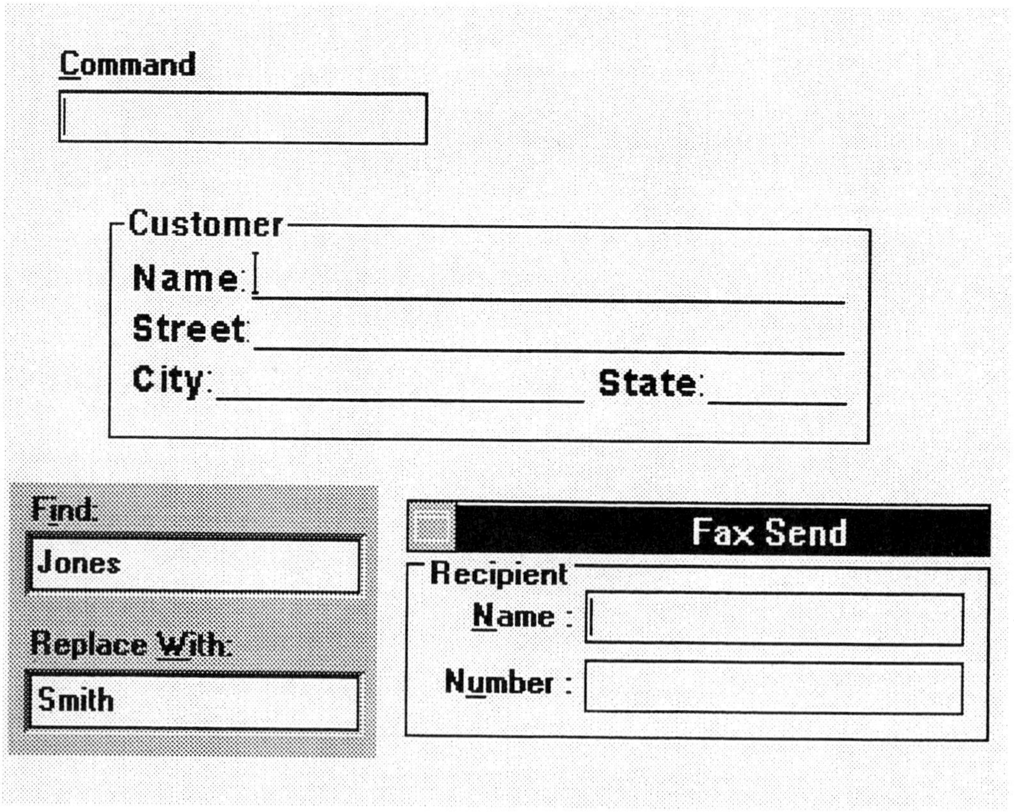

Figure 12-35 Examples of entry fields.

Typical Use

When users are required to make text entries, including commands, field information, and so on.

Guidelines

1. Make the entry field wide enough to show all the data of average length.
2. When several entry fields are displayed, make the length of the fields consistent.
3. Display an entry field with a background color different from the background color of the underlying window.
4. Provide a clear indication when an entry field is required.
5. Provide a clear indication to show that an error has occurred in the information entered in an entry field.

Radio Button

Definition

A control used to select and display mutually exclusive choices. An example is shown in Figure 12-36.

Typical Use

When there is a need to identify one of many (two or more) mutually exclusive choices.

Guidelines

1. Avoid using radio buttons for action, dialog, or cascading choices.
2. Use at least two radio buttons together (never use one alone).
3. If users can choose not to activate any of the choices, provide a choice labeled "None."
4. Arrange related radio buttons in rows, columns, or both.
5. If a choice is currently unavailable, display the radio button and its label with unavailable emphasis.
6. Capitalize the first letter of only the first word of a radio-button choice (unless the choice contains another word that normally would be capitalized).
7. Assign one of the radio-button choices as the default.
8. Avoid using radio buttons for graphical choices (use a value set).
9. Assign a mnemonic to each radio-button choice.

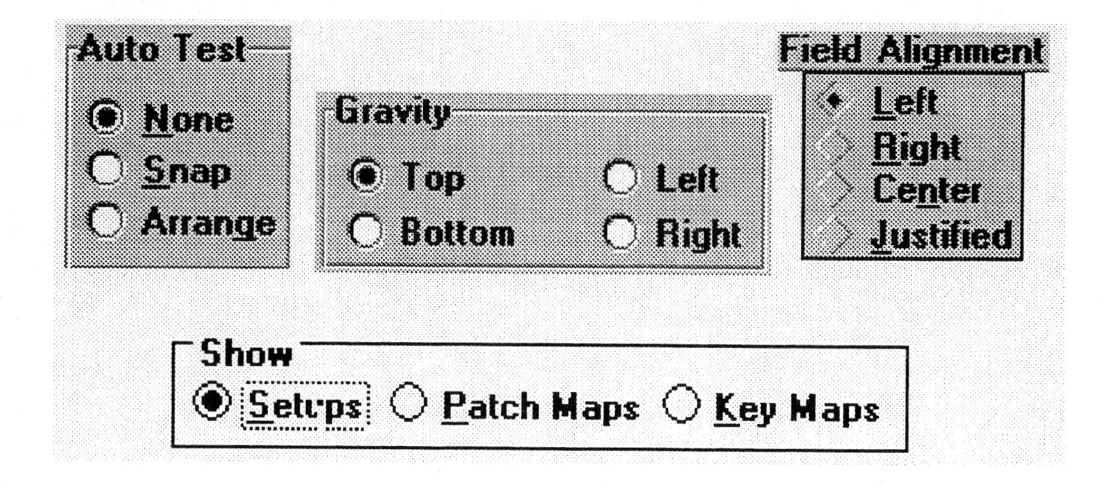

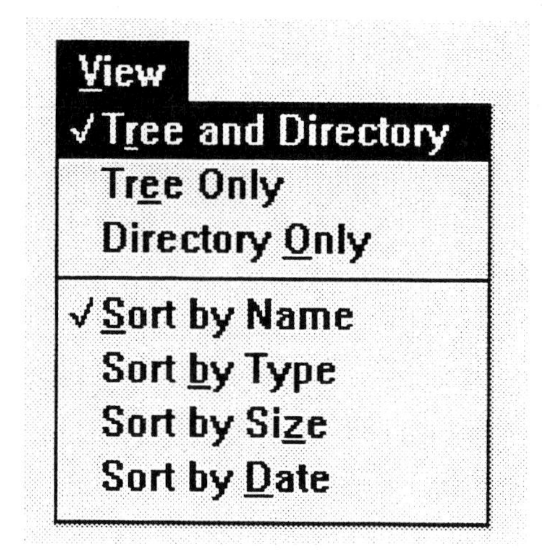

Figure 12-36 Examples of radio buttons.

Spin List

Definition

A control used to display in sequence a series of items, such as days of the week. Users press a small pushbutton to cycle through the options and find at the desired selection. Some spin lists are combo spin lists that allow users to either spin through the options or type into an entry field. An example is shown in Figure 12-37.

Typical use

When potential selections have a logical consecutive order

Guidelines

1. Pressing the up-arrow button should cause the display to increase or move forward, for example, from 14 to 15, from Friday to Saturday, or from low to medium.
2. Pressing the down-arrow button should cause the display to decrease or move back, for example, from 12 to 11, from August to July, or from high to moderate.
3. Allow users to spin through the entire list and back to the beginning by pressing only the up-arrow or down-arrow buttons.

Slider

Definition

A control used for making qualitative settings. Examples are shown in Figures 12-38 and 12-39.

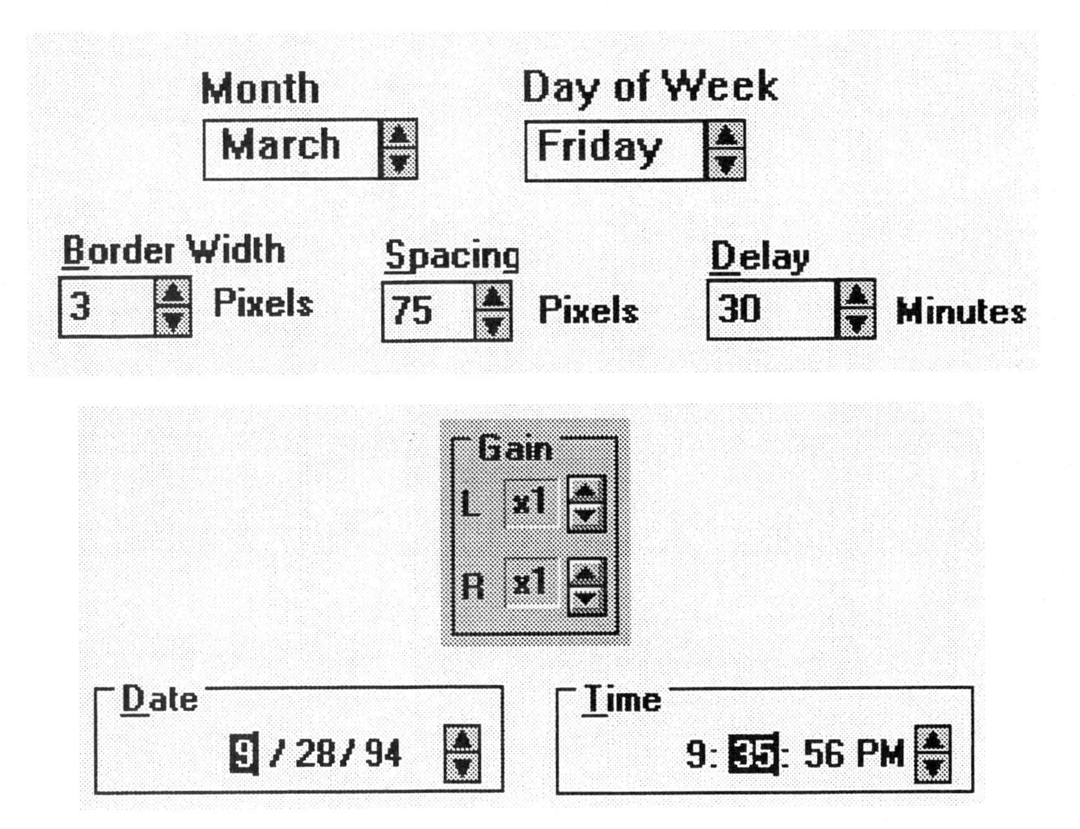

Figure 12-37 Examples of spin lists.

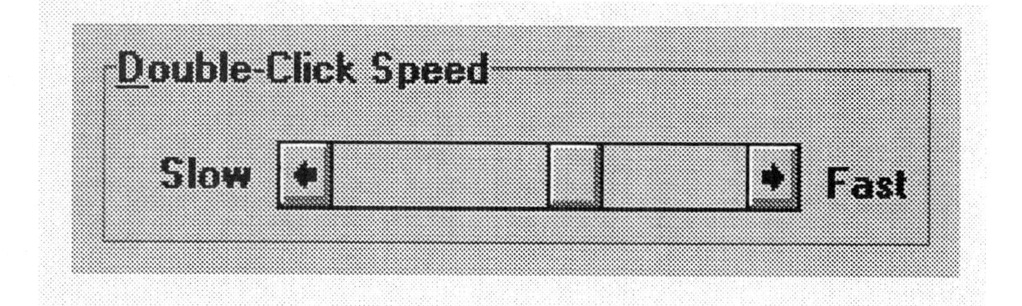

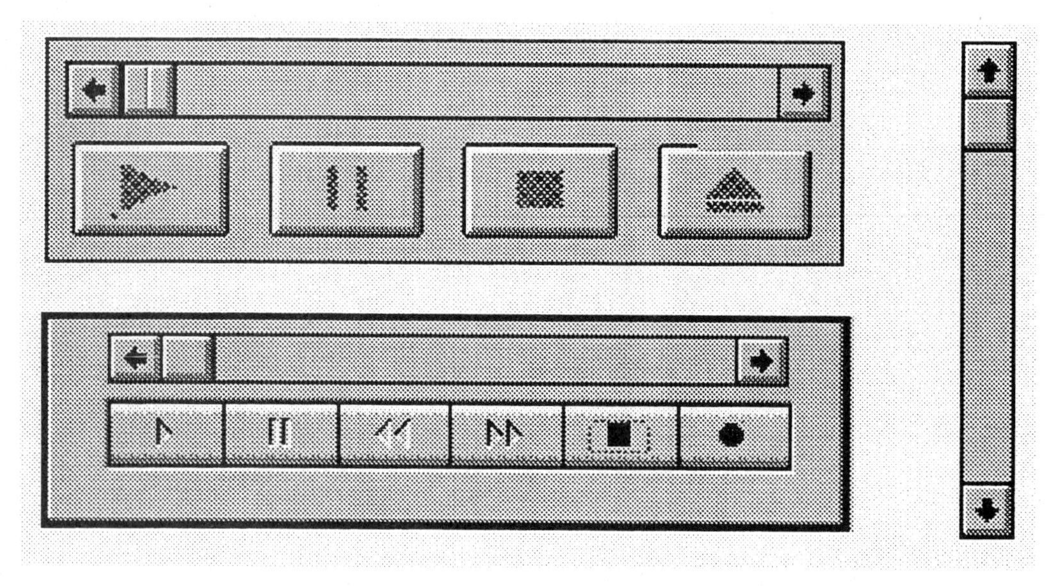

Figure 12-38 Examples of sliders.

Typical Use

When users can benefit from viewing the current value relative to the range of all possible values

Guidelines

1. Identify the slider with a label.
2. Provide a scale to indicate the units of measure represented by the slider.
3. Provide detents on the slider to allow users to easily set values that have special use in a product (if appropriate)

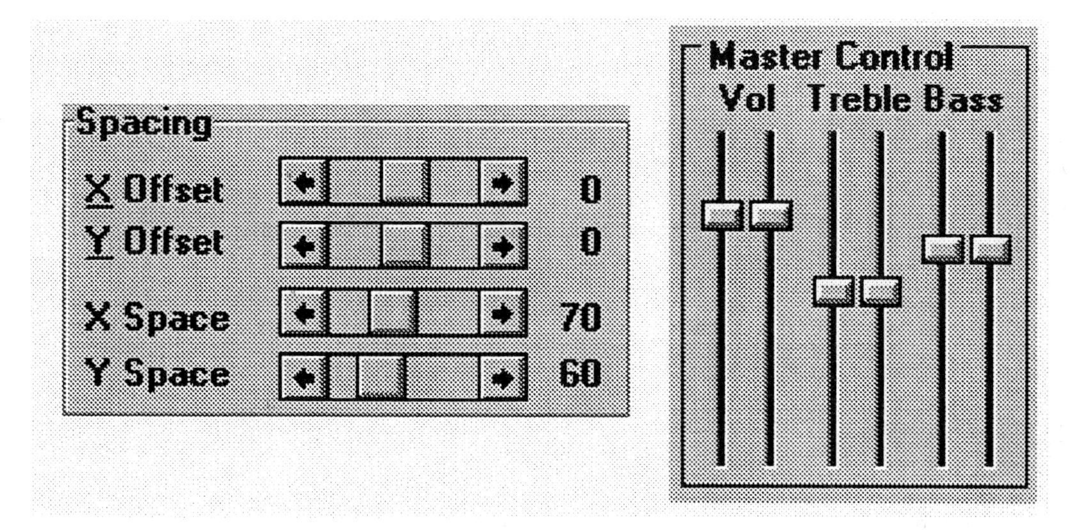

Figure 12-39 More examples of sliders.

Complex Screen-based Controls

There is a growing number of complex screen-based controls. They usually have more than one component and are used in special-purpose situations. Some of these screen-based controls are discussed next.

Notebook

Definition

A control resembling a bound notebook that contains pages separated into sections by tabbed divider pages. The tabs enable users to quickly move from one section to another. Arrow buttons enable users to turn the pages of the notebook. Examples are shown in Figures 12-40 and 12-41.

Typical Uses

When information has only one level of organization (e.g., an alphabetized index or address book)

When data can be logically organized into groups

When dealing with information that users would expect to find in notebooks in real life

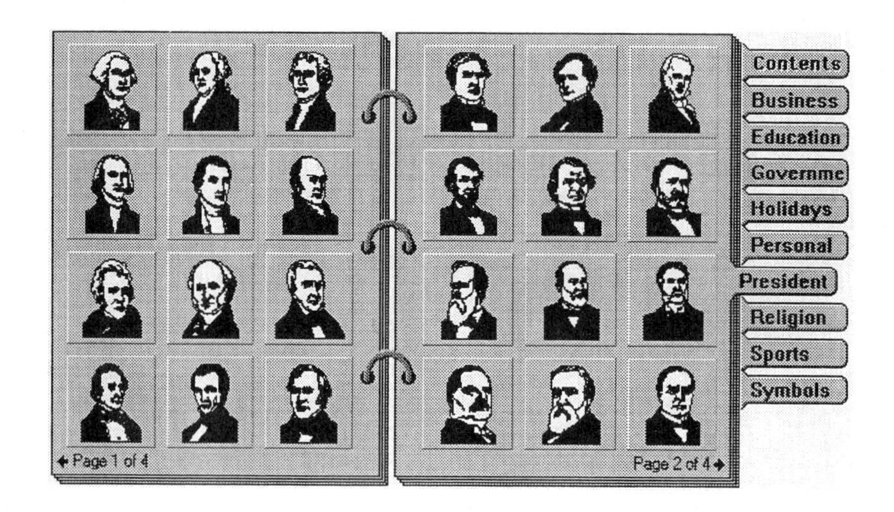

Figure 12-40 Example of the complex screen-based control called a *notebook*.

Figure 12-41 Example of the previous notebook control with the page turned.

Guidelines

1. Place related information within a single tabbed section.
2. Design each page to have approximately the same proportions.
3. Label tabs with either text or graphics (or both).
4. Order vertical tabs from top to bottom and horizontal tabs from left to right.
5. Assign mnemonics to the text label on tabs.
6. Place pushbuttons on the page of a notebook that affect only that page.
7. Place pushbuttons that affect the entire notebook at the bottom of the window (outside the notebook).
8. Do not place a notebook within another notebook or a catalog.
9. Display properties views in a catalog instead of a notebook.

Palette

Definition

An area that provides a place to store commonly used groups of choices or objects. A palette can contain one or more value sets to represent choices. An example is shown in Figure 12-42.

Typical Uses

When users may wish to interact frequently with objects and choices

When it is desirable to have choices that are highly visible and need to be activated quickly

Guidelines

1. Use a value set when the palette contains multiple, related, mutually exclusive choices.
2. Provide a choice within the palette to return the user to a neutral state.
3. When a choice or object on the palette is currently unavailable, display the choice or object with unavailable emphasis.
4. Allow users to choose whether to hide or display the palette.
5. When choices are provided on a palette, also provide access to these choices from a menu.
6. If the palette contains both objects and choices, separate objects from choices using separate panels or group boxes.
7. If activating a choice within the palette presents users with a set of choices, provide a rapid way to view and interact with these secondary choices.

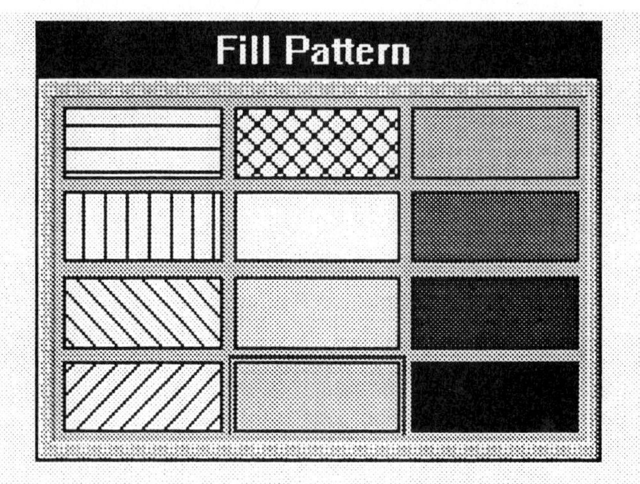

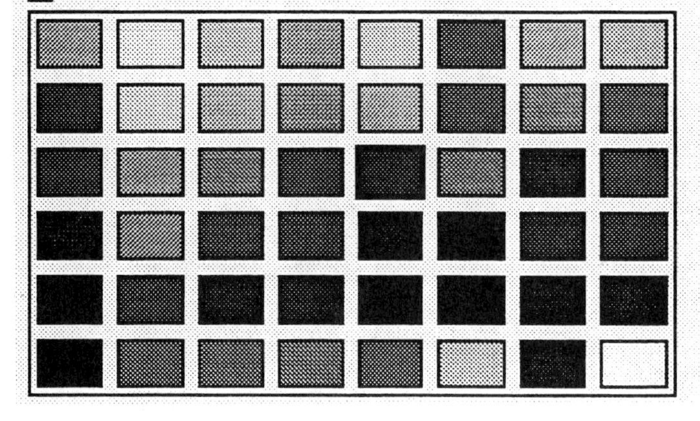

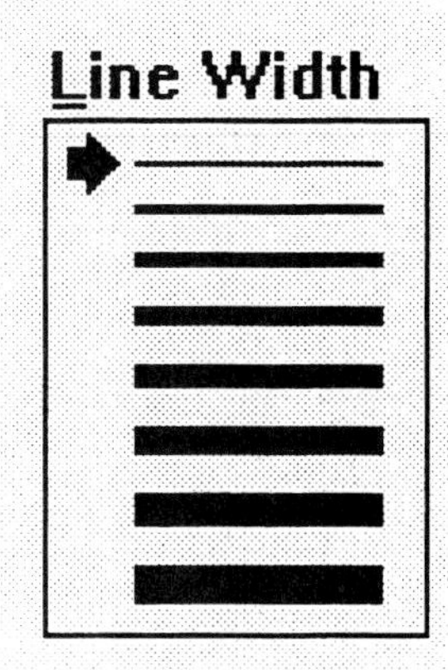

Figure 12-42 Examples of palettes.

Value Set

Definition

A control that allows users to activate one choice from a group of mutually exclusive choices. An example is shown in Figure 12-43.

Typical Uses

When selecting from a set of graphical representations that are mutually exclusive

When selecting from a set of short textual choices that are mutually exclusive

Guidelines

1. Provide at least two choices.
2. Choices can be provided as radio buttons, graphics, or pushbuttons that are grouped together.
3. If choices are displayed as a group of pushbuttons, locate the pushbuttons so that the edges touch (do not overlap).
4. Capitalize the first letter of only the first word of a label for a value set choice (unless the choice contains another word that normally would be capitalized).
5. If a choice is currently unavailable, display it with unavailable emphasis.
6. Assign one choice as the default choice.
7. If choices are text, assign a mnemonic to each choice.

Comparing the Performance of Screen-based Controls

Relatively few published studies provide head-to-head comparisons of even the most used screen-based controls. Two studies have compared the effectiveness of text entry and selection methods for entering dates (Gould et al., 1989) and making airline reservations (Greene et al., 1992). Both studies found *text entry* methods faster and preferred over selection methods.

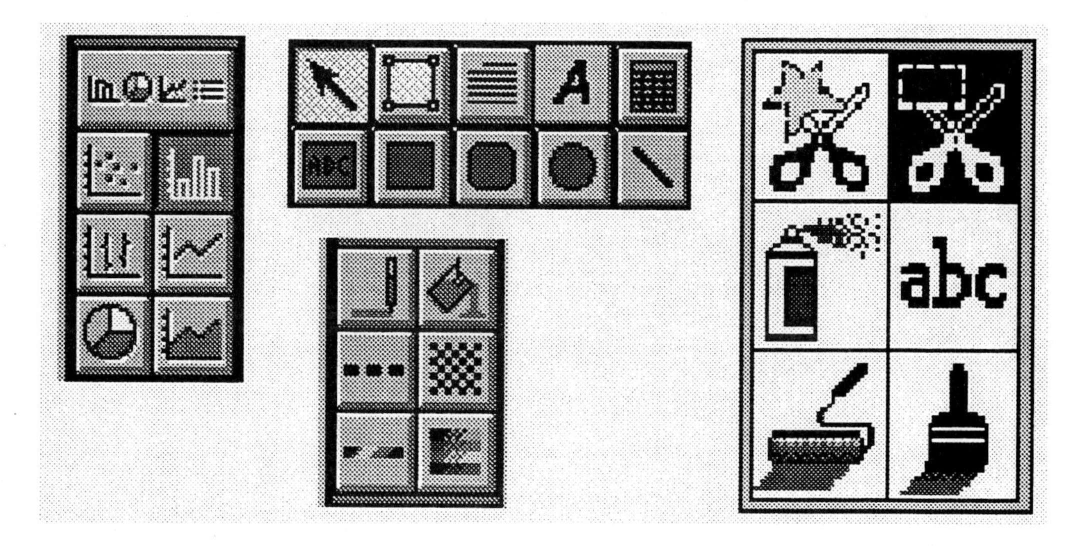

Figure 12-43 Examples of value sets.

Tullis and Kodimer (1992) compared seven screen-based controls for use in reordering fields in a table. They found that

Learning was fastest using a matrix of *radio buttons* and slowest using drag and drop,

Speed was fastest using *radio buttons* and *text entry,* and

The most preferred were *radio buttons, drop-down selection lists,* and *text entry* fields.

Bailey (1993) compared performance making rapid list selections using four screen-based controls. He found that *radio buttons* and *open selection lists* elicited reliably faster performance, typed entry elicited moderate performance, and drop-down selection lists were slower than all other controls.

Johnsgard et al. (1995) compared a set of screen-based controls for making selections (mutually exclusive and nonmutually exclusive), selecting a specific value and setting a value. They reported the following:

For making mutually exclusive selections, *radio buttons* elicited the best performance ($p < 0.05$), and radio buttons and pop-up lists (a pushbutton that shows a selection list when pressed) were preferred. Drop-down selection lists and spin lists were the slowest.

For making nonmutually exclusive selections, *check boxes* elicited the fastest performance ($p < 0.05$) and were the most preferred. Entry fields and accumulators (a control used to move items from a small left window to a right window) were the slowest.

For selecting a specific value, *radio buttons* elicited the fastest performance and were most preferred. Combo drop-down selection lists were the slowest.

For setting a value, *entry fields* were the fastest and most preferred. Spin lists had better accuracy than entry fields. Sliders were the slowest.

Grocery Shopping System Example

Assigning Screen-based Controls

In the grocery store system, at least three windows contained several screen-based controls. These secondary windows and the associated screen-based controls are shown in Figures 12-44 and 12-45.

General Interaction Considerations

Maintaining Consistency

The designer should ensure consistency from display to display, from interaction to interaction, from message to message, and from program to program within the same system. This is particularly helpful for inexperienced users. Experienced users also benefit because it encourages the development of relatively automatic actions and permits the user to transfer principles learned from the execution of well-practiced commands to the execution of new ones.

Name(s) - **Entry field**
Home address - **Entry field**
Home telephone number - **Entry field**
Business telephone number - **Entry field**
Preferred method of payment - **Radio buttons**
    Cash
    Check
    Store charge
    Bankcard charge
Preferred method of obtaining groceries - **Radio buttons**
    Shopping
    Drive-in pickup
    Home delivery
Brand preferences - **Combo drop-down selection list**
Size preferences - **Combo drop-down selection list**
OK - **Pushbutton**
Cancel - **Pushbutton**
Help - **Pushbutton**

Figure 12-44 Proposed screen-based controls for the *customer preferences* window.

Obtaining Groceries - **Radio buttons**
    Shopping
    Drive-in pickup
    Home delivery
        Address for delivery - **Entry field**
        Preferred delivery times - **Check boxes**
            8 a.m. to 10 a.m.
            10 a.m. to 12 noon
            12 noon to 2 p.m.
            2 p.m. to 4 p.m.
            4 p.m. to 6 p.m.
            6 p.m. to 8 p.m.
            8 p.m. to 10 p.m.
OK - **Pushbutton**
Cancel - **Pushbutton**
Help - **Pushbutton**

Figure 12-45 Proposed screen-based controls for the *securing groceries* window.

Doyle (1990) provides a good example of a popular system that was developed with designers focusing only on being consistent in the current product. The study clearly illustrates the confusion and slow user processing that occurred when user expectations were violated (i.e., lack of consistency).

Grudin (1989) observed that ensuring consistency to have ease of learning can conflict with ease of use. He noted that when storing knives the most consistent method (easiest to learn) is to keep them in the same drawer. But the storage strategy that will best facilitate use is to distribute the knives to the location where they will be used. For example, store the table knife in the kitchen, the putty knife in the garage, the hunting knife with the camping equipment, and so on.

Efficient Processing

Users should be able to interact with the computer as efficiently as possible. In some systems, this may require the computer to make adjustments based on the user's experience level. When logging on a system, for example, the user could have the option of stating his or her level of experience. In some systems the computer may make such a decision after determining the experience level of a user from data already collected.

Keeping the User Informed

The designer should provide adequate feedback to inform the user continually about current system status and the options available. For each user action, there should be an appropriate computer reaction. At times there may be a need for an immediate acknowledgment that a request has been received and is being processed, as well as later feedback that the requested action has been accomplished by the computer. When the user is waiting for a computer action to be completed, the user should be informed periodically, using a "progress" indicator, or a "percent done" indicator.

Inexperienced users should be provided with a sensible next step at every point in the development of a transaction. Signaling the possible next steps can take the form of a menu of options; illuminating the next function key or set of function keys; calling attention to items on the CRT through color, blinking, contrasting intensity, or in-verse video (black on white); displaying current options in a special window dedicated to alerting functions; or repositioning a cursor to a new location that suggests the next class of actions available.

Improving User Accuracy and Facilitating Error Handling

The designer should make decisions that encourage user accuracy and improve the quality of data already in a computer. A well-designed system will reduce the opportunity for errors, increase the user's ability to detect errors, and provide users with an immediate opportunity to correct any errors that do occur. Many computer products use numerous error-detection routines that are reasonably effective. These computer detection capabilities can detect as many as half or more of the user's errors.

If these errors are detected by the computer shortly after they are made, this provides an opportunity for users to correct the errors and to take note of what was done to cause the error in the first place. Obviously, the main focus of designers should be to *prevent* errors from ever occurring. However, once an error has occurred, both inexperienced and experienced users should be able to detect and correct it with ease. In some cases, the computer should both detect and *correct* errors.

Natural Interfaces

Computers transfer information *to* users in the form of words and graphics. This form of information is much closer to what takes place in day-to-day communication between people. In fact, in some respects it is an improvement over person-to-person communication because computers have the ability to use full-color graphics sound effects and even full-motion video to portray information.

Self-evident Systems

Some of the best systems are *self-evident,* in that minimal training is needed. Once a user is interacting with a system, each step should be natural and self-evident.

In person-to-person communication, people like the exchange of information to be as natural as possible. For example, if a person is describing a beautiful sunset, a picture of that sunset can transfer information more clearly than the word description. However, if a person is telling how he or she recently solved a problem concerning difficulties with an automobile, words are probably a more efficient medium. Thus, it is important to select the best input language for the activity being performed. Frequently, this language is limited by the available input device.

Exercise 12A: Menu Sorting

Purpose: To determine which menu items should be grouped together to facilitate performance.

Method: Assume that you are developing a new computer product that allows people to quickly access a variety of different types of information on the United States. Once the system is operational, users will access this information using a graphical user interface. Your category names will be the choices on the menu bar, and the associated items will be those found in the pull-down menus.

Do the following:

Step 1. Use the following information on the United States. Write the description and data on cards (one per card).

Step 2. Sort the cards into piles that seem to go together. After sorting the cards, name each pile.

Step 3. Instruct four different people to independently put the items (cards) together that seem to go together (they can have as many piles as they desire).

Step 4. After each person sorts the cards into categories, have the person write the name for each category on a card and place that card on top of each pile.

Step 5. Combine the information from all four subjects.

Step 6. Compare your results with the combined results of the four other people.

Reporting: Prepare a report using the outline discussed in Exercise 1A. Show which items would be on the menu bar and which would be in the pull-down menus for you and for all others combined. Are the two exactly alike? If not, why not? Which represents the best layout? What ideas did you learn from the exercise?

Information on the United States

Average family size (3.2)

Births (4,000,000)

Catholics (25%)

College enrollment (14,000,000)

College graduates (1,000,000)

Deaths (2,000,000)

Divorces (1,000,000)

Employed by small businesses (92,000,000)

Farms (2,000,000)

Federal budget outlay ($1,500,000,000)

Federal budget receipts ($1,300,000,000)

Female life expectancy (79 years)

Female literacy (98%)

Females (51%)

Grades 1–8 enrollment (30,000,000)

Grades 9–12 enrollment (13,000,000)

High school dropouts (1,000,000)

Home computers (35,000,000)

Households (92,000,000)

Households with telephones (85,000,000)

Households with television sets (91,000,000)

Households with VCRs (58,000,000)

Jewish (2%)

Kindergarten enrollment (4,000,000)

Labor force (127,000,000)

Largest area (Alaska)

Male life expectancy (72 years)

Male literacy (97%)

Males (49%)

Marriages (2,000,000)

Married couples (52,000,000)

Median age (32.9)

Median family income ($35,700)

Median single-home price ($103,000)

Most populous city (New York)

Most populous state (California)

Motor vehicles (144,000,000)

No religion (7%)

Owner households (60,000,000)

Population (248,000,000)

Protestants (61%)

Small businesses (6,000,000)

Smallest area (Rhode Island)

Widowers (2,000,000)

Widows (11,000,000)

Exercise 12B: Command Naming

Purpose: To illustrate the many different words and ideas that users try when attempting to use computer systems.

Method: You will need at least four people as subjects (the more you use, the better). A series of command naming questions follow. Make a photocopy for each of your subjects. Give the questions to each subject and have him or her write the one-word command that he or she thinks best. Do not let the subjects know what the others are writing.

In addition, add three questions of your own. Include questions that are similar in style to the ones given but that attempt to secure command names for different actions.

Summarize the responses for each of the eight questions. Determine the best primary command and what would be good secondary commands (i.e., synonyms). Try to have the highest potential "hit rate" with the fewest number of commands.

In other words, show the percentage of people selecting the primary command (e.g., 32%), and combine that with the number that selected the next three alternatives (e.g., 35%). By adding the two numbers (32% + 35% = 67%), you have an estimate of the hit rate expected when the system is operational. In this case, when occasional users enter the command, they will be right about 67 percent of the time.

Test Items

1 Suppose that you are using a software package for the first time and you do not know what to do. What command would you type to remedy this?

2 Sam Jones used to live at 1235 Oakley Avenue. He has recently moved. He now lives at 534 Rupert Street in the same town. What command would you enter to

indicate your desire to have the computer contain the new address rather than the old address?

3. Suppose that you are maintaining a list of telephone numbers of all your friends. Last night at a party you met two new interesting people. You will probably want to call them in the future. What command would you enter to make your telephone list current?
4. Suppose that you are using the computer to make airline reservations for yourself. You have the desired flights and want to use the computer for doing something else. What command would you enter at this point?
5. Suppose that you have used the computer all day and you are ready to go home. What command would you enter to indicate to the computer that you are through working for the day?

Reporting: Prepare a report using the format suggested in Exercise 1A. Include answers to the following questions: How much better is a command language when synonyms are used (quasi-natural language)? Why is this better for occasional users? What is the impact on regular users? Did the wording of the questions bias the results?

Exercise 12C: Designing an Employment System

Purpose: To provide an opportunity to design an object-oriented graphical user interface.

Method: Review the employment system information outlined next. Use this information to (1) conduct a noun analysis and identify potential user interface objects, (2) assign the objects to the appropriate user interface classes, (3) identify primary and secondary windows, (4) develop a menu bar and pull-down menus for each primary window, (5) identify visual displays, (6) identify screen-based controls, and (7) device icons (if appropriate).

Major Steps in the Employment Process

1. *Candidates* indicate that they are seeking a job at the company by
   a. Sending a resume, or
   b. Traveling to the employment office and completing a computer-based job application form.
2. *Managers* at the company who have job openings fill out employment requisitions.
3. *Employment clerks* ensure that candidate applications and manager requisitions are complete, properly coded, and correctly entered into the system.
4. *Employment specialists* repeatedly operate the system, attempting to match candidates with openings, conduct initial interviews, and advertise job openings for which there are no (or very few) candidates.

5 For candidates that pass the initial employment specialist interview, the *employment clerk* notifies both the candidate and manager, and a time is set for a manager's interview.

6 For candidates that pass the manager's interview

  a An *employment clerk* performs a background check to verify candidate information.

  b An *employment specialist* makes a formal offer.

7 After the job opening is filled, an *employment clerk* removes

  a The candidate's information from the candidate database, and

  b The job from the job opening database.

8 *Employment specialists* monitor the employment process and provide periodic status reports to management.

Contents of Application Forms and Employee Requisition Forms

Application forms (completed by job candidates)

  Personal information (name, address, telephone number)

  Education (dates, degrees, and schools)

  Experience (dates, job titles, major job activities)

  Salary requirements

Employee requisition forms (completed by company managers)

  Educational requirements (degrees)

  Experience requirements (type and years required)

  Major activities performed in the job

  Salary range

  Date needed

Tasks for Job Candidates

Find the job categories of most interest (filter)

Find individual job descriptions of most interest

View (read and evaluate) one-page job summaries

Print one-page job summaries

Respond to a request to complete an application form (yes or no)

Complete a computer-based application form

Submit the computer-based application form

Tasks for Company Managers

Prepares an employee requisition

- Completes a computer-based form, or
- Searches to find a requisition for a past opening; then loads and modifies

Sends an employment requisition (request) to the employment specialist

Receives information on candidates from the employment specialist and reviews their qualifications

Informs the employment specialist of possible candidates for a manager's interview

Prepares for a manager's interview

- Provides the employment clerk with times available for conducting a manager's interview
- Reviews a candidate's information, including the results of the initial interview conducted by the employment specialist

Completes the manager's interview

- Summarizes interview results into the candidate's record
- Sends
  - A "not interested" message to the employment clerk, or
  - An "offer" to the employment specialist (how much, desired start date, and so on)

Receives general employment-related information from the employment specialist (e.g., hiring freezes, advertising status, hiring statistics)

Accesses overall employment statistics from time to time

Tasks for Employment Specialists

Runs the system's module that matches candidates and job openings

- Identifies possible matches
- Informs managers of potential candidates for their job openings

Determines the need to advertise openings

- Informs managers of the need to advertise
- Prepares advertising, including newspapers, radio, technical journals, and the like

Evaluates candidate information, and informs managers of highly qualified candidates (even when no openings are available)

Receives feedback from managers on proposed candidates

Provides times available for conducting initial interviews to the employment clerk

Conducts initial interviews and summarizes the results in candidate's files

Receives "offers" (how much to pay, desired start date, and the like) from managers

Instructs the employment clerk to conduct a background check

Makes an official offer to a candidate (after a successful background check)

Receives the candidate's acceptance of an offer and sends the information to managers and the employment clerk

Prepares employment statistics, including

The number of new openings during the month

The number of new candidates during the month

The number of openings filled during the month

The number of candidates on file at the end of the month

The number of openings at the end of the month

Tasks for Employment Clerks

Enters information from résumés or application forms into the system

Ensures that the candidate's information is complete, properly coded, and correct

Ensures that the manager's employee requisitions are complete, properly coded, and correct

Conducts background checks for candidates before offers are made (verifies past employment, education), and records the results of background checks on candidate records

Determines available interview times from managers, employment specialists, and candidates and schedules interviews

Archives information on

Candidates hired

Candidates disqualified by interviews, background checks, or other

Candidates not interviewed by an employment specialist after 3 months

13

Presentation Issues and User Guidance

Computer-to-User Information Flow

The primary considerations when dealing with information flowing from the computer to the human are (1) the types of output devices selected for this purpose and (2) other output characteristics, including screen design, computer messages, and computer response time. The output characteristics are closely related to the output device itself. For example, there are obviously issues associated with using a CRT for output that are much different from those related to using speech output.

Presentation Options

Introduction

There are three relatively distinct presentation options. These include having (1) character-based (text-based) interfaces, (2) graphical user interfaces (GUIs), or (3) multimedia interfaces. Character-based interfaces allow designers the fewest design alternatives, whereas multimedia interfaces offer the most alternatives. Designers should select the presentation option that will elicit the best performance and acceptance.

Character-based Interfaces

Virtually all early computer systems provided only character-based presentation. These first computers dealt only with numeric information, but were eventually able to present both numbers and text. These systems primarily used keyboards for input and CRT monitors and printouts for output.

Graphical User Interfaces

Historical Overview

In the early 1980s, graphical user interfaces began being used. Because of the pixel-based nature of the displays, they were able to display number, text, and graphics simultaneously. These systems primarily used a mouse–keyboard combination for input and CRT monitors and laser printers for output.

Graphical user interfaces were pioneered at the Xerox Palo Alto Research Center (1980), and then popularized by Apple with the Macintosh (1983). They became the favored way to interact with computers when Microsoft introduced Windows 3.0 in 1989. The traditional Macintosh model GUIs are being replaced by object-oriented graphical user interfaces.

Style Guides and Standards

Style Guides

Unlike the earlier character-based interfaces, graphical user interfaces are governed by style guides. Style guides are paper-based guidelines for system developers. The most popular style guides are IBM's *Common User Access* (CUA), Apple's *Macintosh,* the Open Software Foundations' *Motif,* and Microsoft's *Windows.*

Style guides specify the user interface, including the appearance (look) and behavior (feel). They describe the appearance of windows, menus, and screen-based controls. Most style guides also provide design guidance concerning the use of commonly used GUI components. The primary value of style guides is to provide consistency across most graphical user interfaces.

The popularity of GUIs and their accompanying style guides has led to the development of numerous commercial *toolkits.* Toolkits include a library of high-level software routines that enables designers to more quickly develop systems, applications, or products.

User interface design should not take place without first reviewing the large number of guides, standards, and guidelines.

ISO 9241

IS0 9241 is the international user interface design standard. This standard provides a large amount of detailed information on user interfaces. This is the first international user interface standard to be developed and widely used. It contains 19 sections covering the following topics:

1. General introduction
2. Guidance on task requirements

3 Visual display requirements
4 Keyboard requirements
5 Workstation layout
6 Environmental requirements
7 Display requirements with reflections
8 Requirements for displayed colors
9 Requirements for nonkeyboard devices
10 Dialogue principles
11 Usability statements
12 Presentation of information
13 User guidance
14 Menu dialogues
15 Command dialogues
16 Direct manipulation dialogs
17 Form-filling dialogues
18 Question and answer dialogues
19 Natural language

Other Standards

Besides the style guides that include material on multimedia user interfaces, other standards that are closely related to user interfaces include

- CD-ROM (Sony and Philips)
- NTSC (current U.S. and Japan analog video format)
- PAL (current European analog video format)
- SECAM (current French and Eastern bloc analog video format)
- JPEG (for still images)
- MPEG (for full-motion video)

Are Standards of Value?

There are strong arguments both in favor of and opposed to having user interface standards. Those favoring standards point out that standards

Promote ease of learning and ease of use,

Assist in software procurement and product evaluation, and

Facilitate reuse of user interface design and code.

Arguments opposing user interface standards point out that

We do not know enough about usability to standardize,

Standards inhibit innovation in user interface design, and

Only testing can assure usability.

Usability of User Interface Guidelines

Even a good set of guidelines can be difficult to use by designers. Reaux and Williges (1988) had evaluators use a set of user interface guidelines to determine deficiencies in a software product. The investigators knew in advance what concrete and abstract violations existed. The study showed that evaluators found only 50 percent of concrete guideline violations and only 28 percent of abstract violations. Because of these problems, computer-based methods for presenting guidelines are becoming more popular.

Multimedia User Interfaces

Multimedia user interfaces provide more than a simple extension of graphical user interfaces. They provide significant additional alternatives for presenting information to users. Multimedia user interfaces help designers to address an entire new set of issues because they allow and encourage the use of a much broader set of display technologies. Not only do they allow the manipulation of numbers, text, and graphics, but also the creative use of audio, still images (photographs), animation, and full-motion video.

Many early multimedia interfaces included (1) information-on-demand retrieval systems such as encyclopedias and other reference materials, (2) public presentations in art galleries, museums, and zoos, and (3) educational applications such as training courses in music appreciation, history, and geography. Most of the early business-related applications related to training, in-house presentations, or sales demonstrations. Multimedia user interfaces are used with on-line manuals (CD-ROM), storage systems for documents and images (CD-ROM), help and error messages, voice annotation with email, word processing and spreadsheets, and scientific visualization.

Multimedia user interfaces are usually characterized by having (1) a high level of *interactivity* between users and the computer and (2) multiple ways to *display* information.

Interactivity

Interactivity refers to the number of information exchanges in a given period of time. Generally, the more information exchanges between the user and the computer the better (Chapanis, 1975). For example, the typical level of interactivity with television is very low. Users either turn the television on or off or select channels. The viewer is a passive observer, rather than being a participant. Attempts have been made to involve the user in more interaction through the uses of interactive television (ITV), Commodore's Dynamic Total Vision (CDTV), and Phillips' Compact Disc-Interactive (CD-I).

Displaying Information

In computer-based products, displays are generally either visual or auditory. Visual displays include monitors, printers, and paper. Auditory displays include tones, speech, and special sound effects.

Visual Display Characteristics

Content

The following section contains a set of guidelines to help designers to develop an effective interface between displays and their intended users. These guidelines apply particularly to displays; however, many of them, especially those dealing with consistency of presentation, information presentation, and labeling, are also applicable to paper forms. This set of guidelines represents an effort to present what is known or can be reasonably deduced from the present state of knowledge.

Designers are encouraged to apply these guidelines, but should recognize that other considerations may force trade-offs in some instances. One obvious weakness of these and similar guidelines having to do with human–computer interaction is that many of them are based on opinion, judgment, and accumulated wisdom, rather than on research studies. The main problem with this approach is that designers are given much advice, but not told the consequences of ignoring it.

Design Steps

The following steps lead to a systematic design of an effective display. We can divide these steps into two basic groupings. The first set includes activities related to *data collection.* The second set includes activities related to *analysis* and *design.*

Data Collection

The systematic collection of information concerning the use and elements of a display is probably the most overlooked aspect of display design. Yet this is precisely the activity in which it is determined *what* goes on the display. Basically, the designer needs information that will help to determine the proper type and size of display and the compatibility of the new display with other displays in the system (Benbasat et al., 1986). The characteristics of the potential users are also very important. One sequence for the collection of display design information is as follows:

- Determine specific requirements for the display
- Collect information relevant to proposed items and processing
- Collect information on existing displays in the system or adjacent systems

Analysis and Design

There are also guidelines to help in the actual design of displays. The following steps assume that a designer has done an adequate job of collecting the information just discussed. Each step is important to the final product.

- On the basis of data collection, determine the most appropriate grouping of items using grouping principles
- Develop headings or labels for groups and item names as required
- Determine groups of groupings where necessary
- Attempt several alternative layouts and choose the one that will elicit the best human performance
- Determine a meaningful title for the display
- Develop instructions and/or other explanatory material to be placed on the display or to be used as separate instructions
- Determine physical characteristics of the display (e.g., type size and colors)
- Test the proposed layout on people representative of potential users
- Redesign as required

Grouping

Of the nine guidelines just presented, one requires more detailed discussion. Grouping techniques are important in helping to organize information. There are four fundamental techniques for grouping data.

- Sequence
- Frequency
- Function
- Importance

Sequence

Sequential grouping is based on the principle of grouping items in the order that they are transmitted or received. For example, if a designer is developing a visual display that must be used with information that always arrives, then the display should be designed so that the data can be entered in the same sequence. For output data, sequential grouping would dictate that the data be output in the order of use. The first information needed should be the first shown at the top of the display, the second item used should be the second shown, and so forth.

This principle not only applies to item sequence, but also to the sequence of item groups. An example would be a process in which the user is required to verify all historical data before going on to other sections of the displays. The historical data should not only be grouped together, but be placed in a prominent position near the top of the display.

The natural order of data is another basis for sequential grouping. Other factors being equal, some sequences seem more natural than others. For example, name groups are usually alphabetical by last name, and number groups are usually placed in ascending order. One method of establishing sequential groupings is by using a technique called *procedural flow analysis*. Closely analyzing the procedural flow determines what functions a display serves and what processes are performed on the data at each point in the flow of information. With this information the designer could elect to group together those items of information that need to be processed at about the same time. The order of the item groups would then follow the same sequence as the processing steps. The sequence of items within a grouping would also parallel the sequence of processing steps.

Frequency

The frequency-of-use technique is based on the principle that items used most often should be grouped together. If a designer develops a display completely on the basis of this technique, it would be, in effect, a rank order listing of the items according to their frequency of use. The most frequently used item would be at the top and the least frequently used at the bottom. For example, a preprinted mark-sense form for keeping a record of tool use might be designed so that the most frequently used tools are grouped near the top of the form.

Another application might be to arrange items within the groups established by the sequential grouping technique. After sequentially grouping the items, place items most frequently used in the beginning of each group.

Function

A functional grouping technique may also be used. A designer may consider grouping items according to the special needs of the activity being performed. For example, it may be advantageous to group all the items that pertain to inventory in one location and those related to requisition in another location. If the sequence or frequency of use is not too important, then layout based on functional relationships may be the best criterion for grouping. Items grouped on this basis should be identified as such.

Importance

One final technique is to group items according to how important they are to the success of the system. If a certain item or items are critical, then it may be best to place the critical items in the best position on the display so that they are not overlooked.

Grouping Trade-offs

Grouping items on a visual display is a delicate process of elimination, weighting, and judgment. The designer should determine which of the techniques just discussed actually apply to the items. Of those that apply, determine how much weight each should carry in the final decision. And determine weighting on the basis of eliciting the best possible human performance. When determining the best grouping, *give precedence to human performance requirements* over any other requirements, for example, software or hardware requirements. After accomplishing these steps, perform trade-offs in terms of applying one technique in one case and another technique in another case. The object is to arrive at an arrangement of items that has the highest probability of eliciting an acceptable level of human performance.

Specific Guidelines

General Concerns

There are numerous issues concerning how displays should be designed. Issues include (1) what information to put in a window or on a screen, (2) where to put the information, (3) the best way to present information (words, graphics, etc.), (4) use of color, and (5) positive versus negative presentation. In addition, there are issues concerning whether to use the traditional *static* screens (e.g., menus or form filling) or the newer *dynamic* screens (e.g., pointing at icons, opening and closing windows, and overlaying documents) or a combination of both.

Organization

Organizing a screen can be done in a variety of ways, ranging from the use of arbitrary but consistent grouping, to the use of a systematically developed grouping based on the guidelines discussed earlier in the *Grouping* section.

Probably the first consideration should be to standardize the placement of displayed information, including labels and data items, within a specific screen format and to remain consistent. For example, functional areas should remain in the same relative display location on all frames. This means that certain areas of a screen should be reserved for certain types of information. One area may contain computer output only, another may be reserved for user input, a third for reference information, and another for housekeeping messages. Avoid breaking up the screen into too many partitions (*windows*). Numerous windows can be confusing.

But using different screen areas in this way is only the first step. Some way should be used to help the user to know that the screen has been systematically divided. There are several ways of helping a user to understand and appreciate screen divisions. On a large uncluttered screen, each area (window) may be separated by blank spaces in sufficient quantity (three to five rows and/or columns) to indicate clearly that unique information can be expected within the area surrounded by those blank spaces. On smaller and/or more cluttered screens, where extra unused spaces are not available, the user's understanding of the screen divisions, including different areas and/or items on the screen, can be assisted by any one of the following techniques: different colors, surrounding line types (solid, dashed, dotted, etc.), or different intensity levels.

In presenting text on a small screen, there should be a maximum of 50 to 55 characters on each line. On larger screens, break up text into two (or more) columns of 30 to 35 characters per line. Separate columns by at least five spaces if the text is not right justified and by three to four spaces if the text is right and left justified.

Labeling

Labeling is the act of placing a descriptive title, phrase, or word adjacent to a group of related items or information. Good labels provide a quick means of identification and can assist the user in rapidly scanning for an item of interest or can help to ensure that an item is being entered in the proper field. Labels should be highlighted for ease of identification.

If a label is an abbreviation, then it must be meaningful to the users of the display. If there is some doubt about the familiarity of the label, do not use it. Find a label that is meaningful or use complete words.

Labels can be used to identify a single data item, a group of items, or an entire display. For example, each menu should have a label (title) that reflects the question for which an answer is sought. If a good descriptive label for a menu cannot be easily determined, it could mean that the menu does not reflect a logical and consistent set of material and possibly should be redesigned.

Label every item. Do not assume that a user will be able to identify individual items because of past experience. Context plays a significant role. For example, 513-721-2345 may be recognized as a telephone number if it is seen in a telephone directory, but may not be recognized as such on an unformatted display.

Word labels distinctively for data entry fields so that they will not be readily confused with data entries, labeled control options, guidance messages, or other displayed material.

When entry fields are distributed across a display, adopt a consistent format for relating labels to entry areas. For example, the label might always be to the left of the field, or the label might always be immediately above and left justified with the beginning of the field. Such consistent practice will help the user to distinguish labels from data.

Computer Responses

Ensure that the computer makes an appropriate and meaningful message response to each user command.

Not:	But:
User: Remove File A	User: Remove File A
Computer: Ready for next command	Computer: File A removed

Most messages and other text (prose) should be displayed using both upper- and lowercase type. However, where attention getting is needed, as in labels or short titles, use uppercase only.

Not:	But:
ALL UPPERCASE TEXT	Normal reading is
IS HARDER TO READ	easier if
THAN A MIXTURE OF	the text is in
UPPER- AND LOWERCASE	both upper- and lowercase

Put frequently appearing messages in the same place on the screen. Frames should be about the same physical length so that users can depend on finding items in about the same spatial location. Important but infrequent events, such as error messages, may need some enhancement or highlighting to be recognized. Place such messages in the user's central field of view.

Interscreen Considerations

The following guidelines cover situations in which a designer has a display made up of several screens in a series. For the most part, these guidelines reflect the idea that a designer should make decisions consistent with reducing the load on the user's memory.

A message should be available that provides explicit information to a user on how to move from one screen to another or how to select a different screen:

Not:	But:
More . . .	To see more, press the ENTER key

In a hierarchy of screens with different possible paths through the series, a visible trail of the choices already made should be available to a user.

If at all possible, supply all relevant information on a particular topic on one screen. Do not force users to remember data from one screen to the next or to write down information that will be used for subsequent interactions with the computer.

Nontext Displays

These guidelines on data presentation refer primarily to nontext information, including coded information such as part numbers, telephone numbers, or scores on a series of tests. The guidelines attempt to ensure that when information is presented on a screen it will be directly usable. If codes are used, they should reflect the guidelines discussed in Chapter 16. Make use of illustrations to supplement explanations in text.

Strings of six or more alphanumeric characters that are not words should be displayed in groups of two, three, or four with a blank character between them. The grouping should be consistent for all strings of characters. If natural groupings of two to five characters exist, they should be used instead of artificial groupings (e.g., Social Security number: 518-40-1087). In addition, people seem to be able to scan for a certain item of information more quickly and accurately if a tabular format is provided.

Not:	But:
ABBA423675A2	ABBA 423 675 A2
ABBD252389K4	ABBD 252 389 K4
ABCR862534M3	ABCR 862 534 M3
ABRG563487W4	ABRG 563 487 W4
ACGL190537S0	ACGL 190 537 SO

Data items should be presented in some meaningful order, if possible, for ease of scanning and identification. For example, put historical dates in chronological order.

Not:	But:
1215	1215
1941	1492
1975	1861
1861	1917
1917	1941
1492	1975

Identical data should be presented to the user in a standard and consistent manner, despite what it looked like when originally input. The use of dates provides a good example. No matter how they are entered, they should be displayed in a format that allows the speediest and most accurate use.

Entered as:	Displayed in different situations as:
3-5-50	March 5, 1950
	5 Mar 50
	03/05/50
	3-5-50

Justification

With words, alpha, or alphanumeric codes, use vertically aligned lists with *left* justification. Subclassification can be identified by indenting.

Not:	But:
CANCEL (CONTINUE,	CANCEL
END, HELP, SAVE)	CONTINUE
STATUS, SUBMIT	END
	HELP
	SAVE
	STATUS
	SUBMIT

With numeric codes use *right* justification, especially in tables.

Not:	But:
482645	482645
258	258
7295	7295
36890	36890

Natural Formats

Be careful not to change traditionally accepted (natural) formats.

Not: Enter Social Security number, address, and name

But: Enter name, address, and Social Security number

In addition, use standardized (familiar) formats when they exist. For example, some suggested standard data items for American civilian users are

Telephone: (914) 444–0111

Time: HH:MM:SS, HH:MM, MM:SS.S

Date: MM/DD/YY

Highlighting

Highlighting refers to emphasizing some objects, such as label, data item, title, or message, on the screen. This emphasis can be accomplished by the following:

Increasing the intensity of an object (relative to others)

Displaying the item in a unique color

Flashing the object on and off

Underlining the object

Presenting it in a different style or font size (if the object consists of alphanumeric characters)

Pointing to it with a noticeably large flashing object (such as an arrow)

Reversing the image of the object (such as going from light letters on a dark background to dark letters on a light background)

Making a shaded box around the item

Placing some graphics (such as a rectangle composed of a string of asterisks) around or near the object

Whatever the specific technique, the main purpose of highlighting is to attract attention. Highlighting also can be used to provide feedback to the user. For example, when a user is presented with a menu containing a list of mutually exclusive options and one option on the list is selected, those remaining in the list could be dimmed. This is particularly important when the screen is cluttered. Another common use of highlighting is to help the user to detect an item of information. If a user is to perform an operation on some item on a display, highlight that item so that the user is rapidly directed to it.

Screen Format and Visual List Search

Bednall (1992) studied the effect of different screen formats on visual search times of telephone book information (e.g., name, address, and telephone number):

Uppercase letters did *not* slow search time

BORLAN versus Borlan

Spaced screens produced faster search times than nonspaced screens

BORLAN	**BORLAN**
B—64	BORLAN, G—75
G—75	BORLAN, D—93
D—93	BORLAN, L—98
L—98	BORLAN, M—123

Alternating lines produced faster search times

BORLAN

B......64

G......75

D......93

L......98

Adding an extra blank line after the last names produced faster search times.

Reading Text on Monitors

Foreground and Background Colors

Gould et al. (1987) reported that when reading from CRT displays the variables that improve the reading speed most are having (1) high-resolution displays (1000 × 800 pixels), (2) character fonts that resemble paper fonts, and (3) dark characters on light background.

Snyder et al. (1990) had participants perform visual searching and reading tasks. He observed that performance was reliably better using dark characters on a light background (2 to 32 percent improvement) when compared with using light characters on a dark background.

Reading from Computer Displays

Reading from character-based computer displays is up to 28 percent slower than reading from a paper document. Muter and Maurutto (1991) found that to enhance the reading speed of computer text so that it was as fast as reading paper text they had to do the following:

Use a pixel-based display

Use no more than 85 characters per line

Use black characters on a white background

Use bold type

Double-space

Use proportional spacing with a left-justified three-space indention of every other line

Use an eight-space indention of the first line of each paragraph and three spaces (rather than two) between sentences

Skimming speed, which is defined as the user scanning at a rate three to four times faster than normal reading, was not improved by making the changes to enhance reading speed. Skimming from a book was about 41 percent faster than skimming from a computer display.

Using Graphics

Graphics in Decision Making

Jarvenpaa and Dickson (1988) observed that there is little evidence that graphics improves decision making over using tables. The basic tasks when interacting with graphs (Gillan and Neary, 1992) include

Searching for location of an indicator,

Decoding the value of the indicator,

Performing arithmetic operations on the encoded values (when necessary),

Comparing spatial relations (relative heights or lengths), and

Making a response.

When users are not performing well with graphs, designers should evaluate each of these tasks and then determine which one or more need to be improved. Well-done graphs can produce a summarizing effect that helps to facilitate visualization.

Sparrow (1989) determined the best graph for a variety of different task conditions.

Identifying a specific entry

Best: Spreadsheet (table)

Worst: Pie chart

Finding minimums and maximums (year at its peak)

Best: Stacked bar chart and multiple line graph

Worst: Pie chart

Determining where two exemplars intersect (compare two products)

Best: Multiple line graph (better than all others)

Summation of each exemplar (highest one)

Best: Stacked bar chart (better than all others)

Cumulative sum of changes for each exemplar (trend)

Best: Multiple line graph and spreadsheet

Worst: Pie chart

The author recommended

When the task involves retrieval of specific values, use *tables,*

When the task involves retrieval of relational information, use *graphs,* and

For tasks of mixed type, use *tables.*

Tables versus Bar Charts

Another study evaluated the time it took to solve problems using either tables or bar charts. Reading tables took less time ($p < 0.0005$) and produced more correct answers ($p < 0.0005$) than reading bar graphs (Coll et al., 1994). This effect increased with task complexity. Participants preferred using tables to bar graphs.

Color and Graphs

Using color and graphics does not always improve performance or user acceptance (Benbasat et al., 1986). Color can be a positive influence (1) during learning, (2) for certain types of graphical reports, and (3) under time constraints. When color is used, the best presentation alternatives when only using two colors is yellow foreground and black background. In situations where multiple colors are used, the best combinations are red, white, green, cyan, and yellow on black (Hoadley, 1990). Color helps to elicit *faster performance* when reading (1) bar graphs, (2) pie charts, and (3) tables. Color improves *accuracy* when reading (1) line graphs and (2) pie charts.

Three-dimensional Graphics

Using three-dimensional graphics to portray two-dimensional information, when using bar graphs, line graphs, and pie charts, can result in less accurate performance (Carswell, 1991).

Visualization

Visualization refers to when graphics are used to illustrate some aspect of a system. The use of visualization ranges from simple, static visualization to highly interactive, animated visualization (cf. Murray and McDaid, 1993). Unfortunately, many current interfaces simply mimic the way information is presented in paper media. When done well, visualization requires radically different approaches that make better use of human perceptual abilities.

Using Color

Selecting Colors

Page (1993) examined combinations of colors that would allow users to easily identify highlighted buttons in a character-based (text) interface. He found the following:

Best performance

- Gray on blue
- Gray on black
- Gray on dark gray

Users preferred

- White on blue
- White on light blue
- White on red

Lalomia and Happ (1987) reported that the most readable colors on a gray background were black and blue. They found that the most readable colors on a blue background were

- Light cyan
- Green
- Light green
- Yellow
- Light gray

The most readable colors on a black background were

- Cyan
- Light cyan
- Green
- Light magenta
- White

Past research suggests that the colors to avoid on monitors are the following:

Highly saturated blue on red and red on blue can cause chromostereopsis (Helander, 1989).

Red on green and red on blue elicited 10 percent poorer performance on selection tasks (Matthews and Mertins, 1989).

Magenta and green receive the lowest scores in preference tests of computer colors (Taylor and Murch, 1986).

The following color combinations should be avoided for people who are color weak or color-blind (Travis, 1990; Thorell and Smith, 1990).

Cyan and gray

Yellow and light green

Green and brown

Red and black

Bailey and Bailey (1992) conducted a study to determine the preferences of typical users for 16 foreground colors and 8 background colors. A total of 120 color combinations was evaluated. The results are shown in Table 13-1. They found that the most preferred colors (top 5 percent) were

White on blue

Light white on blue

White on black

Yellow on blue

Light white on black

Black on white

They reported that the least preferred colors (bottom 5 percent) were

Cyan on green

Red on magenta

Magenta on red

Green on cyan

Green on yellow

Magenta on green

Red on green

The results of this study are included as an integral part of the Protoscreens software prototyping package (see Bailey and Bailey, 1992).

Table 13-1 Color Combination Preferences*

Foreground Colors	Background Colors							
	Black	**Blue**	**White**	**Red**	**Magenta**	**Cyan**	**Yellow**	**Green**
Light white	Good	Good	Poor	Good	Fair	Fair	Fair	Fair
Black	—	Poor	Good	Fair	Fair	Good	Good	Good
White	Good	Good	—	Good	Fair	Poor	Fair	Poor
Light Yellow	Good	Good	Poor	Fair	Fair	Fair	Poor	Poor
Blue	Poor	—	Good	Poor	Poor	Good	Fair	Fair
Light green	Good	Good	Poor	Fair	Fair	Poor	Poor	Poor
Light cyan	Good	Good	Poor	Fair	Fair	Poor	Poor	Poor
Cyan	Good	Good	Poor	Fair	Poor	—	Poor	Poor
Magenta	Good	Fair	Good	Poor	—	Poor	Poor	Poor
Yellow (brown)	Good	Good	Fair	Poor	Poor	Poor	—	Poor
Red	Good	Poor	Good	—	Poor	Poor	Poor	Poor
Light magenta	Good	Good	Poor	Poor	Poor	Poor	Poor	Poor
Light black	Fair	Poor	Good	Poor	Poor	Poor	Poor	Poor
Light blue	Fair	Poor	Good	Poor	Poor	Poor	Poor	Poor
Green	Good	Fair	Poor	Poor	Poor	Poor	Poor	—
Light red	Good	Fair	Poor	Poor	Poor	Poor	Poor	Poor

**Good* means that the preferences were ranked in the top 25 percent.
Fair means that preferences ranked in the second 25 percent.
Poor means that preferences ranked in bottom 50 percent.

Presentation Alternatives

Sound

Introduction

Computer systems have had the ability to display sound (usually beeps) almost from the beginning. Later, more sophisticated sounds were introduced for use with computer-based games. By the early 1990s, sound had become a recognized part of many new systems. Along with the common use of sound have come new input and output technologies. These include the microphone for input and speakers or headsets for output. Three types of audio are used as part of computer system displays: compact discs–digital audio (CD-DA), waveform sampling (recording), and synthesized sounds (MIDI).

Compact Disks—Digital Audio (CD-DA or CD-Audio)

CD-audio is an optical digital recording–playback technique that was originally designed for continuous playing of high-fidelity audio. This technology rapidly replaced records and cassette tapes for sound. Because CD-audio disks easily fit into the large number of CD-ROM drives found in current computers, the material contained on CD-audio disks can be readily used in computer systems. Designers should find appropriate ways to include this material.

Waveform Audio

Waveform audio can be created, manipulated, and played back by designers. The digitized information is encoded and stored on a hard disk. When the information is played back, it is heard through speakers or a headset.

Recording and Playback

When sound is recorded, the analog sound wave goes through a microphone and is digitized using an analog-to-digital (ADC) converter, converted to a series of numbers, and the numbers are then sampled and stored. When sound is played back, the numbers are sent through a digital-to-analog converter (DAC), converted back into an analog waveform, and the waveform is then amplified and fed into speakers.

Recording and Playback Fidelity

Recording and playback fidelity (accuracy) refers to how closely a recorded or played back sound matches the original sound source. *Sound fidelity* is determined by understanding both the sampling rate and the sample size. *Sampling rate* refers to how often the input is measured, while *sample size* is the number of data bits used to store each sample.

Sampling Rate

The sampling rate is the frequency with which sound is converted from analog waveform to numbers and is reported in samples per second. This governs the highest frequency of sound that can be recorded. Ideally, the sampling rate should be about 10 percent higher than twice the highest sampled frequency.

For example, the range of human hearing is up to about 20,000 cycles per second or hertz, or 20 kilohertz (kHz). Under this condition, the ideal sampling rate would be 44 kHz (2×20 kHz + 4 kHz = 44 kHz). Conventional compact disks (CD-audio) have a 44.1-kHz sampling rate. Assume another example in which all speech sounds are 8000 cycles per second (8000 hertz or 8 kHz) or less; then the required sampling rate would be about 17,600 Hz (17.6 kHz).

Sample Size

The sample size (depth) represents the number of bits used to store a sound. For example, 8-bit sampling can assign each sample one out of 256 values, while 16-bit sampling allows 65,536 possible values for each sample. The more locations to store a sound, the more accurate will be the playback fidelity.

Calculating Audio Storage Requirements

Storage requirements for waveform audio can be calculated by multiplying the duration of the sound in seconds by the sampling rate and the sample size in bytes. If the sound is in stereo, then double the results. For example, for CD-audio quality,

1 second × 44.1 kHz × 2 (16-bit) × 2 (stereo) = 176,400 bytes per second (or 11 megabytes per minute)

Audio Compression

Decreasing either the sampling rate or the sample size decreases the storage requirements. In addition, sound can usually be compressed at ratios up to 3 : 1 (or higher). For example, the storage requirements can be reduced to one-third of the previous calculation. Rather than requiring 176,400 bytes per second, this can be reduced to 58,212 bytes per second.

Musical Instrument Digital Interface

A third way of using sound displays in computer systems is to use an electronic music synthesizer. The Musical Instrument Digital Interface (MIDI) standard was originally developed in the early 1980s. The purpose of the MIDI standard was to provide a consistent way to connect electronic music synthesizers to controllers (e.g., piano keyboards).

MIDI has no innate sound of its own. The sounds are generated when the computer sends a string of commands to a board capable of MIDI audio. Music is created from a palette of preset sounds that is then played through speakers. This means that the storage requirements for MIDI tend to be relatively low.

Comparing Audio Costs

CD-audio provides the highest-quality sound and comes with its own storage media (CDs). These sounds are usually purchased (copyright restrictions must be observed). Waveform sounds can be both recorded and purchased. When sounds are recorded, designers are able to make quality trade-offs relating to sample size, sample rate, and stereophonic sound. Synthesized (MIDI) sounds usually involve sounds that require little storage. They can be either created or purchased.

Sounds in Computer Systems

Auditory icons can help computers convey information to users (Gaver et al., 1991). These audio cues can provide useful information on processes and problems. One main problem is to identify brief, everyday sounds so that they can be used in the operation of computer products. Ballas (1993) has identified many common sounds and determined the time it takes to identify each of them (in seconds) and the accuracy of perception. Both the time it takes to identify an "earcon," and the accuracy of identification make its use questionable in many situations (see Table 13-2).

Table 13-2 Recognition of Common Sounds

Sound	Seconds to Identify	Percent Correct
Telephone ring	1.2	90
Clock ticking	1.6	97
Car horn	1.6	99
Doorbell	1.6	96
Automatic rifle	1.7	68
Water drip	1.8	91
Water bubbling	2.3	76
Bugle	2.4	88
Gunshot indoors	2.4	14
Church bell	2.6	82
Door knock	2.7	73
Toilet flush	2.8	65
Footsteps	2.8	61
Fireworks	2.9	47
Hammering	3.6	36
Power saw	4.1	19
Light switch	6.0	18

Guidelines for Using Sound

Cohen (1993) suggests the following guidelines to help in determining when to use sound in systems.

The origin of the message is itself a sound (voice, music).

Other systems are overburdened (simultaneous presentation).

The message is simple and short.

The message will not be referred to later.

The message deals with events in time ("Your printer is finished").

A warning calls for immediate action ("Your printer is out of paper").

Speech channels are fully employed.

Illumination limits use of vision (e.g., alarm clock).

Receiver moves around.

A verbal response is required.

Users are not age 70 or more (cf. Nielsen and Schaefer, 1993).

Time-compressed Speech

Time compression increases the rate of computer-based speech without altering its pitch. Past research has shown that speech can be compressed by as much as 50 percent with little or no loss of comprehension. DeGroot and Schwab (1993) reported that people over age 60 may have trouble understanding compressed speech.

Presenting Visual Images with Computer Speech

One study examined the contribution of visual images along with speech intelligibility (Greaves et al., 1993). It found that the addition of the head and shoulders of the speaker did not improve listening comprehension.

Hearing Synthesized Speech

People hear more accurately when listening to another person than when listening to a computer making utterances using synthesized speech (Cowley and Jones, 1992; Rudnicky et al., 1994):

	Words	Sentences
Natural speech	99.5%	99.2%
Synthesized speech	96.8%	95.3%

Cowley and Jones (1992) recommend that we use synthesized speech when messages are unique and/or require alteration and use digitized speech for situations when there are a small number of standard messages.

Still Images (Photographs)

Another means of displaying information is through the use of still images or photographs. The images are created by designers using cameras (still or full motion) or scanners. Users tend to access the images using a mouse pointer or touchscreen.

Image Resolution

Number of Pixels

The images on monitors consist of picture elements or *pixels*. Pixels are rectangular, and each pixel is filled with color. Realistic images require a large number of pixels. In fact, the number of pixels in an image is analogous to specifying resolution. Color monitors initially had the ability to display only about 320 × 200 pixels (CGA). This progressed to the common VGA (video graphics array) standard of 640 × 480 pixels. As more pixels are used to create images, the resolution increases. Many monitors are able to display 1664 × 1280 pixels or higher.

Colors per Pixel

Realistic images also require a large number of colors per pixel (bits per pixel). The bits per pixel (bpp) value determines how many colors the video system can display simultaneously. The bits per pixel generally ranges from 1 to 24. Having 1 bit per pixel only allows the display of two colors (e.g., black and white, black and green, or amber and brown), whereas having 24 bits per pixel allows up to 16,777,216 colors. Realistic, but relatively low resolution images can be generated using 8 bits per pixel (256 colors).

Image Compression

Still images can be compressed when stored. Reducing the amount of data needed to reproduce images (1) saves storage space and (2) increases access speed. Compression technologies can reduce storage requirements by a ratio of 15 : 1 (up to about 50 : 1). Obviously, the greater the compression ratio, the greater are the chances of having degraded image quality. The accepted image compression standard is Joint Photographic Experts Group (JPEG).

Much image compression is facilitated by the fact that the eyes are much more sensitive to luminance than chrominance. Thus, the chrominance values in every other pixel can be discarded without affecting human perception of the image. Another favored way of compressing is to discard frequencies outside the range of visible light. This accounts for much data reduction.

Full-motion Video

Video Standards

One of the newest data types available for display in computer systems is full-motion video. The major input source for designers is the camera. The major input technologies for users include the mouse, touchscreen, infrared remote control, joystick, and dataglove. Television was invented in 1927, and the broadcast standard for color television used in North America and Japan was developed in 1952. The broadcast standard was developed by the National Television Standards Committee (NTSC). The NTSC standard dictates that television will have 525 horizontal scan lines, show 30 frames per second, and refresh at 60 times per second (interlaced). Resolution on a television set is equivalent to about 640 × 480 pixel resolution on a computer monitor.

Europe, Australia, and South Africa use the Phase Alternating Line (PAL) standard. France and many Eastern bloc countries use the Sequential Color and Memory (SECAM). In the future, the television standard will become a digital form of high-definition television (HDTV). The resolution level of HDTV will be about 1280 × 1024 pixels and will require a frame rate of 60 frames per second. Another major difference will be the aspect ratio, which will change from the existing 4 : 3 to 16 : 9. Possibly "digital" itself may become the new standard.

Frame Rates

Many frames must be displayed each second to produce the effect of smooth motion. When comparing frame rates, we find the following:

Smooth motion: 15 frames per second

NTSC: 30 frames per second

PAL: 25 frames per second

Film: 24 frames per second

Storage Options

Effective use of full-motion video requires adequate storage. There are two options: (1) use analog video signals directly (television) or (2) convert analog video signals to digital. Analog signals can be displayed directly on computer monitors with their source being from cable, videotape, or videodisk. However, converting analog signals into digital signals allows direct storage, editing, and display (on a monitor). This allows digitized full-motion video signals to be dealt with like other data types.

Optimizing Full-motion Video

There are at least three ways to optimize full-motion video that is being displayed on computer monitors: (1) increase the data transfer rate, (2) use smaller image sizes, and (3) use compression algorithms.

Data Transfer Rates

Video performance is limited by how fast the system can store and retrieve (transfer) data. When compared with a computer's central processing unit (CPU), disk drives can appear to be very slow. Over the past 10 years, CPU performance has increased by a factor of 1000, while disk drive performance has increased by a factor of 3. Data transfer rates tend to range from a slow 500,000 bytes per second to well over 10 megabytes per second.

For example, to show the difficulties with transfer rates, consider a monitor with VGA resolution (640 × 480 pixels). If color is determined using 16 bits per pixel (2 bytes), then one full-screen image would require that 614,400 bytes (640 × 480 × 2) be transferred. If the data transfer rate is 1 megabyte per second, then the fastest possible frame rate is only 1.6 frames per second (1,000,000/614,400). This would not show much motion. If we increase the data transfer rate to 8 megabytes per second, then the fastest possible frame rate is 13 frames per second. This would show fair motion.

Image Sizes

Another way to deal with the requirement for achieving at least 15 frames per second is to reduce the number of pixels per image. This can be done by increasing the size of each pixel, or it can be done by decreasing the number of pixels per image by making the image smaller.

For example, by decreasing the number of pixels to 76,800 (320 × 240), and assuming 2 bytes for color and assuming a data transfer rate of 1 megabyte per second, full-motion video would show at 6.5 frames per second (1,000,000/153,600). This would show jerky motion. It would take a much smaller image of only 16,700 pixels (129 × 129, and 2 bytes for color) to achieve good motion at 30 frames per second.

Full-motion Video Compression

Full-motion video compression makes the use of full-motion video on computer monitors practical. Without compression there would not be enough storage for most uses. Several compression techniques have been proposed. One of the most common standards is the previously mentioned JPEG. JPEG was originally developed for compressing still images; however, it can be used to compress full-motion video on a frame-by-frame basis. The average compression ratio is about 15 : 1.

The Motion Picture Experts Group (MPEG) developed a standard for full-motion video and the associated audio. By using the MPEG standard, it enables full-motion video and associated audio to be easily delivered using existing computers and networks. MPEG allows compression rates up to 150 : 1, with the compressed–decompressed video comparing favorably with VHS.

MPEG full-motion video compression techniques make substantial use of redundancy between frames. When compressing motion video, the first frame is treated as a still image and is then compared against the second frame, where changes are noted. The process continues by comparing adjacent frames. About every six to eight frames, a full reference frame is transmitted. Longer skip rates can result in picture quality deterioration.

Multimedia Development Skills

The knowledge and skills required to adequately develop, test, and deliver multimedia user interfaces can be considerable. Designers will require a general understanding of computing, plus specific capabilities related to using one or more multimedia authoring systems and whatever secondary (support) systems are necessary. There may be a requirement for designers to be good writers, editors, and animators. They may need to understand how to record and edit speech and other sounds. They may need to be able to deal directly with CD-audio and MIDI sounds. In addition, they may need to take high-quality photographs and full-motion video and to edit the photographs and full video as necessary. Other skills that they may need to have available include a musician, actor, graphic artist, photographer, film director, and instructional designer.

Internationalization

Internationalization is the process of isolating the culturally specific elements from a product. Typically, this includes ensuring that a system presents information such as text, numbers, and date formats that are consistent with the culture in which the system is used. *Localization* is the process of infusing a specific cultural context into a previously internationalized product. These can include values, ethics, and the like (Russo and Boor, 1993).

User Guidance

Providing useful user guidance information can make a positive difference in performance when a system is operational. Neerinex and deGreef (1993) reported a study in which poorly designed computer-based help *decreased* learnability and usability for novice users. They redesigned the system, including the computer-based help, using the following method:

Step 1: Specified a "minimal interface" (one that provided sufficient functionality for *expert* users)

Step 2: Had typical users perform using a prototype

Step 3: Only for those functions with which participants had difficulty, they developed computer-based, context-specific assistance

A follow-on study demonstrated that the revised system and computer-based help significantly improved performance.

A complete user guidance procedure has at least four ways of providing assistance to users: (1) encouraging users to ask other users, (2) providing a help facility, (3) providing computer-based documentation, and (4) providing computer-based training. The perceived usefulness of different user guidance approaches is shown in Table 13-3.

Table 13-3 Perceived Usefulness of Computer Information Sources

	Percent Used	Percent Liked
Another user	63	85
Supervisor	54	83
Paper document	58	76
On-line help	61	67
On-line document	51	73
On-line tutorial	41	62

From Granda et al., 1990.

Asking Others

In situations where systems are used, there is a long tradition of asking other people when difficulties are encountered. In these cases, users do not make much use of user guides or other documentation that is provided with the systems. These users tend to ask someone when there is a problem, the system "guru" or "super-user."

Thus, one of the most used methods for providing user assistance is to have someone available who is an expert on using that system. This person should be available to ask questions face to face, by telephone, or perhaps even by sending electronic messages. The main advantage of using people to answer questions is that there is no need to commit resources for trying to anticipate the questions that are going to be asked.

One person asking another person how to do something seems to be the preferred way for many users in many systems. Consequently, it may be difficult to get some users to use the computer to answer questions. This is particularly a problem when computer-based assistance is difficult to access, requires losing what is on the screen, offers incomplete information, or provides no meaningful information at all. It seems that users will seek the guidance that is the quickest and easiest to access and that provides assistance at the appropriate level of detail.

If a super-user is used, then the system should be designed so that he or she receives the necessary training to become expert in the system and that expertise is kept up to date. If this satisfies the user's needs for assistance, then it may be a waste of resources to also provide computer-based user guidance.

It is becoming more popular to have the computer be the primary source of user assistance. If done well, users do not need to rely on others to effectively interact with the computer. Thus, user assistance or guidance comes from the computer, which is causing the difficulty in the first place.

Help Facilities

Designers should provide user guidance that is truly helpful. This usually includes having one or two levels of detail and having a system that is *context sensitive.* For example, if a user is attempting to print a document, he or she may enter an inappropriate command and the computer responds with an error message. The user then asks for help. The computer should assume that the person is asking for help on how to print a document. The system then provides a brief comment, and, if asked for more detail by the user, it is provided.

Intelligent help systems can detect when users are having trouble and suggest solutions for the user (Carroll and McKendree, 1987; Jerrams-Smith, 1987). Such systems can also provide different prompts depending on the level or needs of the user (Mason, 1986).

Computer-based Documentation

A third type of user guidance pertains to computer-based documentation. A computer-based document stores detailed information in the computer, much like the information that has been traditionally stored in paper documents.

A major difference between computer- and paper-based documents is their formatting. Paper documents are usually formatted according to tradition and convention, whereas computer-based documents are formatted in ways that best facilitate user access to information. Computer-based documents usually are not intended to be read like a novel, starting at the beginning and reading to the end. They are laid out to facilitate finding the information necessary for solving a specific problem at a specific time.

Users usually access computer-based documents for more detailed information than can be derived from the help facility. That is, they first use the help facility, and if the help facility does not provide sufficient information, they then access the computer-based documentation. Computer-based documentation tends to be more detailed and usually offers both words and graphics, as well as numerous examples to help clarify issues and concepts.

Computer-based Training

The fourth type of user guidance is referred to as computer-based training. This is usually one or more training modules embedded in the system. This same computer-based training can be used initially to teach users how to use a system. In addition to the information contained in help and computer-based documentation, the computer-based training has the advantage of exercising users to ensure that they truly understand the concepts. Computer-based training could provide a short tutorial and, after the tutorial, provide questions that need to be answered. These could first be knowledge questions, followed by skill questions. Finally, users are then given the opportunity to demonstrate their knowledge and skills by completing exercises for which the computer can check their answers (Mahoney and Lyday, 1984).

Conclusion

Thus, persons having difficulties using a system should first try to resolve the problems using the help facility. If this does not satisfy their needs, they should then access the computer-based documentation. If the answer is still not clear, they should then access the computer-based training.

Human–Computer Communication in the Future

Having an optimum human–computer interface is limited most by the current, somewhat restricted, communication alternatives. Consider the progression of human–computer communication:

The human transfers information to a computer in some noninteractive mode (punched cards, marking, handprinting, or typing on a document and having it read by an optical character reader).

The human transfers information directly to a computer using a keyboard, handprinting, touchscreen, and so on.

The human transfers information directly to a computer by speaking individual words.

The human transfers information directly to a computer by speaking sentences.

The human transfers information to a computer by controlling and changing brainwave (EEG) patterns (e.g., by altering the relative proportion of alpha and beta waves).

The human transfers information to a computer by visualizing individual letters or numbers or nonverbal images.

The human transfers information to a computer by thinking concepts or ideas.

Two or more people are able to communicate knowledge by exchanging thoughts, using a computer as an intermediary.

Two or more people are able to communicate experiences and feelings by exchanging nonverbal impressions and/or images through a computer.

The computer transfers information directly from computer memory to human memory. Theoretically, the user at this point in time will have access to all human knowledge, and the human limitation of forgetting diminishes in importance.

Exercise 13: Designing a Display Screen

Purpose: To help students to apply appropriate window and/or screen design techniques.

Method: This exercise requires you to design and develop the best possible interface for a system. If possible, use a software prototyping package. If you do not have prototyping software, illustrate your ideas using a paper prototype.

Develop a window or screen that will be used by a person working at a hotel. The hotel clerk will use a computer to check in guests. Your user will stand at a computer terminal asking questions and quickly entering the guest's responses.

Assume the following about the system's users (check-in clerks):

They will use the system about 6 hours a day.

They will have about 1 hour of training before beginning to use the system.

They may or may not have touch-typing skills.

They will always have the pressure of the guest waiting for a room while they are completing the registration process.

The window or screen must include the following information:

Name

Address

- Street
- City
- State
- Zip Code

Telephone numbers (include area code)

- Business
- Home

Number of people staying in the room

License plate number (include state)

Payment information

Type of payment (credit card or cash)

Credit card type (Visa, Master Card, American Express)

Credit card number

Room information

Room number

Daily rate

Departure date

Try to make the window or screen as appealing to potential users as possible. In other words, design a useful, esthetically pleasing window or screen.

Part 4

FACILITATING HUMAN PERFORMANCE

Determining Performance Support Requirements

Assuming that the *basic* human performance design decisions are adequate and the *interfaces* are well designed and developed, the only other major human performance consideration concerns *facilitators:* (1) written instructions, (2) performance aids, and (3) training materials. One of the most basic differences between the three is the time that elapses between the time the information is presented and when the performance takes place. Users read (or hear) performance aids and then almost immediately begin performing. Written instructions may also be used in a relatively short time after reading. With training, however, the information may be presented in January and not used until July.

Although many systems rely on all three methods to achieve long-lasting and acceptable human performance levels, many successful systems use none, one, or two of these methods. Designers must decide the proper mix or balance. Too many performance aids and too few training materials may cause user confusion, while too much reliance on training and very limited use of performance aids could result in costly development of training materials that will never be used.

In traditional approaches to system development, the general development sequence was to design, document, and then develop training. Performance aids, if developed at all, were usually developed once the system was operational. Using this sequence, designers rarely had the opportunity to decide whether training or instructions or performance aids were more appropriate for obtaining acceptable human performance.

In addition, traditional approaches frequently resulted in the production of manuals overloaded with useless information that was not referenced on the job and in training courses that contained information that could be conveyed to workers in less expensive and less time-consuming ways. Often, written instructions and training courses covered the same tasks, with the only difference being one of emphasis, because they were developed by different designers. On the job, users found it so difficult to locate data in the mass of unnecessary information that they often dealt with these materials by storing them away and forgetting about them. When workers needed information, they either asked someone for answers or guessed at the solutions.

There are few guidelines to help a designer decide which type of information to provide. However, the best decisions are made when as much as possible is known about potential users, the activity to be performed, and the context in which the performance will take place.

Once developed, system information should be tested and, if necessary, revised. For this process to be most effective, designers should provide the *least* amount of information considered adequate. If representative users are unable to perform adequately, designers can add additional information and retest. If users perform well, materials may be overdeveloped. To determine if this is the case, information should be reduced until representative users begin to experience problems.

Written Instructions

The main question a designer must answer about providing written instructions is whether a set of detailed instructions should be developed at all. If the computer interface provides sufficient support, additional documentation is unnecessary.

Written instructions can be divided into four major types (cf. Schriver, 1986). Designers should decide which type or combination is best for typical users of their system. Each type of document is provided for a different reason.

Tutorials: These are usually step-by-step explanations that cover the *basics* of using a system and are intended for *new system users,* users who are *reading to learn.*

User guides: These assume a task-oriented approach and usually list all activities that can be performed using the system. They are generally intended for *experienced users* who need assistance with *occasionally* or *infrequently* used tasks. Users are *reading to do.*

Quick reference guides or performance aids: These provide skeletal information that serves to explain only the most essential performance-oriented information. They are intended for *experienced users* who are *reading to refresh* their memories.

Full reference manuals: These usually contain highly detailed, encyclopedic information about a system. They can be both performance and topic oriented (descriptive) and tend to provide in-depth information about a system, including a full and detailed discussion of its capabilities and limitations. They are intended for *experienced users* who are *reading to exploit* the full power of a system. Because of their size, they are usually too time consuming to read and too detailed to be of value to the majority of system users.

Written instructions are generally the best solution in the following situations:

- A full description of the work is required (especially if background information or information about where the work fits into the larger system is needed).
- Retention of training information is unlikely because activities are performed infrequently or are complex.
- Step-by-step guidance and detailed reference information are required.
- Complex visual presentations with detailed explanations are required, for example, graphs, flow charts, schematics, and drawings.
- Cross referencing from one set of instructions to another is required.
- The activity is so complex that it is difficult to learn and remember required information.

Performance Aids

Performance aids provide information for immediate use. They may be written or computer based. Familiar examples include fingertip telephone indexes and lists of frequently called telephone numbers. Performance aids are usually characterized by a concise, quick-access format, such as a quick path card or a checklist.

Designers should determine whether performance aids are providing adequate levels of information *before* developing lengthy instructions or training. This bias in favor of performance aids results from the following:

- They are not as subject to problems of forgetting as is training or other types of written instructions.
- They generally cost less to develop and can be developed more quickly.
- They are easier to revise when procedures change.
- They generally shorten training time even when both are required.

Performance aids are generally the best solution in these situations:

- Moderate speed and high accuracy are required at the time of performance, for example, accessing a frequently called telephone number list before making a call.
- Long procedures are required and there is much information to remember. The performance aid can provide key words or pegs for memory.
- Conversion of information is required; for example, from yards to meters or square feet to square yards. A list of conversions can be posted or carried.
- The consequences of error are serious; for example, consider preflight routines performed by flight crews. Trusting this task to memory could prove dangerous.
- The task is performed infrequently and memory needs to be jogged.
- Aspects of the task change frequently and detailed documentation is too expensive to provide.
- Personnel turnover is excessive or only needed on a short-term basis during peak periods.
- Summaries of detailed information are needed. (References to detailed information can be included.)

Training

In many situations, performance should be based on learning situations that preclude the use of detailed instructions or performance aids. Training is generally the best solution in these situations:

- Skill levels must be attained and demonstrated before performance begins. For example, telephone installers should know their jobs *before* they try to install customers' telephones. In such cases, training is usually necessary.
- On-the-job speed is so critical that there is no time to use performance aids or instructions. For certain critical and emergency activities, it is not desirable for workers to lose time looking through written material.
- Activities are too complex to perform without training.
- Workers need preparation (explanations, demonstrations, practice) to adequately use performance aids or instructions.

- Training is the quickest way for users to achieve and maintain an acceptable performance level.
- Training that is already available for other purposes will produce the desired performance if supplemented.
- It is more cost effective to train people who do not have the necessary skills than it is to select people who do.
- Skills and knowledge must constantly be enhanced or updated.
- Tasks are performed frequently. For example, a user must conduct an inspection of computers on an average of three times per shift. The frequency of these inspections tends to support the use of training rather than performance aids or instructions.
- The required performance is hard to communicate in written form.
- The use of performance aids is impractical or inappropriate. For example, performance aids are usually not too useful for users who must perform troubleshooting under confining or poorly illuminated working conditions.

Generally, training is more expensive to develop, provide, and maintain. Furthermore, the most expensive elements of training are often the combined costs of student salaries, training facilities, instructor salaries, and training administration.

14

Written Instructions

Introduction

No matter how well designed are the software or hardware in a system, people still may have difficulty performing in the absence of clear, accurate, and complete instructions. Even small changes in the way these instructions are written and presented can have sizable effects on human performance. The designer must ensure that when information is required it can be accessed easily and used efficiently, whether it is stored in a 3-inch-thick user manual, on the face of a 1-meter square road sign, or in an integrated circuit measuring only 3 micrometers in a computer.

Understanding Information Sources

When system users require additional information to effectively use a system, three major information sources are typically available to them.

- *People.* Users first ask other people for the needed information. They prefer asking someone face to face, but will use the telephone if necessary.
- *Paper.* Hard-copy documents (e.g., user manuals, performance aids, or tutorials) are second. Paper documents still offer many advantages for users.
- *Computer.* Computer-based information, such as help systems and on-line documentation, is generally used only when other sources fail or the computer system offers significant advantages, such as easy and comprehensive search capabilities.

Designers must account for this hierarchy as they prepare system information for users. It is also important for a designer to consider carefully the exact amount of information needed, to determine the best way of storing the information until it is needed, and to devise a method for presenting the information (when needed) so that it can be used as efficiently and effectively as possible. Verbal instructions are flexible, but the person who is giving the instructions may not consistently include complete information. Written instructions do not change over time. Written and verbal instructions are not merely an adjunct to a system, but an integral part that deserves careful consideration as early as possible in the design of a new system.

Understanding Users

Sanders (1976) in *The Tomorrow File* gives an account of the use of instructions that, although fictional, applies to many real-life situations. In his society of the future, a precisely timed sleeping pill or "hypnotic" called Somnorific is developed and marketed as an over-the-counter drug. Somnorific initially fails because, after unwrapping the substance, people fail to wait the required 10 seconds before inhaling. Clear *instructions* to wait are on the wrapper. Sanders' human engineering (ergonomics) team recommends that a more tenacious adhesive be attached to the product package. The adhesive sticks to users' fingers when they open the package and 10 seconds pass before they can roll it into a little ball and completely flick it off. People are constantly bombarded with instructions and, like the people who used Somnorific, ignore many of them, posing a problem for designers: even though the instructions are available, users don't read them.

Designers must also overcome a long-standing tradition related to instructions. We have assumed that, given sufficient time, a person assigned a task can figure out what to do and how best to do it even if instructions are inadequate or even nonexistent. A classic case is the Adam computer for children developed by the Coleco Corporation. Coleco advertised the computer as simple enough for kindergarten children to operate. However, even adults could not properly use the machine, primarily because the documentation was too difficult to understand. The Adam computer lost the Coleco Corporation millions of dollars and was eventually withdrawn from their product line. The *Wall Street Journal* had a headline that read, "Hundreds of Coleco's Adams are returned as defective; *Firm blames user manuals*" (November 30, 1983).

In too many situations, instructions are provided as an *afterthought.* A designer who understands how the system operates often assumes that users have that same knowledge and fails to give enough attention to developing the documentation.

Providing Adequate Information

When a user does not understand an instruction and consequently does not perform in accordance with that instruction, the resulting degraded performance is the direct responsibility of the designer. Chapanis (1965b) reports on such a problem in a large building in Baltimore. This building has elevators on each floor. The wording of signs placed near the elevators is shown in Figure 14-1. Chapanis conducted a limited study to

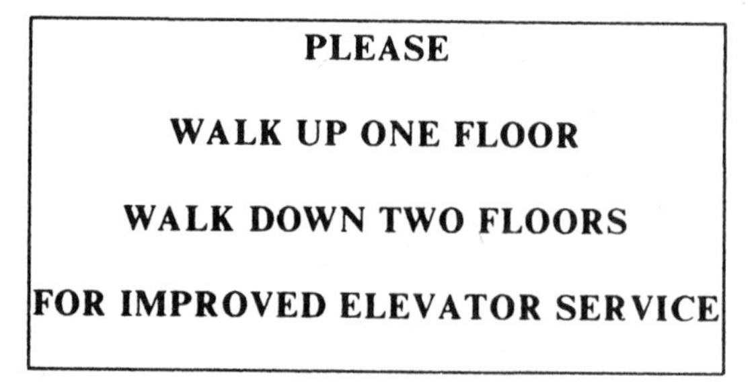

PLEASE
WALK UP ONE FLOOR
WALK DOWN TWO FLOORS
FOR IMPROVED ELEVATOR SERVICE

Figure 14-1 Sign placed between two elevators.

determine what the signs meant to people and found that most people thought something like "This must be one of those fancy new elevators that has something automatic. The elevator doesn't stop at this floor very often. If I want to get on the elevator, I'd better go up one floor or down two floors." And this is exactly what people were doing. Unfortunately, after they had trudged up or down the stairs, they found the same sign. The instructions did not convey what the designer intended. Chapanis proposed some alternative wordings that would communicate the message more clearly (see Figure 14-2).

IF YOU ARE ONLY GOING

UP ONE FLOOR

OR

DOWN TWO FLOORS

PLEASE WALK.

IF YOU DO THAT WE'LL ALL HAVE

BETTER ELEVATOR SERVICE.

Figure 14-2 A clearer version of the sign (adapted from *Words, Words, Words,* by A. Chapanis. Reprinted by permission of the Human Factors Society, Santa Monica, California).

Instructions that are not clear are sometimes worse than no instructions at all. Conrad (1962) tells of a tragedy that occurred during very cold weather in England. A small boy left school in the afternoon and did not return home. The following morning a search was organized. He was found dead from exposure at a place nearby, where he often played. An inquest disclosed that the search was not organized sooner because neither the boy's mother nor any of her neighbors knew how to use the public telephone at the end of the street. Conrad reviewed the instructions for using the telephone and was impressed by their difficulty. He conducted a study to determine how well people could read and understand them. He found that only 20 percent were able to use the telephone correctly. When another group received a shortened and rewritten version of the instructions, there was a marked increase in the number who could perform the task correctly.

Thus, even something as simple as using a telephone is becoming more sophisticated. Users must deal with numbers with 10 or more figures; they must know how to dial direct, reverse charges, use a credit card, and charge to a third-party number. Some telephone systems require people to know how to transfer calls, set up three-way calls, and put calls on hold. Telephones with sophisticated CRT displays are in common use.

Even though the technological complexity of the telephone and other systems has increased, this does not mean that the requirements placed on users of these systems should become more complex. Usually, the more complex a system, the harder a designer needs to work to ensure that the user population be able to adequately perform in the system. Clear and concise instructions help simplify performance requirements.

Problems with providing adequate instructions have been well documented for several years (Bailey et al., 1975). In some cases, documents did not give the user enough information. In other cases, instructions contained omissions and were inaccurate and poorly organized. At times, poorly designed tables of contents, indexes, section headings, and cross-referencing schemes (or their omission) led to degraded performance. At other times, the information was at the wrong level of detail.

Testing for Performance

The best way to ensure adequate instruction is to *test* for completeness and *performance,* preferably by having people similar to the target group read the instructions and then do what they say (without any outside help).The design process is not completed until the designer is assured that the instructions will elicit the appropriate actions.

Instructions should be developed as early in the design cycle as possible and tested at the same time as other portions of the system. As the designer develops a procedure, he or she should ask, "Is this something I can readily communicate to a user with a written instruction?" The skill required to provide a good set of instructions is every bit as complex as that required to engineer the software or hardware of a system.

Writing for a Specific User Audience

It cannot be emphasized enough: *prepare instructions for the people who will use them.* Elegance of style, clarity of organization, and simplicity of presentation are wasted if instructions are prepared for someone other than the ultimate user. Preparing good instructions becomes almost impossible if the designer does not have a clear set of reader characteristics in mind or a precise definition of the target audience for each set of instructions. Occasionally, a set of instructions is intended for such a diverse audience that the only alternative is to reduce the instructions to the lowest common denominator—the least capable person who will be required to use them.

Reasons for Using a Document

Past studies suggest that about 14 percent of users used the manuals delivered with new systems (Lazonder and vanderMeij, 1993). When documents are used in computer-based systems, Dillon (1991) found that they were used as follows:

	%
For reference (how to do)	73
Getting started	67
When in trouble	53

For a summary of features	33
Aid for exploring software	13
As a guide before buying	7
For detailed technical information	7

Guidelines for Developing Documentation

Writing instructional materials is much different than writing for the *New York Times,* an article in *Scientific American,* or a novel. As we will discuss in more detail later, instructional materials should convey information completely, accurately, concisely, and objectively. An appropriate writing style is usually terse and relatively flat (humor and creativity are not essential and are usually unappreciated).

Numerous studies related to presenting instructions are available to designers to help them to make good decisions. The guidelines that follow are only a small subset and are taken primarily from Ronco and others (1966), MacDonald-Ross (1977), Wright (1977), Hartley and Burnhill (1977), Dever and others (1978), and Schriver (1986). More detailed information can be found in each of these resources.

Typographical Features

Typographical characteristics that affect legibility should be carefully selected and evaluated by the designer. The characteristics that have received the most attention are size, case, style of typeface, and use of capitals. Numerous studies have also been conducted to determine optimal spacing or leading, line length, and the contrast between paper (or screen) and print. Formats have also been studied to determine their affect on reading comprehension.

Legibility

Legibility has been studied under both favorable and unfavorable reading conditions.

Favorable conditions. The aim generally has been to determine an optimal combination of typographical characteristics for normal reading.

Unfavorable conditions. The aim usually has been to determine how small or how dimly illuminated a letter, word, or phrase can be and yet be correctly perceived.

Both situations are of interest to designers. And, in fact, a designer should carefully consider all possible situations in which an instruction will be read. It is obviously not a good idea to design instructions to be used with high illumination only to find that they are used with very low illumination a good portion of the time. For example, highway signs that can be easily seen during the day might be seen only with great difficulty after dark unless care is taken to adjust for this difference.

Favorable reading conditions exist when the illumination is acceptable and other degrading influences such as vibration or viewing from long distances are absent. Unless stated otherwise, the conditions we will discuss will focus on legibility under favorable reading conditions.

Type Size

Three terms are useful when discussing the physical characteristics of type.

Point size is a unit of measure equivalent to 0.01384 inch (72 points approximate 1 inch). Points are generally used to indicate type size and interlinear spacing. In terms of letter dimensions, point size can be estimated by measuring from the top of the highest letter (a capital or a lowercase b, for example) to the bottom of the lowest letter (such as a lowercase p or y). The type size in books and magazines usually ranges from 7 to 14 points, with the majority being about 10 to 11 points. Probably the optimum range is from 9 to 11 points; sizes smaller or larger can slow reading speed (Tinker, 1963).

Pica is another unit of length. One pica is equal to 12 points and 6 picas equals about 1 inch. Picas are used to specify typographical dimensions such as line width, column depth, and column spacing.

Leading refers to spacing between lines and is measured in points. Type set with no leading appears crowded but is usually readable because the letters do not occupy all the vertical spacing available on the type body.

Type Styles

Most typefaces or type styles in common use are about equally legible, but designers should keep the following in mind:

Serif versus san serif. Serif styles were once considered best, but with most modern typefaces, both are equally readable.

Italics. Italic print tends to slow reading speed.

Uppercase. Text printed in all uppercase letters is read more slowly than text in lowercase letters.

Boldface. Boldface type is read as rapidly as ordinary lowercase type. Users prefer boldface type for emphasis.

Number of typefaces. Text presented in too wide an assortment of typefaces retards reading speed.

Line Widths

According to a study by Tinker (1963), line widths in common use fall in a relatively small range:

Double column	17 to 18 picas

Single column	21 to 26 picas
Reader preference	moderate length (14 to 36 picas) and two-point leading

Increased leading particularly improves the legibility of smaller types (6 and 8 point), although the improvement is not generally sufficient to make them more legible than 10- or 11-point type set without leading.

Margins

Readers seem to prefer ample margins and believe that wider margins increase legibility. However, studies have shown that text set with no margins is just as legible as text set with large margins. It is interesting to note that legibility may be adversely affected by page curvature such as exists near the inner margin of some books. This problem can be dealt with by using a larger inner margin. Readers tend to prefer multiple columns per page. Spacing between columns can be accomplished in a number of ways without adversely affecting legibility.

Print and Background Colors

For ordinary reading, black print on white background is more legible than white print on a dark background. In addition, black print on a white background is generally preferred by readers. However, the use of tinted papers with black print does not seem to impair legibility if the reflectance of the paper is high (70 percent or greater) and the type size is at least 10 point. Colored inks on colored papers can adversely affect legibility. Combinations that provide good brightness contrast (e.g., a dark-colored ink on a light-colored paper) tend to be the most legible.

Impact of Typographical Features on Performance

Designers should be aware that even after applying the best of all typographical characteristics only small differences in performance usually occur. That the differences tend to be small is not surprising when we consider the refinement process in the preparation of printed pages that has taken place over the last few centuries. Even so, any increase in speed or accuracy in interpreting the meaning is worthwhile.

Finally, designers should constantly keep in mind where the reading of the instructions will take place. Characteristics that are nearly optimal under favorable conditions still may lead to degraded performance under less favorable conditions. A maintenance manual that is well designed for use in a well-lighted office may be almost totally unreadable in an equipment room where the light is dim.

Language

Words

When a word or concept does not mean the same to a designer as it does to a user, the intent of an instruction is altered and its effectiveness impaired. Using words that convey the intended meaning is therefore critically important to designers. In short, designers should use words that the audience will understand:

Familiar words. Use words that have high frequency in everyday use.

Short words (relative). Many experts have long contended that using short words improves comprehension and makes text more readable (see section on readability indexes). However, familiarity, rather than length, is probably the more critical factor. Short words don't always convey the appropriate meaning and technical terms have no (short) substitutes. In addition, difficulty is not confined to long words—it is the concept, not the word, that is difficult (an *erg,* however short the word, is not an easily understood unit of measure).

Little jargon. Consider the audience. If documentation is designed for experts, it should take advantage of the expert jargon common to the audience (consider medical documentation, for example); if designed for the inexperienced, jargon should be avoided when possible and explained when it must be used.

Few abbreviations. Only *standard* abbreviations should be used and then only sparingly; designers should not make up abbreviations for their own particular situations.

Sentences

Sentence structure and style can also influence the effectiveness of instructions. The following principles have been summarized from several sources (see Gough, 1965; Slobin, 1966; Broadbent, 1977; Miller, 1962; and Rude, 1991). Sentences should

Use clear, unconvoluted syntax, usually subject–verb–object (or complement).

Use action verbs (move, speak, do, notify). Consider restructuring sentences that use "to be" verbs (is, are, was, were) and wimpy verbs.

Be complete, well organized, unambiguous, and concise.

Include correct grammar and appropriately placed modifiers.

Be relatively short (20 words or less). Long and complex sentences are usually more difficult because readers have to sort relationships. However, for this reason it may be more important to reduce sentence *complexity* than to reduce sentence *length.* Consider the following:

> Two clauses, appropriately coordinated or subordinated, may be better than two separate sentences just because they *do* establish needed relationships.
>
> Structuring all sentences as simple sentences can make the text seem childish.
>
> Expert users often appreciate the information they receive from complex sentences, while inexperienced users need to get information one chunk at a time.
>
> Some studies indicate that there is little correlation between line lengths and readability. A 1987 study conducted among 360 readers by the *Journalism*

Quarterly, using various line lengths and type sizes, showed no significant relationship between readability and line length. In fact, the longest line used in the test got the highest readability scores.

Avoid nominals; change *ions* and *ents* to *ings.*

Use positive rather than negative constructions. Tell users what *to do* rather than what *not* to do.

Use active voice ("flip the switch" instead of "the switch should be switched by someone").

Active versus Passive Voice

Active and passive sentences deserve additional attention. Miller (1962) compared active sentences with their passive forms to determine which would elicit the best performance. On the average, people took 25 percent more time to understand simple passive sentences like "The small child was warned by his father" or "The old woman was liked by her son" than the corresponding active forms: "The father warned the boy" and "The son liked the old woman."

The passive voice, used without cause, tends to weaken writing. However, passive does have a place; it is probably best to manipulate active and passive sentences to emphasize or subordinate information. Consider the following guidelines.

Use active

For *instructions,* when users are expected to perform an action (as opposed to reading about an action).

To add energy to writing. Active voice emphasizes the doer and the action.

When the actor is more important than the receiver.

Use passive

For *descriptions* of tasks (usually in introductions and in reading-to-learn sections that describe concepts without expecting user action). Much of this book, for example, is written in passive voice. In usability tests, users frequently ask, "Am I supposed to be doing something?" when active voice is used inappropriately for descriptive information.

When it places the most important information in the subject position ("All the children were hurt in the accident").

When it eliminates an unknown, unimportant, or "understood" actor and emphasizes the result of an action ("Recent events clearly indicate that this bill should be passed").

To mask responsibility for action ("A decision was reached to fire you").

Generally speaking, switching back and forth between active and passive voice is distracting and should be avoided.

Negative versus Positive Statements

While positive statements are usually easier to understand, designers must constantly evaluate to determine whether the person reading an instruction is likely to have an assumption or presupposition as to what the instruction will ask them to do. If this is likely, then it may be a good idea to use a negative sentence to deny the assumption. This underlines the importance of considering not only the logical meaning of an instruction, but also of thinking about the effect the instruction will have on the person at the very time the person will be reading it.

Writing Guidelines

Thompson (1979) provides a short set of easy to understand guidelines for writing (p. 42):

1. Outline. Know before you start where you will stop.
2. Write to your readers' level. Remember your prime purpose—to explain something, "not to prove that you're smarter than your readers."
3. Use familiar combinations of words. For instance, "We're going to make a country in which few are left out," instead of "We are endeavoring to construct a more inclusive society."
4. Stick to the point. Use your outline to judge the value of each instruction.
5. Be as brief as possible. Tighter writing is easier to read and understand.
6. Present your instructions in logical order (the order in which they will be performed).

   Poor: "Before turning off the calculator, clear the memory."

   Better: "Clear the memory before turning off the calculator."
7. Do not tell people what they already know.
8. Eliminate common word wasters.

Word Wasters	Cut to
at the present time	now
in the event of	if
in the majority of instances	usually
green in color	green
actually installed	installed
sufficient quantities	enough

9. When you have finished, stop.

Strunk and White's *The Elements of Style* (1972) contains a more detailed, but also highly readable set of guidelines. Designers who are preparing instructions should have a good understanding of style and grammar. It is beyond the scope of this book to cover these areas in more detail.

Illustrations (Graphics)

As designers prepare instructions, they must constantly ask whether key ideas can best be communicated by words or illustrations or both. Because most people are taught to communicate only with words, they are not as familiar with the power obtained when illustrations and written information are combined.

In many situations, illustrations can be used as the *primary* source of information, with words used as a secondary source. Illustrations generally should not be included if they only repeat in some graphical form what has already been said adequately in words. Thus, the decision to use illustrations should depend on whether the illustrations add value or clarity. More specifically, illustrations should be used to show details that are difficult to describe verbally. For example, tables generally present detailed numerical information in a compact and organized way.

Illustrations can also make instructions more interesting and serve to make a presentation of rather dull material more lively. Whenever an illustration is used, the narration should provide directions on reading and interpreting it. Probably the best approach to incorporating illustrations in instructions is to determine the illustrations first and then write around the illustrations with whatever narrative is then needed.

If it has been determined that when graphics are necessary several decisions must be made:

- What form of illustration best conveys the information
- What the relative position of text and graphics should be
- How many illustrations are needed
- How much material should be included in each illustration
- What size and proportions are best
- What colors, shading, and contrast are optimal

Type of Illustration

One of the most difficult problems with providing adequate illustrations is determining the most appropriate illustration for a particular type of information.

Tables work well when large amounts of specific numerical data must be presented.

Maps could be included when distances are to be calculated, routes chosen, or instructions for getting from one point to another are indicated.

Photographs are useful for providing an exact impression of an appearance of an object and to illustrate objects in three dimensions.

Drawings are helpful for showing the simple dimensions of an object and for emphasizing certain detail. If cross-sectional diagrams are used, the designer should indicate from what part of the entire object the cross section was taken. Exploded-view drawings are frequently used when providing instructions on assembly and disassembly of mechanical objects and for showing relationships among mechanical parts. Artistic drawings attract attention and may help motivate individuals to read the text.

Cartoons can be used to highlight points that call for special attention. They should, however, contain the idea to be communicated and not serve merely as decoration since they might detract from the main intent of a set of instructions.

Location

Illustrations should be placed close to related textual material. If an illustration and accompanying text will not fit on the same page, the illustration should be placed on the facing page. Generally, illustrations should not be accumulated and then presented as a set at the end of a text. A possible exception might be illustrations that need to be consulted throughout a set of instructions. It may be convenient to place such illustrations in an appendix or, if possible, on a foldout.

Symbols

Any symbols used on illustrations should be few, familiar, and precise. The symbols should be compatible with one another and consistently used throughout the instruction. Straight lines and dashes tend to be somewhat difficult to interpret as symbols, whereas arrows indicating movement and serrated edges indicating cutaways are two of the most easily interpreted symbols.

Color

Color can be used to highlight specific ideas or objects. However, when color is used for coding purposes, no more than eight different colors should be used. A designer should ensure that the illustrations will be viewed under acceptable lighting or else the colors may not be seen as the colors that they are supposed to represent. If colors are to be viewed in low-level illumination, changes in color tend to be less when the colors are on a white background. The designer should also keep in mind that about 8 percent of the population has some kind of color weakness or color blindness.

Contrast, Shading, and Texture

These devices can be used to advantage in presenting certain types of illustrations, including tables, and tend to draw attention to the illustration's content. Designers should exercise caution in using shading or crosshatching, particularly when they cross each other or when their patterns are extremely divergent. Crosshatching textures that are fine enough not to impair legibility of overprinting are likely to be overlooked when an illustration is used with low illumination.

Titles and Captions

All illustrations should have titles or captions. Titles should be placed above tables and below graphs or other figures. Tables and graphs particularly should be numbered and have titles that are descriptive of the content. In addition, words used in titles and labels should be consistent with the words used in the text. Titles should be short; between 5 and 11 words tend to be remembered best. If a subtitle is used to help clarify the title, the subtitle should be placed beneath the title and in a smaller font than the title.

Text and Annotation

Any annotation or text that appears on graphics should be kept to a minimum, usually to a few words or phrases. Care should be taken to avoid too many typefaces and sizes and elaborate or decorative fonts. Font sizes and styles should correspond to those used for regular text and do not need to be larger. If possible, text should be oriented horizontally, not vertically (except for the wording in vertical scales).

Learning Styles

Designers need to be cautious in deciding when to use illustrations. Reading a picture is probably a learned activity that is easier for some than others. Some users skip the pictures; others read *only* the pictures. Designers must also recognize that visual conventions are not universal and that individuals develop their own mental schema and expectations in interpreting visual information. For example, most adults have formed a schema for reading graphs and charts. If designers violate that schema, confusion results. Social and cultural values, visual experience, and explicit training also play a role. For these reasons, usability testing with the intended audience to determine levels of understanding, accompanied by redesign where indicated, is particularly important with graphics.

For a detailed discussion of guidelines pertaining to illustrations, consult Tufte (1983) and MacDonald-Ross (1977).

Organization

The ways instructions are organized can obviously influence users' speed and accuracy. All too often, writers fail to review organizational patterns and instead offer users little more than "Here's-everything-I-know-in-the-order-I-thought-of-it." Users should not have to deal with "Agatha Christie" organizational patterns that force them to gather clues and then draw their own conclusions or to loop back to previous steps or decisions because information is not presented in the order it will be used. For example, warnings about possible problems are sometimes presented to users *after* they have made mistakes.

How instructions are best organized probably depends on the purpose:

Training (reading to learn). The most typical organizational pattern for training materials includes (1) preorganizers to provide an overview of an entire process before instruction begins, (2) a section providing instructions, and (3) a summary of the completed activities.

Performing (reading to do). Users who are focused on completing a task rarely make use of introductory materials, previews, and summaries; they tend to skip to the point where the instructions begin and return to introductory information only when they encounter problems with the procedure. For this reason, proponents of "minimalist manual" theories (cf. Carroll et al., 1987) suggest eliminating introductions and summaries and further suggest that users need instructions for only critical or typical tasks—just enough to introduce the features of the interface. Users can then strike out on their own. Carroll explains that users are "apt to plunge into a procedure as soon as it is mentioned or will try to execute purely expository descriptions. This is a strong preference, and the fact that it often leads to trouble does not seem to curtail it." Anyone who has watched usability tests consistently will agree with Carroll's conclusions that users

Come to the task with their personal agendas

Jump the gun, ignoring prerequisites

Skip (what cannot be executed can be skipped, as one user commented, "it's just information")

Reason instead of read

Ignore the screen or the documentation and strike out on their own

Make errors (but we assume that they will proceed from step to step with no problem)

Want to do real work, not read

The minimal manual concept

Focuses only on real activities,

Slashes the verbiage (attempts to reduce the first draft by half),

Enables users to use the manual to deal effectively with errors, and

Is constructed using an iterative approach with usability tests.

The minimalist approach to tutorial documentation involves the following:

Task orientation allows users to get started fast and focuses on the basic goals that the software can help satisfy

Adequate use of text uses as little text as possible (no preface, index, summaries, etc.), simple sentences (averaging 14 words, presented in subject–predicate order), and little or no jargon or technical terms

Modularity each chapter should be self-contained (totally independent from each other) and require no cross-referencing

Providing information to detect and correct errors

Lazonder and vanderMeij (1993) compared the minimal manual approach with a commercially available manual. They found that those using the minimal manual (1) completed training about 25 percent sooner ($p < 0.01$), (2) made fewer errors ($p < 0.01$), and (3) completed 9 percent more items ($p < 0.05$). There was no difference among novices, beginners, and intermediates in the time taken to complete training.

Those who design instructional materials must take these user characteristics into account. As with other aspects of good instructions, how information is organized should depend on the potential readers. Although there are a few basic rules to keep in mind, the designer should be willing to adapt any set of instructions to fit the reading habits of the intended audience.

Organizational Patterns

Many different approaches can be used for presenting a set of instructions. Among them are the following:

Narrative or top-down organizational patterns. Probably the most frequently used organizational pattern is a narrative one in which instructions are presented in a chronological sequence. The narrative approach is obviously very useful for procedural instructions, although it frequently does not allow a designer to show the relative importance of various items. Designers must make sure that they deliver first things first, provide logical tracking (a road map), and then deliver what they promise. There should be no surprises. Although this sounds obvious, putting information in sequential order is not always easy, especially when several actions must occur simultaneously or when a sequence depends on which of several alternatives a user may have chosen in a previous step. Because designers are usually familiar with the process, they also tend to fill in information from their experience that the user has not been given. For these reasons, all sequences need to be tested with users to make sure that important steps have not been missed and that alternatives have been accounted for.

Inductive patterns. Inductive patterns lead users from specific instances to a generalization that encompasses all the instances. This approach is used frequently when instructions tend to be complex.

Deductive patterns. Deductive patterns, sometimes called *given–new contracts,* move from known (the given) to unknown (the new). For example, an equipment description might begin with nontechnical terms but move the user toward the technical jargon used by experts.

Processing Information

The primary purpose in organizing instructional materials is to help users to find and process information. Users should not have to put all the pieces together to arrive at an appropriate conclusion. That is the designer's responsibility. Each element of the process should contribute.

Hot Spots

Each document, page, and paragraph has "hot spots" that the eye is attracted to. These hot spots should be reserved for the most important tracking information on the page. In American culture, the top, leftmost position (on the page, in a paragraph) is primary. In addition, material at the beginning and end of a list, a sequence, or a set of instructions tends to be learned most rapidly and recalled more easily. This places great emphasis on choosing what material is presented early, in the middle, and toward the end of a set of instructions (Reder and Anderson, 1979). Hot spots, prioritized in order of importance, include the following:

Title or heading (first)

The first material in the sentence, paragraph, or page (second)

Summary or lower rightmost position (third)

Up and left (cultural)

Charts and graphics

Introductions

When using introductions and previews is appropriate, designers must make sure that they are well worded to facilitate comprehension, learning, and retention. Typically, introductions outline what is to come and the most important facts to be learned. They help readers call up appropriate schema (relevant past experiences), allowing them to pull together any appropriate information that they already know.

Headings

Headings should reflect the user's point of view not the software's. Using active verbs usually helps to ensure a task-oriented approach. Consider these examples:

Poor: Appointment Time Options Preferences

Better: Schedule an Appointment

The first (poor) heading invites the writer to explain the time options that the software allows; the second invites the writer to explain the user's task. Designers should remember that the user's goal is not to use or understand the software but to complete a task.

Effective headings can also help in look-up tasks and to give a task orientation to index entries.

Selecting and Abridging Information

To communicate instructions effectively, the designer must carefully select which information should be presented. Attempting to transmit *all* available information to a user would in most systems produce only confusion. A designer should first consider all available information, decide what information is relevant to the user, and then determine how best to organize the information. Unfortunately, some designers feel that the best way to achieve an acceptable level of performance is to dump everything known about the task on an unsuspecting user.

After the core set of information has been selected, it should be abridged, winnowed, culled—put into its most concise form. Abridging goes far beyond eliminating a few unnecessary words or tightening the writing style in a single sentence or paragraph. It often means providing the same level of technical detail in fewer words, a difficult but not impossible task given the versatility of the English language. It can mean cutting whole sections or even a well-designed graphic if it is not necessary to successful task completion. Designers are well advised to simplify the presentation of instructions to their essential points, to avoid unnecessary detail, and to maximize their ability to identify and understand which material is critical. Good designers must have the courage to delete information readers don't need, no matter how carefully constructed or time consuming its preparation. Testing prototypes with actual users in realistic situations is the best way to identify which information can be confidently deleted.

An instruction may be complex due to the large amount of information that a designer must fit into it. However, the longer the instruction, the fewer the readers. William Horton calls this the "curve of diminishing returns." Conrad (1962) found that people simply "close up their minds" when faced with a large mass of printed instruction, because it just seems too much to understand. The greater the amount of information that needs to be processed within a given period of time is the poorer the performance. Conrad rewrote a set of instructions, including only those things that were directly related to performing the task. Given the revised instructions, test participants were better able to read and understand and substantially improved their performance. This unwanted effect may be intensified still further if relevant information is not clearly differentiated from the irrelevant.

Researchers have found that shorter paragraphs tend to facilitate comprehension better than longer paragraphs. A good rule of thumb is to organize material in paragraphs that contain between 70 and 200 words; each paragraph should deal with only a single, logically presented idea. A good technique for applying this principle is to insert a space after every third line of text and then work with the three lines to form a complete, unified paragraph.

Providing Redundancy

Redundancy is not necessarily a negative attribute, especially when training is the objective or the information must be delivered to several levels of users. Humans learn best when materials are presented to them several times. Redundancy may be achieved in two ways: (1) by repetition of the same material or (2) by providing alternative ways of looking at the same information (e.g., with examples or with a mix of text and graphics rather than one mode alone).

Because simple repetition is likely to bore adult readers, using examples or other alternative modes is the better choice and has the added advantage of making the concept more concrete.

Examples. Examples should be presented as closely as possible to the initial introduction of the concept. Examples increase the reader's ability to generalize the concept and understand its application to new situations. For the designer, this suggests that presenting information in terms of various frames of reference and providing numerous illustrative examples of critical points will aid greatly in the understanding of instructions. Positive examples are generally more effective than negative ones.

Mixing modes. Making judicious use of illustrations, graphics, and tables provides users with alternative ways to get at information than just providing text alone. For example, the following checklist is an alternative (redundant) way to present the information in this section. Not all adults learn using the same methods. Using a mix of modes helps designers to ensure that no member of the intended audience will be left out. However, designers must maintain a realistic balance between enough and too much. They must make sure that no one mode is overused and that too many graphics, diagrams, or tables are not used in too short a sequence. Too many can be worse than too few.

Organizational Checklist

One way to check for organizational details is to check text elements—headings, bullet lists, paragraphs, graphics, concepts—for the following:

- Is it in the right place?
- Is it in the right order?
- Does it deliver what it promises?
- Is it grouped with the right things?
- Does it need to be "unpacked" (broken out into more manageable pieces)?
- Are emphasis and subordination easy to pick out (are part–whole relationships clear)?
- Are instructions clearly distinguished from background materials?
- Is the balance right (insignificant details are not given more attention than major concepts)?
- Is it concise (the user has not been inflicted with words)?
- Is there enough redundancy to make the point memorable and to accommodate different levels of experience and learning styles without being overdone?
- Is the most important information on the page in one of the hot spots?

Formatting Procedural Instructions

It is important to recognize that the purpose of procedural instructions is to cause the reader to *do* certain things, *not* to cause the reader to be trained or entertained. Good procedural writing is not the same as good narrative writing. A good set of instructions has little in common with a volume of Shakespeare; the different styles have different objectives.

In most situations, written instructions should *not* be presented as a series of prose paragraphs—the traditional text format. Many of the suggestions in this section deal with breaking the paragraph mold and formatting pages in ways that are easy to scan.

Text organization and formatting go hand and glove. In fact, one of the primary purposes of text formatting is to make the underlying organizational structure *visible.* Because of this, many of the principles in the preceding section must be applied to formatting principles as well.

Locating Information

If designers are to make instructions easy to use, they must make it easy for users to locate and process the information that they are interested in. Well-designed look-up features provide ways to access needed information without reading other materials. When rapid access is a requirement, a table of contents, indexes, and tabs are primary. Presenting information in modules can also be highly beneficial.

Table of contents. The TOC should show structure so that readers can see the overall plan in a single glance. The TOC should fit on one page or, at most, two facing pages, even if the levels of detail have to be restricted. If necessary, include more detailed indexes at the beginning of each chapter.

Indexes. Indexes should be task specific and anticipate the key words users might look up. It may be useful for both the traditional hierarchical and the "key-word" types of indexing to be included. An index that merely lists system nomenclature is next to useless. Since it is usually not possible for designers to anticipate all requirements for looking up material in a set of instructions, carefully conducted tests of the new system should give designers a good idea of the majority of words that will be used as key words.

Tabs. Large manuals should include tabs on all major sections. Tabs divide even the most ponderous manual into manageable sections and narrow the user's field of search.

Modules. Where possible, divide large manuals into manageable sections or modules that are functionally independent of other sections. Using modules also narrow's the readers search field. Again, each module should be tabbed.

Making Instructions Easy to Scan

One principle that underlies making information easy to scan is processing as much information as possible for the user. The designer must work hard so that the user's job is easier. To improve scanning follow these guidelines:

Abstracting. It is important to pull the main idea or topic of sentences, paragraphs, and sections to the top, leftmost position in the element, a process sometimes called abstracting. This simplifies the reader's job. The processing has been done for them so they don't have to read an entire paragraph to find the main idea. The paragraphs in this section are an example of this principle. Abstracting facilitates scanning. One glance at the page should allow readers to locate the information that they want to read.

Chunking (grouping). Presenting information in manageable chunks so that the user can deal with one part at a time instead of the whole page also facilitates scanning. Users should be able to see at a glance what goes with what. Bullet list, charts, headings, indentation, font or font size changes are examples of chunking devices. Using white space strategically to separate groups of information is a must. The reader should be able to see, without reading, which items belong in a group or chunk. It should be visually apparent where one kind of information ends and the next begins.

Queuing. Queuing involves providing visual cues that allow readers to recognize types of information that are similar, whether they are positioned next to each other or separated by several pages. Typically, warning formats, command syntax, bullets, numbers, and the like, are queued to distinguish them from regular text. Queuing devices include different levels of indentation, icons, color, or different sizes and weights of fonts (but not all at once) to create a visual hierarchy so that the reader can distinguish the relative importance of ideas. In other words, *typography* becomes *topography.* It is important to achieve this differentiation and contrast using the fewest possible means.

Passing lanes. Put different types of information into separate columns (steps in one column and concepts or background information in another, for example). Using columns has the added advantage of keeping lines short enough that they can be read while scanning vertically. Users don't have to scan both horizontally and vertically at the same time.

Tables. Put complex information into a table. This makes it easy for users to find the one piece that they need without reading about everything else.

Layering Instructions for Multiple Audiences

One dilemma facing designers is how to format instructions for audiences that include users with different levels of experience. For example, *experienced users* seem to perform better when given more general guidelines. Their experience enables them to take a specific instance or example and generalize principles to other, often only loosely related situations. To explain every instance not only increases the complexity of the instructions but is counterproductive. *Inexperienced users,* on the other hand, need detailed instructions that include nearly every keystroke. Unfortunately, instructions that are too specific seem to impede generalization to other activities. *Occasional users* (experienced but infrequent users) need only a reminder.

One way to accommodate audiences made up of several types is to provide layered information so that users can read at the level best suited for their purpose. Consider the following, simple example.

---

10. Log out and reboot.

A. Log out of your application by selecting *Close* from the *File* menu.

B. Close Windows by double-clicking on the icon in the upper left-hand corner of the *Main* window.

C. Press <Ctrl>+<Alt>+<Delete> simultaneously to reboot.

If this instruction were intended for only experienced users, only the headline step (Log out and reboot) is needed. If, however, the audience includes users who are unfamiliar with computers, the substeps might be necessary. By including both the headline and the detailed substeps, the instruction is layered for both audiences. Experts can read the headline and skip to the next step; the inexperienced still have all the detail they need to allow them to successfully log out and reboot.

A designer must also provide instructions that are neither too concrete nor too abstract. Concrete, specific instructions seem best suited to situations requiring precise performance of a set of well-ordered activities. More general instructions encourage response flexibility. A general type of instruction may be preferable in situations demanding originality or creativity on the part of a user.

Formatting Checklist

- Can the reader skim the material and still get information?
- Is the visual hierarchy the same as the underlying information hierarchy? (If a word or idea is highlighted, is it really more important than other information that is not highlighted?)
- Does the page provide the framework—a map or grid with landmarks—to help users navigate through the information?
- Are there landmarks to help readers find what you want them to find?
- Is there enough variety? Is text mixed with well-chosen tables, photos, or diagrams, or is everything a bullet list?
- Is there too much variety? Do too many typefaces, sizes, weights, indents, graphics, or photos make the material difficult to navigate and understand?

Choosing a Format

There are many ways to format instructions that elicit good performance; only a few are presented here since other contemporary texts cover these methods in greater detail. Whatever the method, moving instructions from traditional paragraph format (see Figure 14-3) to numbered lists, logic trees and tables, tree diagrams, matrices, or structured writing methods is almost always better. Again, using visual cues to represent underlying organizational structures is the primary goal.

Probably the most traditional and well-known methods are numbered lists and flow charts. A *numbered list* format describes the activity using steps numbered sequentially and arranged from top to bottom down the page (compare Figures 14-3 and 14-4).

Instructions for Logging In

Users will login with the login name assigned by their course manager. For simplicity, the password for a given login will be the same expression as the login itself. A successful login will place the user in the appropriate student directory of the file system. During login, the unconditional execution of profile will solicit a response to this prompt:

TERMINAL?

The valid responses are hp, ti, and dasi for Hewlett-Packard, Texas Instruments, and Data Access Terminals, respectively.

Figure 14-3 Example of the traditional paragraph-style format for instructions.

Instructions for Logging In

1. Obtain login name and password from course manager.
2. Login with login name and password assigned by course manager.
3. Wait for prompt word "TERMINAL?"
4. Use the following table to determine what response to make to the prompt word "TERMINAL."

IF	THEN
Hewlett-Packard	hp
Texas Instruments	ti
Data Access	dasi

Figure 14-4 Example of a numbered list format for a login instruction.

Description versus Instruction

Most instructions can be divided into two distinct treatments: *descriptive* (this is what the system is, how it works, and why) or *performance related* (this is the process—the how to do it). Descriptive treatments try to describe *all* about a system for *any* person who wants to know, whereas performance-related treatments are usually intended for a restricted audience whose primary objective is to complete a specific task. When the two types are combined in the same document, it is best to clearly distinguish between them. Descriptive treatments are usually reserved for introductions and overviews and typically use passive voice (these tasks can be done by someone, but not necessarily now). Performance-related treatments are usually presented in step-by-step processes and typically use active voice (do this, now).

Process versus Rationale

A fundamental dilemma in formatting procedures within numbered lists is determining how to handle instructions (process) and related supporting information (the rationale—why this way?). Experienced users don't need the "why," but it is often critical for inexperienced users. Designers have run the gamut from providing every keystroke, system response, and concept in one grand mix to relegating support materials to entirely separate sections and manuals. In general, however, it is probably more effective to present all the information related to a process in one place. This helps to ensure that related information will be read without the interruptions introduced by flipping pages and consulting other documents.

Maintaining a distinction between process and rationale can be accomplished, within numbered lists, using a number of page formatting techniques or *grids.* Grids divide the page into logical units so that each element (title, instruction, system response, background information) has the same physical location on every page. Options for both double- and single-column grids are given next.

- **Double-column Grid:** General background information is presented before the steps begin. Steps and system responses are displayed in the left-hand column, each with its distinctive format; appropriate syntax and task completion criteria are explained on the right.
- **Single-column Grid:** Step information is provided first, followed by any system response. Step and response can be distinguished by boxes; different size, style, and weight of fonts; or other means. Rationale is indented below the system response and numbered if more than one is needed; again, different fonts and weights can be used to visually distinguish information types.

The following discussion assumes the use of either numbered lists or flow charts for procedural instructions. These guidelines were derived from Poller et al. (1981).

Sequencing

Correct sequencing of procedural steps produces effective procedural documents. The ordering of events in documents must

Be consistent with the natural sequence required by the work environment,

Be clear to the user, and

Be presented to minimize both writer and user errors.

A good task analysis helps to specify the preferred order of actions and decisions in a given work environment. If the task analysis indicates that event A precedes event B, then present them in that order in the procedure. Although common sense and psychological research dictate this, sometimes it is overlooked.

Overviews

Use overviews at the beginning of tasks to supply meaning to the work activities contained in a procedural document. They give users an overall idea of the activities involved and how these activities fit together. Even though users typically skip overviews, particularly if they are of any length, they should be provided for reference as needed.

Use the following guidelines to write an effective overview.

Keep it short. The more words, the less likely it will be read.

Describe work activities in general terms.

Describe inputs to the work activities or the conditions likely to be present when that set of procedures is applicable.

Describe the outputs or ensuing conditions that result from successful execution of the work activities.

Include any cautions or warnings relevant to the set of procedures to alert users before they start the procedures. However, repeat warnings at the appropriate point in the step list since users often skip overviews. Warnings should *precede* any action that might cause a critical situation and should be clearly distinguished from other text.

Identify, when appropriate, the destination of the outputs or subsequent operations that result from correctly performing the procedures.

Operational Statements

An operational or action statement describes an *action* that the user must perform. Both numbered list and flow-chart formats can contain operational statements, with or without support information.

Use the following guidelines when writing an operational statement.

Start the operational statement with a verb. Do not use phrases such as "should be" or "must be" to indicate an action. Rather, state the action positively in an imperative sentence.

Place the object of the verb close to the verb.

Place relevant text after the object.

Keep the statement short, 20 words or less if possible.

In list formats, underline key words (verb and object) with a continuous line. If boldface type is available, print key words in boldface with no underlining.

Contingencies

This section describes rules for procedures that require the user to make a decision by choosing the appropriate answer to a question. In these situations, the performed *actions* are contingent on the *answer* the user chooses.

Decision Statements

Decision statements typically appear as questions and answers or as conditional "if..., then..." statements. Even when these statements are not in question form, they always imply a question.

Write decision statements that ask exactly one question each. Look for the word "and" in the decision statement; this often indicates that two or more questions have been lumped together in one statement. For example, "Are the fire alarm and the water sprinkler on?" clearly contains two separate questions. Be sure to include all possible answers to the question asked by the decision statement. Also, state each answer explicitly. Do not use words like "else" or "otherwise."

The possible answers to the question asked by the decision statement should be obvious to the user of the document. Answers that are not obvious require the user to complete a series of operations or steps to determine the correct answer. If the answer is not obvious, include the procedural step(s) necessary to make the answers obvious before presenting the decision statement.

Formatting Decisions

Decision display structures present decisions and the actions to be taken depending on the decisions. Options include IF–THEN lists, IF–AND–THEN contingency tables, logic trees, and decision tables.

The structure depends on

- The number of decisions involved
- Whether the answers to the questions are obvious
- The number of actions contingent on the answers
- The amount of support material required
- Logical limitations imposed by the page size
- Characteristics of the user (many users have difficulty using decision tables effectively; for this reason, decision tables are a less preferred option)

IF–THEN Lists

Use IF–THEN lists when only *one* decision must be made. Lists can also be used for flow charts when one decision has more than seven answers. See Figure 14-5.

IF–AND–THEN Contingency Tables

Use contingency tables when *two* consecutive decisions must be made. See Figure 14-6.

Use the following table to decide which action to take for different colors of traffic lights.

IF	THEN
light is red	stop.
light is green	go.
light is yellow	proceed with caution.

Figure 14-5 Example of IF–THEN list format.

IF	AND	THEN
The traffic light is red		stop.
the traffic light is green		go.
the traffic light is yellow	there is heavy traffic	go with caution.
	there is no heavy traffic	go.

Figure 14-6 Example of IF–AND–THEN contigency table.

Logic Trees

The logic tree, a treelike arrangement of decisions and the actions taken as the result of each combination of answers, directs the user from one decision to the next with connecting lines. Logic trees differ from flow charts in several ways:

No flowchart symbols are used.

Actions occur only at the end of a sequence.

Each item, except the first decision, has only one line leading to it.

They must fit on one page.

Decision Tables

Decision tables display several decisions compactly using a table format. However, since many users have difficulty using decision tables, it is usually best to reserve them for audiences that have special training in reading them. See Figure 14-7.

If your client-server...	Load this driver...
Uses token ring topology	TOKENSHR.LAN
Uses IBM communication protocols	TOKENLNK.LAN. You must be on a token ring network and have ODINSUP set up.
Uses IPX/SPX protocols on a non-token ring network	LANSHARE.LAN

Figure 14-7 Example of a decision table.

Care must be taken to

Maintain strict parallel structure for each statement and each action

Include any repeated phrases in the heading line so that they will not have to be included in each statement and action

Use as few words as possible

Keep the format uncluttered (reduce font changes and other topographical features to a minimum)

Place keywords (topics), particularly in the statement column, as close to the beginning of the sentence as possible to aid in scanning.

Support Information

Support information explains, clarifies, or expands the meaning of operations in procedural documents. It enhances the user's ability to perform a procedure, but is not an operational (action) statement or decision statement. Although support information never directs the user to do a particular procedure, it

Alerts the user to possible consequences of performed actions

Increases the user's knowledge about a procedure

Provides the user with feedback on machine, equipment, or environmental responses

References other sources of information that might be needed

Support information should be placed with the action or decision it supports. This prevents users from having to search for important information. In addition,

Do not include required actions or decisions in support statements.

Place support information in the sequence of procedures to correspond with the behavioral sequence.

Provide feedback to the user about the correct outcomes of actions, when the user may not know the correct outcomes. This feedback should *immediately follow* the action statement.

Check Statements

Check statements keep users on the right track and prevent them from performing incorrect actions. Users of procedural documents often make errors when the instructions tell them to go to a step other than the one immediately following. In some cases, the user continues with the step immediately following anyway. In other cases, the user carries out the GO TO instruction, but goes to an incorrect step. Using check statements can reduce these errors. Place the check statement *immediately before* the action to which the user was directed by the procedure. See Figure 14-8.

1. Is the form completed?
2. Write today's date in section A.
3. Write your telephone number in section B.
4. CHECK: You should be here only if you are working with a completed form.
5. Mail form to local administrator.

Figure 14-8 Example of a check statement.

Computer-based Documentation

Overview

The traditional ways of developing and distributing paper documents have not kept pace with the increased need for information in systems. With the ever increasing availability of computers with text editing and storage capabilities, a considerable amount of documentation is becoming computer based.

The main advantages of computer-based documentation are the ease with which items can be found and the ease with which it can be updated. Paper documents are "frozen" when they are written. Since they are not easily changed, they must be complete when issued. This frequently leads to having large, cumbersome documents that contain more information than any one person might want or need. In contrast, computer-based documents can be easily changed.

Well-organized computer-based documents, particularly hypertext or hypermedia documents, can be as complete as paper documents without containing nearly as much information. Because information can be selectively displayed, the need for redundancy is reduced. In addition, a record can be kept of the sections accessed most frequently (or not at all), allowing the designer to make the necessary adjustments (add, subtract, or change information) while the system is in use.

It is important not to overlook the desirable properties of paper documents; many of these advantages are not possible with computer-based documentation. For example, paper documents give the reader the option of underlining important points in the text, adding marginal notes, making "dog ears," or using paper clips to mark pages for later reference. It appears that even the most technologically advanced computer-based documentation will be supplemented by paper documents, at least for the foreseeable future.

The important message here is that designers should *not* automatically provide computer-based documentation and then think that they have satisfied all documentation-related needs. At least for the next few years, many systems will probably require some mix of computer-based and paper documents that takes full advantage of the benefits provided by each approach.

Evaluations of Hypertext versus Paper-based Documents

Egan et al. (1989) conducted studies using two hypertext products (Hyperties and SuperBook). When compared with a paper document, *Hyperties* resulted in longer searches, hypertext and paper elicited the same accuracy, and users preferred hypertext. *SuperBook* resulted in faster searches when a text button existed, but paper resulted in faster searches when there was no text or heading cues. SuperBook produced fewer errors and was preferred by users.

Fox (1992) reports on a study for which a hypertext tool was used to select design guidelines. Participants used a paper-based textbook and a hypertext design aid (Dynamic Rules for User Interface Design). Users selected more relevant guidelines when using the paper book, but preferred using the hypertext document.

Nelson and Smith (1990) evaluated four alternatives:

Original hard copy text

Improved hard copy text

Hypertext version of the original text

Hypertext version of the improved text

Users performed best using the improved hard copy, but preferred using hypertext.

Dexter Hypertext Model

Most current hypertext systems are based on the Dexter hypertext model (Halasz and Schwartz, 1994). This model attempts to capture the important abstractions found in a wide range of existing and future hypertext systems.

Information Allocation

Because paper- and computer-based documents each have distinct advantages, a new activity is being performed by designers in systems. It is called *information allocation.* After collecting all information that needs to be communicated to users, designers must then decide which medium is best for each type of information. These decisions are based on the current strengths and weaknesses of each approach.

Paper Based

The advantages of paper-based documentation include the following:

Legibility. The same information can be read up to about 28 percent faster from paper documents than from a character-based CRT screen (Gould et al., 1987; Muter et al., 1982; Wright and Lickorish, 1984).

Maintaining orientation. Users of computer-based documentation frequently complain of lacking a "sense of text" (Haas and Hayes, 1986). Frequently, this requires the effective use of navigation aids to help users answer the question, Where am I?

Maintaining context. With the large number of systems that use only one monitor, users must lose the image on their CRT screen or manage several windows on a CRT screen to read a computer-based document.

Ability to easily cross-reference. Frequently, needed information is found in bits and pieces in several different documents. This requires much cross referencing, which is difficult in many computer-based documentation systems.

Portability. Obviously, it is still much easier to carry a book than even the most portable of lap-top computers.

Reading comfort. When people read, they have a tendency to lower reading materials at least 35 to 40 degrees down from the horizontal. Many CRTs are placed so that readers are required to sit upright, with the CRT placed no more than 15 degrees down from the horizontal and allowing little opportunity for movement. In addition, the computer cannot be easily read while snuggled comfortably in a warm bed on a cold and snowy winter evening.

Computer Based (Hypertext)

Some major advantages of computer-based documentation (over paper-based documentation) are as follows:

Easier to update and revise. This is particularly true for mainframe systems, but has been done successfully in microbased systems by circulating updated disks.

Easier to retrieve information. Well-designed computer-based documentation allows fast and efficient access to material by using sophisticated search capabilities.

Allows effective user interaction. The computer-based document is not a lifeless page; it can encourage an interaction between users and the information that frequently leads to finding more information in shorter periods of time.

Allows animated graphics. Computers can be connected to videotape and videodisk players, which allows the use of animation, sound, and the like, when presenting graphics (hypertext).

Easier to transmit information. Information can be transmitted electronically thousands of miles (or to the next office) in a few seconds. The need to physically move a paper document from one place to another, using a runner or modern mail services, is reduced.

Preferred Characteristics for User Manuals

The most preferred characteristics for user manuals were studied by Angiolillo et al. (1990). They found that users prefer the following:

- Good table of contents
  - Much white space
  - Spread out, not too compact
- Colored and labeled tabs
- Clear system of section headings
  - High ratio of headings to text
  - Larger or different type font (distinctive)
- Headings outdented into margins or indented
- Wide margins
- Color to highlight content information (e.g., light gray screens)
- The more figures (graphics) the better
- The more examples the better

Simplified English

Kincaid et al. (1990) proposed the use of Simplified English for international technical documentation. Simplified English is a subset of English with a limited vocabulary (about 1500 words) and a limited set of writing rules (less than 40). Simplified English offers the advantage to foreign users of not requiring them to learn a 100,000-word vocabulary to read English user manuals. In addition, with a limited vocabulary and relatively few writing rules, it is much easier to develop software to assist in the writing process.

Thomas et al. (1992) reported that Simplified English is easier to read for users whose first language is not English, but harder to prepare by technical writers. A software product designed to assist with writing Simplified English (Simplified English Analyzer, or SEAN) was able to identify the percentage of problems shown:

	%
Verb tense or form	79
Multiple instructions	74
Needs article	73
Passive voice	71
Sentence too long	63
Needs imperative	36
Missing commas	28

Readability Measures

It is amazing that so many sets of instructions are written at levels too difficult for the majority of readers in the intended user population. Although long sentences using a difficult vocabulary with multisyllable words may be preferred by certain readers, most users in systems prefer instructions made up of *short sentences* with *familiar* words. Readability measures have been developed to assist designers in preparing instructions that make better use of word and sentence characteristics.

Readability measures provide a means for estimating the difficulty a reader may have with a set of instructions, whether paper or computer based. Readability scores depend on the writing style rather than the technical content of a set of instructions. These stylistic features are under the control of the designer and range from the specific words a designer may choose to the way a set of instructions is organized into major topics.

A number of readability formulas has been developed over the last several years. The first true readability formula was published in 1923. However, the prototype of modern readability formulas was not published for another 5 years. These early formulas and those developed since have continued in their attempt to quantify stylistic difficulty. Readability measures are most useful as *predictors* of reading difficulty. Faced with two texts covering the same subject matter, a designer can predict that the more readable text is likely to be more quickly read and easier to understand.

Even though readability measures attempt to *predict* reading difficulty, they do not always reveal *why* the material is difficult. This is because many important attributes of style that contribute to difficulty cannot be quantified. But, fortunately, many of these unmeasurable attributes are highly correlated with stylistic features that *can* be measured. Klare (1974, 1975), in a review of readability formulas, concluded that "as long as predictions are all that is needed, the evidence that simple word and sentence counts can provide satisfactory predictions for most purposes is now quite conclusive" (p. 98).

Before a designer accepts or rejects a set of instructions on the basis of its readability, a number of other things should be considered. For example, the instructions should be read to see if they make sense. A document classified as highly readable solely on the basis of a formula could be a disorganized disaster. In addition, it is possible that the readability of a document may be very important when it is used for step-by-step instructions or in training, but less important for a reference document. Consider, for example, that Coke (1976b) found some evidence that readability was not as important when readers looked for specific information as it was when they had to remember that information.

The user and the conditions under which they use documents should be carefully considered. Klare (1975) found that highly motivated readers were unaffected by readability in many cases. In circumstances where time is not crucial and readers are highly motivated, the readability of a document may be of less importance. Finally, other features of a document that are not related to content should be examined. A difficult document, according to a readability formula, might be made more readable by changing its format rather than its writing style (Frase and Schwartz, 1979; Hartley, 1977).

In using a readability measure to evaluate instructions, a decision has to be made as to which constitutes an acceptable level of readability for the material being prepared. This decision should be based on information about the reading abilities of the intended audience. As a general rule, it is better to write to a readability level that is *below* the reading skills levels of the intended audience (Kulp, 1976). This reduces the risk of developing a set of instructions that is too difficult for users. The reading ability of a group of potential users can be measured with a standardized reading test, like the *Nelson–Denny Reading Test.* A designer should avoid an extreme mismatch between the reading demands of a set of instructions and the reading skills of the intended user.

As used here, reading grade level is defined as a score on a standardized reading test. For example, the score of a high school graduate who reads at an eighth-grade reading level would be about the same as that of the average eighth grader.

If it is not practical to use a reading test, some idea of reading skill levels can be obtained by looking at the users' educational levels. In general, people with more education have better reading skills than people with less education. However, the *actual* reading ability of a person should not be confused with his or her educational level. A high school graduate does not necessarily read at the twelfth-grade level. Coke and Koether (1979) found that for a group of over 200 craft-level telephone company employees with an average education of 12 years 95 percent had reading test scores above the tenth grade reading level. In fact, the group as a whole averaged at the fourteenth-grade reading level. The readability measures that follow use only two text features to predict readability: a measure of word difficulty and a measure of sentence difficulty.

Several readability measures were evaluated by Coke (1978). Based at least partially on her recommendations, two formulas have been selected and will be presented. The first is the Kincaid (Kincaid et al., 1975), which uses the average length of a word in syllables to measure word difficulty. The second is the Automated Readability Index (Smith and Kincaid, 1970), which uses the average number of letters per word to measure word difficulty. Both use the average words per sentence as a measure of sentence difficulty.

Using these two formulas, readability calculations can be made with or without a computer. When a formula is being computed by hand, it is easier to use the Kincaid method because syllables are counted rather than letters. When a formula is being computed by a computer, letters are easier to count than syllables, which is what the Automated Readability Index does.

Readability Formulas

To calculate the readability of a document using the Kincaid formula, a designer should

1. Select five or more 100- to 150-word samples.
2. Determine the average number of syllables per word in each sample.
3. Determine the average number of words per sentence in each sample.
4. Apply the following formula to make the calculation:

$$\text{Reading grade level} = 11.8S + 0.39W - 15.59$$

where

S = average syllables per word

W = average words per sentence

The process is almost the same for using the Automated Readability Index. The only difference is that the average number of letters per word is used, rather than the average number of syllables per word. Using the average number of letters per word facilitates the use of this formula with computers. The Automated Readability Index formula is given next for designers who want to computerize the readability calculation process.

$$\text{Reading grade level} = 4.71L + 0.5W - 21.43$$

where

L = average letters per word

W = average words per sentence

Often it is not practical to use the entire set of instructions when making the needed counts. A common practice is to take a number of short selections from the document. The average readability of these selections is then used to estimate the readability of the set of instructions. Each selection should be from 100 to 150 words long. Selections should be taken from the beginning, middle, and end of a document. Enough selections (at least five) should be chosen to obtain a good overall estimate of the readability level.

Abbreviations, acronyms, and numbers can often be a problem when syllable counts are made because these words usually do not conform to the usual rules for determining syllables. When counting syllables by hand, Flesch (1949) has suggested dropping all words whose pronunciation is in doubt. Since most words in a set of instructions ought to be pronounceable, omitting a small percentage of the words should have little effect on average word length or the average number of words per sentence. This same reasoning holds for computer programs that make syllable counts. Coke (1976a) has recommended that words that do not conform to a program's syllable-counting algorithm should be dropped from the analysis.

Usually, the only difficulty encountered when determining the average number of words in a sentence is knowing for sure what a sentence is. Sentences commonly end with a period, question mark, or exclamation point. However, a writer may separate complete thought units with a semicolon or other punctuation. In analyzing readability, a decision must be made about whether to treat these units as separate sentences. Kincaid et al. (1975) suggest that these units be counted as sentences. But determining when such units have occurred can be a problem, especially when readability is analyzed with a computer. It is much simpler and little accuracy is lost by defining a sentence as a set of words terminated by one of the punctuation marks commonly used to end sentences (Coke, 1976a).

Readability Example

As an example of how to use the Kincaid readability formula, consider the following paragraph. It is a 127-word section from a much larger document.

> Readability formulas predict reading difficulty, but they do not reveal why a document is difficult. Many important attributes of style that contribute to difficulty cannot be quantified. These unmeasurable attributes are often highly correlated with stylistic features than can be measured. For example, a poorly organized document might contain a lot of long sentences because the writer is forced to reference topics he or she has or will discuss. In this case, the document would be classified as difficult by most readability formulas since average sentence length is often used in these formulas. But shortening sentences to produce a more "readable" document would not get at the source of difficulty. A readability formula should be used as a *predictor* of difficulty and not as a *diagnostic* tool.

This selection contains 230 syllables. The number of syllables (230) is divided by the total number of words (127), which produces an average syllable per word count of 1.81. The 127 words are contained in 7 sentences. The number of words (127) is divided by the number of sentences (7), which produces an average word count per sentence of 18.14 ($W = 18.14$). These results are then entered in the formula.

$$\begin{aligned}\text{Reading grade level} &= 11.8S + 0.39W - 15.59 \\ &= 11.8 \times 1.81 + 0.39 \times 18.14 - 15.59 \\ &= 21.36 + 7.07 - 15.59 \\ &= 12.84\end{aligned}$$

The reading grade level for this one section is 12.84. A designer should take similar readability counts for at least four other sections (the more the better) and then average them. For example, assume that the following readability counts were taken for a document:

Section	Readability Count
1	12.84
2	11.03
3	10.65

Section	Readability Count
4	13.48
5	14.72
6	11.64
7	9.87
8	13.52

The average of these counts, which is 12.22, is the readability score or reading grade level for this document.

Assume that the designer knows that this particular document is to be used by people who read at about the fifth-grade level. The designer found this out by having the potential users take a standardized reading test (cf. *Nelson–Denny Reading Test,* 1973). The designer then rewrites the document without changing its meaning. The section reviewed before now reads as follows:

> There are a number of formulas that predict how hard a document is to read, based on average word and sentence lengths. They may show *that* a passage is hard to read, but not *why.* Writing style cannot be measured, but some effects of style can be. For example, a writer may need to use long sentences because his writing is not well organized. The long sentences are the result of the problem, not the cause. The cure is not to make the sentences shorter, but to organize the material so that shorter sentences can be used.

This section now contains 97 words, 134 syllables, and 6 sentences. Applying the Kincaid formula,

$$\begin{aligned}\text{Reading grade level} &= 11.8S + 0.39W - 15.59\\ &= 11.8 \times 1.38 + 0.39 \times 16.17 - 15.59\\ &= 16.28 + 6.31 - 15.59\\ &= 7.0\end{aligned}$$

The reading grade level is now down to 7.0, which is much closer to the fifth-grade reading level of the potential users. It could be written on even a lower level if desired.

By comparing the readability of a document to the reading abilities of its users, a designer can determine whether a document will convey its information effectively. A designer should keep in mind that a readability formula is a useful *predictor* of the difficulty that a particular group of users may have with a document, but the formula does not tell how to make instructions more comprehensible. Incidentally, the average reading grade level for this book is 12.3.

Grammar Checkers

Over the past several years, electronic tools to check grammar have been created and refined. These can be effective tools for evaluating written materials, but, again, should not be used in isolation. They can only check against whatever rules are programmed into the system, so they have their limits. One of the first grammar checking tools was created at Bell Laboratories in the late 1970s. It is called the *Writer's Workbench* (MacDonald et al., 1980, 1981, 1982).

The Writer's Workbench is actually a system that consists of over 20 separate computer programs that evaluates and suggests improvements for written documents. The programs detect features that characterize poor writing, including problems with punctuation, spelling, split infinitives, long sentences, awkward phrases, and passive sentences.

Although the Writer's Workbench contains many short programs, users need to know only one or two short commands to start the editing process. For example, if a designer has the Writer's Workbench available, he or she can make it run by using a keyboard to enter the command "wwb" and the filename where the text is stored. Users can also run any of the 20 programs individually if they choose.

The original Writer's Workbench evaluated a document and then provided a printout of possible problems. Types of issues dealt with include

Possible spelling errors

Sentences that appeared to be incorrectly punctuated

Double words

Sentences with wordy or misused phrases (followed by suggested revisions)

Split infinitives

Many such programs are on the market today. These programs provide numerous statistics and editorial comments to help in critiquing the written material. Grammar checking programs should be used to help to develop documentation that elicits acceptable levels of human performance.

Performance Aids

Another type of documentation that should be considered is performance aids (quick reference guides). Generally, they are condensed versions of more detailed instructions that are designed for on-the-job use when manuals and on-line computer programs would be impractical. They are essentially memory joggers that include enough information to allow trained users to accomplish tasks successfully. They help by reducing cognitive processing so that users don't have to remember large numbers of details.

Using performance aids to help to achieve acceptable levels of human performance is not a new idea. The Incas of fourteenth-century Peru used a device called a *quipu* (pronounced *kee-poo*) for a similar purpose. Runners, who carried messages about trade of livestock, goods, births, and deaths, used quipus as they moved from town to town. This device was simply a colored length of rope with a series of knots in it. Depending on its shape and position along the rope, each knot enabled the runner to recall information such as who sent the message, the subject of the message, and the number of items involved (Lanning, 1967).

Benefits

The four most significant benefits of performance aids are

1. Reduction of errors (the user relies less on long-term memory)
2. Increased speed in performing tasks (reduced uncertainty can lead to faster responses)
3. Reduced training requirements (although users might be taught to use the performance aid, they do not have to learn and remember all the information contained in the aid)
4. Reduced skill-level requirements (in many situations, a well-designed performance aid allows the work to be done by a person with fewer skills and less knowledge)

Types of Performance Aids

Two types of performance aids (PAs) are used frequently. Both assume that (1) the user knows how to use the aid, and (2) the user possesses the skills and knowledge to use the information the aid contains. The first kind assists in *remembering specific details.* For example, rather than memorizing the 15 items that we need to buy at the food market, we prepare a written shopping list to use while at the store. The second kind provides step-by-step *guidance* in performing an activity or executing a set of procedures. Instructions for assembling new bicycles or model airplanes are examples of this category.

Identifying the Need for Performance Aids

Performance aids are valuable for almost any task, but are especially needed if

The task is critical.

The task is too lengthy, complex, or infrequently done to be remembered.

Verbal explanations or other documents are less effective.

Consider this scenario: A 500-pound unexploded bomb is sticking out of the ground in an army camp. From a bunker 30 feet away, a nervous officer is shouting defusing instructions to a terrified private standing next to the bomb.

Officer: Turn the tailpiece counterclockwise and lift it off slowly.

Private: Yes, sir.

Officer: You will have exposed two wires, one red and one black; these are to be cut.

Private: Yes, sir.

The private carefully cuts the red wire and begins to cut the black one.

Officer: Oh yes, cut the black wire *first* or the bomb will explode.

In this situation, a well-designed performance aid may have been more helpful than step-by-step verbal instructions from the nervous officer.

A designer should carefully consider all activities to determine where performance aids can best help and whether they are needed. As a minimum, designers need to know as precisely as possible (1) the type of person who will perform the task, (2) how and when the task will be performed, and (3) under what conditions or context the task will be performed. A well-done task analysis helps to answer all these questions.

General Considerations

Designers should consider the following process as they design performance aids.

Analyze. During system development, anticipate needs.

Observe. Watch and talk with system users. Especially, look for "cheat sheets," notes pasted on equipment or the wall or written in margins; these are candidates for PAs.

Consider. Picture yourself in the user's place. Consider *how, when,* by *whom,* and *where* the PA will be used. Tailor it to the *users'* (not your) skills, needs, and physical environment.

Test. Perform the task yourself under actual conditions. Hold design reviews. Personally *observe* the PA's use and *talk* with users. Ensure that they interpret the PA correctly and perform the task accurately and efficiently.

Human (User) Characteristics

To design an effective performance aid, a designer should first determine the human–activity–context characteristics of the situation in which the performance aid is to be used. A designer can then make other design decisions in order to develop a performance aid that is as responsive as possible to user needs. There are at least three major considerations.

Reading Ability

Users obviously should be able to read and understand the performance aid. For example, a performance aid to help prepare a federal income tax return should be written clearly and at a level that is easy to understand (especially if the material in the lengthy preparation instructions is confusing). This applies whether the information is written on a form or on a CRT screen.

Effect of Past Experience

Designers should determine whether users have had (or will have) training and/or experience in the activity. Prior or similar experience does not always promote successful performance. For example, a person may have considerable experience dialing telephone numbers using a push-button phone; however, the new activity may require the extensive use of a hand-held calculator whose keys are arranged in a different order. This change may cause a degradation in performance because of the differences between the two keyboards. Any new performance aid for using the calculator must take the previous telephone experience into account.

Resistance to Use

A designer should determine whether the user is likely to be opposed to using performance aids and, if so, determine the reason. For example, certain highly skilled people may view a performance aid as a crutch to be used only by the less skilled. The aid may also be perceived as an inconvenience; many key operators do not like to interrupt their keying to access even the best designed performance aids.

Activity Characteristics

When considering whether a performance aid is a good choice for a given activity, the designer should ask the following:

Will the information in the aid ever need to be updated and, if so, how often; can changes be made easily?

Is the activity performed often? If so, will the aid be able to withstand frequent use?

Will the aid need cross referencing for other closely related activities?

Will the context in which the aid will be used interfere with using it appropriately?

Performance aids are typically either a single page or divided into several pages called *frames*. An example of a single-frame performance aid is a card used by automobile assembly line workers to install a windshield. Dictionaries and telephone books are multiple-frame performance aids. Each page or frame should contain all the information necessary to perform a single occurrence of the activity.

Designers must (1) concentrate on how to make the information in the performance aid communicate clearly and effectively and (2) focus closely on how each piece of information should be presented. In so doing, designers must attempt to optimize at least five quality standards:

Accessibility

Arrange the information in such a way that it can be scanned easily.

Make the aid clear and simple to use. Also make it *appear* clear and simple at first glance so that people will *want* to use it.

Use short words, sentences, paragraphs, and sections.

Use active voice and present tense.

Be informal. Use *you, we,* and so on.

Do not feel constrained by rules of grammar. Use short phrases, even if they are not complete sentences.

Use diagrams, checklists, lists, charts, and the like, instead of or in addition to text.

Use extra spacing, horizontal and vertical lines of differing widths, and perhaps color to set off and highlight data.

Use a standard format to specify information for similar items.

Accuracy

The information must be correct and be presented in a way that ensures a correct and accurate response from the user. Edit and reedit. Ask others, especially future users, to edit the aid. Test the performance aid with typical users.

Clarity

The information must be directly usable; no judgments or interpretations should be needed.

Completeness and Conciseness

The aid must contain *all* the information users need and *only* the information they need.

Include enough identifying, instructing, and detailed information so that a trained person can use the aid without further explanation or references.

For unusual or complex procedures that would be impractical to cover in the aid, refer to the manual. But do not just ignore them.

Procedures to be used just once should not be in the aid.

Use as few words as possible to convey as much meaning as possible.

Concentrate on the difficult and unfamiliar, not the obvious.

Legibility

The information must be legible, even under the worst possible environmental conditions.

Use larger, bolded, clearer printing than usual if the PA will be used by people in a hurry or under other stress and/or with poor lighting or other unfavorable conditions.

Use proportional spacing to improve legibility.

Use regular lowercase (initial caps only) for text.

Do not use all italics for text or for all cap headings. Italics, especially in all caps, are hard to read.

Run all text and charts the same way. Readers should not have to turn the aid sideways.

Write numbers in digit, not word, form.

Allow adequate spacing for margins and between words, lines, paragraphs, and sections. Keep the spacing uniform.

If color is used, get expert advice on visibility, contrast, and readability.

Context (Environment) Characteristics

To be effective, performance aids must be suited to the context or work environment. The characteristics of the work area should be identified for each activity. Consider the following:

Is the illumination sufficient or is it abnormally low or high?

Will the activity be performed on a horizontal work surface (desk or table)?

Can the aid be placed on a wall, yet still be readily accessible and legible at a distance?

Will the aid be used with equipment to which it may be attached?

Will the task be performed at one location or several?

Will there be enough room at the work space to accommodate the aid?

Will the work area be dirty? If so, a way must be provided to keep the aid clean enough to be read.

Will the work area be subjected to adverse conditions (heat, snow, or high humidity) that would require the performance aid to be constructed of special materials?

Testing Performance Aids

It is worthwhile to state again that once a prototype performance aid is developed it should be carefully tested. A designer should evaluate the aid for poor design, such as the heavy use of underlining and excessive color coding. Legibility, accessibility of information, clarity, accuracy, completeness and conciseness, and compatibility with the work conditions should also be evaluated.

The evaluation should include discussions with several potential users to help to identify any flaws in the design. Finally, each performance aid should be tested in a simulated situation very close to the real one. Users should use the prototype aid as it will be used on the job. This will help a designer to make necessary final modifications before full-scale production begins.

Exercise 14A: Performing a Readability Analysis

Purpose: To provide an opportunity to use a readability formula for evaluating written text.

Method: Use either the Kincaid or the Automated Readability Index to calculate the readability level of the following paragraph. Then make whatever changes are necessary to the paragraph so that it elicits a *seventh grade* reading level (or lower).

It may be helpful to know that this paragraph contains 9 sentences and 258 words. The average sentence length is 28.7 words, and the average word length is 5 letters. Also, if you are not good at counting syllables, the number of syllables can be estimated by counting the number of vowels in the text and applying the following formula:

$$\text{Syllables} = 0.98V - 0.34W$$

where

V = total number of vowels

W = total number of words

Errors

In most computer systems, errors can be detected either by people or the computer. For this reason, a good system designer will use both means of error detection in order to detect and correct as many errors as possible. Obviously, the less attention given to error prevention the more attention a designer will have to give to error *detection.* But even when many errors are prevented from ever occurring, the designer should seek to detect those few errors that are still being made or that may be inherited from other systems. People are able to detect many of their own errors right after they make them, while many other errors can be detected at a later time by proofreaders. A properly programmed computer can detect certain types of errors, particularly those difficult ones that are extremely tough for inexperienced people to catch when they are trying to do all that they can to find and correct all the errors in any written text. If we were to rely only on manual proofreading to catch the unpreventable errors in computer systems, our systems would contain a substantially greater number of errors. Fortunately, numerous computer detection techniques have been developed, and the computer has greatly increased the capacity for detecting human errors because there is no question that most people have difficulty detecting errors made by others. However, even the most sophisticated and elaborate computer error detection techniques cannot locate and indicate to people using a system all the many different varieties of errors that people are capable of making.

Reporting: Turn in a one-page report showing (1) your readability calculation for the original paragraph, (2) your revised paragraph, and (3) your readability calculation for the revised paragraph. Also, discuss the types of changes that you made. If you substituted certain words, what were they?

Exercise 14B: Designing a Performance Aid

Purpose: To provide an opportunity to design a useful performance aid.

Method: The final exam for this course will be coming up soon. To assist your performance on that test, design and construct a performance aid. Follow the guidelines shown in the textbook. The only restriction is that your performance aid can be no larger than one 3-inch by 5-inch card.

Turn in the card at the next class or when it is due. Either your performance aid or one from another member of the class will be given back to you at the beginning of the final. In general, the better you design the performance aid, the better your performance should be on the test.

15

Training Development

Introduction

Training is the systematic acquisition of *skills, knowledge,* and *attitudes* that will lead to an acceptable level of human performance on a specific activity in a given context. The primary purpose of training is to improve (change) the user in some task-related way, not merely to add to his or her store of knowledge. For example, when developing training for a new computer system, a designer could ensure that terminal use *skills* are developed, that a general *knowledge* about the system is provided, and that users *attitudes* about the new system change be positive.

The central goal of training is to achieve acceptable performance. Keep in mind, however, that training development is only one part of a much larger whole that encompasses a variety of human performance considerations. Some designers mistakenly feel that, once the hardware and software are developed, all that is necessary to ensure an acceptable level of human performance is a set of good training materials. This is simply not true. In systems that are *operational,* for example, many human performance problems (perhaps even the majority) have little if anything to do with training.

Costs and Benefits

Training development should not be initiated automatically, casually, or haphazardly. An often quoted estimate for training costs is 200 hours of development time for each hour of training. This can be a hefty investment for a population characterized by significant turnover.

With such great amounts of money, time, and resources at stake, the development of training must be approached with considerable care. A few years ago a major communications company developed and presented a 10-week training course. The course was given to new employees who were hired to install and service equipment. A later analysis of the course indicated that at least 5 weeks of the course were being spent teaching irrelevant material. The course was redesigned as a 2-week self-paced course. The redevelopment costs (a substantial $350,000) paid off. Since nearly 2000 employees receive the training each year, reducing the course from 10 weeks to an average of 2 weeks resulted in savings of over $4 million a year. Making the course self-paced enabled some students to finish the course early or, if necessary, take extra time. This company reported that in one 6-year period there was a total savings of $37,800,000 from this course alone. Perhaps even more important, a follow-up evaluation indicated that performance on the job had improved. This savings and improved performance could have been realized from the *beginning* if the initial course had been designed properly.

Consider another example for which improved training reduced costs. Many telephone systems now have features such as

Automatic callback: you can be called back automatically when a busy telephone that you are trying to call becomes free.

Call forwarding: all your incoming calls will ring at another telephone.

Three-way conference: you can set up a three-way conversation without operator assistance.

Transfer: you can transfer a caller from your telephone to someone else's without operator assistance.

The problem here is to train new users to take full advantage of these features. Originally, this was done for business customers by having a telephone company instructor explain and demonstrate each feature to groups of eight or ten people seated around a conference table. Following a step-by-step description and demonstration of each feature, one or two people from the group would be given the opportunity to try out the feature. This procedure would be employed with different users so that each had an opportunity to practice a few of the features. Each session took about 2 hours. With some customers having hundreds of telephones, this type of training proved to be very expensive for the telephone company.

One approach to reducing training costs focused around the use of hands-on training. Originally, it seemed that learning to use a complex set of features would definitely require hands-on practice by users. Ellis (1977) and others conducted studies in which half the people had training sessions that included an opportunity to practice and half had training sessions with *no* practice. Ellis found that there was no improvement in user performance for this type of task (requiring little movement skill) when people were provided an opportunity to practice during training. Karlin (1977) estimates the savings realized by eliminating hands-on training in this situation was about $2.5 million for 1 year.

Another approach taken by designers to reduce the cost of training was to develop a one-page card performance aid that described the procedure for using each feature. In the new training sessions, users were given an overview of the features, and the details were given in the performance aid (Ellis and Coskren, 1979). Again, training time was reduced, with no decrease in feature use or performance.

It should be noted that designing a system that is easy to learn does not automatically ensure that the system will be easy to use. The Phelps tractor was introduced as a replacement for the horse in 1901 (Gentner and Grudin, 1990). To take full advantage of what farmers already knew and to reduce training, they made the tractor so it could be hitched to a carriage or a wagon. In addition, the farmers could control it with reins: they could loosen the reins to go forward and pull back on the reins to slow or stop. The tractor was not successful.

Learning versus Training

A distinction should be made between learning theories and training. *Learning theories* usually describe the conditions under which a behavior is acquired. Such theories are *descriptive;* they provide a theoretical base that can be modified for application in specific practical situations. *Training theories* specify the most effective and efficient ways to obtain knowledge, skills, or attitudes at identified levels and under particular conditions. Such theories are *prescriptive.* They suggest principles of instruction, criteria for learning, and the conditions that are likely to ensure that learning will take place.

While inroads have been made in the direction of developing research-based guidelines for training development, the job has by no means been completed. Considerable experimentation over the past 100 years has yielded only a few principles that have received broad acceptance (Hilgard and Bower, 1975). For example,

- Keep the trainees *active* (skill can be best developed by doing, not just listening or reading).
- Make use of *repetition* (practice makes perfect).
- Make use of *reinforcement* (reward correct responses).
- Have trainees *practice* in many different situations so that they are able to generalize.
- *Organize* the presentation of information in some meaningful way.
- Provide for learning with *understanding.*
- Encourage *divergent thinking* (urge students to develop creative solutions and explore alternative solutions).
- Consider the trainee's *ability to learn* (some people learn quickly, some learn slowly).

Developing Skills

Training courses usually help to develop new skills, knowledge, or attitudes. Probably the most critical to human performance is the development of *skills.* There are two basic ways to acquire a new skill. One approach is simply to have the person perform, and over a period of time the skill develops. Frequently, there is a model of some kind to imitate, but no specific instructions are given. Most of us learned to walk, talk, and ride a bicycle in this way. The second way is to have another person communicate in some more or less systematic way what is to be done (i.e., suggest a strategy). These instructions can be verbal, written, or contained on a CRT display.

For the most part, designers should not allow skills simply to evolve. Instructions should be provided to make the learning process as efficient as possible. The following discussion focuses on a skill development process that usually begins with and uses instructions.

Developing Skills with Instructions

Skills seem to reflect a set of internalized instructions (or plan) that was originally voluntary but that has become relatively inflexible, involuntary, and automatic. Once the internalized instructions that control a sequence of skilled actions become fixed through overlearning, they function in much the same way as instinctive behavior in animals. The conditions under which various skilled components are triggered, or released, is much the same in both cases.

By consciously following a good set of verbal or printed instructions, a beginner may achieve the same objective as the automatic following of internalized instructions by a skilled performer. In a sense the performance is the same. But the beginner's performance is carried out in a way that is voluntary, flexible, and communicable, whereas the expert's performance is automatic, inflexible, and usually locked in. The development of skill appears to free the expert to automatically implement larger and larger behavioral units.

When a person sets out to acquire a new skill, particularly in a system, he or she usually begins with a set of instructions of some kind. But just having the basic strategy in verbal form does not mean that the learner can correctly perform on the first try. For example, when an individual learns to fly an airplane, a set of instructions like the following may be provided:

> To land this plane you must level off at an altitude of about ten feet. Then, after you have descended to about two feet, pull back on the elevators and touch down as you approach stalling speed. You must remember that at touch-down the control surfaces are less sensitive, and any gust may increase your airspeed. That may start the plane flying again, so be prepared to take corrective measures with the throttle and elevators. And if there is a cross-wind, lower the wing on the windward side, holding the plane parallel to the runway with the opposite rudder. (Miller et al., 1960, p. 83)

These are the instructions for landing the plane. When skillfully executed, they serve to get the pilot and airplane safely back to earth. It is a short paragraph and could be memorized in a few minutes, but it is doubtful whether the person who memorized it could land a plane, even under ideal weather conditions. In fact, it seems likely that someone could learn all the individual acts that are indicated in the instructions and still be unable to land successfully. Even given the description of what to do, the trainee still faces the major task of converting the *knowledge* into *separate skills* and the separate skills into an *integrated skill.*

Skill Development Stages

Dreyfus and Dreyfus (1979) present an example of stages through which a pilot passes as he or she develops the flying skill. First, the novice pilot focuses all attention on a list of memorized instructions (procedures) to be appropriately applied. In doing so, the pilot trainee is so absorbed in details that he or she is unaware of most surrounding events and experiences little sense of actual flying.

With further experience the pilot trainee acquires the ability to recognize and learn the importance of such situations as being in the landing envelope and such sensations as acceleration and characteristic sounds and vibrations. The pilot analytically determines the appropriate actions by applying rules, such as determining whether the aircraft is in the landing envelope or returning to base when vibrations are abnormal. This intermediate pilot trainee begins to feel that he or she is flying the plane.

Finally, a pilot's repertoire of flying experiences becomes so extensive that each *whole situation* is recognized as similar to a previous typical situation, and this previous situation elicits remembered appropriate responses. Furthermore, associated with the memory of each of these past experiences are other associated experiences. For example, suppose that the current situation is a normal landing, and hence location in the center of the landing envelope is a crucial aspect. If the pilot perceives that he or she is very high in the landing envelope, the associated past experience suggests a "go around and try again" situation. Analytic, conscious control becomes almost completely bypassed and replaced by an automatic (skilled) response. The pilot now feels that he or she is flying.

Dreyfus and Dreyfus (1979) suggest that the same type of phenomenon shows up whenever a person acquires a complex skill, be it highly intellectual, like chess, or largely associated with movement control, like tennis. For example, in chess a beginner uses instructions to learn simple rules, such as to trade pieces to maximize material balance (calculated by adding up the values of the individual pieces involved). In tennis, a player first learns several independent movements, such as swinging the racket at the proper speed or transferring weight from one foot to the other while making a stroke. With experience, a chess player learns to follow rules, such as to exploit a weakness on the king's side or avoid an unbalanced pawn structure. And in tennis an experienced player may be advised to use top spin on a return shot.

When truly proficient, a chess master, immersed in the world of the game, immediately perceives the forces and tensions on the board as similar to those previously experienced in actual play or in the involved study of previous games. The world class tennis champion no longer thinks about shifting weight or using top spin. Bypassing analytical control of performance, he or she enters into an almost unconscious series of movements that are appropriate responses to the other player's actions and, furthermore, leave the player free to develop new offensive strategies as the game goes on.

Skill development, then, probably takes place for most activities roughly as a three-stage process. First, the performance is almost totally under conscious control. Second, the performance is under shared control; some activities require conscious deliberation and others are automatic. Finally, in the third stage, performance is totally under automatic control, perhaps leaving the person free to engage in other performance or to monitor and improve the performance presently taking place under automatic control.

Developing Skills without Instructions

Although it at first seems necessary to begin all training with a set of clear, precise instructions that will eventually become internalized as skills, this is not always possible. Almost no one, for example, can communicate the instructions for a new bicyclist to maintain balance. According to Miller et al. (1960), the underlying principle would not really be much help even if a person did know how to express it: "Adjust the curvature of your bicycle's path in proportion to the ratio of your unbalance over the square of your speed." For most people this instruction is almost impossible to understand, much less to do. In such cases, a designer uses the other option for building a skill and simply has someone run along beside the bicycle, holding it up until the trainee "catches on."

A careful review of the tasks on which skills are to be developed should suggest which of the tasks require explicit instructions and which do not. Keep in mind that small children frequently acquire skills without first memorizing verbal descriptions of what they are supposed to do.

Another problem with relying too heavily on an initial set of instructions when developing skills is that even the finest set of instructions may be too general from a user's point of view. Training instructions in many situations tend to deal more with overall strategy than with the movement-by-movement or thought-by-thought details required by the trainee.

Miller et al. (1960) observed that the general strategy provided by the designer usually says little or nothing about the activities of individual muscle groups. The designer knows these interrelated acts because he or she knows how to perform, but they are implicit, rather than explicit and communicable. Thus, designers are usually working from the *general* (actions that need to be done) to the specific when attempting to communicate training instructions, while the trainee is working from the *specific* (thoughts and muscle movements) to the general when trying to carry them out. The designer and trainee may not see the required performance in the same way.

General versus Detailed Instructions

In fact, in many situations it is probably best for a designer to provide general information and let the trainee work out the best specific way of doing it. In this way a trainee is free to develop specific thought and movement patterns that best enable him or her to meet the performance requirements.

In some systems, however, designers have provided very detailed instructions. On an assembly line in a factory there may be a task that consists of, let us say, assembling three washers on a bolt. The analysis of this task into micromotions will specify the exact time at which each hand should move and the operation it should perform. For the left hand, the instructions may read "Carry assembly to bin," "Release assembly," "Reach for bolt," and so on, while at the same time the right hand is instructed to "Reach for lock washer," "Select and grasp washer," "Carry washer to bolt," and so on. For each of these motions, a fixed duration is specified. This is about as near as anyone can come to *writing programs* for people that are as detailed as the programs we write for computers.

With instructions as specific as these, designers attempt to find a sequence of motions that achieves the result most efficiently, with fewest movements and in the least time. Following these rules, designers may be able to develop chains of responses that can be executed with very high efficiency. But, unfortunately, many users may object to being so tightly regimented. When people have time to develop the skill themselves, they are able to determine the interposed elements that produce the skill. Once the skill has been developed, alternative modes of action then become possible, and the person is free to make changes that can help to improve performance.

Developing Efficient Skills

We need to make one final note on skill development. Designers should seek to develop skills that lead to performance with the fewest errors and shortest time frame.

Many skills tend to be developed in such haphazard ways that the resulting performance is inefficient. Most people perform numerous activities every day without considering whether they are performing in an efficient or inefficient manner. Morehouse and Gross (1977) provide an example that helps to illustrate the problems we frequently encounter with inefficient skills.

> The first time you made a bed, you probably followed someone's instructions—your mother's or your sergeant's. Like a rat in a maze who finds a route to the food and then repeats that route over and over without searching for a better one.
>
> The next time you make a bed, observe how many times you go from one side to the other before you're finished. The first time I tried this experiment I was surprised to learn that I was making six trips around the bed. I decided to see whether I could make the bed in one trip. I could, but it was a pretty sloppy-looking bed. So I modified my objective, and found that in three trips I could often make a bed as neatly as I could in six (adapted from *Maximum Performance,* by Laurence E. Morehouse, Ph.D., and Leonard Gross. Copyright © 1977 by Laurence E. Morehouse, Ph.D., p. 48. Reprinted by permission of Simon and Schuster).

To develop skills that produce efficient performance is no simple matter, but it can be done with effort and by knowing what to do. Keep in mind that simply *passing out information* is not the same as training a user. Good training concentrates on *building skills* and, in many cases, *changing attitudes.*

Learning Preferences and Training

Raban (1988) asked people to indicate their preferred learning preference, either using a step-by-step tutorial or using guided exploration. Half of the participants were then given tutorial training, and the other half were given guided exploration training. Participants performed better when assigned to their preferred learning method. Overall, *guided exploration* was more effective than traditional tutorial training.

Skill Degradation

Rullo and McDonald (1990) reported that skill degradation occurs differently for different human information processes:

- Little degradation
    - Attitudes
    - Motor skills
- Moderate degradation
    - Perception related
    - Decision making and problem solving
- Considerable degradation
    - Memory related

The factors most closely related to skill degradation are as follows (Rullo and McDonald, 1990):

Infrequent use of the skill

Training-to-performance lag is too long

Control-display incompatibility (nonoptimal procedures)

Limited time allowed to perform a task (speed degrades more rapidly than accuracy)

If one or more of these conditions exist, designers should consider using refresher training for successful performance of critical activities.

Using Reinforcement

Studies on the systematic modification of behavior supply additional information for training developers. The concept of behavior modification states that people will act according to a set of rules if these rules are reinforced in a direct, immediate, and consistent manner. The emphasis is on the interaction between what the person does to the world and how the world reacts (Margolis and Kroes, 1975). *Reinforcers* are positive or rewarding consequences that increase the likelihood of a certain set of performance behaviors recurring.

Possible reinforcers include

Attention and praise (social reactions)

Prizes

Status

Privileges

Awards

Money

Punishers are negative consequences, such as

Reprimands

Ridicule

Fines

Deprivation

In training, the appropriate use of reinforcers enhances the possibility that the desired performance skills will be acquired and not forgotten. The use of punishers is usually inappropriate because their consequences are difficult to predict. Ignoring undesired performance is often aversive enough to decrease the likelihood of the undesirable performance being repeated.

Another relevant behavior principle is *modeling*. A person's performance can be modified by his or her observations of reinforcers or punishers received by others. People learn by watching and imitating others. An observer is more likely to imitate the behavior performed by a person with prestige, particularly if the behavior was reinforced. Finally, if an observer sees a behavior being punished, he or she is less likely to imitate that behavior (Mager and Pipe, 1970).

A designer should consider the following when scheduling consequences for behavioral changes during training.

There should be sufficient opportunities for reinforcement.

Responses to correct performance behavior should be *positive, immediate,* and *consistent.*

Specific desired behaviors should be recognized and rewarded even if the overall or final performance is less than adequate.

Reinforcers that support *undesirable* performance behavior should be identified and removed.

Training Development Process

A systematic approach to training development enables designers to produce courses that are relevant, effective, and efficient. A *relevant* course has high validity; that is, the knowledge, skills, and attitudes users learn in the course match the job requirements. A course is *effective* when students who complete the training achieve acceptable scores on criterion examinations and case problems, thus indicating that they have achieved course objectives. An *efficient* course is lean and does not waste the time of the training developer, students, or instructors. A main goal of training is to bring about the greatest amount of change in performance capability in the shortest amount of time.

The training development process can begin as soon as a designer determines that training can help to ensure an acceptable level of human performance (Kearsley, 1985). Ten major considerations in the training development process are shown below:

1. What are the specific performance requirements?
2. What performance requirements are to be met through training?
3. What skills, knowledge, and attitudes do trainees bring with them?
4. What skills, knowledge, and attitudes should the trainees have at the end of training?
5. How can trainees best learn the needed skills, knowledge, and attitudes?
6. What assurance is there that the instruction will be effective?
7. Exactly what performance is desired?
8. What performance level is acceptable?
9. Is *training* a good way to achieve the needed performance level?
10. What are the training alternatives for ensuring acceptable performance?

Most experienced designers know that training cannot overcome problems caused by such things as inadequate function allocation, poor work design or interface, incomplete instructions or performance aids, or inappropriate personnel selection.

Training may not be the only or even the best way of achieving adequate human performance. Other options include personnel selection, instructions, and performance aids. Since training is expensive to develop, deliver, and maintain, it is usually selected only after other approaches are ruled out.

Analysis

The foundation of any training development enterprise is the analytic work that precedes it, such as function allocation, task analysis, and work module design. There are no shortcuts.

Actual training development efforts cannot begin until the analysis work is complete. A critical part of the analysis deals mainly with identifying the skills, knowledge, and attitudes necessary for acceptable performance on a particular activity. Clearly, the better the analysis and initial design are, the less likely that performance problems will develop.

The results of task analysis and work module design should lead to a detailed work description that includes identification of the skills, knowledge, and attitudes required to perform each task in the work description. This becomes a list of the skills, knowledge, and attitudes that training must produce or enhance.

There should also be well-developed user profiles for each work module. These will provide a good description of the target population (e.g., the skills and knowledge that they already possess). A well-defined target population provides the starting point for training development. The performance expected of that population at the end of the course is the finishing point. The content of the course is established by subtracting what trainees can do from what one wants them to be able to do (Mager, 1975). No matter how carefully a target population is specified, the analysis will always reflect a spectrum of abilities.

When the appropriate analyses are complete, the design portion of training development begins.

Determining Objectives

Well-stated, detailed, and measurable objectives are the logical first step in the training design process. Objectives control subsequent course design by providing a framework for the rest of the development process. A good set of objectives should specify skills, knowledge, and attitudes not already possessed by the target population and that, therefore, the course must produce or enhance. In addition, objectives should specify the required performance levels of the target population at the completion of the course (Mager, 1975).

The two major types of objectives are

Terminal or end-of-course objectives: These reflect the performance levels expected of the trainee upon course completion.

Lower-level, subordinate, or enabling objectives: These include all levels and/or kinds of performances required to reach the terminal or end-of-course objectives.

All objectives should lead to one ultimate goal: an acceptable level of human performance. The objectives should track closely with all levels and kinds of abilities required of the trainee.

A clear objective includes the following three elements:

A given: A statement of the conditions, limitations, aids, or tools that affect the performance involved

An action: A description of the overt performance expected of the student

Criteria: The standards for that performance (time allotted for performance, number of items, degree of accuracy)

The development of objectives is iterative; as development progresses and more is understood about the new system, the objectives should be adapted to reflect the new level of understanding. Objectives ordinarily should not be changed just because they are difficult to reach.

Developing Tests

Tests serve a number of purposes. In-class and end-of-course tests *measure performance.* They also *provide feedback* to trainees and instructors on the attainment of objectives. By providing feedback, the test results potentially affect both learning and motivation. By providing instructor feedback, test results suggest needed revisions.

As soon as all levels of objectives have been completed, the designer should begin to develop test items that match the objectives. Good test items depend on well-thought-out objectives.

There are several types of tests. Two of the most important are prerequisite and entry-level tests. *Prerequisite tests* are administered to determine whether a trainee has the skills and knowledge necessary to begin a course. *Entry-level tests* can help to determine what material can be omitted from the course because it is already known by the trainee. Both kinds of tests reveal what need *not be* taught to ensure competent performance.

Two other important tests are those administered *during* a course (in-session tests) and at its *termination* (end-of-course tests). These tests measure student success in accomplishing the objectives.

A designer may discover that some objectives cannot be tested. If so, such objectives should be redone. An objective should be measurable. The designer must ensure that all objectives truly reflect the expected end-of-course performance.

When the designer has developed a clear set of *objectives* and is satisfied that the *test items* measure the objectives, he or she should move on to the next step in the design process.

Developing Training Materials

A basic document outlining key elements of the course should be produced at this point. Ideally, it includes the following elements:

Introduction and course outline: A general description of the course, including the overall objectives, an analysis of the target population, and an outline of the lessons in the course

Unit outlines: Includes objectives for each unit, specifications for training strategies (e.g., unit sequencing, practice strategies, review and feedback points, testing), and specification of materials and media

Implementation materials, facilities, and procedures: Includes a description of administrative roles and materials, instructor or course manager roles and materials, trainee materials, audiovisual materials, training aids, training facilities, instructor training requirements, and pre- and postcourse procedures and materials

This design work is done first on a broad preliminary basis and then in a more detailed fashion. When these efforts are completed, drafts of the material are ready for trial and the system designer moves to evaluating the course.

An important consideration is whether to develop group-paced or self-paced training. There are many variations or methods of each mode. For example, *group-paced* includes lecture, illustrated lecture, lecture and discussion, or discussion. The outstanding characteristic of the group-paced mode is that trainees constitute a group and move at the *same* pace.

The predominant characteristic of the *self-paced* mode is that each student proceeds at his or her own pace. In some self-paced courses, the trainee learns from previously prepared or programmed materials:

The materials provide for continual student activity (responding).

Frequent quizzes are provided for self-checking.

Materials are divided into modules.

The student demonstrates mastery of a module before he or she moves to the next module.

In many self-paced courses, a course manager or mentor is provided to carry out managerial and tutorial functions as needed. There are several variations of the self-paced mode. Figure 15-1 lists common methods of training classified under the two modes.

Instructor-led, group-paced courses may have small inefficiencies or gaps because of inadequate training development that can be overcome by a good instructor. Self-paced training, on the other hand, is often used by trainees working alone, with little support except for occasional assistance from a course manager or mentor. Since self-paced instruction materials must stand alone or bear the whole, or major, burden or delivering the training, their design and development must be particularly thorough and well done.

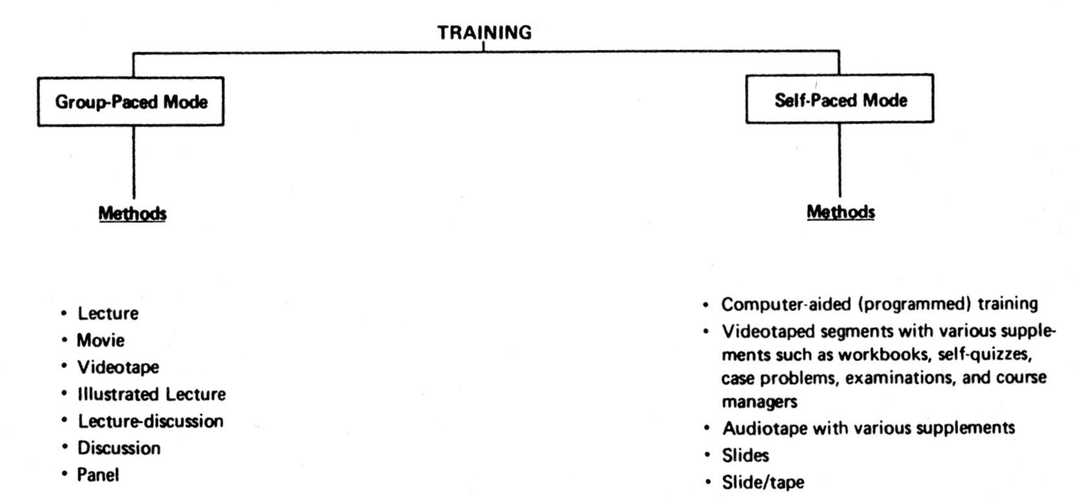

Figure 15-1 Representative methods of training used in group-paced and self-paced modes.

Although the ideal self-paced training course is *effective,* it is not always the most *efficient* in the use of available resources, and it is seldom as economical as traditional lecture-based training. Three conditions conducive to the development of self-paced training are learning situations in which:

There is a need for quality control of the output of instruction; the organization requires *identical* responses from all learners to standard work conditions.

The *scheduling constraints* (availability of trainees) for training do not permit the organization of group-paced classes of instruction.

The *location constraint* of users does not permit the organization of group-paced classes of instruction.

Three conditions *not* conducive to development of self-paced training are learning situations in which:

The subject matter is still too volatile (considerable changes are still being made to the new system).

The potential benefits, including number of trainees, do not exceed the break-even cost of developing and implementing self-paced training courses.

Group learning exercises are essential to simulate actual work conditions.

Evaluation: Tryout and Revision

Throughout the entire training development process, the course is tested. Two concerns are paramount for the designer:

Internal validity: Does the course elicit the type of performance identified in the objectives?

External validity: Will a person who completes the course perform at an acceptable level on the job? The final test of a course's effectiveness is a follow-up study of trainees after they are on the job.

Four things to consider when conducting *training tryouts* are the following:

1. Were *preliminary tryouts* done with first-draft materials?
   a. What were the results?
   b. What revisions were made?
2. Was another tryout done that included a *sufficient number* of trainees to establish confidence that the course is effective?
   a. Do the data summaries indicate high achievement (80 to 90 percent of objectives achieved)?
   b. Were indicated revisions made based on the tryout data?
3. Were the tryouts done with *representative* users of the actual target population?
4. Were the tryouts done in an *environment* that resembles as closely as possible the actual training environment?

Each unit of the course should be evaluated separately; then the entire course should be tested as a package. Trainee comments are a primary source of formal and informal feedback, not only during the tryout but throughout the life of a course. Each tryout will suggest revisions. A number of tryouts or revisions is inevitable before a finished product is ready.

Training courses should be kept up to date, even after they are tested, revised, and in regular use. Designers should keep courses current by exploring changing needs in the user population and updating subject matter.

Conclusions

Designers who are developing training should keep several important principles in mind. First, training is a means of improving *human performance.* If a training package is to be effective, it must be based firmly on the analysis of the desired performance. Training is only one viable option that may be chosen to increase the likelihood of competent performance. Because the development, presentation, and maintenance of training is expensive, it is not an option to be chosen lightly.

Second, when developing training, the designer should keep it lean, that is, as efficient as possible. It seems to be simpler to add missing pieces to a course than to spend too much development time covering all aspects of everything.

Finally, whatever presentation strategies are chosen (e.g., instructor led or computer based), it is good to remember that learning is, in a sense, *always self-paced.* A designer must support the reality of people progressing at their own speed.

Computers and Training

A discussion of training development would be incomplete without a look at the impact of computers. Computers affect training in at least two ways:

1. Their growing presence in homes, schools, and the workplace creates a mushrooming need for training in their use.
2. Computers may be used as instructional or learning tools in training design and delivery.

The first point is self-explanatory. Few of us remain untouched by computers and more of us are actual users. The second point, the utility of computers as instructional and/or learning tools, requires more discussion. All the power of computers is available to training developers.

To understand the potential of the computer for training, let's look first at some computer-related possibilities.

The opportunity to provide demand scheduling: A trainee can be scheduled for a course when and where he or she wants it by using a computer.

The opportunity to reduce training costs:

The use of the computer can reduce travel costs by providing training or segments of training at the trainee's home base (in fact, in a trainee's home).

Time away from the job and its attendant costs are significantly reduced if onsite training is available.

There is a reduction of such costs as those for instructor, facilities, or lab equipment.

The opportunity to provide more flexible training design: Many training courses have become so complex that they can be difficult to control. The computer can perform many information management activities for the training system, thus freeing trainers to spend more time on creative design activities.

Several names are currently in vogue to describe the role of computers in training. For example, one might speak of computer-based training, computer-based instruction, computer-assisted instruction, or computer-managed instruction. In the following discussion, we will use the term *computer-based training* or *CBT.*

Computer-based Training

Advantages

Many systems training needs can be best met using computer-based training (CBT). CBT can present explanations of course content, exercises with feedback, tests, and diagnostic information to trainees (Faust and O'Neal, 1977). Other advantages include having *interactive* training. Frequent responding and immediate, tailored feedback are important training advantages. CBT also enables considerable adaptation to individual needs.

Centralized control is an advantage because all users of a particular CBT course can depend on receiving *exactly the same course.* If there is a human performance problem associated with training, it can frequently be corrected by making the right change in just the one course. With centralized course material, the necessary changes can be accomplished at all locations simultaneously. This reduces costly reproduction, storage, and distribution of printed materials.

Training delivery can be made when and where it is needed. Trainees need not wait for a scheduled class. They can take a course or even portions of a course as work schedules permit. This helps to provide training just before it is needed. In addition, many system users can work at their own job-related terminals.

CBT courses can be designed to monitor and record trainee progress and performance. A trainer at one location can monitor and direct the work of trainees at a number of locations. Data about the achievement of objectives and the efficiency of a course can be maintained and easily accessed by trainers. Although the original CBT courses can be developed by the system designer, it can be left to trainers in the new system to maintain and improve the courses.

Disadvantages

Some disadvantages are also associated with using CBT. Expensive hardware and specialized software are required. Acquisition, development, and maintenance of the software may turn out to be much more costly and difficult than other forms of training. Computer downtime and slow or irregular response time may also prove to be a problem. Because its special qualities are often not understood, CBT can be misused. For example, there is a tendency by less experienced designers to use the computer to simply move from one page of text to the next (i.e., an electronic page turner).

Designers should recognize that CBT course development often requires more time than do other types of course development—up to 200 or 300 hours or more per hour of instruction. However, estimates of CBT development time vary greatly depending on how they are computed. Medsker (1979) suggests that factors that may affect development time calculations include whether a systematic design process has been used, whether time spent testing and revising lessons has been included, the developer's familiarity with CBT, his or her subject matter expertise, and the complexity of the CBT lesson format. Usually, the more sophisticated the software is (i.e., the better it uses computer capabilities), the longer the preparation time. For systems that change fairly rapidly, long development time may not be feasible.

Evidence for Computer-based Training

CBT is being used more frequently in new systems, either to deliver all the training or as a supplement to training delivered in other ways. Existing CBT approaches range from courses to teach a user to operate a computer terminal, to courses used to train aircraft pilots to fly an airplane. Trollip (1979), for example, provides an evaluation of American Airlines flight crew training using a CBT simulator. He reports that, when compared with more traditional simulator methods, CBT simulation results in more effective and less expensive training.

Keep in mind that the quality of CBT projects tends to be quite variable. Imaginative and sophisticated applications of CBT produce the best results. It is not enough to simply have the computer present information and have a user respond while at a terminal. The novelty of using a computer soon wears off, and the basic problems associated with knowledge learning and skill building continue. Consider, as an example, the teaching of computer programming using CBT. Results of studies tend to follow the pattern established in many other subject matter areas in which usually no significant differences are found between effectiveness of CBT courses and their alternatives (cf. Danielson and Nievergelt, 1975; Montanelli, 1977; and Gilbert, 1979).

Deciding When to Use CBT

To select the most appropriate means for delivering a training course (whether it be computer based or not), the designer should consider the available and reasonable alternatives and match them to the needs of the trainee. Mager and Beach (1967) suggest some principles for selecting the best delivery approach. Briefly, they suggest choosing the approach that

Most closely approximates performance conditions on the job

Causes the learner to perform in a manner closely approximating the performance called for on the job

Requires frequent interaction by the learner

Medsker (1979) suggests that CBT is a good candidate when individualized, self-contained training with low trainer involvement is appropriate. To help determine if this is the case, she suggests that the designer see if these conditions are present:

Mastery of specific, measurable objectives by each learner is required.

Active responding by the learner (practice) is needed.

Job conditions are constant or predictably variable, which minimizes the need for on-the-spot adaptation to local conditions.

Long development time is acceptable.

Subject matter is fairly stable.

A large number of people will take the training.

Flexible scheduling of trainees is desired.

Instructors are expensive or difficult to find.

CBT is *not* as desirable when

Training must adapt to rapidly changing subject matter or job conditions.

Courses teach interpersonal skills.

Limited time and/or resources exist.

Courses will not be needed over a long period of time.

Within the category of individualized, self-contained instruction with low instructor involvement, CBT is a particularly good choice if the system designer already possesses the required course development skills, including experience with CBT and programming ability or training on how to author a CBT course.

Exercise 15A: Describe Your Most Effective Teacher and Course

Purpose: To provide an opportunity to reflect on an effective teacher and course.

Method: Think back on the elementary school, high school, college courses, and training courses that you have had. Pick out the one teacher that had the greatest influence on you.

Reporting: Prepare a one-page report that outlines why the teacher was so effective. What was there about the training that made you want to learn more? What was there about the teacher that made you remember the experience? How could this experience be applied in the design and development of computer-related training?

Exercise 15B: Developing a Training Course Outline

Purpose: To provide an opportunity to design a small computer-based training package.

Method: In Exercise 12C, you designed an employment system. A primary user of that new system was the *employment clerk.* Consider the employment clerk's user profile and the way you designed the clerk's activities to be performed with the new system. Assume further that because of the very high turnover rate of employment clerks the training for the job must be totally computer based.

Prepare an outline that describes the content and format of the computer-based training. Follow the steps outlined in this chapter.

Reporting: Prepare a report that includes a complete set of course objectives, a complete set of test items, and an outline of the course materials.

16

Usability Optimization

Introduction

This may be a good time to formally introduce myself: my name is Robert Bailey. My parents, as far as I know without the help of any usability research, devised a six-character code, "Robert," to go with the six-character code, "Bailey," the latter having been handed down through many generations. I found as I grew older, however, that many of my associates rejected the six-character code, "Robert," preferring the shorter code, "Bob."

One day I realized that there were some who even rejected my three-character code. To my auto insurance company, for example, I wasn't Bob or Robert. I was 950424F1130B. My driver's license code, a terribly long 15 characters, is B81811658863405. Blue Cross/ Blue Shield also rejects the Robert and Bob codes and gave me one of their codes. I recently called the bank, and they wouldn't do business with me until they knew both my name code and my bank number code. Exxon, preferring my credit-card code, does not even know my name code.

For my employer's records, I found that I was no longer Robert Bailey but 270471. Furthermore, I couldn't become 270471 until they verified that the government had me on file as 518-42-1887. The telephone company, not to be outdone, has given me a 10-character code that I am supposed to memorize. Aside from the phone number, the telephone company also has much more information, all in the form of codes, that must be directly or indirectly generated, processed, or updated to offer me telephone service (Table 16-1).

Codes

Codes are supposed to be a shortened way of representing longer messages, and for that reason it is desirable to have each code convey as much information as possible. My name, Robert Bailey, has 12 characters. It is a very inefficient code because the 12 characters convey only one message—the identity of a single individual. We might, however, assume two other messages. The name Robert suggests that I am male, and both names together may suggest that I was born in the United States, England, or Ireland. Nevertheless, taking 12 characters to convey my identity is not an efficient use of the code. The United States post office's ZIP code is an example of a more efficient code. The first character refers to a sectional center, the next two to a post office, and the remaining characters to a specific delivery area.

Table 16-1 Codes Associated with Customer's Telephone Service

Cable/wire number = 79	Low binding post = 105
Apparatus number = 18	Cable/wire size = 404
Signaling arrangement = SX	Building = 13
Type set = HCK-BLK	Equipment type = DLCID
Pair number = 1511	Customer code = 660
Building suffix = A	Class of service = A
Numeric address = (201) 745-3621	Wire center = 348
Unit = 3	Terminal capacity = 24
Street code = ABAB	Even/odd/both indicator = E
Multiplicity = 3	Terminal location code = 16
Vertical row = 130W - 020	Terminal ID number = 650032-101
Class of service = 2FR	Engineering plate number = 1475
Bunch block number = 31	Terminal type = F
Section of plant length = 1562	Cable number = 13
Complement type = 3	USOC code = CJJER
Connect to section of plant = 136784	Distribution frame side = H
Facility type = 2	Bridge lifter number = 190
Type of plant = MUL	Section of plant number = 3381
Taper code = 110100	Bay relay rack = 0103-02
Binder color = 660	Distribution frame type = MDF
Wired out of limits = 1	Pair present condition = 13

As it is used here, a *code* is a shortened representation of a word or group of words. Codes are usually made up of all letters (alpha codes), all numbers (numeric codes), or a mixture of letters and numbers (alphanumeric codes). Some codes use special symbols that are not alpha or numeric, such as $, #, or ;. Codes are commonly used to define people or items, such as equipment parts, locations, and facilities. Thus, rather than spell out California, we write "CAL"; rather than say that a person lives on the left side of the road, in the 23rd house from the corner, starting from where the brick barber shop is located, we say the person lives at 1444 Oakley Avenue; and rather than ask for the person who is 5'9", weighs 175 pounds, and has dark hair with hazel eyes, we ask for "Robert."

There are at least two reasons why words by themselves are not convenient. The first is that they are not necessarily unique in situations where it is essential not to confuse one identification with another. There are many Adams, Baileys, and Smiths and even with the first and middle names added, there is no guarantee that two individuals will not have the same name. The second reason is that the full words can be very lengthy, especially if they must be unique. This is a handicap for human use when the words must be read and copied by hand or typewriter. It is also a handicap for computer processing, particularly in systems characterized by large-volume files.

Zipf (1935) observed that in a number of diverse languages a word's frequency of use and its length were negatively correlated. Furthermore, this relation appeared to be causal; as a word or phrase increased in frequency of use, it became shorter. Examples of this sort of word abbreviation (or code making) are plentiful in English. Consider for example, that *television* becomes the code *TV, cathode ray tube* becomes the code *CRT,* and *video display terminal* becomes *VDT.*

For numerous reasons, then, not the least of which is the advent of computers, communication takes place with information in a shortened form. There is no question that the use of codes has proliferated to the extent that virtually everybody uses them. Coded information can help a person communicate faster and more accurately.

Types of Codes

There are two broad types of codes, *arbitrary* and *mnemonic.* Arbitrary codes aim to provide a unique identification. The numbers or letters of these codes hold little or no special significance to the user; examples are a Social Security number, passport number, or ticket number. Mnemonic codes tend to convey information that is meaningful in some way to users.

Arbitrary codes usually bear no direct relationship to the word or groups of words that they represent. Most telephone numbers are arbitrary codes. Usually, knowing a person's seven-digit telephone number or nine-digit Social Security number tells little about the person.

Mnemonic codes are purposely designed to have some association with the word or groups of words that they represent. Mnemonic codes frequently, although not always, consist of alpha characters and tend to be easier to recall than arbitrary codes. A person's name is a mnemonic code and represents a shorthand way of referring to a person without a lengthy description of age, height, weight, type of nose, and color of hair. Many mnemonic codes are abbreviations of the words that they represent: NJ for New Jersey, Chevy for Chevrolet, or Fri. for Friday.

Certain arbitrary codes, even those that are all numeric, may become mnemonic codes if used frequently enough for an association to develop between the code and what it represents. Consider, for example, a system in which each type of computer-detected error is assigned a specific error code. An omitted name may be error type 1, a misspelled name, error type 2; a wrong address, error type 3; and an incorrect telephone number, error type 4. After a relatively short period of time, the people working in this system will begin communicating by making statements like, "yesterday I corrected twenty-seven 3's and nineteen 4's." Translated, this means that yesterday 27 wrong address errors and 19 incorrect telephone numbers were corrected. In this case, *meaningless* arbitrary codes were converted to *meaningful* mnemonics.

Coding Errors

Errors made while using codes may be classified as *clerical* or *procedural.* Generally, clerical errors are related to code design, whereas procedural errors are more closely related to other aspects of the system. A procedural error occurs when a user selects and enters the wrong code or the right code in the wrong location. It can be an error related to

- Code usage, such as failure to encode a street address when operating procedures call for a code to be assigned.
- The information content of the message unit, such as entering the code for one equipment type when a different equipment type should have been entered.
- The formal setup of the message unit, such as entering the code in the wrong field.

A clerical error occurs when a user correctly selects the code, but incorrectly enters one or more characters while printing or keying. The definition of clerical errors excludes omission of a whole code or substitution of one code for another. In general, when using codes, about one out of every five errors is clerical (Bailey, 1975).

Character-level clerical errors can be classified in many ways. Probably the most common classification scheme is shown in Table 16-2. A richer classification scheme for clerical errors, oriented toward the stage in cognitive processing at which an error probably occurred, is shown in Table 16-3. This scheme suggests that clerical coding errors occur due to faulty processing in one of the three major cognitive stages: perceptual, intellectual, or movement control. Within each stage, different types of errors can occur. For example, in the intellectual (including memory) stage, it is suggested that there can be either anticipation or perseveration errors. Whether anticipation or perseveration errors only originate in the intellectual stage is still not clear. We have more confidence in identifying errors from the perceptual and movement control stages than the intellectual stage. Nevertheless, this classification scheme provides a more useful way to consider character-level coding errors than the older, purely descriptive method in Table 16-2. This approach suggests ways of reducing errors once they are identified. For example, if errors are found to be perceptual, then the way to reduce these errors is to make perception-related changes. We will review briefly each category.

Table 16-2 Common Classification Scheme for Character-level Clerical Errors

	Example	
Category	**Correct**	**Error**
Substitution	ABC	A*K*C
Transposition	ABC	A*CB*
Omission	ABC	AC
Addition	ABC	AB*B*C

Perceptual Errors

Perceptual errors probably occur while sensing, storing, or encoding stimulus information. We suggest that the majority of errors in this stage will occur from misreading

Table 16-3 Revised Classification Scheme for Character-level Clerical Errors

Category	Example: Correct	Example: Error
Perceptual		
Visual	ABCDE	AB*G*DE
Auditory	ABCDE	*H*BCDE
Intellectual and memory		
Anticipation	ABCDE	AB*D*E
		AC*C*DE
		A*C*BDE
Perseveration	ABCDE	AB*B*DE
		AB*B*CDE
		ABC*A*DE
Movement control		
Adjacent Key	ABCDE	*S*BCDE

visual stimuli or mishearing auditory stimuli. Confusion matrices are the best available source of commonly confused characters. Numerous confusion matrices have been published showing which uppercase or lowercase English characters tend to be most frequently confused with one another. A list of *visual* confusions for uppercase letters and numbers taken from several studies is shown in Table 16-4 (Neisser and Weene, 1963; Owsowitz and Sweetland, 1965; Fisher et al., 1969; Bailey, 1975). The visual confusion "T for Y" is also an eligible combination, but is not included because it also qualifies as a movement control confusion. A list of *auditory* confusions taken from Conrad (1964) and Hull (1973) is shown in Table 16-5.

Intellectual Errors

The second category in Table 16-3, intellectual, involves all the processes required for choosing a response, that is, translating perceptual information into a plan for action. Even though all three stages make use of long-term memory, the intellectual stage also makes considerable use of short-term memory. Intellectual errors are defined operationally as errors in the sequence in which required characters are entered. Two major types of errors are included: *anticipation* and *perseveration.*

Anticipations

Anticipation errors occur when a character is keyed before its proper time in a sequence. There are three subclasses of anticipation errors. In the first the erroneous (anticipated) character is one or possibly two spaces ahead of its position in the original

Table 16-4 Visual Confusions

Response	Stimulus	Response	Stimulus
B	R	P	D
C	F	S	J
C	G	T	I
D	O	T	J
D	P	U	J
F	E	U	V
F	P	V	U
G	C	V	Y
G	Q	W	N
K	Z	X	Y
L	Z	Y	V
M	H	Y	X
N	H	1	I
O	C	2	Z
O	D	5	J
O	G	5	S
O	Q	6	G

stimulus. For example, ABCDE, is keyed ABDE, or ABCDE is keyed as ABE. The old classification scheme emphasized characters not entered between the last correct character and the anticipated character by calling this type of error an omission error. What was omitted was emphasized instead of what character was anticipated.

Table 16-5 Auditory Confusions

A–J	C–E	E–V	H–8	N–A	T–2	4–5
A–K	C–P	F–S	I–R	O–A	V–3	6–8
A–L	C–T	F–X	I–4	P–Q	Z–7	
A–N	D–B	G–P	I–5	P–T	0–4	
A–O	D–E	G–Q	I–9	P–V	0–8	
B–C	D–T	G–T	J–2	Q–T	1–7	
B–D	D–V	G–U	K–N	Q–U	1–8	
B–E	E–G	G–V	L–O	Q–E	1–9	
B–G	E–P	H–S	L–R	R–4	3–8	
B–P	E–T	H–X	M–7	S–X	4–1	

In the second subclass, a character is again omitted, but the erroneous or anticipated character is correctly entered in its original character position: ABCDE, but keyed as ACCDE. A third subclass of anticipation errors is commonly known as character transpositions. Transposition errors exist generally, but not exclusively, when the positions of two adjacent characters are interchanged: ABCDE keyed as ACBDE. Exactly why transpositions occur in human cognitive processing is still controversial. For example, Murdock and Von Saal (1967) argue that transpositions only occur after characters are stored, whereas Conrad (1964) believes that transpositions occur as responses are being executed.

Perseverations

Perseveration errors occur when a character is entered in its proper position and again in an incorrect position. There are three subclasses of perseveration errors. In the first subclass, the wrong character is entered immediately after the correct character. It is always a repeat of the first character (a "stutter") and an omission of a subsequent character: ABCDE keyed as ABBDE. In the second subclass, the persisting character again is entered immediately after the correct character. It is always a repeat of the first character, but without an omitted character. In fact, all the original stimulus characters end up in their proper order in the response: ABCDE keyed as ABBCDE. The third subclass of perseveration errors involves a character perseverated in a character position at least two away from its original position. Again, all the original stimulus characters are in their proper order, but not in their proper positions: ABCDE keyed as ABCADE. Because an extra character shows up in the latter two subclasses of perseveration errors, these errors used to be termed commission or addition errors.

Movement Control

The third category in Table 16-3 is movement control errors. These errors occur on a keyboard when a subject attempts to depress a key for a properly perceived and translated response but hits the wrong key. These errors, by definition, result from typing a letter on the keyboard immediately adjacent to the one required by the stimulus, for example, by depressing a key to the left or right. A list of keyboard confusions is shown in Table 16-6. The character combinations of N for H and U for J are also visual confusions. In addition, movement control errors may result, again by definition, from using the correct finger, but the wrong hand (should key an E but strikes an I). We assume that both of these error types occur after a stimulus is correctly perceived and the proper action is decided. In typing and keying tasks, these movement errors traditionally have been called *motor confusions*.

Error Control

Errors are costly; therefore, it is important to design codes that elicit the fewest errors. This is particularly true for errors that are not self-detected. When a proofreader or computer detects errors, the point where the error was originally made is frequently

Table 16-6 Motor Confusions for a QWERTY Keyboard

Response	Stimulus	Response	Stimulus	Response	Stimulus
A	S	L	K	X	C
B	V	M	N	Y	T
B	N	N	B	Y	U
C	X	N	M	Z	X
C	V	O	I	1	2
D	S	O	P	2	1
D	F	P	O	2	3
E	W	Q	W	3	2
E	R	R	E	3	4
F	D	R	T	4	5
F	G	S	A	4	3
G	F	S	D	5	6
G	H	T	R	5	4
H	G	T	Y	6	7
H	J	U	Y	6	5
I	U	U	I	7	8
I	O	V	C	7	6
J	H	V	B	8	9
J	K	W	Q	8	7
K	J	W	E	9	0
K	L	X	Z	9	8
				0	9

obscure. This frequently requires costly and time-consuming analysis to determine the correct information. However, the most costly errors may be those *not detected;* they can be proliferated in printouts and files and eventually could result in a general lack of integrity for the total system.

One of the best ways to eliminate coding errors comes through having well-designed codes. Designers should know the general characteristics of codes that enhance human performance. They learn these by carefully considering issues related to the user, activity performed, sources of information, type of response required, and type of format used. These general considerations and the specific guidelines that follow are derived from Sonntag (1971); Hodge and Field (1970); Jones and Munger (1969); Field and others (1971); Hodge and Pennington (1973); and others.

User Characteristics

Codes should be designed for human use. We have dealt with many user characteristics in great detail in earlier chapters. Nine characteristics that are closely related to code use are reviewed next.

1. Users tend to make more errors working with codes that sound or look alike than with codes that do not (*perceptual*).
2. Users should make use of the code as soon after seeing or hearing it as possible (*short-term memory*).
3. If users have trouble speaking a code, they will have trouble rehearsing and remembering it (*short-term memory*).
4. Users prefer short codes (one to three characters). The longer the code is, the greater the user rate of error in copying tasks (*short-term memory*).
5. Long strings of code characters become more manageable when they are grouped into three- or four-character elements (*short-term memory*).
6. Users tend to make more errors at the end of short codes or to the right of center in longer codes (*short-term memory*).
7. Users react more quickly and more accurately to familiar codes than to unfamiliar codes (*long-term memory*).
8. Users best recall codes when the context (environment) at the time of recall is similar to the context at the time that the code was learned (*long-term memory*).
9. Users have more difficulty keying with the middle and little fingers than with the index and ring fingers (*movement control*).

Except for activities for which the same codes are frequently used and that rely heavily on long-term memory, most critical design issues are associated with perception and short-term memory.

Activity Characteristics

The activity being performed by users helps to determine the type of code. Frequently used codes should be committed to memory. For infrequently used codes, it is usually best to design a system in which users look up codes in either a code book or a computer file. If users are employing a large number of different codes and the code books or computer files are poorly designed and awkward to consult, users may attempt to guess the correct code, and many errors are likely to result. In one telephone company study, users were correct only 73 percent of the time when trying to remember the correct code, although they indicated that they thought they knew *all* the codes.

In some applications, imposing codes that are difficult to memorize may actually improve accuracy by requiring users to access code books rather than to rely on memory.

This technique should be used only when accuracy is more important than performance rate (it takes time to look up codes). The fastest performance comes when people have memorized the codes.

Response Considerations

The design of an activity determines the way users will make *responses.* Responses may be verbal, printed, mark-sense, keyed, dialed, and so on. Speaking is the most familiar and usually the easiest way of responding for most people. In computer systems, keying is commonly used. Certain types of codes elicit fewer responding errors. A small survey was made in one company to determine the frequency of different types of responses when codes are used. The results showed that in 12 jobs 3 emphasized verbal responding, 8 emphasized handprinting, and 1 emphasized keying. A designer should determine the types of responses to be made with codes before developing the codes. Certain codes, for example, can be keyed with fewer errors, some can be spoken more easily, and some speed up handprinting.

Guidelines for Constructing Codes

A designer should consider the following guidelines when developing a new set of codes. If these considerations are appropriately applied, the probability of having code-related errors once the system is operational is reduced.

Secure Basic Information

Once the need for a code is established, the designer should secure the following information:

- Description of the information to be coded
- Purpose of the code
- Total number of items the code is to encompass when the system is first operational
- Growth potential of the coded items (the number of different items over the life of the system)
- Media on which the items are to appear (CRT, paper, etc.)
- Primary tasks performed by the people using the code
- Accuracy requirements, including error detection and correction
- Any computer limitations that will be imposed on the code
- Whether there is a code in use similar to the one to be designed and, if so, what is that code

Establishing Rules

Some designers create code sets worthy of military intelligence. Some of these codes, unfortunately, are almost impossible to break without a code book. When designing systems, we should have just the opposite objective. Construct codes so that users, by knowing the rules for construction, can easily determine code meaning (and even make up new codes if necessary).

Develop a set of rules before designing a set of codes. The choice of rules used for developing a code set depends on various factors, including how and by whom the codes will be used. The rules, once clearly defined, should be followed consistently. A definite set of rules helps to provide users with a means of decoding (converting the code into the information that it represents) and serves to facilitate both code learning and performance in the use of codes. Encoding schemes that are not based on systematic and consistently applied rules may lead users to make incorrect code associations, thereby increasing the probability of error.

User Experience

Codes should be designed for the least skilled users within the expected population of code users. Codes that are designed for the newest users do not usually degrade the performance of more highly skilled personnel. When the possibility exists that experienced users will have their performance penalized by codes designed for less experienced users, the designer should consider using two or possibly three different versions of the same code. In some computer systems, for example, less experienced people may prefer codes that are descriptive, such as "ed" for "edit," "sub" for "substitute," "wrt" for "write," whereas experienced users may prefer much briefer codes: "e," "s," or "w."

Preference versus Performance

Most users eventually accept and use a set of codes as a second language and become quite comfortable using even poorly constructed codes. For example, many experienced users see no problem with using long codes, but performance tests consistently show that the proportion of errors increases as code length increases.

When asked to rank codes of 2 through 11 characters, users seem to show no preference for either short or long codes. Judgment of the ease or difficulty of use seems to be based on other characteristics. Performance data, on the other hand, give a clear indication that users have difficulty with codes that have 7 or more characters. Users do rank those codes for which internal patterning is readily recognized as easier, which is consistent with performance data.

One telephone company study compared user code preferences and user performance. Three conclusions interest us.

1. Users showed high agreement about which codes were easy and which were difficult to use.
2. Performance data tended to support the idea that usually codes were not more difficult for different types of users.

3 Where major differences in performance were found, the differences were more related to the overall design of the activity, rather than the design of the codes.

When users express dissatisfaction with a set of codes, usually it is because of inconsistent rules for code generation, lack of logic in codes (little relationship of code to meaning), difficult formatting requirements, or difficulties in finding codes in code dictionaries.

Most users believe that alpha codes are easier to work with than numeric or alphanumeric codes. Among the alpha codes, mnemonic codes tend to be rated higher than those that are nonmnemonic. On the other hand, performance data indicate that users make fewer errors with alpha codes only when the alpha codes convey special meaning, and the tasks for which they were used are complex.

The length and complexity of a dictionary definition appears to be associated with the judged difficulty or preferences for codes. When the relationship between the code and its definition is not obvious, codes tend to be rated as more difficult. For example, the C911 code with a dictionary definition of "audible ring overflow" was rated as extremely difficult by all who had to use the code.

Uniqueness

The demand for code uniqueness means that codes close in meaning and that are auditorily or visually similar should not be used. A code should uniquely represent the element of information that it defines. Keep in mind that when codes are verbally transferred there can be considerable confusion among those that sound alike. Failure to provide uniqueness leads to confusion and errors.

Meaningfulness

Meaningfulness refers to those attributes of a code that lead the user to readily associate the code with the item object, instruction, or action that the code represents. A meaningful code is known as a *mnemonic.* Meaningfulness should be built into codes whenever possible.

Users tend to perceive codes more accurately and quickly if they are meaningful. Both complex and simple activities seem to benefit from meaningful codes. Characters that are grouped to provide meaningfulness and that appear logical to the code user elicit far fewer errors than characters that are drawn at random.

Mnemonic alpha codes almost always produce better performance (both increased speed and reduced errors) than numeric codes of the same length. Even in typing tasks, fewer errors occur when the text is meaningful than when codes made up of random letters are used. The reduction of errors when keying mnemonic codes appears related more to learning a code pattern than to learning the code's meaning. Certain alpha characters appear in combination much more frequently than others. Mnemonics also permit a significant decrease in keying errors when the task involves looking up a code in a code book and then keying.

The following attributes are related to the meaningfulness of codes.

Similarity to the English Language

Part of the concept of meaningfulness is associated with the order of characters within a code (cf. Howe, 1970). For example, a random code such as QVIL would be more subject to error than QUIL, since there is no general habit pattern in the English language using QV in sequence. While it is true that users can, over a period of time on the job, adapt to letter combinations that do not occur or occur infrequently in the English language, the design of code sets for the least experienced user dictates avoiding unusual combinations whenever possible. Codes that are pronounceable (i.e., rehearsable) tend to be more meaningful, more easily learned, and better remembered than those that are not pronounceable.

The material in Table 16-7 (letters), Table 16-8 (digrams), and Table 16-9 (trigrams) was prepared to assist designers in selecting common characters and character combinations. In each table, the most frequently used letters, digrams, or trigrams are shown first. Designers should construct codes that contain frequently used character combinations.

Number and Type of Different Vocabularies (Alpha, Numeric, or Mixed Alphanumeric) Used

All alpha codes such as CRT tend to be more meaningful than all numeric or mixed alphanumeric codes. Codes drawn from only two vocabularies (for instance, the alphanumeric P38) are more meaningful than those drawn from three vocabularies (for example, the alphanumeric and symbolic MC*237$5).

Frequency of Code Use

Frequently used codes generally are considered more meaningful than moderately or infrequently used ones.

Context of Code Use

Codes that are placed consistently in the same location in a message tend to have higher meaning than those that appear randomly, since the location serves as an additional cue for code meaning. Codes that are part of a longer code phrase tend to have higher meaning than codes that stand alone, since cues from the total set contribute to the meaning of any one code in the set. This is similar to the idea that the meaning of a word may be more apparent in a sentence than when the word stands alone and out of context.

Similarity between the Code and the Full Description of the Item, Object, Action, or Instruction for Which It Stands

Codes that are similar to their basic words (abbreviations) that contain some of the same characters in the same order as they appear in the definition can be expected to have more meaning than those with characters that differ and/or appear in a different order.

Table 16-7 Frequency of the 26 English Letters Based on a Sample of 20,000 English Words

Letter	Percent of Total
E	13.3
T	9.8
A	8.1
H	7.7
O	6.6
S	6.1
N	6.0
R	5.9
I	5.1
L	4.5
D	4.3
U	3.1
W	2.9
M	2.5
C	2.4
G	2.2
Y	2.1
F	1.8
B	1.6
P	1.5
K	1.1
V	1.0
J	.2
X	.1
Q	.1
Z	.1
	100.0

Adapted from Mayzner and Tresselt, 1965.

Table 16-8 200 Digrams Based on 20,000 English Words

TH	3774	CA	368	IE	189	EW	106
HE	3155	NO	349	FR	188	EF	103
AN	1576	LO	344	EM	187	WN	103
ER	1314	YO	339	TR	187	FT	102
ND	1213	KE	337	EC	181	AP	100
HA	1164	OO	336	CK	178	NA	100
RE	1139	EL	332	AM	177	BL	98
OU	1115	LA	332	SU	175	GR	98
IN	1110	TO	331	EV	172	NC	98

HI	824	SH	328	PL	169	PI	97
OR	812	IL	324	SS	168	GI	96
AR	802	AI	322	HT	165	DS	95
EN	799	AY	319	IV	165	GA	94
AT	785	RS	318	MI	165	HR	91
NG	771	ET	316	CT	154	EP	90
ED	767	RI	309	FE	154	RU	89
ST	754	AC	308	TT	154	BR	88
AS	683	IC	304	YE	152	IO	88
VE	683	US	299	PO	149	OI	88
EA	670	CO	298	DI	148	AU	87
AL	664	GE	289	NS	148	EX	87
ES	630	LD	289	UG	148	UM	87
SE	626	MO	289	AK	146	FF	86
ON	598	RA	289	FA	145	IK	86
WA	595	GH	288	RY	145	MU	85
LE	591	CE	285	AB	144	TW	84
TE	583	WE	285	PA	142	DR	83
IT	558	PE	280	AG	138	KN	82
LL	546	UN	278	OP	138	LU	81
ME	530	LY	276	DO	137	YS	81
NE	512	IR	272	BA	136	NL	80
RO	504	WO	264	OV	136	OF	80
UT	492	ID	260	GO	135	BI	78
HO	487	TA	259	NI	135	MP	77
IS	484	BU	256	RD	133	HU	75
WH	472	IM	255	TU	132	TL	75
EE	470	TI	252	EI	127	LT	74
FO	429	UL	247	KI	127	CR	71
OM	417	BO	240	OK	123	RL	71
BE	415	AV	233	LS	121	UE	71
OT	415	IG	233	TY	121	FL	69
CH	412	OL	218	OD	119	RR	69
UR	402	SI	214	NY	115	PU	68
OW	398	SO	213	UC	114	AF	67
MA	394	TS	209	PR	110	CI	67
LI	390	FI	205	VI	109	OB	67
AD	382	SA	196	CL	108	QU	66
NT	378	OS	195	SP	107	OA	65
DE	375	RT	190	DA	106	RM	65
WI	374	EY	189	RN	106	UI	65

Adapted from Mayzner et al., 1965.

Rule Used to Construct the Code Is Known by Users

Codes that are consistently derived by established rules tend to be more meaningful than codes with inconsistent or obscure design rules, since the code user, when in doubt, tends to decode by application of the rule.

Table 16-9 200 Trigrams Based on 20,000 English Words

THE	2565	WER	119	ACE	80	EAT	65
AND	959	ATE	118	AID	80	ERY	65
ING	526	HOU	114	IND	80	HOW	65
HAT	479	OVE	114	URE	80	NIN	65
THA	438	NOW	112	COU	79	OSE	65
HER	414	WHO	112	LEA	79	RES	65
HIS	354	OUN	111	TUR	79	STE	65
FOR	353	COM	110	IDE	78	TIM	65
YOU	326	EVE	110	TIN	78	ION	64
WAS	304	HIN	109	ART	77	OOD	64
ALL	270	OUG	109	EAS	77	PLE	64
THI	259	USE	109	EST	77	RIE	64
ERE	255	ERS	108	VEN	77	WIL	64
ITH	238	AKE	107	HEM	76	NGE	63
ARE	228	MOR	107	LON	76	THR	63
WIT	227	WAY	106	ANT	75	TTE	63
OUT	225	INT	103	END	75	CHA	62
VER	221	STA	101	LED	75	HES	62
OUR	209	ABO	99	MEN	75	SHO	62
ONE	205	HIC	99	YEA	75	TEN	62
EAR	197	UND	99	HEA	74	DAY	61
AVE	194	AIN	98	PLA	74	ILE	61
NOT	191	ICH	98	ACK	73	MOS	61
OME	191	OWN	97	ARD	73	NEW	61
TER	179	OLD	96	GET	73	ONL	61
BUT	178	UST	96	ROU	73	ACT	60
HAD	173	ONG	95	SAI	73	BEC	60
GHT	163	WOR	94	ARS	72	LAS	60
IGH	161	BOU	93	LES	71	ANG	59
ORE	153	AME	91	PER	71	ICE	59
HAV	147	AST	91	UCH	71	ROV	59
ILL	146	CAN	91	AYS	70	ECT	58
OUL	145	HAS	89	EIR	70	EET	58
IVE	143	OST	89	RED	70	FIN	58
MAN	143	WOU	89	CAR	69	STO	58
SHE	143	ANY	88	HEI	69	AIR	57
ULD	143	KIN	88	SOM	69	EVE	57
OTH	140	WHA	88	THO	69	ITT	57
ENT	139	REE	87	NTO	68	MAR	57
FRO	138	BEE	86	TOO	68	ACH	56
HEN	133	IKE	86	AGE	67	EED	56
HEY	131	TED	86	CAM	67	RSE	56
WHE	131	ELL	85	NCE	67	TLE	56
ROM	130	LOO	84	ORT	67	IRS	55
EEN	128	OOK	84	CAL	66	ITE	55
HAN	128	SEE	84	DER	66	OIN	55
REA	128	EAD	83	FTE	66	PRO	55
UGH	128	LIK	83	IME	66	WAN	55
HIM	126	ITS	81	LLE	66	ADE	54
WHI	125	KED	81	NLY	66	DOW	54

Adapted from Mayzner et al., 1965.

Alpha versus Numeric Characters

The two most popular character vocabularies from which codes are made are letters and numbers. Since it is usually easier to convey meaning with letter or alpha characters than with numbers, tasks requiring complex cognitive processing of a code should use alpha character codes as much as possible. Frequently, alpha characters can be used to develop codes that have an easily associated meaning. Building meaningfulness into codes permits the code user to make maximum use of past learning. For many activities, meaningfulness may be the most critical characteristic in the code design.

Give numeric characters preference when designing codes for simple transfer tasks, for example, when a code is transferred from one form to another or keyed from a form to a computer using a terminal. These are tasks for which past learning may not be as critical. Numeric codes are also superior to random alpha codes in "listen–recall–write down" tasks, as long as the user is not required to hold the code in memory for more than a few seconds (Conrad and Hille, 1957; Conrad, 1959).

Gallagher (1974) pointed out that one advantage of using a combination of numbers and letters for codes is that they can be arranged in a pattern to mark off different fields without having to resort to other symbols such as hyphens or virgules. A code of the type B12A05P38 would create natural groups of B12, A05, and P38. It is likely, however, that the groups would be even better recognized if spaces were also used.

Hull (1975) compared the performance elicited by codes constructed using different code lengths, vocabularies, and groupings. The codes that elicited significantly fewer errors were those that had the least characters (six in this study), contained both letters and numbers, and had no more than five characters in a group (e.g., 1 E2NVL or QVTEK 4). The most errors were elicited by eight-character codes made up of all uppercase letters that were ungrouped (e.g., JBDEBCJD). Codes that had eight ungrouped numeric characters (e.g., 64195897) and those that were constructed of nine alpha characters with consonants and vowels, grouped in a consonant–vowel–consonant format (e.g., DOL FIK TAG), elicited a moderate number of errors.

Meaningfulness and Frequency of Use

When meaningfulness cannot be provided for all codes within a code set, preference should be given to those codes that will be used the most. In a system that requires a large number of codes, an equal amount of meaning cannot be built into all codes. This problem has its origins in the high redundancy in the English language. For example, consider a situation when the words "subscriber," "substitute," and "suburban" all need to be encoded for a user. Probably the most meaningful code for all three is SUB. To keep the codes unique, the designer must either make the code longer or give up some of the meaningfulness. As a general rule, meaning should be stressed in the design of frequently used codes even if it means making them slightly longer.

Special Symbols

Codes should be composed of characters already in the user's vocabulary. Alpha and numeric characters are familiar to all workers. Even though many other symbols are available for building codes (e.g., $, !, #), most of them are not as commonly known by users. When special symbols are required, the most common ones should be given preference. They should have, wherever possible, the same functions or meanings that users would expect them to have: # means pound, ? means question, $ means dollar. When using a keyboard to input data, requiring a user to use the shift key for special symbols will substantially increase errors (Hammond et al., 1980).

Compatibility of Code Sets

The total set of different code types (all alpha, all numerics, etc.) developed for users of a new system should be taken into consideration. A tendency to independently make up different code types occurs when more than one designer develops a system. One designer may use only numerics, another only alphas, and still another may mix alphas and numerics with special symbols. Users tend to form habit patterns or preferences for the types of codes that they use most often. When numeric characters dominate in a new system, more errors will be made against alpha characters than against numeric characters. When alpha characters dominate, fewer errors tend to be made.

If other factors do not provide a clear indication of the type of code to be used in a new system, consideration should be given to using the same type of code already in use in the system being replaced. If that set of codes is not acceptable, a designer should at least ensure that the new code type does not severely conflict with codes being used in adjoining systems.

Limited Vocabulary

The vocabulary (characters used to make up codes) for a given code set should contain the fewest amount of different characters that is consistent with providing meaningfulness, unique coding, and the ability to expand. The larger the vocabulary base from which a code is drawn, the more difficult it is to remember the characters that could be legally contained in a code. If the lowercase L and alpha O characters are not included, for example, and this is known by users, then symbols having the characteristics of a 1 and 0 have a better chance of being correctly interpreted as a one and zero.

If all codes are made up of only five alpha characters arranged in different ways, people and computers could immediately recognize that any character other than these five is illegal. Codes made up using a single vocabulary (all alpha) are less subject to error than codes made up using multiple vocabularies (mixed alpha and numeric, or numeric and special symbols, etc.).

Code Length

Codes should be short, preferably containing six or fewer characters, consistent with providing meaningfulness, unique coding, and anticipated growth. Code length is one of the most important considerations in code design. In general, a long code is more difficult to use and elicits more errors than a short code. Users work best with codes of three to five characters. The more complex the task, the shorter the code should be. Studies clearly indicate that in complex tasks errors increase substantially when codes contain more than six characters; in relatively simple tasks, errors increase rapidly when codes are longer than eight characters. There is at least one exception to the code length rule. One-character codes tend to elicit more errors (generally errors of omission) than do codes of two, three, or four characters.

Shorter codes elicit fewer handprinting and keying errors and, as a bonus, require less memory storage in a computer. Even so, examples of long codes abound. Consider the New Jersey Division of Motor Vehicles code for a driver's license. It is *15* characters long. The New Jersey driver's license number, like many codes, is read, hand printed, and keyed numerous times by a variety of different people, for example, by police when issuing a traffic ticket, court clerks while preparing dockets and collecting fines, and banks and other places of business when trying to establish identification for cashing checks.

The number of possible 15-character combinations, with only the 10 numbers zero through nine is 10^{15} or 1,000,000,000,000,000. But the New Jersey driver's license number also includes one character position for an alpha character, which increases even further the total number of possible 15-character combinations. When using all 10 numeric characters in 14 character positions and all 26 alpha characters in 1 character position, the number of possible different driver's license numbers is 2.6×10^{15} or 2,600,000,000,000,000.

New Jersey only has about 4 million licensed drivers, which means that a code of seven characters is all that is required. In fact, every person in the world, if we assume a population of about 6 billion, could be assigned his or her own number by using only a 10-character code. By using a 15-character code, unique numbers could be assigned to every man, woman, and child on earth, plus every person on over 700,000 other worlds with the same population as earth. The question is: From a human performance point of view, is it really necessary to use a 15-character code?

One of the first steps in developing any code set is to determine the code length. Once the length has been tentatively established, another early step is to determine how many codes of this length will be required for the life of the system. Depending on the nature of the system, the number of codes may range from two to a very large number. For example, only two codes would be required to respond to a computer's request to save a file, yes or no. On the other hand, the number of unique Social Security numbers required in the United States may someday exceed 500,000,000.

Once the code length and the total number of codes are determined, a third step is to consult Table 16-10. This table shows the minimum number of different characters necessary to develop a character vocabulary when the number of codes and the code length are known. For example, if the system requires 500,000 codes and each code is to contain six characters, the character vocabulary will require only nine different characters.

Trade-offs may be required when selecting a character vocabulary. For example, if a system requires a total of 50,000 codes and the original decision was to have a code length of four characters, the total number of different characters that would be required is 15. This requirement precludes using only numeric codes. The best performance may be achieved, depending on how the codes are used, by using codes that contain 5 characters (rather than 4 characters) and thus using only numeric characters.

Table 16-10 Number of Different Characters Necessary to Develop a Set of Codes*

Total Number of Codes Required by the System	Desired Code Length Number of Characters per Code 2	3	4	5	6	7	8	9	10
100	10	5	4	3	3	2	2	2	2
200	15	6	4	3	3	3	2	2	2
300	18	7	5	4	3	3	3	2	2
400	20	8	5	4	3	3	3	2	2
500	23	9	5	4	3	3	3	3	2
1,000		10	6	4	4	3	3	3	2
5,000		18	9	6	5	4	3	3	3
10,000		22	10	7	5	4	4	3	3
50,000			15	9	7	5	4	4	3
100,000			18	10	7	6	5	4	4
500,000			27	14	9	7	6	5	4
1,000,000				16	10	8	6	5	4
5,000.000				22	14	10	7	6	5
10,000,000				26	15	10	8	6	6
50,000,000					20	13	10	8	6
100,000,000					22	14	10	8	7
500,000,000					29	18	13	10	8
1,000,000,000						20	14	10	8

*Use this table when code length is between two and ten characters and the total number of codes required over the life of the system is between 100 and 1 billion.

Lengthening Codes Artificially

Since error potential increases as codes get longer, the addition of one or two characters to enable computer detection of errors (e.g., using check digits) actually increases error potential. Cost trade-offs between preventing errors and using the computer to detect errors should be considered before codes are artificially lengthened to provide for computer checks. The cost of computer detection of errors might exceed the cost of preventing the errors in the first place. Such trade-offs should be carefully evaluated before doing something that is *guaranteed* to increase errors, for instance, artificially lengthening codes.

Growth Requirements for a Code Set

Few code sets remain the same size throughout their operational use. New equipment or modifications evolve, new instructions are written, and new locations and buildings are developed, all of which require expanding an existing code set. Designers must realistically estimate the requirements for new codes and allow for expansion. When, for example, there is some doubt about whether to use a six-character versus a seven-character code, the best decision may be to go with the six-character code, recognizing that another character can be added in 15 or 20 years (if needed). The costs for adding one character at some later date may be far less than the costs associated with detecting and correcting code-related errors over a 20-year period.

Visual and Auditory Confusions

Characters that may be easily confused, either visually or auditorily, should be avoided. For example, confusion brought about by visual similarity of characters is very common for

- The alpha character O and the numeric character 0
- The alpha l and the numeric 1

Visual and auditory confusion also can be a problem with entire codes. An example of similar visual codes is HAY and HAP, and an example of similar auditory codes is GOME and GOAN. The alpha Z and the numeric 2 are often confused (Bailey, 1975; McArthur, 1965; Neisser and Weene, 1963). The accepted standard is to slash the Z (Z) to differentiate it from the numeric 2. Figure 16-1 illustrates a proposed method for handprinting all 36 alphanumeric characters. Notice particularly the techniques used to differentiate between the l and 1, alpha O and numeric 0, S and 5, and Z and 2.

Certain characters made up of 5 × 7 or 7 × 9 dot matrices on character-based screens are difficult to discriminate on many CRT screens. Those that tend to cause the most confusions are the alpha O, alpha D, and numeric zero; the alpha l, lowercase L, and numeric 1; and the alpha S and numeric 5. If the CRT screen for the new system does not provide a high degree of differentiation for these and possibly some other characters, the designer should avoid these characters in the new code set. If they are used, point out potential difficulties in reading these characters to users.

The same kinds of visual confusion problems result from using many computer printers. For example, one printer that is widely used presents a lowercase alpha L that is identical to the numeric 1. This same printer presents an alpha O that is slightly "fatter" than the numeric zero, but for all practical purposes, a user cannot tell the difference between an alpha O and numeric zero when reading a code. The same is true of lowercase q and g.

We advise designers to carefully inspect devices (and even the handprinting of users) that will be generating characters in a new system and to try to identify characters that could cause confusions.

Consistency

Special symbols should be used consistently for the same meaning or function. The use of like symbols for the same function should be consistent within the new system and

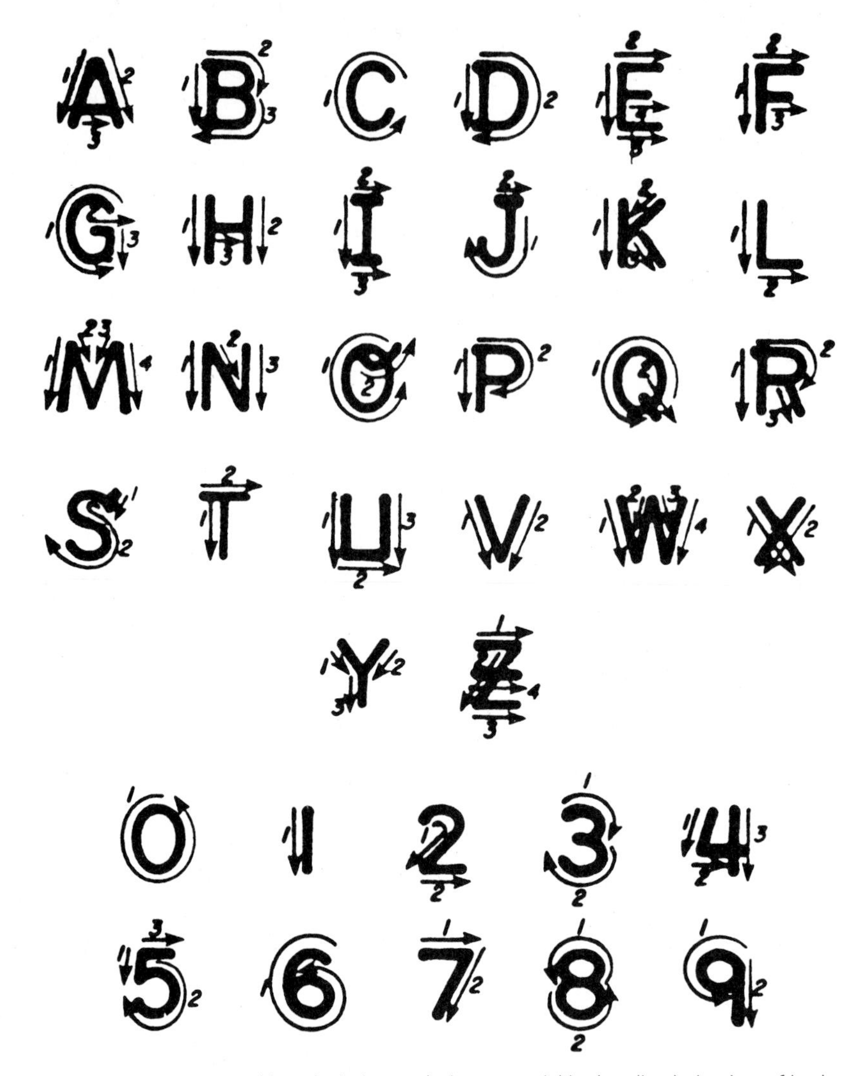

Figure 16-1 Proposed handprinting techniques to aid in the discrimination of look-alike characters.

with all other systems that a user is exposed to. For example, a designer could use either a dash or virgule as a separator in a code phrase (e.g., month-day-year), but should not use a dash one time and a virgule the next: 11-9-39 versus 11/9/39.

Formatting

Formatting of codes and code books should facilitate scanning for accuracy and completeness. Codes with easily detected patterns are more meaningful and easier to scan. If users are accustomed to a certain pattern in codes, an inconsistent code pattern may create errors. For example, one group of codes observed in the same department had inconsistently structured codes involving dates. In one instance the date was *bracketed* by alpha characters: X10–10BET; in a second, the date was *followed* by alpha characters: 10-11-CA; and in a third, the alpha character prefixed the date X10–7. If these codes had been more consistent, users would have made fewer errors.

Evidence suggests that formatting can affect the ability of users to detect their own errors. For example, a format that has exactly six spaces provided for a six-character code would make it relatively easy to detect when only five characters are entered. Formatting should also facilitate scanning an output for error. When using a properly designed format, a user should be able to easily scan for errors in a check for internal consistency, for example, to ensure that a white cord is provided for a white phone.

Formatting, as applied to coding, refers to (1) how pages are formatted in a code book, (2) how codes are formatted, and (3) how codes and code elements are arranged on a printed form or CRT screen. We will address guidelines related to the last two considerations later in this chapter.

A code dictionary or code book (whether printed or computer based) should be formatted to reduce errors and to minimize time for use. Also, take into consideration the density of a page; performance appears to suffer with high-density material. Keep in mind that the format of a code book should be divided into two parts: one for encoding and the other for decoding.

Grouping

When people set out to remember a series of numbers or letters, they break the series up into natural groups that enable them to remember the codes better. Grouping codes in a pattern encourages the best performance. A designer or user can do the grouping. When left to users, grouping may be synonymous with syllabication, such as dividing the code ALTECEP into AL, TE, and CEP. In other cases, patterning is associated with redundancy, such as mentally grouping 4LLL4 to three Ls between 4s. Users may also break codes by associating the parts with specific meanings, such as remembering BISCUS as BIS for "business" and CUS for "customer."

There is no question that grouping codes into patterns facilitates remembering, but the ways of grouping vary widely with individuals. One telephone company study found that, when a given grouping technique was defined and people instructed on that technique, those with instruction made significantly fewer coding errors than those without instruction.

It is usually preferable for a *designer* to group codes rather than the user. The designer should take advantage of available information on grouping. For example, the familiar telephone number 2019816425 is a long character string containing three distinct codes. The first code (201) is the area code, the second code (981) a district within the area, and the third (6425) is a subscriber's number. Breaking code phrases into the individual codes facilitates the use of both short- and long-term memories. The telephone number, for example, is generally broken up as (201) 981-6425.

It is almost always preferable to use short codes, rather than long codes or code phrases. (A code *phrase* occurs when two or more codes are used together.) Even when symbols are used to separate individual codes, there still can be confusion.

A considerable amount of work has been carried out in the area of grouping (cf. Klemmer, 1969). The most natural subgroup size is three or four units. Thorpe and

Rowland (1965) report on a study in which people were asked to speak codes containing nine numbers. They found that about 90 percent of people used three groups of three. At the same time the authors showed that the people who did not use the most common natural groups had many more errors.

Gallagher (1974) reports that when long lists of series are copied, more errors result when they vary in structure. For example, in the following two lists, a list of type (a) will produce more errors than (b).

(a)	(b)
68-51381	685-1381
924-1861	924-1861
4-5921-37	459-2137

Abbreviations

Code design frequently requires rules for abbreviations since abbreviations are considered to be mnemonic codes. There are many methods for creating abbreviations. Four of the most common methods are listed next. In each case the proper abbreviation for the word "command" is shown as an example.

1. *Truncation* (the user quits keying when he or she chooses or when the computer starts processing): c, co, com, comm, comma, comman, or command
2. *Contraction* by dropping all vowels: cmmnd
3. *Phonics* (keep only the first vowel): comnd
4. *System unique* (either the designer or the user decides the valid abbreviation for each item): cmd

The best abbreviation method for encoding tasks appears to be truncation (Ehrenreich, 1985). The system should be designed to have the computer decide when enough characters have been entered to designate a unique item and then start processing. After a few instances, users learn the minimum number of characters needed.

A less acceptable alternative is to have users enter their preferred truncation and then press the Enter key. If the computer has enough characters for uniqueness, it processes the entry. If it does not have enough characters, it presents a message that so indicates, and the user must try again with a longer entry. Research does not show a clear-cut abbreviation method for decoding tasks.

Exercise 16: Evaluating Coding Errors

Purpose: To provide an opportunity to become familiar with different types of coding errors, with the goal of finding ways to reduce them.

Method: An error matrix provides a convenient way to display coding errors. Look at the matrix below. These errors were made by 20 people as a result of either handprinting, keying, or speaking five-character codes that they read from a paper form. The people were entering information into a computer. About 98 percent of the characters were entered correctly. The characters that were incorrectly entered are shown in the error matrix.

The characters shown across the top of the page are the ones that *should have been* entered. Those down the left side are those that *actually were entered.* For example, on 10 different occasions the G was supposed to be entered, but the people entered a C; 25 times the W was supposed to be entered, but they entered a Q.

After analyzing the error matrix, what activity do you think the people were doing when the errors were made (handprinting, keying, or speaking)? Why? Based on your evaluation of the numbers and types of errors that were made, what could be done so that people would make fewer errors while performing this activity? What could be done so that people would self-detect more errors? What could be done so that the computer would detect more of the errors?

Reporting: Prepare a report using the format presented in Exercise 1A.

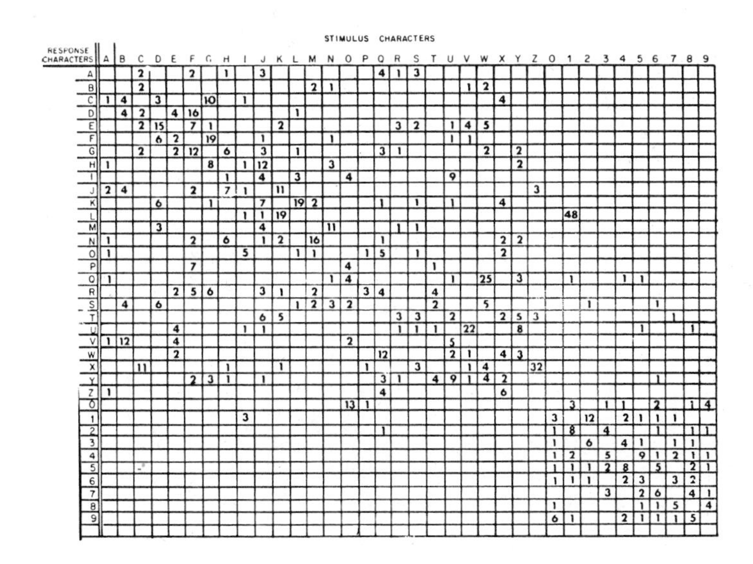

STIMULUS CHARACTERS

RESPONSE CHARACTERS	A	B	C	D	E	F	G	H	I	J	K	L	M	N	O	P	Q	R	S	T	U	V	W	X	Y	Z	0	1	2	3	4	5	6	7	8	9
A			2			2		1		3							4	1	3																	
B			2										2	1								1	2													
C	1	4		3			10		1															4												
D		4	2		4	16						1																								
E			2	15		7	1				2							3	2		1	4	5													
F				6	2		19			1				1							1	1														
G			2		2	12		6		3		1					3	1					2		2											
H	1						8		1	12				3											2											
I								1		4		3			4						9															
J	2	4				2		7	1		11															3										
K				6			1			7		19	2				1		1		1			4												
L									1	1	19																	48								
M				3						4				11				1	1																	
N	1					2		6		1	2		16				1							2	2											
O	1								5			1	1			1	5		1					2												
P						7									4					1																
Q	1													1	4						1		25		3			1			1	1				
R					2	5	6			3	1		2			3	4			4																
S		4		6								1	2	3	2					2			5						1				1			
T										6	5							3	3		2			2	5	3								1		
U					4				1	1								1	1	1		22			8							1			1	
V	1	12			4										2						5															
W					2												12				2	1		4	3											
X			11					1			1					1			3			1	4			32										
Y						2	3	1		1							3	1		4	9	1	4	2									1			
Z	1																4							6												
0															13	1												3		1	1		2		1	4
1									3																		3		12		2	1	1	1		
2																	1										1	8		4			1		1	1
3																											1		6		4	1		1	1	
4																											1	2		5		9	1	2	1	1
5																											1	1	1	2	8		5		2	1
6																											1	1	1		2	3		3	2	
7																														3		2	6		4	1
8																											1					1	1	5		4
9																											6	1			2	1	1	1	5	

Figure Ex 16A

17

Physical and Social Environment

Introduction

The environment in which a system is used can have a major impact on performance in that system. An extreme *physical* environment, including loud noise, low light level, or hot room temperatures, can lead to degraded performance (e.g., reduced speed and increased errors). Out-of-tolerance *social* environments also can cause deteriorization in performance.

Designers may have little control over the environment in which a system is used. Nevertheless, they should be aware of existing environmental conditions and take steps to provide a system that fits with current environmental constraints.

All human activity takes place within some environmental context (see Figure 17-1). The contexts in which people live, work, and spend their leisure hours are varied. Some environments expose users to physical extremes such as the temperatures near the equator or in the Arctic. Other contexts include the weightless environment of outer space or the watery environment of the ocean.

Figure 17-1 Human performance model.

The natural environment is no longer so natural, however. The light level, temperature, and sounds in many modern buildings are predominantly human made. In addition, the contamination and pollution of both the natural and work environments place new demands on people's ability to resist toxic and irritating substances.

The context in which people work can enhance or degrade their performance. This does not necessarily mean that the environment always controls performance in totally predictable ways. Only extreme environments can do this.

Whether entirely human made or natural, the environment always seems to exert some influence over system users and at times may become one of the most crucial factors affecting human performance.

Example of an Altered Environment

Designers should be aware of how environment may affect task performance and have sufficient understanding to design, prescribe, and maintain environments that reflect human needs and foster acceptable levels of human performance. Consider the following example.

Reliable Knitwear, Inc., is a medium-sized knitting mill that makes sweaters and sportswear. The company employs over 200 machinists, sewers, steam pressers, and maintenance personnel. Productivity is measured quantitatively (number of items produced) and qualitatively (percentage of seconds and rejects). In reviewing the books, the company's president saw that company profits were declining. She diagnosed the problem as resulting from excessive overhead costs. It seemed to her that it was costing too much to operate and maintain the plant. She announced to her employees that changes had to be made to cut down on overhead.

Her policy was instituted. Cooling and heating units in the steam press room were put on half-power. Lighting was reduced in storage areas, as well as in the knitting room itself. One of the two sewing rooms was closed and the workers all placed in the remaining room, making the room crowded and noisy. As a final economy measure, two of four maintenance people were dismissed.

Unfortunately, the company did not begin to save money. To start with, the overhead costs did not decrease nearly as much as she had estimated. Individuals turned lights back on and reset the controls on the heating and cooling equipment. In addition, production went down and the percentage of rejects and seconds increased. The workers began complaining; more began calling in sick. The production manager could not pinpoint the problem and suggested that the company president call in a consultant.

After an investigation, the consultant's report was received. The workers were not producing at their old rate because they felt abused by the company's policy of reducing the overall comfort level of the work environment. They felt that the company did not care about them, only about profits.

But a negative attitude was not wholly responsible for low productivity. The poor lighting and the extreme temperatures in the press room seemed to be causing workers to make more errors and become fatigued sooner. The crowded and ill-maintained conditions in the sewing room caused further discomfort and also impeded the work process.

The president of this company had treated her work force as if they were machines indifferent to the environmental conditions of the workplace. Her actions could make only a small difference to the overhead costs, while bringing about a major problem resulting in a considerable loss of revenue for the company and the loss of the goodwill of many employees.

As this example suggests, the working environment is very important to human performance. There is a mistaken idea that designing acceptable environments consists of wrapping the worker in luxury and providing a soft reclining chair. Ensuring acceptable system environments is not a matter of pandering to the workers. Rather, it is a way of providing a work situation or context in which an acceptable level of human performance is most likely to occur.

Goals of Environmental Design

Performance Support

Just as an astronaut is provided a life-support system in order to live during space flight, so we can think of the context of workspace in which an activity is performed as a *performance support system.* The designer requires and expects a certain level of performance from users as they undertake and complete various tasks. The environment in which the tasks are performed has to support the required performance level, which takes into account meeting the basic physical, physiological, and psychological needs of users. For example, a driver should fit comfortably into his or her car and be within easy reach of the controls (physical needs). The automobile suspension should reduce vibration as much as possible, and the exhaust system must be functioning properly to avoid carbon monoxide poisoning (physiological needs). If the trip is a long one, a radio may alleviate boredom (psychological needs).

At the very least, environments should do no harm. That is, the environment should not drain off energies or be a point of dissatisfaction in itself. The environment should provide for the user's basic needs, such as adequate light, air temperature, workspace, and safety. But more than that, the environment should *support* the performance of work so that work is easily organized, resources are readily accessible, and communication is facilitated.

Considering the environment as a performance support system also includes aspects of the social environment. When designing an environment for a research group, one would wish, in general, to provide each person with his or her individual space, at the same time encouraging interaction, group problem solving, sharing of ideas, and a sense of group identity.

Organizational Support

Not only should the environment support user performance, but it should also transmit such messages to users as the nature of the organization, its structure, its values, and the nature of the work. People are very good at picking up these cues and tend to behave and react accordingly.

The issue is not that one organizational structure is better than another or that one environmental design is better than another. The issue is rather that the environmental design portrays or reflects the organizational structure and, at least indirectly, the kinds of performance behaviors that are acceptable in that particular environment or context. Designers should recognize that the environmental recommendations that they make will help create an organizational image. Careful consideration of design issues associated with the environment is needed to ensure that the image is the one desired and that the performance elicited by the environmental cues is acceptable and desirable.

A Normal Environment

A third consideration when designing a work environment is that all critical conditions should fall within a normal range. People are quite tolerant of environmental diversity, but not infinitely so. For any condition in the environment, there is a normal range wherein people perform best. In environmental conditions outside that normal range, human performance may be degraded. The more abnormal the environment is, the greater the probability of having degraded performance.

Guidelines have been developed that include information concerning workspace size, noise, lighting, and temperature. Special applications may require designers to consider additional aspects of the environment, such as vibration, acceleration, toxic substances, electrical shock hazard, or radiation hazard. System designers should conform to available guidelines wherever possible. It is particularly important that designers recognize the environmental conditions outside the normal range and take appropriate steps either to change the environment or minimize its effects on users.

Human–Environment Interaction

Someone once observed that people who chop their own wood get warmed twice. The human and the environment interact in some very interesting ways. Consider working in a cold environment (cf. Fox, 1967). The body responds in a variety of ways to conserve heat (e.g., withdrawal of blood from the periphery) and to generate heat (e.g., shivering). The person may also do exercises (such as slapping his or her arms), put on warmer clothing, limit exposure time to the cold conditions, or turn up the thermostat. People and their environments frequently interact in a chain of events. The environment affects people; people then make changes either to lessen the effects of the environment or to alter the environment. A designer should provide the user with as much control over a work environment as possible.

In the real world, the effects noted may be more or less pronounced, depending on the nature of the work activity and the particular people involved. Delicate, fine work performed by sedentary, highly skilled people is more affected by adverse environmental conditions than less skilled, muscular work. In addition, highly motivated people usually tolerate inferior environmental conditions better than those who are less motivated.

These variations emphasize the interactive nature of the situation that we are studying and are in keeping with the model presented in Chapter 1. Human performance in any situation is a three-way interaction among the human, the work activity, and the environment. For example, studies show that heat affects Morse code reception (the more difficult task) more than Morse code transmission. Yet highly skilled telegraphers have been found to maintain their performance even in heat conditions that seriously diminished the performance of less skilled persons.

We must not take too lightly the potential effects of environmental conditions on human performance. The role of the designer is to understand the ranges of human adaptation to various environments and, most importantly, to design the environment to be a positive factor in promoting an acceptable level of human performance.

Adaptation and the Optimal Environment

Poulton (1972) compiled a list of environmental conditions and the resulting effects on performance, which is shown in Table 17-1. Designers should take care that the characteristics of their work environments fall within the ranges shown. An inappropriate amount of heat, light, noise, or vibration, for example, could result in degraded human performance.

Designers should not be misled by the apparent adaptation capabilities of people. The concept of adaptation refers to changes that take place in an individual to enable him or her to deal adequately with an adverse environmental condition. Adaptation may occur when people are exposed to a totally new environment or an existing environment that has changed considerably. There are limits to adaptation imposed by physical, physiological, or psychological characteristics. When a person reaches the limits of adaptation, he or she will experience stress, which in turn may result in degraded performance. Designers should consider the limits of human adaptation and design realistic environmental conditions that do not exceed these limits. In so doing, designers will be creating performance supportive environments.

Environmental Components

A designer should think of the environment as having both a physical component and a social component. The *physical* environment includes such factors as noise, lighting, temperature, vibration, air pollution, and radiation. The social environment includes such considerations as social facilitation, conformity, privacy, personal space, territoriality, and crowding. Designers usually have more control over the physical environment than the social environment. This by no means suggests, however, that a designer should not attempt to make design decisions to enhance the social environment.

Physical Environment

Human systems are vulnerable to extreme conditions in the physical environment. In aviation, many accidents have been related to such things as poor visibility, wind, or thunderstorms. Automobile accidents frequently are associated with sun glare or icy roads. These conditions, however, seldom directly cause the accidents. Rather, an environment with difficult and hazardous conditions may severely stress the person who is performing a particular activity. The environment can alter a person's ability to perform. Excessive cold or heat over long periods of time cause physiological changes (e.g., frostbite and heat exhaustion) that not only lead to seriously impaired performance, but also may result in injury or even death.

Table 17-1 Environments Just Severe Enough to Produce a Reliable Deterioration in Performance

Environmental Condition	Condition Just Severe Enough to Degrade Performance	Task	Kind of Deterioration
Heat	Air Temperature 27°C (80°F)	Making knitted clothes	Reduced output
		Reading	Reduced speed and comprehension
		Making munitions	Increased accidents
Cold	Air temperature 13°C (55°F)	Tracking	Reduced time on target
		Making munitions	Increased accidents
Dim light	7 to 10 foot-candles	Reading 7-point newspaper type	Reduced speed
		Reading 6-point italic type	Reduced speed
Glare	Depends on the angle of the glare source and of the line of sight, as well as on its brightness	Inspecting cartridge cases	Reduced speed
		Typesetting by hand	Reduced speed
Noise (continuous)	100 decibels (dB)	Tapping 1 of 5 targets in response to 1 of 5 lights	Increased errors
		Threading photographic film through a machine	Increased broken rolls and shutdowns
Noise (intermittent)	95 dB for 1 second	Reporting differences between pairs of cards	More omitted reports
	Noise varying randomly between 90 and 65 dB	Recording digits	Increased variability in rate of work
Noise (interference with speech communication)	70 dB (600 to 4800 Hz)	Speaking at 3 feet and telephoning	Reduced comprehension

Adapted from *Environment and Human Efficiency,* by E. C. Polton, 1972. Reprinted by permission of Charles C Thomas, Publisher, Springfield, Illinois.

Controlling the environment to make it more compatible with human physical, physiological, and psychological limits has become commonplace. A person leaves an air-conditioned home to drive an air-conditioned car to an air-conditioned restaurant for dinner. Except for brief periods, the temperature is maintained at a comfortable level. Divers explore and do useful work in the oceans as a consequence of the development of underwater breathing gear. The use of fast, high-altitude aircraft in commercial aviation is possible only because the passenger cabins are maintained at a low-altitude pressure. These are some of the many ways designers have controlled people's environment to enable adequate performance.

Noise

Noise may be defined as unwanted sound. This is not an entirely satisfactory definition because a sound wanted by one person may be unwanted by another person. Think, for instance, of a noisy party next door, which is not considered noisy to the party goers. Thus, some sounds will be "noisy" because they are unwanted, intrusive, irritating, or distracting. The more control a user has to start or stop a noise, the less the annoyance value of the noise (Kryter, 1970; Glass and Singer, 1972).

People constantly experience a wide variety of noises. These noises range from the gentle hum of a fluorescent light to the deafening roar of a jet at takeoff. Due to human adaptation, we tend to grow accustomed to most noises and pay little attention. Even so, the single most often voiced complaint in work environments is that there is too much noise and that it is distracting. The individuals most likely to be affected are those engaged in complex tasks requiring close, undivided attention and concentration (Glass and Singer, 1972; Finkelman, 1975).

Sound has at least two characteristics related to human performance: frequency and loudness. There is no single frequency at which noises produce harmful effects; however, low-frequency noises, when sufficiently loud, can produce the most serious hearing impairments. On the other hand, high-frequency noises (over 1200 Hz) usually interfere most with work and are considered the most annoying. Noises in the intermediate frequencies tend to interfere most with the intelligibility of speech.

Loudness is directly related to the mechanical pressure it exerts on the ear. Table 17-2 shows intensity levels of various common sounds and their effects on people. In general, a loud noise (90 dB or more) maintained over at least 30 minutes will cause degraded performance. Intermittent and unpredictable noises are worse than steady sounds. When unpredictable, even low-intensity noise can be annoying.

The total absence of noise may also impair performance. Noise tends to have an arousal value with many people. Particularly for dull or uneventful tasks, it is important to have some sounds in the environment to help to keep people alert (Warner, 1969; Eschenbrenner, 1971; Poulton, 1972).

The aim of a designer should be to have a noise level of about 50 dB in the environment. This is equivalent to the hum of voices and quiet machinery. This noise level is also information bearing in that it keeps the hearer apprised of what is going on in the environment, but it is not loud enough to be distracting.

Lighting

Although the sun, and sometimes the moon, provide adequate lighting outdoors, people require artificial light to perform activities inside. Lighting requirements range

Table 17-2 Intensity Levels and Effects of Various Common Noises

Common Sounds	Noise Level (dB)	Effect
Air-raid siren	140	Painfully loud (blurring vision, nausea, dizziness)
Jet takeoff (200 feet)	130	Begin to "feel" the sound
		Thunderclap
Loud disco	120	Hearing becomes uncomfortable
		Auto horn (3 feet)
Pile drivers	110	Cannot speak over the sound
Garbage truck	100	
Heavy truck (150 feet)	90	Very annoying
		City traffic
Alarm clock (2 feet)	80	Annoying
		Hair dryer
Noisy restaurant	70	Telephone use difficult
		Freeway traffic
		Man's voice (3 feet)
Air-conditioning unit (20 feet)	60	Intrusive
Light auto traffic (100 feet)	50	Quiet
Living room	40	
Bedroom		
Quiet office		
Library	30	Very quiet
Soft whisper (15 feet)		
Broadcasting studio	20	
	10	Just audible

from the low, softly illuminated environment of a commercial airline cockpit to the bright surroundings of a hospital operating room. The lighting in both these situations has been designed to elicit acceptable levels of human performance.

The optimization of artificial illumination has long been a concern to researchers and designers. In particular, a great deal of attention has been given to work situations where there is too little light, low brightness contrast, or glare. The Hawthorne studies began in 1929 to determine the effects of illumination on performance. A short time later, Wyatt and Langdon (1932) and Weston (1949) found a deterioration of performance (reduced speed) in inspection and typewriting tasks when subjects performed these tasks under lighting conditions that were too bright. Tinker (1943, 1963) has reported decreases in human reading speed in dimly lighted environments.

Individual workers require different levels of lighting. A common mistake with both noise and illumination is to assume that all people are the same. Thus, we erroneously provide an average level of illumination to accommodate the average person. But the average person does not exist. Guth and Eastman (1955) suggested that, although it is perhaps idealistic to provide for individual lighting requirements in the workplace, it should be obvious that supposedly average illumination levels automatically provide insufficient lighting for a certain percentage of the work force. Designers should establish human performance requirements for each activity in their particular systems and then provide sufficient illumination levels to ensure the fulfillment of these requirements for *all* users.

Blackwell (1959, 1964, 1972) studied a variety of factors related to poor visibility and his work has served as the basis for specifying illumination levels for various work activities. In general, research indicates that insufficient illumination, glare, reflectance, shadows, low brightness contrast, and flickering adversely affect human performance and should be controlled by system designers. Specific design recommendations are summarized in Tables 17-3 and 17-4.

Temperature

Another condition of the physical work environment that system designers should be aware of is temperature. Extreme outdoor temperatures, while largely beyond the control of the system designer, can be compensated for by cooling equipment to control heat and proper clothing to counteract cold. Indoor temperatures, on the other hand, may be under the control of designers and should be modified to support an acceptable level of human performance. Like many other conditions in the physical environment, extreme temperatures may degrade the performance of work activities.

The preferred ambient temperatures (average and range) for a group of seated young adults is given next in degrees Fahrenheit (Grivel and Candas, 1991):

Men	Morning: 78 (72–87)
	Afternoon: 82 (73–90)
Women	Morning: 80 (68–85)
	Afternoon: 82 (75–85)

Most people readily adapt to moderate fluctuations in environmental temperatures but not to extremes. The designer should recognize that there are *limits* to this adaptation process and design the system environment so that these limits are not exceeded.

Heat and Performance

The effect of heat on human performance has a considerable research history (Baron and Bell, 1976; Bell et al., 1964; Griffiths and Boyce, 1971; Provins and Bell, 1970). Bell et al. (1978), in a review of laboratory studies dealing with the effects of heat on performance, report that research has found (1) detrimental effects, (2) improvements in performance, and (3) no effects at all. They suggest that these results are best interpreted in the light of the arousal theory discussed in Chapter 9. For example, effective temperatures

Table 17-3 Recommended Illumination Levels (Footcandles)

Assembly	
Rough, easy seeing	30
Rough, difficult seeing	50
Medium	100
Fine	500
Extra fine	1000
Auditoriums	
Assembly only	15
Exhibitions	30
Social activities	5
Banks	
Lobby	
General	50
Writing areas	70
Teller's stations	150
Posting and keying	150
Conference rooms	
Critical seeing tasks	100
Conferring	30
Note-taking during projection	30
Hospitals	
Anesthetizing and preparation room	30
Lobby (or entrance foyer)	
During day	50
During night	20
Medical records room	100
Nurses station	
General, day	70
General, night	30
Patients' rooms	
General	20
Reading	30
Observation (by nurse)	2
Surgical suite	
Operating room, general	200
Operating table	2500
Inspection	
Ordinary	50
Difficult	100
Highly difficult	200
Very difficult	500
Most difficult	1000
Library	
Reading areas	
Reading printed material	30
Study and note taking	70
Card files	100
Circulation desks	70
Offices	
Reading handwriting in pencil, active filing, mail sorting	100
Reading handwriting in ink, intermittent filing	70
Reading high-contrast or well-printed materials	30
Conferring and interviewing	30
Corridors	20
Residences	
Specific visual tasks	
Dining	15
Grooming, shaving, makeup	50
Food preparation	50
Ironing	50
Sewing	
Dark fabrics	200
Medium fabrics	100
Light fabrics	50
General lighting	
Conversation, relaxation	10
Visual tasks	30

(effective temperature is an index of perceived comfort level influenced by temperature and humidity) of about 90°F (32°C) increase arousal and tend to help the performance of simple tasks and degrade the performance of complex tasks.

Much research dealing with the effects of heat on performance has been conducted in actual job settings. Tichauer (1962), for example, studied the relationship between temperature and the performance of cotton pickers. He found that workers produce optimally between 75° and 80°F, whereas workers slowed down when the temperature was above

Table 17-4 Recommendations for Glare Reduction

Avoid intense light sources within 60° of any commonly used line of sight
Use shields or hoods where appropriate
Use indirect lighting
Use several lower-intensity lights instead of one very bright one
Use dull, matte surfaces instead of polished or glossy

From Van Cott and Kinkade, 1972.

90°F. Wyon (1970) found reduced speed and comprehension in readers at temperatures greater than 80°F (27°C). Research with mineworkers (Wyndham, 1969) and students (Pepler, 1972) has reliably shown performance deterioration with increasing temperatures.

It seems that the actual level of heat necessary to degrade performance varies from situation to situation and from individual to individual. What is important from a system designer's point of view, however, is the fact that excessive heat does affect human performance. Crockford (1967) and Hill (1967) have discussed the problem of excessive industrial heat in factories (near furnaces or boiler rooms) and have suggested that workers be provided with an adequate intake of water and salt, have frequent rest breaks, and, when new to the job, have adequate time to adapt to working conditions. Other possible ways to help alleviate the effects of heat include the use of increased ventilation, fans, and air conditioners. See Figure 17-2 for recommended exposure times to heat.

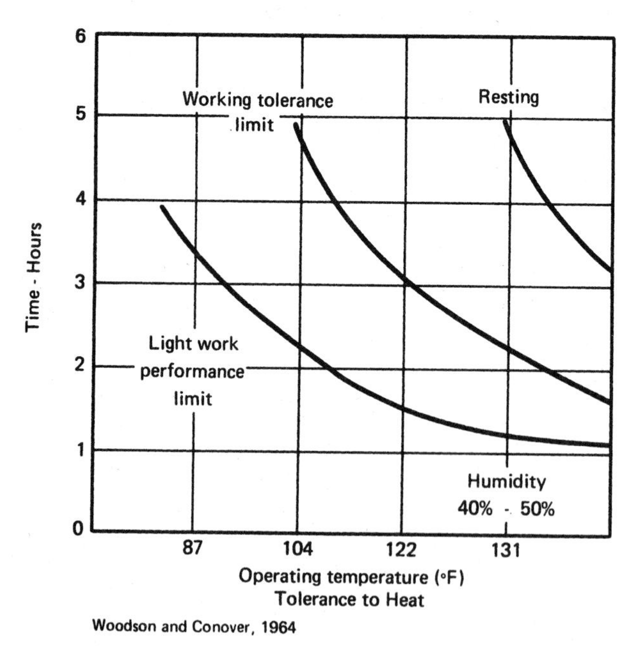

Figure 17-2 Recommended exposure time to heat as a function of activity and temperature.

Designers should also consider the comfort levels of workers. Comfort levels are usually measured in the effective temperature units discussed above. See Figure 17-3 for a graphical summary of comfort levels. It should be pointed out that comfort levels for feeling cool are generally below those at which performance deteriorates. However, a sustained experience of discomfort can lead to distraction, which may produce performance decrements.

Cold and Performance

It is relatively rare to find indoor environments sufficiently cold to seriously impair performance. However, some machines and storage areas require an ambient temperature below 65°. This creates a chilly environment for sedentary workers. If activities require the use of cold rooms for storage, people need to be clothed in ways similar to those specified for working in cold outdoor conditions. Energy conservation policies will most likely cause marginally cold conditions for many workers during winter months. It is important to remember that people vary widely in their ability to adapt to moderately low temperatures. For example, older people and those with poor vascular circulation could feel much colder than others.

Long exposure to cold temperatures as high as 55°F may cause hands and feet to become chilled and lose some of their strength, touch sensitivity, and ability to make fine manipulations (Fox, 1967). These losses increase with lower temperatures and longer exposures. The individual becomes progressively more awkward. In time, psychological abilities are affected; it becomes difficult to concentrate and memory lapses occur.

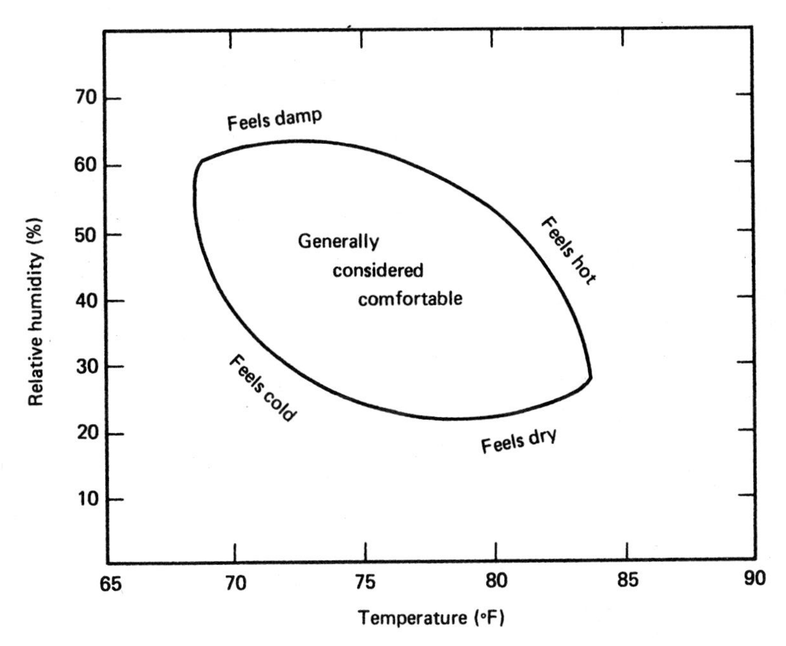

Figure 17-3 Comfort zone as a function of relative humidity versus temperature.

To guard against possible degradation of human performance in extremely cold temperatures, designers should make a variety of preventive decisions. First, they must consider the amount of protective clothing that is necessary when workers perform in environments of different severity. Second, selection procedures should be used to screen out individuals who cannot tolerate extreme cold. A third consideration would be to ensure that workers are given sufficient time to adapt to the environment. Finally, design decisions should be made that help to guarantee that reasonable work and rest schedules are implemented.

Designers should also take steps to develop machine controls such as handles, push buttons, levers, or switches with oversized dimensions to make them easier to operate with gloved hands. In addition, they should consider entry and exit problems for people who may be slower and more awkward in their heavier clothing.

Other Conditions of the Physical Environment

Aside from the major characteristics of the physical environment just discussed (noise, lighting, and temperature), designers should be aware of numerous other conditions that may affect human performance in certain systems. *Vibration* when using heavy machinery, motor vehicles, and other equipment has been linked to performance decrements in tracking tasks (Lovesey, 1970), reading (Dennis, 1965), and tasks requiring considerable eye-hand coordination (Guignard, 1965). The effects of vibration can be reduced, for example, by selecting machines that vibrate less, incorporating spring suspensions where possible, and increasing letter and number size in displays for easier readability.

Anyone who has ever attempted to perform an outdoor activity on a windy day can attest to the fact that *wind* can be detrimental to performance. Poulton and others (1975) exposed a group of people to varying degrees of turbulence and found, as expected, that high wind (20 mph) and gustiness produced decreases in performance over a variety of tasks.

Wind has the greatest potential effect on performance when it occurs together with cold temperatures. The *wind chill index* (see Table 17-5) is important to designers who must calculate maximum exposure times and clothing requirements for system users performing work activities in cold environments. For example, telephone line workers making emergency repairs in –4°F temperatures with a steady wind of only 15 mph are

Table 17-5 Wind Chill Index

		Wind Speed			
		5 mph	**15 mph**	**25 mph**	**35 mph**
	32	29	13	3	–1
Actual	23	20	–1	–10	–15
Temperature	14	10	–13	–24	–29
(°F)	5	1	–25	–38	–43
	–4	–9	–37	–50	–52

actually being subjected to a temperature that feels like –37°F. This type of environment will require shorter work periods, more frequent warm-up sessions, and better insulated clothing.

Designers should be aware of the effects of *air pollution* (particularly carbon monoxide) on workers. Breisacher (1971), in a review of studies dealing with air pollutants, reports that exposure to relatively small amounts of these pollutants adversely affects human reaction time, attention, and manual dexterity. One study exposed subjects to a level of carbon monoxide similar to what one would find on a moderately traveled freeway at rush hour. After 90 minutes of exposure, subjects evidenced a significant impairment on a time-judgment task. With heavier concentrations of carbon monoxide, less exposure time was required to produce performance decrements. Workers such as automotive mechanics, heavy outdoor machinery operators, and others exposed to even moderate levels of exhaust fumes or cigarette smoke (cigarette smoke contains relatively large quantities of carbon monoxide) are subject to the adverse effects mentioned over a period of time. Designers should provide proper ventilation. Reducing or even eliminating the pollutant at its source by selecting vehicles and machinery that emit less carbon monoxide is an even better solution.

Designers of some systems should be aware of the performance correlates of prolonged exposure to other environmental conditions, for example, weightlessness and acceleration in space capsules and radiation in nuclear power plants. For a discussion of these topics, readers may refer to Poulton (1972).

Combined Environmental Stressors

When two or more adverse environmental conditions exist at the same time, the effect on human performance is not easy to predict. The effect may be greater than the sum of the environmental conditions taken singularly, less than each individual condition, or equal to their sum. For example, Harris and Shoenberger (1970) found that the combination of excessive noise and vibration produced an additive decrement in a compensatory tracking task. Again, many of these effects are best understood by considering the arousal concept discussed in Chapter 9. A discussion of research on combined environmental conditions can be found in Poulton (1972).

Environmental Preferences and Esthetics

Although little research has looked directly at the relationship between human performance and user preference for a given work environment, the issue of environmental preference itself should be important to system designers. If the work area or environment is not pleasing to users, it may lead to job dissatisfaction and possibly degraded performance. Hertzberg and his fellow researchers reported (1959) that a pleasing work area would not necessarily lead to greater satisfaction, but one that is nonpleasing would lead to dissatisfaction. S. Kaplan (1974), R. Kaplan (1974), and Wohlwill (1974, 1976) have investigated the concept of environmental preference. These researchers suggest that, although people generally prefer environments that are complex, spacious, identifiable, coherent, and of smooth texture, individuals may have different levels of preference depending on past experience.

The issue of environmental colors and environmental esthetics has also been of concern to researchers. Ertel (1973), for example, has shown that when school children were tested in rooms that they thought were "beautiful" (light blue, yellow, yellow-green, orange) their IQ scores were significantly higher than when tested in rooms they considered "ugly" (white). Follow-up studies showed similar findings.

Social Environment

Presence of Others

The user's social environment could include one or more co-workers, management people, customers, and possibly even spectators. The total effect of these people on a user often affects human performance. As an example, consider one of the first research studies on the effects of the social environment, conducted in the late 1800s by Triplett. While examining the official records of bicycle races, Triplett (1897) noted that a rider's maximum speed was approximately 20 percent greater than when he was paced by a visible multicycle. Desiring to learn more about the matter, he set up an experiment with children in the age range of 10 to 12, giving them the task of winding fishing reels. Alternating situations *alone* and *together,* he found that when working together 20 of his 40 subjects *exceeded* their own solitary record, while 10 did less work and 10 were essentially unaffected. All in all, he concluded that the group situation must be thought of as producing greater achievement. He felt that the children's competitive spirit, together with the facilitative effects of others watching the competition, caused the increase in performance.

Since Triplett's early studies, hundreds of other studies have been conducted on the relationship between human performance and the social environment. The general conclusion of investigators has been that the presence of others enhances performance of well-learned skills, but inhibits the acquisition of new skills, such as memorizing a list of nonsense syllables. It appears that the presence of others increases arousal in such a way that the dominant (or most probable) response becomes even more dominant.

The effect of others is more dramatic when *pressure* is placed on the individual to conform. People tend to feel uncomfortable when they are alone in their actions or out of step with the crowd. Two studies of interest in this regard are those conducted by psychologists Asch (1951) and Milgram (1965). Asch showed that the judgment of a subject was influenced by several other people who had first purposely made incorrect responses. That is, there was a tendency for the person to respond like other members of the group even when he or she thought that the response was wrong. Many people follow directions from their supervisors and co-workers in much the same way.

Milgram, in another study of conformity, instructed subjects to administer what they were led to believe would be a painful electrical shock to another person. They did this because they were expected (by the experimenter) to do so, even though it appeared to result in extreme pain to other people.

The implications for designers from these and other similar studies are relatively straightforward. Most people are strongly influenced by social pressure, competitive forces, and even the mere presence or absence of others in the workplace. Particularly in situations where important judgmental or evaluative decisions need to be made, designers should develop a system in such a way that decisions are made as independently as possible. For example, if an engineer needs to decide how much sheet metal to purchase for a given work project, he or she should make an estimate alone and then individually consult with other engineers. Only after a decision is made should it be passed to a supervisor or others for approval. Such a procedure is preferable to a situation in which an engineer and two or three levels of management provide their views prior to the making of an initial decision. In the latter case, the engineer may be influenced against the *best* decision because some advisors (including members of management) gave their opinion first.

Personal Space and Privacy

People have very definite feelings about their control over privacy, being isolated or crowded, having a certain amount of personal space, and having a certain territory that is their own. For example, a person seated on a train or bus when there are many empty seats is likely to feel uncomfortable if another passenger chooses to sit next to him or her rather than in a vacant seat. Similar feelings are held by people in their workplace. Designers can improve human performance by understanding this aspect of the social environment.

Privacy

Altman (1975) has suggested that people are constantly struggling to obtain a certain desired level of privacy. People attempt to achieve a desired privacy level through the regulation of personal space boundaries and through claiming territory as their own. There are three possible outcomes when a person attempts to regulate privacy:

1. The achieved privacy is less than the desired level, in which case they experience crowding.
2. The achieved privacy is more than the desired privacy level, in which case they experience isolation.
3. The desired level of privacy is achieved.

Opponents of the open office or landscaped office concept, for example, point out that lack of privacy is a major factor behind employee discontent with such layouts. Restaurants, public rest rooms, and private homes all are designed in part with a consideration for privacy. Designers need to consider privacy when making decisions for their systems if they want to ensure an acceptable level of human performance.

Personal Space

The term personal space was first coined by Katz in 1937. The term has come to refer to an area with an invisible boundary surrounding a person's body. Encroachment on an individual's personal space may result in feelings of displeasure, anger, or withdrawal.

Hall (1966) investigated human spacing behavior in several different cultures and developed a system of studying how people use space as a communications vehicle. Hall discussed four spatial distances or zones that are used in social interaction. The first zone, from 0 to 18 inches, is called the intimate zone and generally is used only for communication between intimate friends. The next personal zone extends from 18 inches to 4 feet and is used for contacts between close friends and everyday acquaintances. The third social zone, from 4 to 12 feet in distance, usually is used for impersonal and businesslike contacts, while the public zone extends from 12 feet out and is appropriate for contacts between politicians, actors, lecturers, and the general public.

Research on personal space indicates that the amount of personal space one maintains with another depends on the individual and the activity being performed. In general, people find it unpleasant when they are forced to interact with others at inappropriate distances (either too close or too far) or when they must intrude on the personal space of others. Unpleasant effects tend to be heightened when the other person is either a stranger or disliked or when the encroachment takes place in a formal setting.

System designers should be concerned with minimizing personal and interpersonal problems that may lead to human performance degradation. They also should be aware of the consequences of inappropriate personal spacing behavior. Although designers cannot interfere with users' preferences as to how close or how far away they choose to interact with others, certain system design features—the size of a conference room or the distance between desks—should not inhibit an individual user from selecting the proper personal space for achieving the desired performance. Users should have the flexibility of choosing the level of personal space that they wish to maintain about them; designers should make decisions to accommodate such flexibility.

Crowding and Density

Crowding is closely related to personal space and will be distinguished from density in accordance with a definition provided by Stokols (1972). Stokols defined density as a strictly physical variable—the number of people per unit of space. Crowding, on the other hand, is a subjective feeling of too little space. Density is a necessary though not sufficient condition for the feeling of being crowded. It is entirely possible, for example, for someone to be among many co-worker friends and not feel crowded at all.

Crowded working conditions are sometimes unavoidable. From a designer's point of view, one of the most critical questions that can be asked about crowding is whether the crowding affects human performance. Performance effects of crowding are dependent on the task in question. More specifically, simple tasks do not seem to be affected by crowding (Bergman, 1971; Freedman, 1975), while more complex tasks may be adversely affected (Dooley; 1974; McClelland, 1974).

Territoriality

The concepts of personal space and crowding help to predict how closely individuals like to interact in the environment. The concept of *territoriality* helps to predict with whom the interaction will take place. Territorial behavior usually is defined as a personalization of or establishing ownership of a place by a person. The designer should keep in mind that defensive responses may occur when territorial boundaries are violated.

Human territorial behavior, at least in part, is a response or reaction to some need that people have to identify a place that they can call their own. Whether we are talking about a small child reminding a friend about whose sandbox they are playing in or a computer operator who is quite particular about who steps into the computer room, the basic idea is still the same. One person lays claim to a particular location that is off limits to unwanted others.

Studies suggest that the use of personal markers and objects, such as nameplates, is effective in defining and maintaining territories. Territory holders seem to be more influential in their own territory as opposed to elsewhere. Based on these findings, it would seem to make good design sense to allow system users maximum flexibility in personalizing their given work area and to provide boundaries or barriers to support individual territorial needs.

Isolation

While it is important to take into account the effect of crowding on human and system performance, the effects of isolation should also be considered. Freedman (1975) reviewed several isolation studies and concluded that most people perform adequately under isolated conditions. Freedman concluded that, although most people can apparently endure seemingly unpleasant conditions of isolation for long periods of time with relatively few negative performance effects, they find little satisfaction in being so isolated.

Workspace

Space allocation is particularly important to system designers in large, growing companies. Although, ideally, the space allocation process is done on an objective basis, the end result frequently is that some workers end up with too little space and others with much more than they need. Often space allocation problems arise because the process does not take place on a functional basis (work requirements), but rather on the basis of seniority or status with the company. With increasing seniority and status in the company, workers are afforded larger offices, just as they are afforded more pay and more vacation. This may start to degrade performance when employees are clustered so close together that they do not have adequate workspace to perform their work effectively.

Designers should assume the responsibility of specifying the minimum amount of workspace required for a user to adequately perform. For example, designers planning to place several individuals together in one area should provide at least 70 square feet per individual. Some well-designed systems do not maintain an adequate level of human performance simply because the designer did not specify the necessary amount of workspace.

Exercise 17: Evaluating the Home-court Advantage

Purpose: To demonstrate the potential effects of the social environment on human performance.

Method: It has been observed that in certain team sports there seems to be a home-court advantage. As used here, the home court advantage refers to a slight advantage a *home* team seems to have over a *visiting* team. This advantage is frequently exemplified by teams winning more games when playing at *home* than when playing *away.*

Evaluate the following information from the United States professional basketball season that ended in 1995. Can you identify any evidence supporting the home court advantage?

	Overall		Home		Away	
Team	**Won**	**Lost**	**Won**	**Lost**	**Won**	**Lost**
San Antonio	62	20	33	8	29	12
Utah	60	22	33	8	27	14
Phoenix	59	23	32	9	27	14
Seattle	57	25	32	9	25	16
Orlando	57	25	39	2	18	23
New York	55	27	29	12	26	15
Indiana	52	30	33	8	19	22
Charlotte	50	32	29	12	21	20
LA Lakers	48	34	29	12	19	22
Chicago	47	35	28	13	19	22
Houston	47	35	25	16	22	19
Portland	44	38	26	15	18	23
Cleveland	43	39	26	15	17	24
Atlanta	42	40	24	17	18	23
Denver	41	41	23	18	18	23
Sacramento	39	43	27	14	12	29
Dallas	36	46	19	22	17	24
Boston	35	47	20	21	15	26
Milwaukee	34	48	22	19	12	29
Miami	32	50	22	19	10	31
New Jersey	30	52	20	21	10	31
Detroit	28	54	22	19	6	35
Golden State	26	56	15	26	11	30
Philadelphia	24	58	14	27	10	31
Washington	21	61	13	28	8	33
Minnesota	21	61	13	28	8	33
LA Clippers	17	65	13	28	4	37

One explanation for having an advantage at home is that the local crowd tends to provide more and better support to the home team. They give extra encouragement in many ways, such as cheering, clapping, and yelling. In other words, the *social environment* seems to be more favorable and tends to elicit better human performance.

Reporting: Prepare a report. If you find evidence for the home-court advantage, make sure that you provide a discussion of why you think the social environment is more favorable. If you find no evidence of a home-court advantage, discuss why so many people seem to think that one exists.

18

Conducting Comparison Studies Using Statistics

Common Sense versus Fact Finding

Too often designers rely on commonsense guessing to solve human performance problems. Although useful at times, common sense should not be relied on solely. Consider that just 90 years ago it was only common sense that human flight was impossible. Common sense also seems to dictate that hearing in a very noisy environment is unlikely without the use of a microphone or earphones.

> Nearly everyone knows that boiler factories are very noisy and that workmen in boiler factories have difficulty hearing and understanding speech because of this noise. Does it seem sensible that you can improve the intelligibility of speech by putting earplugs in the workmen's ears? Research proves . . . that . . . workmen can hear better with earplugs! (adapted from *Research Techniques in Human Engineering,* by A. Chapanis. The Johns Hopkins Press. Copyright 1959, p. 7).

Product design has, for years, depended on common sense. This has resulted in products with many different shapes, sizes, and colors. A good example is the control burner linkages on stoves. The best design causes the fewest burns and least damage. Using only common sense, which configuration in Figure 18-1 is best?

Designers tend to make many decisions they feel are "obvious." Unfortunately, basing design decisions related to human performance on knowledge that seems obvious may result in degraded performance. Designers also rely on two other, usually unprofitable,

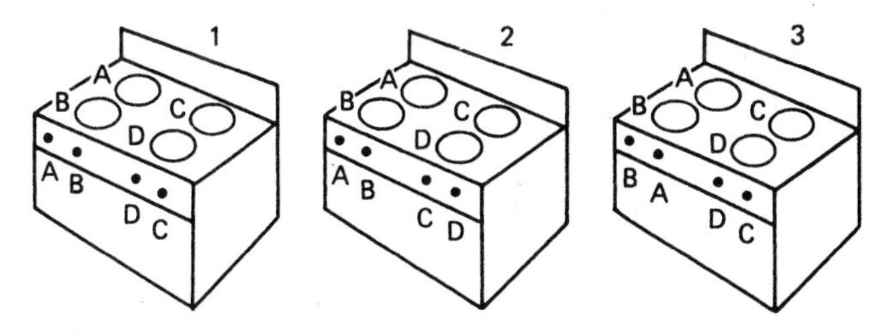

Figure 18-1 Control-burner linkage designs (Courtesy Johns Hopkins University Press).

approaches to decisions. First is the idea that something is true because someone says so. The "someone" could be a member of management or a well-known authority in the particular area. Many designers depend on authorities for much of their knowledge. However, an authority's judgment should only be accepted if the authority can also provide factual evidence. Remember, Aristotle was considered an irreproachable authority for hundreds of years after his death, even though undeniable evidence contradicted his views. Authorities are useful and certainly can aid the design process, but they should not be used as an easy substitute for conducting meaningful studies.

The third approach to design decisions is following habit or tradition. Continued and unwavering belief in something does not make it true. Taking the time to conduct a study to compare the traditional approach with a suggested new one could result in a better design. Incidentally, a study was done to see which of the stoves in Figure 18-1 was best. They found that configuration 1 resulted in the fewest errors (Chapanis, 1959).

Reliable information for critical decisions can come only from careful studies. Designers can conduct many of these studies without the help of other specialists. The techniques covered in this chapter will help designers use methods that are much better than commonsense guessing.

Study Design and Statistics

Measuring People

Human complexity makes the study of human performance difficult. When you measure a desk's height, you can assume that the measurement will not change. However, after measuring a single human dimension, such as speed of performance on a difficult task, you may find it *often* changes. The simplest human measurements are of length of limbs, weight, or height. Even these can vary with time. For example, height can vary up to 1 inch during the day. Also, the measuring devices and techniques themselves may give different readings from one measurement to the next.

Most human performance studies attempt to measure cognitive (brain) functioning in terms of either an attitude (e.g., user satisfaction) or some aspect of performance. This includes how long it takes to learn or to perform an activity or how accurately an activity can be performed. These are much more difficult measurements than height or weight because people do not perform consistently, and available measurement devices are imperfect.

Artificial Performance

A good study will have results that can be duplicated in a second or third study. A good study takes into account that the situations being evaluated are deliberately created and that the people performing in them may behave in unnatural (artificial) ways.

We can easily study the natural performance of microbes through a microscope or galaxies through a telescope, but the minute we turn to look at people performing, we find that they make adjustments to accommodate the person doing the evaluating. When human performance is studied, some elements of the situation tend to be somewhat artificial. Thus, a designer must work hard to make a study situation as *real world* as possible and recognize, also, that the same performance may be slightly different once a system is operational.

This problem was observed and reported as early as the Hawthorne studies conducted many years ago (Roethlisberger and Dickson, 1939). Although questioned since that time, the *Hawthorne effect* (the tendency of people to work harder when experiencing a sense of participation in something new and special) occurs frequently enough to make almost all human performance studies somewhat artificial.

Establishing Relations

Human performance studies look for the existence of relations: A certain performance aid is related to fewer errors, certain training materials are related to having shorter training time, certain activity characteristics relate to greater job satisfaction, or a certain activity configuration relates to shorter manual processing times. In all these relations we assume some type of cause-and-effect relationship. The performance aids cause greater accuracy, training materials cause shorter training time requirements, or activity configuration causes greater job satisfaction. Establishing that one is closely related to the other in some causal way is the essence of a good human performance study.

When things go together in a systematic way, a relation occurs that usually can be stated statistically. Most of the relations that interest us concern the effect that design decisions have on the four standards: errors, processing time, training time, or job satisfaction. Appropriately using statistics will help to determine the best way to optimize the critical variables. For example, a designer must decide which of two training approaches will result in a lower on-the-job error rate. A study measuring the errors elicited by the two different approaches will help the designer to choose the best approach. The time invested in such a study often results in a payoff of greatly improved human and system performance.

Average and Range

Probably the most common statistic used to describe a set of study results is the *average*. The average is determined by adding up all the scores and dividing the result by the number of scores. For example, the average of the scores 1, 2, 3, 4, 5 is 3 (1 + 2 + 3 + 4 + 5 = 15; and 15/5 = 3).

But to know the average of a group of scores only tells part of the story. Frequently, we are also interested in the scatter or spread of the separate scores around the average. This can be indicated for any group of scores by reporting the *range*.

Consider a situation where two groups are involved in a study to see if using a color CRT for a map-reading task results in fewer errors than using a monochrome CRT. The average number of errors for the color group is 23.6, and the average errors for the monochrome group is 23.9. As far as the averages are concerned, there is little meaningful difference in the performance of the two groups. However, the color group's errors ranged from 3 to 41 (3, 10, 15, 20, 24, 28, 33, 38, and 41), while the monochrome group's errors ranged only from 20 to 28 (20, 21, 21, 23, 24, 24, 27, 27, and 28).

This difference in range indicates that the color group is more *variable* than the monochrome group. This greater variability in one group could mean that the color CRT was, in fact, affecting the error performance of people differently than the monochrome CRT, but in a way that did not show up in the average number of errors. By taking a closer look at the individual error scores of people in the color group, we see that the color CRT tended to elicit the fewest errors from some subjects in the study, but it also elicited the most errors from other subjects. This could be important information when evaluating the two types of CRTs.

Consider another performance situation in which we have two basketball players who each average 22.5 points a game. Suppose that one is consistent (exhibits low variability) and almost always scores from 20 to 25 points a game. The second player, on the other hand, is erratic (exhibits high variability); this player scores over 30 points in some games and fewer than 10 in others. There are important performance differences between these two players that are best indicated by the range and not the averages.

Thus, a rough guide to the spread or variability of a set of study results (scores) can be given by indicating the range of scores (i.e., by noting the low and the high score). You can see why it is usually a good idea to report results giving both the *average* and the *range*. Sometimes the range is summarized in a statistic known as the *standard deviation.*

Establishing Controls

In most studies conducted to aid in making design decisions, one object or method is compared with other objects or methods. No matter how limited a designer's background in statistics, numerous good comparison studies can be conducted if a few rules are closely followed. The most crucial feature of a comparison study is the need to *control* conditions that might cause an ambiguous test.

A main question we pose is "How much of the observed performance can be attributed to differences in the objects or methods being compared (e.g., performance on two CRT displays) and how much of it can be attributed to other sources?" A well-developed study tries to minimize the effects of "other" sources by

- Ensuring that all subjects perform the same activity in the same context
- Conducting the study in a place with no unwanted distractions (unwanted distractions are those that one would not expect to occur in a real-world situation)
- Using a consistent (standardized) set of instructions
- Ensuring that subjects are as similar as possible to the larger group that they represent

Controlling *all* conditions is usually very difficult when using people as subjects. Consider a situation when we must decide which of two CRTs to use. Both appear acceptable and some people favor one, some the other. We decide that the most important performance measure is how quickly people can read the material on one CRT versus the other. We select a *person* to view the material on both CRTs and time the subject with a stopwatch. We find that the person can read the material on CRT A in 18 seconds and the same material on CRT B in 15 seconds. These results are shown in Figure 18-2. Based on these results, CRT B seems superior, but we feel that the study was too brief and extend it to include *two* people. The second subject reads the material on CRT A in 16 seconds and

TIME FOR ONE SUBJECT TO READ

CRT A: 18 seconds

CRT B: 15 seconds

Figure 18-2 Comparison of CRT performance using one subject.

CRT B in 19 seconds. For both subjects, the *average* time to read the material on CRT A is 17 seconds and the average time on CRT B is also 17 seconds. Figure 18-3 shows that the average difference between CRTs is 0, while the average difference between subjects is 3. Since the difference between subjects is substantially greater than any difference between CRTs, we have little confidence in using these results to select the better CRT.

	Subject 1	Subject 2	Difference Between Subjects	Average Time to Read CRT's
CRT A	18	16	2	17
CRT B	15	19	4	17
			2 ⌊ 6	

Average difference between subjects= 3 seconds

Average difference between CRT's= 0 seconds

Figure 18-3 Comparison of CRT performance using two subjects.

Assume that two other subjects are asked to participate in a study of two other CRTs and they provide the data shown in Figure 18-4. In this case, the average difference between the two subjects (2) is much less than the difference found between the CRTs (5). Thus we have more confidence that the difference found between the CRTs is *reliable.* The results of a study are said to be reliable if, in a second study using different subjects, we find very similar results. By measuring the hard-to-control difference between people and comparing it with the difference between objects or methods, we have an idea of how reliable our results are.

	Subject 3	Subject 4	Difference Between Subjects	Average Time to Read CRT's
CRT C	18	20	2	19
CRT D	15	13	2	14
			2 ⌊ 4	

Average difference between subjects= 2 seconds

Average difference between CRT's= 5 seconds

Figure 18-4 Comparison of CRT performance using two new subjects and two different CRTs.

Using Statistics to Determine the Reliability of a Study

When preparing a study, it is important to try to provide a means for determining whether or not the difference between subjects is *more* or *less* than the measured difference between objects or methods. Modern statistical techniques help in making this determination (cf. Hays, 1973). Four useful techniques for designers are the *t* method, *F* method, *b* method, and chi-square method. These statistical methods, which all provide a quantitative estimate of reliability, will be discussed later.

Avoiding Bias

A second consideration when preparing to conduct a study is to *avoid systematic bias*. Look at Figure 18-5 and determine what can be concluded from this study. The person who prepared this study made a mistake. In each case the subjects always read from CRT A first and CRT B second. There may have been a *transfer* effect; that is, the experience with CRT A may have affected performance on CRT B. We know that the subjects were always slower reading CRT A than CRT B, but we do not know if this is because CRT A was always read first or because CRT A is more difficult to read. In this case, a systematic bias was included in the study.

		Time to Read (seconds)
Test 1:	Subject 1 reads CRT A	19
Test 2:	Subject 2 reads CRT A	18
Test 3:	Subject 1 reads CRT B	16
Test 4:	Subject 2 reads CRT B	15

Figure 18-5 Comparison of CRT performance using four tests.

Reducing Transfer Effects

It is important to be aware of the possible existence of *transfer effects* in any study. People change in the course of a study. Experience gained while performing changes them, even if only slightly. Positive transfer occurs when the second trial benefits from the practice gained in the first. This could have occurred in Figure 18-5. Negative transfer occurs when people have an experience in the first example trial that interferes with performance in the second. Positive transfer occurs when people have an experience in the first trial that actually improves performance in the next trial. Two ways of controlling transfer effects are to counterbalance or randomize. In the example shown in Figure 18-5, the performance could be *counterbalanced* by having subject 1 perform on CRT A first and CRT B second and subject 2 perform on CRT B first and CRT A second.

Another approach, particularly useful when dealing with a larger number of subjects, is to *randomize* the order of presentation for all subjects. Some would see CRT A first, others CRT B first. Besides having a random order of presentation, each subject's assignment within a study group can also be randomized.

Consider the following example. An advertisement appeared in the company newspaper asking for 20 people to participate in a study. The person conducting the study decided in advance that the first 10 people to volunteer would be one group and the second

10 to volunteer would be put into the second group. Twelve people volunteered the first day of the ad, while the other 8 took up to 2 weeks to respond. Unfortunately, the study was related to motivation, and the motivation of the people who responded first was different from the motivation of those who responded later. By assigning the first 10 to one group and the second 10 to a second group, the person conducting the study systematically biased the results before the study had been conducted. In this case, after all 20 had volunteered, the subjects should have been randomly assigned to the two groups. Exactly how this can be accomplished will be discussed in the next section.

Determining Groups of Subjects

The groups in a comparison study are inevitably composed of people who differ in a variety of ways, such as manual dexterity or ability to make decisions. To reduce the chance of these differences between people affecting the final results, it is important to use participants that all have similar traits. This increases the chances that the results will be due to true differences in the objects or methods being compared (e.g., CRTs), as opposed to basic differences of people in the groups.

The groups in question might be composed of those who will perform on CRT A and those who will perform on CRT B. If a total of 10 subjects is to be used, 5 could be assigned to view CRT A (group 1) and 5 to view CRT B (group 2). To ensure that the groups are approximately equal in all relevant characteristics, designers frequently use one of two techniques when assigning subjects to groups. These techniques are *random* or *same subject.* In the random situation, each subject is assigned to one group, while in the same-subject situation, each subject is used in all groups.

Random Groups

In random assignment, each subject should have an equal opportunity to be assigned to each group. One simple way to assign group members to one group or another, when there are only two groups, is by flipping a coin. Another quick and easy method is to put the names of all subjects into a hat and draw them out one at a time. The first name drawn is put in group 1, the second name in group 2, the third in group 1, and so on.

In some instances, it is convenient to use a random number table like Table 18-1. Consider another study that you are conducting using 20 subjects to test which of two new computer products elicits the fastest performance. First, list the participants and assign each a two-digit number.

Sam	01	Joe	11
Pat	02	Rog	12
Lou	03	Pam	13
Bob	04	Ken	14
Sue	05	Rob	15
Jon	06	Jaq	16
Cid	07	Jill	17
Ron	08	Sar	18
Lin	09	Mic	19
Ali	10	Don	20

Table 18-1 Random Numbers

1143919283	5091114209	3959467368	9774236252	2767155091
9697119968	3170902197	1631381020	0158821654	5032804577
0777947712	3384684716	4987059671	4694671716	5062338681
7167595993	0879013241	7126016658	8331668482	1029445137
3280472742	1623772550	1057031470	9261294917	4882279794
1483556263	5306271543	6763230337	2873917582	4092332434
1554414327	0758048813	3016110756	9647060680	4350714435
9223041243	9076508867	0803805038	1090800633	2174055450
3356893563	1077010595	7132384243	0940262877	4976256151
8446156618	4057072906	3079441144	6523921788	3828829180
9164542451	8377699246	4554802457	7480449536	8981574285
7830563797	2699523146	5607197081	2237609819	5685597424
9788855122	6554502904	4004270653	2448331258	9647577668
6728609001	0971867231	5403324185	5209778713	5951084400
5361059459	8994572102	6659502198	2696888467	4693952318
5296576189	2889245419	0222509603	5930438179	7592080486
2533639735	2559450557	9625759700	2771542432	2765288151
7307844321	7761649296	5588271507	3016831876	2828353424
8174722244	8354437800	4845443304	1425674281	8227928882
4717252798	3691039986	3403339868	2400997123	5915127583
5415370832	3757531898	3921263993	0541977565	7315098537
9374599871	3712955032	9444417884	2708223502	0613689476
8167651330	5882874199	8721413727	8053995037	7353616862
7978802193	3325005865	5301862394	5699741534	0195313763
9211261235	6876461201	0218909424	2415610368	2652789107

Second, pick any line of random numbers in the table. Part of the tenth line in Table 18-1 is

8446156618 4057072906 3079441144 6523921788...

Combine adjacent pairs of numbers to form two-digit numbers. If there had been 9 or fewer subjects, you would use one-digit numbers; and if there had been 100 or more subjects, you would use three-digit numbers. When converted to two-digit numbers, the tenth line becomes:

84 46 15 66 18 40 57 07 29 06 30 79 44 91 44 65 23 92 17 88...

Third, pick the numbers that are part of the original list of participants. The first one should be put in group 1, the second in group 2, the third in group 1, and so on.

84 46 **15** 66 **18** 40 57 **07** 29 **06** 30 79 44 **11** 44 65 23 92 **17** 88...

The people begin to be assigned as follows:

Group 1	Group 2
15 (Rob)	18 (Sar)
07 (Cid)	06 (Jon)
11 (Joe)	17 (Jill)

Continue looking for two-digit numbers that previously had been assigned to the subjects, using each successive line until all subjects are assigned to either group 1 or group 2.

Same-subject Groups

A second commonly used approach is to have each subject participate in all groups. In this way, we keep to a minimum the differences between subjects (but we have to be very careful about *transfer* effects).

This method cannot be used in all studies because it does not always make good sense to have the same subjects involved in two or more conditions, for example, teaching the same person to use a new computer system by training them with two different methods. We also must avoid subject fatigue or boredom. Remember, people change as they spend time participating in a study.

Reliable Differences

After subjects are assigned to groups, the study completed, and the data collected, the results may be analyzed *statistically* to determine the *reliability* of differences. Statistical methods are most useful when there seems to be an important difference, but it is difficult to tell if the difference in performance is large enough to be reliable. If it is reliable, the same results should occur if the study were done again using different people with backgrounds similar to the original group. It is not always necessary to use a statistical method. If, for example, it is evident that, say, 10 out of 10 subjects performed with fewer errors using CRT A than CRT B, then a statistical method is not necessary.

Whether or not a difference is reliable depends on the probability that the cause of the difference is chance. In this book, we consider a finding reliable if the observed difference would have occurred by chance less than once in 20 studies (i.e., less than 5 percent of the time). The aim is to accept as many genuine differences as possible, without accepting chance differences. Remember that, even though a set of results is found to be reliable, there is still a small possibility that they are due to chance. Stating that a difference is reliable only means that if the study were repeated we would probably get the same result.

Selecting a Statistical Method

Consider the following when attempting to decide which of the four statistical methods discussed in the following sections to use:

Whether the study data are reported as scores or averages (of group performance) or as frequency counts

The *number* of different groups or conditions

The way that people are *assigned* to the different conditions being studied (random, same subject)

Probably the most difficult decision is determining how the study data are reported. Virtually all studies collect their primary data in the form of a *score* or a *frequency count.* A typical score could be the time taken by each person to perform a task or the number of errors made by each subject. Scores also include the average time to perform all tasks or the average number of errors across all tasks. Speed and error scores are the result of user performance and tend to be the most common scores reported in usability tests. Scores tend to be *performance* related (i.e., the results of user performance).

Frequency counts, on the other hand, reflect the number of occurrences or votes for one option versus another. If 100 people are asked to indicate whether they prefer using a personal computer or typewriter for doing their word processing, and 98 prefer personal computers and 2 prefer the typewriter, these numbers (98 and 2) are frequency counts. They are simply the number of people preferring one over another. Frequency counts tend to be *preference* related.

Table 18-2 can help in deciding the appropriate test for a set of data. As an example of how to use this logic tree, assume that a designer wants to compare the number of errors made on the exact same activity performed in crowded versus isolated conditions. Ten people are selected as subjects and randomly assigned to one of the two conditions (crowded or isolated). The designer conducts the study and ends up with the number of errors made by each subject under each condition.

There are three consecutive decisions to be made. First, are the study results reported as scores or frequency counts? In this case they are *scores,* reported as the number of errors made by each subject. The second decision has to do with the number of conditions evaluated. There were *two* (crowded and isolated). The third and final decision concerns how the subjects were assigned to each condition. They were *randomly* assigned to one group or the other. Following these branches on the logic tree leads to the conclusion that the t method is the appropriate statistical method. In some situations, more than one method can be used. In these cases, designers can use whichever method they prefer.

Statistical Methods

The following sections introduce and demonstrate the application of statistical methods for determining whether the differences between groups are large enough to be considered reliable (i.e., statistically significant). The designer does not need an in-depth understanding of statistics to use these techniques. If none of these methods seems appropriate or difficulties are encountered, the designer should consult a statistical specialist.

Table 18-2 Decision Table for Selecting the Most Appropriate Statistical Method

What is the purpose of the evaluation?

- To determine performance
  - How many conditions?
    - 2
      - How were subjects assigned?
        - Randomly → t-method (A)
        - Same subject → t-method (B)
    - 3 or more
      - How were subjects assigned?
        - Randomly → F-method (A)
        - Same subject → F-method (B)
- To determine preferences
  - Number of participants?
    - 20 or less → b-method
    - Any number → chi-square

t Method

The *t* method is a commonly used technique that compares two groups to determine if there is a reliable difference between them. The number of subjects in each group should be the same. The following two examples illustrate different ways of applying the *t* method. The first example is used when subjects are *randomly assigned.* The second example is used when subjects are either *matched* or in *same-subject* groups. The formulas are slightly different, so be careful to use the correct one. Remember, the correct formula depends on *how the subjects were assigned* to the different groups.

t Method A: Randomly Assigned Subjects

A designer wants to find out if different methods of training result in reliably different numbers of errors on an end-of-course performance test. He randomly assigns 24 subjects to two groups. He teaches group 1 using *videotapes* and group 2 using *lectures* by highly experienced workers. At the conclusion of the training, he obtains the following error scores:

Group 1 (Videotape)		Group 2 (Lectures)	
Subject	Score	Subject	Score
1	27	13	24
2	28	14	28
3	20	15	31
4	27	16	27

(continued)

Group 1 (Videotape)		Group 2 (Lectures)	
Subject	Score	Subject	Score
5	30	17	28
6	28	18	23
7	19	19	31
8	25	20	29
9	31	21	26
10	24	24	33
11	23	23	29
12	27	24	28

The following computational steps are used to calculate *t:*

Step 1. List the error data by group as shown in the table. No particular order is necessary within the groups.

Step 2. Add the scores for group 1 to obtain P_1.

$$27 + 28 + 20 + 27 + \cdots + 27 = 309 = P_1$$

Step 3. Square every score in group 1, and add these squared values to obtain S_1 (squaring a score means that the score is multiplied by itself, e.g., 3^2 is 3×3 or 9).

$$27^2 + 28^2 + 20^2 + 27^2 + 30^2 + 28^2 + 19^2 + 25^2 + 31^2 + 24^2 + 23^2 + 27^2 =$$

$$729 + 784 + 400 + 729 + 900 + 784 + 361 + 625 + 961 + 576 + 529 + 729 =$$

$$8107 = S_1$$

Step 4. Count the total number of subjects in group 1. Call this value N_1.

$$N_1 = 12$$

Step 5. Square the value obtained in step 2 (P_1) to obtain the value $(P_1)^2$.

$$309^2 = 309 \times 309 = 95{,}481$$

Step 6. Divide the value obtained in step 5 by the value obtained in step 4.

$$\frac{95{,}481}{12} = 7956.75$$

Note: Two decimal places is sufficient for all calculations performed in this chapter.

Step 7. Subtract the value obtained in step 6 from the value in step 3.

$$8107.00 - 7956.75 = 150.25$$

Step 8. Add the scores in group 2 to obtain P_2.

$$24 + 28 + 31 + \cdots + 28 = 337 = P_2$$

Step 9. Square every score in group 2 and add these squared values to obtain S_2.

$$24^2 + 28^2 + 31^2 + 27^2 + 28^2 + 23^2 + 31^2 + 29^2 + 26^2 + 33^2 + 29^2 + 28^2 =$$
$$576 + 784 + 961 + 729 + 784 + 529 + 961 + 841 + 676 + 1089 + 841 + 784 =$$
$$9555 = S_2$$

Step 10. Count the total number of subjects in group 2. Call this value N_2.

$$N_2 = 12$$

Step 11. Square the value obtained in step 8 to get $(P_2)^2$.

$$337^2 = 337 \times 337 = 113{,}569$$

Step 12. Divide the value obtained in step 11 by the value obtained in step 10.

$$\frac{113569}{12} = 9464.08$$

Step 13. Subtract the value obtained in step 12 from the value obtained in step 9.

$$9555 - 9464.08 = 90.92$$

Step 14. Add the values obtained in steps 7 and 13.

$$150.25 + 90.92 = 241.17$$

Step 15. Add together N_1 and N_2, and then subtract 2.

$$N_1 + N_2 - 2 = 12 + 12 - 2 = 22$$

Note: This is also the number that you will use later to access the t table (Table 18-3).

Step 16. Divide the value obtained in step 14 by the value in step 15.

$$\frac{241.7}{22} = 10.96$$

Step 17. Divide 1 by N_1 and 1 by N_2; then add the results together.

$$\frac{1}{12} = 0.08 \qquad \frac{1}{12} = 0.08 \qquad 0.08 + 0.08 = 0.16$$

Step 18. Multiply the value from step 17 by the value obtained in step 16.

$$0.16 \times 10.96 = 1.75$$

Step 19. Use a calculator to calculate the square root of the value obtained in step 18.

Square root of 1.75 = 1.32

Step 20. Calculate the average for group 1 by taking the value from step 2 and dividing by N_1.

$$\frac{309}{12} = 25.75$$

Step 21. Calculate the average for group 2 by taking the value from step 8 and dividing by N_2.

$$\frac{337}{12} = 28.08$$

Step 22. Subtract the value obtained in step 20 from the value obtained in step 21.

25.75 – 28.08 = –2.33

Note: Only the absolute difference is important. Therefore, a negative sign in the answer (if there is one) can be deleted (e.g., –2.33 = 2.33).

Step 23. Divide the value obtained in step 22 by the value obtained in step 19. This yields the t value.

$$t = \frac{2.33}{1.32} = 1.77$$

Step 24. Determine the tabled t by consulting Table 18-3. In the first column, find the number closest to the value calculated in step 15. The tabled t is the number shown to the right of it.

Step 25. Compare the calculated t with the tabled t. If the calculated t is the same as or larger than the tabled t, the findings are reliable (statistically significant).

Because the calculated t (1.77) is smaller than the tabled t (2.07), the designer must conclude that the difference between the two group averages is *not* large enough to suggest that one training method is reliably better than the other. Thus, the results do not show a reliable difference between the groups.

t Method B: Same-subject Groups

The following is an example using matched subjects; this technique can also be used with studies using the same subjects in each condition.

A designer wanted to know if having a printer at each user's desk would increase the number of customer transactions completed each week. Ten customer representatives were recruited to participate in the study. Each person had his or her own printer for 4 weeks and shared a printer with one other person for 4 weeks. Half of the subjects started

Table 18-3 Values of t

	Tabled t value
1	12.71
2	4.30
3	3.18
4	2.78
5	2.57
6	2.45
7	2.37
8	2.31
9	2.26
10	2.23
11	2.20
12	2.18
13	2.16
14	2.15
15	2.13
16	2.12
17	2.11
18	2.10
19	2.09
20	2.09
21	2.08
22	2.07
23	2.07
24	2.05
25	2.06
30	2.04
40	2.02
60	2.00
120	1.98
Over 120	1.96

To conclude that there is a reliable difference between the two groups, the calculated t must be equal to or greater than the tabled t (adapted from Fisher and Yates, 1963; courtesy Cambridge University Press).

with having their own printer, and half started with sharing a printer. After 4 weeks, they switched conditions. Obviously, in this study the same subject had a chance to be in both conditions.

Step 1. List the data in the form of a table, with the scores for each subject listed side by side.

Subjects	Average Number of Customers per Day	
	Own Printer	**Shared Printer**
1	8	4
2	6	6
3	9	7
4	7	6
5	9	3
6	8	5
7	8	7
8	12	8
9	10	7
10	7	6

Step 2. Obtain the difference (*D*) between each pair of scores.

Subject	Difference between Conditions
1	$8 - 4 = 4$
2	$6 - 8 = -2$
3	$9 - 7 = 2$
4	$7 - 6 = 1$
5	$9 - 3 = 6$
6	$8 - 5 = 3$
7	$8 - 9 = -1$
8	$12 - 8 = 4$
9	$10 - 7 = 7$
10	$7 - 6 = 1$

Step 3. Square each difference and add these squared values.

$$(4^2) + (-2^2) + (2^2) + (1^2) + (6^2) + (3^2) + (-1^2) + (4^2) + (7^2) + (1^2)$$
$$= 16 + 4 + 4 + 1 + 36 + 9 + 1 + 16 + 49 + 1 = 137$$

Step 4. Count the number of pairs of scores. Call this value *N*.

$$N = 10$$

Step 5. Multiply the value obtained in step 3 by the value obtained in step 4.

$137 \times 10 = 1370$

Step 6. Add together the differences from step 2.

$(4) + (-2) + (2) + (1) + (6) + (3) + (-1) + (4) + (7) + (1) = 25$

Step 7. Square the answer from step 6.

$25^2 = 625$

Step 8. Subtract the value obtained in step 7 from the value obtained in step 6.

$1370 - 625 = 745$

Step 9. Take the value from step 4 (N) and square it.

$10^2 = 100$

Step 10. Take the value from step 4 (N) and subtract 1.

$10 - 1 = 9$

Step 11. Multiply the value in step 9 by the value in step 10.

$100 \times 9 = 900$

Step 12. Divide the value obtained in step 8 by the number obtained in step 11.

$\frac{745}{900} = 0.83$

Step 13. Use a calculator to obtain the square root of the number obtained in step 12.

Square root of 0.83 = 0.91

Step 14. Obtain the average of group 1 by adding the scores in column 1 and dividing by the value obtained in step 4 (N).

$8 + 6 + 9 + 7 + 9 + 8 + 8 + 12 + 10 + 7 = 84$

$\frac{84}{10} = 8.4$

Step 15. Obtain the average of group 2 by adding the scores in column 2 and dividing by the value obtained in step 4 (N).

$4 + 6 + 7 + 6 + 3 + 5 + 7 + 8 + 7 + 6 = 59$

$$\frac{59}{10} = 5.9$$

Step 16. Subtract the value in step 14 from the value in step 15.

$$8.4 - 5.9 = 2.5$$

Note: The negative sign (if one shows) can be ignored.

Step 17. Divide the value obtained in step 16 by the value obtained in step 13 to obtain the *t* value.

$$\frac{2.5}{.91} = 2.75$$

Step 18. Determine the tabled *t* by consulting Table 18-3. In the first column, find the number closest to the value calculated in step 10. The tabled *t* is the number shown to the right of it.

Value from Step 10	***t* Value**
9	2.26

Step 19. Compare the calculated *t* value with the tabled *t* value. If the calculated *t* value is the same as or larger than the tabled *t* value, the findings are reliable (i.e., statistically significant).

In this example, the calculated *t* value of 2.75 is much larger than the tabled *t* value. The difference then is reliable. Therefore, we can conclude that having access to one's own printer is related to better performance. Having one's own printer seems to lead to reliably more customer interactions (average = 8.4) than does sharing a printer (average = 5.9). There is a slight possibility, about 5 percent, that the difference in the number of customer interactions in this study is not due to printer availability, but is due to some other factor.

F Method

The *F* method provides a way to compare two or *more* groups simultaneously to help to decide if there are reliable differences among them. Usually, if you have only two groups, the *t* method is used.

Determining if differences among groups are reliable by computing an *F* value and consulting a table is very similar to that used with the *t* method. If the *F* value is large enough to be reliable, it means that real differences probably exist among the groups and can be expected to reliably occur again in a study with similar circumstances. Again, having an equal number of people in each group is important.

F Method A: Randomly Assigned Groups

A designer wishes to see if people make fewer errors according to the type of training that they receive on an activity. The three training types are (1) formal classroom training, (2) on-the-job training where they receive some assistance from those already working, and (3) "pickup" learning, where they receive no help and learn the activity on their own through trial and error. The number of errors made during a 1-week period was collected on a sample of 30 people who were trained with these three methods.

Step 1. Start with three columns of error scores, one column for each condition (i.e., each training type).

	Type of Training	
Formal	**On-the-Job**	**Pickup**
20	21	29
22	24	28
25	26	29
21	22	31
19	20	34
17	19	32
18	20	21
24	29	35
28	28	29
19	20	28

Step 2. Add the scores for each column and then the total of each column together. This total is *T*.

$$20 + 22 + 25 + 21 + 19 + 17 + 18 + 24 + 28 + 19 = 213$$

$$21 + 24 + 26 + 22 + 20 + 19 + 20 + 29 + 28 + 20 = 229$$

$$29 + 28 + 29 + 31 + 34 + 32 + 21 + 35 + 29 + 28 = 296$$

$$T = 213 + 229 + 296 = 738$$

Step 3. Square the *T* value obtained in step 2.

$$T^2 = 738^2 = 544,644$$

Step 4. Square each score and then add these squared scores for all groups together. This squared quantity is *S*.

$20^2 + 22^2 + 25^2 + 21^2 + 19^2 + 17^2 + 18^2 + 24^2 + 28^2 + 19^2 =$

$400 + 484 + 625 + 441 + 361 + 289 + 324 + 576 + 784 + 361 = 4645$

$21^2 + 24^2 + 26^2 + 22^2 + 20^2 + 19^2 + 20^2 + 29^2 + 28^2 + 20^2 =$

$441 + 576 + 676 + 484 + 400 + 361 + 400 + 841 + 784 + 400 = 5363$

$29^2 + 28^2 + 29^2 + 31^2 + 34^2 + 32^2 + 21^2 + 35^2 + 29^2 + 28^2 =$

$841 + 784 + 841 + 961 + 1156 + 1024 + 441 + 1225 + 841 + 784 = 8898$

$S = 4645 + 5363 + 8898 = 18{,}906$

Step 5. For each column, square the value obtained in step 2, and count the number of scores in each column (call this n).

Column 1 = $213^2 = 45{,}369$, $n = 10$

Column 2 = $229^2 = 52{,}441$, $n = 10$

Column 3 = $296^2 = 87{,}616$, $n = 10$

Step 6. Divide each value by n.

Column 1 = 45,369 / 10 = 4536.9

Column 2 = 52,441 / 10 = 5244.1

Column 3 = 87,616 / 10 = 8761.6

Step 7. Add the results from step 6. This value is Z.

4536.9 + 5244.1 + 8761.6 = 18,542.6

Step 8. Count the total number of scores on the table in step 1. Call this quantity N.

$N = 30$

Step 9. Use the results of steps 3 and 8 to calculate T^2 / N.

$$T^2 / N = \frac{544{,}644}{30} = 18{,}154.8$$

Step 10. Subtract the results of step 9 from step 7.

18,542 – 18,154.8 = 387.2

Step 11. Subtract the results of step 9 from step 4.

18,906 – 18,154.8 = 751.2

Step 12. Subtract the results of step 11 from step 10.

$$751.2 - 387.8 = 363.4$$

Step 13. Count the number of columns on the table in step 1. Call this C.

$$C = 3$$

Step 14. Subtract 1 from the value of C.

$$C - 1 = 2$$

Step 15. Divide the quantity from step 11 by the value in step 14.

$$\frac{387.8}{2} = 193.9$$

Step 16. Subtract the value of C in step 13 from the value of N in step 8.

$$N - C = 30 - 3 = 27$$

Step 17. Divide the value from step 11 by value in step 16.

$$\frac{363.4}{27} = 13.46$$

Step 18. To find the value of F, divide the value from step 15 by the value from step 17.

$$F = \frac{193.9}{13.46} = 14.41$$

Step 19. Determine the tabled F by consulting Table 18-4. In the top row of numbers (1 through 10), find the value from step 14. In the first column find the number closest to the value from step 16. Where the two (row and column) intersect in the table is the tabled value of F. In this case, it is between 3.39 and 3.32.

Step 20. If the calculated F is equal to or greater than the tabled F, the difference among the groups is reliable. Otherwise, there is no reliable difference.

Since the calculated F is larger than the tabled F in this example, we can conclude with some confidence that there are real differences among the groups (i.e., the differences are not due to chance). By inspecting the results, we see that the most errors occurred with the subjects who were trained using the pickup method.

F Method B: Same Subjects in Each Group

A designer wishes to determine whether illumination level will affect how people feel about their work area. She measures satisfaction using a questionnaire. She selects nine people from a group of employees and has each person work for 1 month under three

Table 18-4 Values of F

	1	2	3	4	5	6	7	8	9	10
2	18.51	19.00	19.10	19.25	19.30	19.33	19.35	19.37	19.38	19.40
3	10.13	9.56	9.28	9.12	9.01	8.94	8.89	8.85	8.81	8.79
4	7.71	6.94	6.69	6.39	6.36	6.16	6.09	6.04	6.00	5.96
5	6.61	5.79	5.41	5.19	5.05	4.95	4.88	4.82	4.77	4.74
6	5.99	5.14	4.76	4.53	4.39	4.28	4.21	4.15	4.10	4.06
7	5.59	4.74	4.35	4.12	3.97	3.87	3.79	3.73	3.68	3.64
8	5.32	4.48	4.07	3.84	3.69	3.58	3.60	3.44	3.39	3.36
9	5.12	4.26	3.86	3.63	3.48	3.37	3.29	3.23	3.18	3.14
10	4.96	4.00	3.71	3.48	3.33	3.22	3.14	3.07	3.02	2.98
11	4.84	3.48	3.59	3.36	3.20	3.09	3.01	2.95	2.90	2.85
12	4.75	3.89	3.49	3.26	3.11	3.00	2.91	2.85	2.80	2.75
13	4.67	3.81	3.41	3.18	3.03	2.92	2.83	2.77	2.71	2.67
14	4.60	3.74	3.34	3.11	2.96	2.85	2.76	2.70	2.65	2.60
15	4.54	3.68	3.29	3.06	2.90	2.79	2.71	2.64	2.59	2.54
16	4.49	3.63	3.24	3.01	2.85	2.74	2.66	2.59	2.54	2.49
17	4.45	3.59	3.20	2.96	2.81	2.70	2.61	2.55	2.49	2.45
18	4.41	3.58	3.16	2.93	2.77	2.66	2.58	2.51	2.46	2.41
19	4.38	3.52	3.13	2.90	2.74	2.63	2.55	2.48	2.42	2.38
20	4.35	3.49	3.10	2.87	2.71	2.60	2.52	2.45	2.30	2.35
25	4.24	3.39	2.99	2.76	2.60	2.49	2.40	2.34	2.28	2.24
30	4.17	3.32	2.92	2.69	2.53	2.42	2.33	2.27	2.21	2.16
40	4.08	3.28	2.84	2.61	2.45	2.34	2.25	2.18	2.12	2.08
60	4.00	3.16	2.76	2.53	2.37	2.25	2.17	2.10	2.04	1.99
120	3.92	3.07	2.68	2.45	2.29	2.17	2.09	2.02	1.96	1.91

To conclude that there is a reliable difference between groups, the calculated F must be equal to or greater than the tabled F (adapted from Hays, 1973).

different light levels. To help reduce transfer effects, three of the people begin in the lowest light level (moderately bright), three in the middle (bright), and three in the highest light level (very bright). Each month the people are assigned to different light levels. The satisfaction score for each person in each group is shown in step 1. Because each person performs in each group, a different F method is used.

Step 1. List the subject's satisfaction scores by the illumination levels to which they belong (a higher score means the person was more satisfied).

	Illumination Level		
Subject	Moderately Bright	Bright	Very Bright
1	18	20	19
2	17	21	14
3	23	20	27
4	28	23	26
5	16	17	15
6	15	19	17
7	19	20	18
8	22	23	21
9	24	25	22

Step 2. Add the scores for each column. Then add the totals of the columns. This total is T.

$$18 + 17 + 23 + 28 + 16 + 15 + 19 + 22 + 24 = 182$$
$$20 + 21 + 20 + 23 + 17 + 19 + 20 + 23 + 25 = 188$$
$$19 + 14 + 27 + 26 + 15 + 17 + 18 + 21 + 22 = 179$$

$$T = 182 + 188 + 179 = 549$$

Step 3. Square the value obtained in step 2.

$$549^2 = 301{,}401$$

Step 4. Count the number of columns containing scores in step 1. Call this C. Count the number of rows in step 1. Call this R. Find the product of C times R.

$$C = 3, \qquad R = 9$$
$$3 \times 9 = 27$$

Step 5. Divide the value from step 3 by the value in step 4.

$$\frac{301{,}401}{27} = 11{,}163$$

Step 6. Square each score in step 1, add the squared scores for each group, and then add the scores for all groups.

$$18^2 + 17^2 + 23^2 + 28^2 + 16^2 + 15^2 + 19^2 + 22^2 + 24^2$$
$$= 324 + 289 + 529 + 784 + 256 + 255 + 361 + 484 + 576 = 3828$$

$$20^2 + 21^2 + 20^2 + 23^2 + 17^2 + 19^2 + 20^2 + 23^2 + 25^2$$
$$= 400 + 441 + 400 + 529 + 289 + 361 + 400 + 529 + 625 = 3974$$

$$19^2 + 14^2 + 27^2 + 26^2 + 15^2 + 17^2 + 18^2 + 21^2 + 22^2$$
$$= 361 + 196 + 729 + 676 + 225 + 289 + 324 + 441 + 484 = 3725$$

$$3828 + 3974 + 3725 = 11{,}527$$

Step 7. For each column, square the value obtained in step 2, and count the number of scores in each column (call this n).

Column 1 = $182^2 = 33{,}124$, $n = 9$

Column 2 = $188^2 = 35{,}344$, $n = 9$

Column 3 = $179^2 = 32{,}041$, $n = 9$

Step 8. Divide each value by n.

$$\text{Column 1} = \frac{33{,}124}{9} = 3680.44$$

$$\text{Column 2} = \frac{35{,}344}{9} = 3927.11$$

$$\text{Column 3} = \frac{32{,}041}{9} = 3560.11$$

Step 9. Add the values from step 8.

$$3680.44 + 3927.11 + 3560.11 = 11{,}167.66$$

Step 10. Add the scores for each row from step 1.

$$18 + 20 + 19 = 57$$
$$17 + 21 + 14 = 52$$
$$23 + 20 + 27 = 70$$
$$28 + 23 + 26 = 77$$
$$16 + 17 + 15 = 48$$
$$15 + 19 + 17 = 51$$
$$19 + 20 + 18 = 57$$
$$22 + 23 + 21 = 66$$
$$24 + 25 + 22 = 71$$

Step 11. Square each of the row totals from step 10; then add the results.

$$57^2 + 52^2 + 70^2 + 77^2 + 48^2 + 51^2 + 57^2 + 66^2 + 71^2 =$$
$$3249 + 2704 + 4900 + 5929 + 2304 + 2601 + 3249 + 4356 + 5041 = 34{,}333$$

Step 12. Divide the value in step 11 by the value of C from step 4.

$$\frac{34{,}333}{3} = 11{,}444.33$$

Step 13. Subtract the value in step 5 from the value in step 9.

$$11{,}167.66 - 11{,}163.00 = 4.66$$

Step 14. Add the value in step 5 to the value in step 6.

$$11{,}163 + 11{,}527 = 22{,}690$$

Step 15. Add the value in step 9 to the value in step 12.

$$11{,}167.66 + 11{,}444.33 = 22{,}611.99$$

Step 16. Subtract the value in step 15 from the value in step 14.

$$22{,}690 - 22{,}611.99 = 78.01$$

Step 17. Subtract 1 from the value of C, and subtract 1 from the value of R from step 4.

$$C = 3 - 1 = 2$$

$$R = 9 - 1 = 8$$

Step 18. Divide the value in step 13 by the new value for C in step 17.

$$\frac{4.66}{2} = 2.33$$

Step 19. Multiply the values in step 17.

$$2 \times 8 = 16$$

Step 20. Divide the value from step 16 by the value from step 19.

$$\frac{78.01}{16} = 4.88$$

Step 21. To find F, divide the value from step 18 by the value from step 20.

$$F = \frac{2.33}{4.88} = 0.48$$

Step 22. Determine the tabled F by consulting Table 18-4. In the top row of numbers (1 through 10), find the value for C from step 17. In the first column, find the number closest to the value for R in step 17. Where the two (row and column) intersect in the table is the tabled F of 3.63.

Since the calculated *F* (0.48) is less than the tabled F (3.63), there is no reliable difference between the different illumination levels as it relates to user satisfaction with the light.

Chi-square Method

When using the *t* method and the *F* method, groups were compared in terms of the *amount* of a given characteristic. If the results consist of frequency counts, we use a different method for evaluation, such as the chi-square method. Usually, these problems consist of counting the number of people who prefer a particular alternative. The question asked by a designer dealing with frequency counts is "Can the differences between the observed frequencies and the expected frequencies be attributed to chance or not?"

The chi-square method is used to compare observed and expected frequencies. The observed frequencies are the results obtained in the study. The expected frequencies are usually calculated. If the differences between the observed frequencies and the expected frequencies are small, chi square will be small. The greater the difference between the observed and expected frequencies, the larger will be the chi square. If the differences between observed and expected values are so large collectively as to occur by chance only 5 percent or less of the time, we conclude that the differences between the observed and expected frequencies are reliable.

Consider this example: A designer develops three alternative ways of interacting with a new computer system. These included using a keyboard and mouse, using a touchscreen, and using speech. Wondering whether they all have an equal potential for success, he asks a random sample of 28 potential users which of the three interaction devices they preferred.

Step 1. List the frequencies (number of people preferring each alternative) in one row. These are the *observed* frequencies.

Interaction Methods		
Keyboard and Mouse	**Touchscreen**	**Speech**
6	15	7

Step 2. Determine the *expected* frequencies by dividing the total of observed frequencies by the number of observations (28 / 3 = 9.33). This suggests that we expect about the same number of people to prefer each of the three alternatives.

	Interaction Methods		
	Keyboard and Mouse	**Touchscreen**	**Speech**
Observed	6	15	7
Expected	9.33	9.33	9.33

Step 3. For each column, subtract the expected frequency from the observed frequency.

Observed	6	15	7
Expected	9.33	9.33	9.33
	–3.33	5.67	–2.33

Note: Ignore the minus signs.

Step 4. Square the values from step 3.

$-3.33^2 = 11.09$

$5.67^2 = 32.15$

$-2.33^2 = 5.43$

Step 5. Divide the quantities in step 4 by the expected frequencies.

$11.09 / 9.33 = 1.19$

$32.15 / 9.33 = 3.45$

$5.43 / 9.33 = 0.58$

Step 6. Chi square is determined by adding the values in step 5.

$1.19 + 3.45 + 0.58 = 5.22$

Step 7. Count the number of columns and subtract by 1.

$3 - 1 = 2$

Step 8. Determine the tabled chi square by consulting Table 18-5. In the first column, find the value from step 7. The tabled chi square is the number shown to the right of it.

Step 9. Compare the calculated chi square with the tabled chi square. If the calculated chi square is the same as or larger than the tabled chi square, then the findings are reliable.

Since the calculated chi square (5.22) is smaller than the tabled chi square (5.99), we conclude that there is no reliable difference between the preferences for the three different interaction methods. Although on the surface it appears that using a touchscreen for this system was preferred over the other two approaches, the difference is not large enough to preclude the finding that the results may change if the study is repeated.

Table 18-5 Values of Chi Square

	Chi-square Values
1	3.84
2	5.99
3	7.82
4	9.49
5	11.07
6	12.59
7	14.07
8	15.51
9	16.92
10	18.31
11	19.68
12	21.03
13	22.36
14	23.68
15	25.00
16	26.30
17	27.59
18	28.87
19	30.14
20	31.41
21	32.67
22	33.92
23	35.17
24	36.42
25	37.65
26	38.89
27	40.11
28	41.34
29	42.56
30	43.77

To conclude that there is a reliable difference between groups, the calculated chi square must be equal to or greater than the tabled chi square (adapted from Fisher and Yates, 1963).

b Method

Another means for determining the reliability of a set of data is the *b* method. With this method, a designer can very quickly determine whether one item is reliably different from one to four others. Consider the following example.

A designer wants to know which of four different monitors is preferred by a group of bank clerks. The monitors are of four different sizes: A = 10 inches, B = 13 inches, C = 15 inches, and D = 17 inches. Eleven clerks agreed to evaluate each of the four monitors and to indicate which one they prefer. For example, subject 1 preferred terminal C, subject 2 preferred B, and so on. The results of their evaluations are shown next:

	Most Preferred Monitor Size			
Subject	**A**	**B**	**C**	**D**
1			×	
2		×		
3			×	
4			×	
5	×			
6				×
7		×		
8			×	
9	×			
10			×	
11			×	

Monitor C was preferred by six of the subjects, while monitors A, B, and D were preferred by two or fewer subjects. The question is "How many times must monitor C be selected in order to confidently conclude that the preference for monitor C is *not* due to chance?" Note that this method makes use of *frequency counts* and not the *scores* used with the t and F methods.

By looking at Table 18-6 for 11 subjects and four categories, we see that six or more votes for a specific monitor are required for there to be a reliable difference. Thus, the results from the study are reliable. If another group of bank clerks with similar backgrounds was asked the same question, the result would most likely be the same.

The same method can be used with two or three different groups. Consider this example: A designer wants to determine which of two computer screens is better for a particular task. Screen A uses few graphics, while screen B uses many graphics. In this study, 17 people performed a variety of tasks and were timed on how long it took them to find items on the screens. Each person used both screen A and screen B during the study. Half of the people used screen A first, and half used screen B first. The results are as follows:

Average Time to Find Items (Seconds)		
Subject	**A**	**B**
1	2.5	**2.3**
2	3.2	**3.1**
3	2.1	2.1
4	**3.6**	3.7
5	4.7	**4.5**

(continued)

Average Time to Find Items (Seconds)		
Subject	**A**	**B**
6	**2.1**	2.5
7	1.9	**1.7**
8	3.2	**3.0**
9	3.6	**3.4**
10	**2.7**	2.9
11	**2.6**	2.9
12	3.1	3.1
13	4.5	**4.2**
14	3.2	**3.0**
15	**2.8**	2.9
16	2.7	**2.4**
17	2.5	**2.3**

Note that the average times for subjects 3 and 12 were the *same* for both screens. These will be discarded, leaving 15 subjects. The rule is that if ties occur on potential hits we discard all the responses for that subject. The majority of people (10) performed better with screen B than with screen A. According to Table 18-6, at least 12 people must do better with one of the screens when there are 15 subjects who are performing in order for these findings to be considered reliable. On the basis of this study, we would have to conclude that there is *no* reliable difference between the two screens.

Table 18-6 Values for the b Method

Minimum Number of Hits Required to Be Reliable				
Number of Subjects	**Number of Categories**			
	2	**3**	**4**	**5**
5	5	4	4	4
6	6	5	4	4
7	7	6	5	4
8	7	6	5	5
9	8	6	5	5
10	9	7	6	5
11	9	7	6	5
12	10	8	6	6
13	10	8	7	6
14	11	9	7	6
15	12	9	8	7
16	12	10	8	7
17	13	10	8	7
18	13	11	9	7
19	14	11	9	8
20	15	11	9	8

Exercise 18: Conducting Statistical Evaluations

Purpose: To provide an opportunity to apply statistical methods for evaluating the reliability of study and test results.

Method: Use the appropriate statistic to determine whether the following questions are reliable.

1 A designer wants to know if the time it takes to read text on a monitor is different, depending on whether the information is black on a white background, white on a black background, or red on a blue background. He selects 12 subjects and randomly assigns each person to one of the three groups. The average times (in minutes) taken by the subjects to read eight different documents were as follows:

Minutes to Read Text Using Different Foreground–Background Colors

Black on White		White on Black		Red on Blue	
Lou	8	Sam	10	Pam	13
Pat	7	Bob	11	Tim	12
Ray	5	Liz	13	Tom	15
Ron	6	Fay	12	Mei	16

Are the results reliable? Is it correct to conclude that the selected color combination can affect reading speed?

2 A designer has developed a computer product that can make full use of either a two-button or three-button mouse. She selects a sample of 76 employees. Each participant uses the two-button mouse with the product for 3 months and then the three-button mouse for 3 months. Half of the subjects begin using the two-button mouse and half begin using the three-button mouse. At the end of 6 months, she administers a questionnaire to find out which is preferred. The results are as follows:

Number Preferring Each Mouse Type

Two-button Mouse	Three-button Mouse
52	24

Are the results reliable? Is it correct to conclude that one mouse type is reliably preferred over the other?

3 A designer conducts a usability test to determine which of two training methods is best: step-by-step tutorial or assisted exploration. She selects 20 potential

users and randomly assigns each person to one of the two training groups. The first group receives all training only with the step-by-step tutorial. The second group receives training only with the assisted exploration method. The participants were timed to see how long it would take to complete training and pass a difficult end-of-course test. The results were as follows:

Minutes to Complete Each Training Method	
Step-by-step Tutorial	**Assisted Exploration**
Subject 1: 7	Subject 11: 7
Subject 2: 7	Subject 12: 5
Subject 3: 8	Subject 13: 6
Subject 4: 9	Subject 14: 5
Subject 5: 8	Subject 15: 4
Subject 6: 9	Subject 16: 5
Subject 7: 6	Subject 17: 3
Subject 8: 7	Subject 18: 6
Subject 9: 8	Subject 19: 3
Subject 10: 7	Subject 20: 5

Are the results reliable? Is it correct to conclude that one of the training courses results in faster learning?

4 A designer is interested in determining which of four screen-based controls (widgets) is most preferred for a particular activity. The screen-based controls are a selection list, radio buttons, text entry, and spin lists. He selects 16 subjects and has each person use each of the four screen-based controls to complete a series of activities. The total test time is about 1 hour. He makes sure that the subjects use the screen-based controls in different orders. At the end of each test session, he asks the subjects to indicate which of the screen-based controls they most preferred. The results are as follows:

Number Preferring Each Screen-based Control			
Selection List	**Radio Buttons**	**Text Entry**	**Spin List**
5	8	1	2

Are the results reliable? Is it correct to conclude that one screen-based control is reliably preferred over all the others?

Appendix A

Human Performance Engineering (Usability) Resources

Professional Societies

The following professional societies are concerned with human performance engineering and usability issues.

Primary Societies

Human Factors and Ergonomics Society
Box 1369
Santa Monica, California 90406
(310) 394-1811

Ergonomics Society
4 John Street
London WC1N 2ET

Special Interest Group on Computer and Human Interaction (SIGCHI)
Association for Computing Machinery (ACM)
11 West 42nd Street
New York, New York 10036
(212) 869-7440

Other Societies

Usability Professionals' Association (UPA)
10875 Plano Road #115
Dallas, Texas 75238

American Institute of Industrial Engineers
25 Technology Park
Norcross, Georgia 30092

Society for Information Display
654 N. Sepulveda Boulevard
Los Angeles, California 90049

Society for Technical Communication
815 15th Street, NW
Washington, D.C. 20005

National Society for Performance and Instruction
1126 16th Street, NW
Washington, D.C. 20036

Institute of Electrical and Electronics Engineers
Systems, Man and Cybernetics Society
345 East 47th Street
New York, New York 10017

Journals

There are numerous journals published by professional societies or privately that contain information on human performance engineering. These journals are generally available in university libraries. The journals or primary interest include the following:

Applied Ergonomics

Behaviour and Information Technology

Communications of the ACM

Ergonomics

Human–Computer Interaction

Human Factors

International Journal of Human–Computer Studies (previously the *International Journal of Man–Machine Studies*)

Major Conferences (each conference publishes its proceedings)

Human Factors and Ergonomics Society Annual Meeting (usually in the fall)

CHI Conference (usually in the spring)

Ergonomics Society Annual Conference (usually in the spring)

HCI International (usually in the summer)

Usability Professionals Annual Meeting (usually in the summer)

Usability Training Courses

Computer Psychology, Inc.
960 South Donner Way #350
Salt Lake City, Utah 84108
(801) 582-2100

Appendix B

Guidelines for Developing Questionnaires

Self-administered Questionnaires

Special care must be exercised in framing the questions for self-administered questionnaires. Because the self-administered questionnaire must play the role of interviewer, all the rules of interviewing apply. For example, terms must be defined, instructions must be clear, the questionnaire must be easy to fill in, and it should look important to win the respondent's cooperation. In effect, the interviewing function is programmed into the content and makeup of the questionnaire.

Also, when self-administered questionnaires are used, the response rate is apt to be low unless some form of supervision is applied. When large groups of people are to be sampled, as in an attitude survey among workers in a plant or office, it may be best to bring the employees together in groups.

Question Content

Once the type of questionnaire has been determined, the development of question content should begin. Designers must decide which questions are most likely to elicit information that will satisfy the objectives and requirements of the survey. An effective way to start the assembly of questions is to simply write down as many possibilities as come to mind. Careful attention to exact wording is not important at this point.

Emphasis on the topics to be covered is more important. If the questionnaire is to include more than a few questions, it may be more convenient to write each on a separate card. This makes it easy to rearrange or eliminate questions as development of the questionnaire progresses. Once all questions are listed, the final selections and wording of questions should be considered. As with other steps in the data collection process, pretesting is important for individual questions, as well as for the full set of questions.

Wording of Questions

Here are several rules or guides that will help in the wording of questions:

Use words that are as simple and familiar as possible. While technical language assists clarity if the respondents are equipped to deal with the technical terms, it can be a liability. Thus, if there is any doubt about the technical understanding of respondents, use only simple words.

Avoid "loaded" terms. Terms may have an emotional value and can mean more to most people than their simple dictionary definitions. The use of such words can bias the results of a data collection effort.

Avoid the use of words that suggest the desired response. "You do read a newspaper every day, don't you?" almost forces the respondent to say yes. Responses to such a question would be worthless.

Be careful to word questions so that they do not embarrass respondents. Questions that make the respondent appear poorly educated or inadequate at his or her job can produce distorted responses and in some cases outright refusal to participate in the survey. For example, don't say "only a clerical worker."

Do not place too great a burden on the respondent's ability to remember. Ask only for details that you are reasonably certain that the respondent will know. If more detailed information on specifics is desired, another type of data collection may be indicated, for instance, observation, mechanized measurement, or file searches.

Anticipate the context of the responses. For example, if the respondent is asked for an average, be sure that "average" means the same to all respondents. Otherwise, define the term in the question context.

Use words that have precise meaning.

Types of Questions

There are two main types of questions:

1. *Open-ended* questions are unstructured and permit free response.
2. *Closed-ended* questions are structured, offering only specifically stated response alternatives that can be easily tallied.

Open-ended questions are characterized by little or no restriction on the range of responses. This kind of question is particularly useful when opinions or attitudes are being probed or when it is difficult or undesirable to suggest possible answers to the respondent. Open-ended questions are often used in the exploratory phases of a study as a method of establishing alternatives for final questions that can be tallied quantitatively. In such situations, the open-ended question helps establish final wording in terms used by the respondents themselves.

The following is an example of an open-ended question and some typical replies used in an interview where users are asked about the time that they spend each day handling computer-detected errors.

Question: On a typical day, about how much time do you spend handling computer-detected errors?

Replies: "Hardly any time at all." (Depending on whether the respondent liked or disliked this part of his or her job, such a response could mean almost anything, two hours, twenty minutes, five seconds.)

"Twenty percent of my time."

"Four to five hours."

"It varies."

"Hard to say."

Obviously, most of these responses would be virtually useless because of the lack of definite information. Thus, in addition to being an example of an open-ended question, this is also an example of a poorly designed and/or misapplied question. The question is poor because it fails to indicate the kind of answer required. Better responses might have been obtained with a closed-ended question.

Assuming that the respondents were qualified to answer, a more appropriate application for open-ended questions might be as follows:

What could be done to reduce the time it takes to handle computer-detected errors?

While open-ended questions can be a great help in the early phases of a study, unstructured responses may present problems in tabulation and analysis.

Closed-ended questions can take several forms, all carefully structured for quantitative tabulation and analysis. For example,

On a typical day, about how much time do you spend handling computer-detected errors?
Less than 1 hour _____ 1 to 5 hours _____ More than 5 hours _____

A point to remember, however, is that the question should not ask for more detailed information than the respondent is likely to have. While the answers actually will be estimates or guesses, the detail will imply a degree of accuracy that does not exist.

Another form of closed-ended question asks the respondent to evaluate alternatives or checklists according to rating scales. These scales may be numerical or verbal. When such checklist questions are used, allowances must be made for biases that result from typical respondent patterns. For example, if a list contains more than six or seven items, the respondents may tend to concentrate on the first few alternatives and pay less attention to the others. Such a survey could erroneously lead to conclusions favoring the earlier items. For example, a questionnaire might have a question like the following:

Would you please indicate how each of the following words describes your feelings about the majority of customer complaints? If the word describes your feelings very well, give it a rating of 7. If the word does not fit your feeling at all, give it a 1. If the work falls between the extremes, please choose the number that best describes where you think it belongs.

An interview's questionnaire may have a similar question:

> As I mention each of these words, please tell me how well it describes your feelings about the majority of customer complaints. Would you say each word describes your feelings very well, fairly well, not too well, or poorly?

One way to offset such bias is to prepare more than one version of the checklist to be used in the survey. The simplest method for doing this is to print two lists, one in reverse order of the other. Then, when the results are tabulated and combined for analysis, the bias factor is somewhat balanced. This simple solution, however, has limitations when the list is long or when the alternatives are complex and require a lot of thought on the part of the respondent.

The reverse-order method could still tend to bias the results against the items that appear toward the middle of both lists. In such cases, four orderings of the items could be printed, increasing position balance for alternatives. To illustrate, assume that there are 12 items in a response selection list. The items are grouped by quarters, so each item appears only once in each quarter. This will help to average out any possible effects of the order of presentation on the responses.

Organization

The sequence and interrelationship of the questions and their parts are also an important part of questionnaire design. Elements should be sequenced so that they carry the respondent through the questionnaire in a logical, easy manner. Organization of a questionnaire actually involves at least three separate levels of activity:

1. Logical ordering of questions according to topical groups
2. Logical sequencing of questions within topical groups
3. Ordering of response alternatives within individual questions

Organization of the questions into logically ordered, topical groups is done so that all the questions covering a specific area are asked while the respondent is thinking about that subject. Topics should be ordered so that they start with easier, more readily discussed subjects and lead gradually into more difficult areas. For example, in conducting a survey of teen-agers about telephone use, starting with questions about how the telephone is used in dating or with personal questions about grades in school could hamper cooperation. However, starting with neutral questions such as those about frequency of use and length of use could help establish credibility. Later in the interview the more personal questions are less likely to be a problem.

The order of the topics in the questionnaire should be developed so that ideas flow easily from topic to topic. When the questionnaire switches from one topic to another, a statement should be inserted that announces that the change is being made. Such statements are usually simple bridges: "Now I'd appreciate getting a few answers from you on the subject of..." or "Let's turn to another subject."

Arrangement of the questions into logically ordered sequences within the topic groups is important because illogical question order could confuse respondents. To be successful in capturing the interest and cooperation of the respondent, both the topic groups and the questions themselves should follow some logical pattern.

In addition, question order can have a significant effect on the bias of responses. To illustrate, consider the next two statements, which might be included on a public opinion survey about telephone service charges.

A seven percent increase in charges for telephone service seems

______ reasonable at this time

______ too high

______ unacceptable

______ I'm not sure

Do you consider telephone service

______ a good value

______ reasonable

______ too expensive

______ no opinion

If the questions were presented in this order, the response to the first could bias the selection of alternatives on the second. It would be more logical to put the second question first. This situation also illustrates an important general principle about presentation sequencing: It is usually best to begin with the general and proceed to the specific. The second question could be referred to as an *establishing question.* It indicates an interest in the subject of telephone service charges and helps respondents to establish a general attitude framework in preparation for a response to a specific situation.

Another device for channeling respondent attention logically is the *filter question,* an item designed to establish respondent qualification for participation in a survey response sequence. An example of a filter question on a service charge questionnaire follows:

Do you have a telephone in your home?

______ Yes

______ No

Since a rate increase would affect telephone users only, it might be decided not to seek responses on the service charge questions from persons giving a "no" response to the filter question. A filter question could then be followed by a *branching instruction,* which directs respondents to questionnaire sections that they are qualified to answer and away from areas for which they have disqualified themselves. For example, a branching instruction on the filter question about whether the respondent has a home phone might direct attention to another question aimed at finding out why the respondent does not have a phone. In such a case, a separate branching instruction would be needed to direct respondents who do have phones around the question on reasons for nonuse.

For self-administered questionnaires, it is best to minimize the use of filter questions and branching instructions. If used on self-administered questionnaires, instructions to respondents must be very explicit. The same principles apply to the sequencing of response alternatives within individual questions.

Questionnaire Length

As part of developing a questionnaire, a certain target length should be set. The questionnaire length relates to a number of factors within the data collection effort.

- Amount of time respondents can reasonably be expected to devote to a questionnaire
- Available time designers can devote as data collectors
- Actual data requirements for the study
- Costs and available funds for the study

These factors, obviously, must be considered in the development of a questionnaire. After questions have been polished and sequenced effectively, the constraints listed are brought to bear in determining the final design of the questionnaire and the plan for using it.

Layout

The physical layout of a questionnaire depends in part on whether it is to be self-administered or administered by an interviewer. In both cases, the important thing is to make the instructions clear to the person using it. Although this requirement is more critical for self-administered questionnaires, the following seven general rules apply to both:

1. The instructions on the questionnaire should be as complete as possible, even when the survey is conducted by an interviewer.
2. The typeface for the instructions should be distinct from the one used for the questions. A common procedure is to use italics or all capital letters for the instructions.
3. Questions should be listed and numbered in an orderly manner so that the reader is easily led from one question to the next. Any complicated arrangement invites the omission of parts of the survey and/or errors in the recording of responses.
4. Where the question sequence calls for skipping questions under certain conditions, the branching instructions must be clearly written and, where possible, supplemented by arrows or other devices to assist the reader.
5. Provide adequate space for recording all responses. This is particularly true for open-ended questions. If the space is too small, only partial responses may be recorded.

6 All questionnaires should include entry and exit statements. These are to assist the interviewers or to enlist the cooperation of the respondent in the case of self-administered instruments. The following are simple examples of entry and exit statements:

a. "Hello, my name is ______. I'm involved in the design of a new system that I think you will find very useful. We're conducting a survey on computer usage in your department and I would like to ask you a few questions."

b. "Thank you very much for your help."

7 Ensure compatibility with techniques to be used for tallying, tabulating, and analyzing results.

Testing

Before a questionnaire is used for full-scale collection of data, it should be pretested on selected, reasonably typical groups of respondents. If the questionnaire is to be administered through an interview, the pretest should be done by experienced interviewers who can pinpoint trouble spots. Particular attention should be paid to such points as the following:

- Are the instructions clear and complete?
- Do the respondents understand the questions?
- Are the response categories clear and distinct?
- Can the interview or questionnaire be completed in the planned time?
- Is there adequate space for entry of responses?
- Is the questionnaire usable in all respects?

When a self-administered questionnaire is pretested, a small group of actual respondents should complete the questionnaires. This affords the opportunity to truly evaluate real-world aspects of the questionnaire. Some of the most important features of question evaluation are listed next. As a first step in testing, a designer should carefully evaluate each question on a questionnaire against the following criteria.

For all questions consider:

- Is the issue fully defined?
- Does each question reflect a clear understanding of the issue?
- Is the issue meaningful to potential respondents?
- Is there reason to suspect that the issue being dealt with by the questionnaire is still not sufficiently well understood by all potential respondents? Consider ways of changing certain questions or eliminating the uninformed.
- What is the stage of development of the issue? It may be a mistake to ask a closed-ended question on a topic if the opinion of many respondents is still hazy

on the subject. Conversely, if opinion is well crystallized or falls into definite patterns, using an open-ended question may be a waste of time.

For each open-ended question consider:

- Is it necessary to ask open-ended questions? Remember that the analysis of thousands of verbal replies represents a good deal of work.
- Which questions, if any, can be converted to a closed-ended type? If the different points of view on the issue are generally well known, then present them as alternatives (multiple choice), rather than leaving them to the respondents to articulate in their various ways.
- Are open-ended questions sufficiently directive? This type of question can be too broad and leave respondents free to give answers from every direction and in every dimension. By carefully wording each question, however, the answers can be confined to a particular frame of reference.
- Do respondents know the number of ideas expected on each question? If one idea is accepted from one and five from another, it will be difficult to tell whether respondents are being weighted according to their ability to articulate or their weakness of conviction.
- Is a probe question needed? If it is desired to extract all possible thoughts on the subject, it may be advisable to add a probe question.
- Are check boxes needed? Even though the question is in an open-ended form, it may be possible to provide check boxes for the answers. This is especially likely if asking for amounts or figures.

For two-way questions consider:

- Are all alternatives included?
- Are alternatives complementary? In some cases, however, it is wise to take account of the realities of the situation rather than to use the literal opposites as complementary.
- Are the choices mutually exclusive? If they cannot be made so, an answer box should be added for a "Both" category.
- Are implied alternatives avoided? No fault can be found with stating the alternative, while some harm may result from leaving it to be carried by implication.
- Are "negatives" detailed enough? The phrase "or not" may not be enough to convey the full implications of the negative side.
- Are "Don't know" or "No opinion" answers provided?
- Is a reasonable middle-ground position available for respondents? If so, it must be decided whether to allow respondents this option.
- Are double-choice-type questions such as "better–worse" or "now–then" avoided?

- Are fold-over questions used? Many two-way questions are easily converted to the fold-over type in which an expression of opinion and its intensity can be obtained in case a designer is interested in both.

For multiple-choice questions consider:

- Are choices mutually exclusive?
- Are choices exhaustive? That is, none of the alternatives should be overlooked. If combinations of the alternatives are possible, those combinations should be included.
- Are some choices intentionally restricted? It may be all right to restrict the choices, but this restriction should be stated in the question.
- Are choices balanced? The choices should be well balanced within a realistic framework. The number of alternatives presented on one side or another of a central point can affect the distribution of replies.
- Is it clear if respondents are expected to express one choice or more than one?
- Are lists provided for respondents? If a questionnaire is used with an interview, give respondents a card list if the question has more than three alternatives.
- Are numbers listed in logical order?

It is important for designers to put themselves in the position of the person responding to the questionnaire. The following items should be considered:

- Avoid the appearance of talking down or otherwise insulting the intelligence of respondents.
- Do not make questions sound stilted.
- Do not use slang.
- Do not try to be "folksy."
- Help the respondents; do not confuse them.
- If there can be any possible question about the meaning, restate it.
- Keep away from wordings that result in ambiguous answers. A "Yes" that means "No" is worse than a "Don't know."
- Each question should be specific without being elaborate.
- Fine distinctions are often not understood by respondents.

When evaluating the use of particular words, the following items should be considered:

- Use as few words as necessary. Most questions can be asked in 20 words or less.
- Use simple, frequently used words.

- Familiar words are the most useful, but only if they do not have too many meanings in context.
- When a word with more than two syllables is used, make sure it is a familiar word.
- Ensure that each word actually does have the intended meaning. Use a dictionary to check what other meanings the word might have that could confuse the issue.
- If it is to be used in an interview, make sure the word has only one pronunciation. Also, consider the possibility of homonyms (*pear* and *pair*).
- If a synonym is used, make sure that it actually is synonymous with the idea at hand.

When evaluating a question by an interviewer, the following items should be considered:

- Misplaced emphasis can be minimized by underscoring the words that should be emphasized.
- To reduce "jumping the gun" on the part of respondents, hold back the alternatives until all conditions have been stated.
- When used in interviews, eliminate unnecessary punctuation because a pause may be taken as the end of the question.
- With interviews, indicate correct pronunciation of difficult words.
- Tongue twisters have no place in interview survey questions.
- With interviews, spell out all abbreviations so that an interviewer will say them correctly.
- Instead of the indefinite "how much?" approach, indicate the denomination in which the answer is desired, for example, percentages, dollars, miles, or pints.

Appendix C

Guidelines for Designing Forms

Introduction

Forms are centrally important in most systems. If well designed, they aid the human–computer interaction. If poorly designed, they can take far longer than is desirable to complete and/or produce large numbers of errors. Designers may or may not be interested in the *time* people take to complete a form. For example, forms completed by people outside the system boundaries—by customers—may take considerable time to complete without having a negative impact on system performance. Customers complete these forms on their own time. The only potential negative effect on system performance is that, if customers feel it takes too long to complete a form, they may not do it. Or they may do it hurriedly, making many mistakes. A well-designed form collects the necessary information in the shortest possible time and with the fewest errors.

The *accuracy* of information collected by having people fill in forms is important whether the information is collected within or outside the system boundaries. Mistakes made by people completing forms are as much the responsibility of the designer as the form filler. In fact, most errors can be attributed to designers.

A designer has the responsibility to develop forms that elicit accurate information in the shortest time. As Wright and Barnard (1975) pointed out, we are no longer bound by the traditional, arbitrary conventions for writing that have evolved over the past few hundred years. We now have the results of considerable research on many human performance-related characteristics of form design. Many of these results already have been discussed. Other guidelines specifically related to form design are summarized in the following paragraphs as general rules or important considerations.

Instructions for Completing Forms

The instructions for some forms leave users confused. They cannot understand what is wanted. We will review some of the more common problems designers encounter when preparing instructions for using forms. Much of the following discussion is adapted from Wright and Barnard (1975).

General Considerations

In general, all instructions necessary to complete a form should appear *on the form.* The main exception to this rule is when the same individual fills in the same form over and over again. The instructions should include

- A brief description of why the form is used
- A sample of a completed form
- A complete list of valid codes to be used on the form (when codes are necessary)
- Specific instructions for filling out the forms that cover all contingencies
- Specific instructions for transmitting the form
- Procedures for correcting errors

Make the instructions as brief as possible and use clear, unambiguous words and sentences. Generally, the instructions should follow the same sequence used for completing the form. Caution the user, in turn, to complete the form in the same order as the instructions.

Placement of Instructions

Whenever possible, put all instructions on the front of a form. Place general instructions at the top of the form and specific instructions just before the corresponding item(s). Instructions that pertain to an action to be taken with the completed form should appear at the bottom of the form. Forms in which instructions are presented and then have to be remembered until the appropriate section on the form is reached can generate numerous problems. In fact, Wright (1969) found that even the reading of a question between the reading of a set of instructions and making an entry is sufficient interruption to increase errors.

Error-correction Procedures

Spell out error-correction procedures. This becomes especially important with machine-processed forms such as those used in mark-sense or optical-scan processes. Unless instructions are provided, the initiator may assume that he or she can simply cross out errors and put another mark in the correct column.

Testing Instructions

Always test instructions on a group of people representative of the group for whom the instructions are intended. Do not simply ask the people if they understand the instructions or if the instructions are clear; rather, ask them to perform using the instructions (with no help from the tester) and then evaluate their performance.

Additional Considerations

Other guidelines for instructions used with forms are discussed in the following paragraphs. Chapter 14 contains a more detailed discussion of some of these topics.

Headings

Appropriate headings can greatly benefit a reader. There is good reason to expect similar benefits from the use of headings within forms. Headings appear to provide a context that assists understanding of both words and sentences. Some methods for displaying headings are better than others. For example, headings printed sideways to bracket several rows of questions all relating to the same topic will be less effective than a heading written horizontally (Wright and Barnard, 1975).

Words

Familiar words are easier to read and remember (cf. Loftus et al., 1970). People also find it easier to think about and draw inferences from sentences using familiar words (Wason and Johnson-Laird, 1972).

One difficulty with familiar words is that they may not mean the same thing to everyone. For example, "income" may mean income before any tax deductions to some people; to others it may mean after tax deductions.

A designer should check words used on a form against their frequency of use using such sources as a study by Kucera and Francis (1967). Familiarity seems to apply to abbreviations as well.

Sentences

Short sentences seem to be more easily understood than long ones. Sentences with more than one clause, with the clauses nested inside one another, are particularly difficult to understand. One alternative to long sentences may be to convert them into a list. Wright and Barnard (1975) give an example of a long, unwieldy sentence that appeared in an instruction on a form:

> It should be noted that a rent allowance cannot be granted to a tenant of a local authority or to a person with a service tenancy or occupying a dwelling partly used for commercial or business purposes but he or she may qualify for a rate rebate.

Ignoring the fact that this was printed in all capital letters, a 47-word sentence is bound to be difficult to understand. They suggest that a sentence such as this might be more easily understood if written as a list:

> The following people cannot have a rent allowance, but they qualify for a rate rebate:
>
> tenants of a local authority
>
> people with a service tenancy
>
> people occupying a dwelling partly used for commerce or business.

Blumenthal (1966) and Hakes and Foss (1970) have shown that sentences with subordinate clauses are more easily understood if the clauses are introduced by relative pronouns (e.g., *which, that*) than if these pronouns are omitted. For example, people grasp the meaning of "The dog that the milkman found chased the cat" more readily than "The dog the milkman found chased the cat."

Active sentences are easier to understand and remember than the equivalent passive forms (Greene, 1972; Barnard, 1974). For example, application forms should not instruct "The notes should be read" when they mean "Read the notes." Passive voice seems to focus attention on the agent, whereas the active is relatively neutral, showing no strong differentiation between two nouns.

Positive sentences are best. Sentences with negative elements are more difficult to understand than positive sentences (cf. Herriot, 1970; Greene, 1972). Words such as *not, except,* and *unless* all have negative elements. The form that said "Do not delay returning this form simply because you do not know your insurance number" could as easily have said, "Return this form at once even if you do not know your insurance number" (Wright and Barnard, 1975).

Even when single negatives have a clear interpretation, performance will be better if an alternative affirmative can be used (Wason, 1961; Jones, 1968). People more easily understand the instruction to "Leave this box blank if you already receive a pension" than the negative version, "Do not write in this box if you already receive a pension." Similarly, when people must answer yes or no questions, evidence suggests that instructions to check, underline, or circle what *does* apply are more easily followed than instructions to delete what does *not* apply (Wright and Barnard, 1975).

Some of the studies examining the terms *more* and *less* have the greatest relevance to designing forms (Palerno, 1973; Wright and Barnard, 1975). These studies show that people most easily understand the term *more.* Yet forms still tell applicants that they need not declare their interest from savings if the amount is *less* than $100.

Finally, designers should avoid double questions. Wright and Barnard (1975) pointed out that double questions, such as "Are you over 21 and under 65? Yes/No," cause difficulties for many people over 65 who appear to take each part of the question in turn. They want to answer "Yes" to over 21 and "No" to under 65. In one study, over one-third of the responses to this question were either incorrectly answered or not answered at all. The problem was solved by making it two separate questions.

Physical Characteristics

It does not take a Rembrandt to produce a thoroughly useful form. In fact, it is usually much better to spend time dealing with performance-related considerations than esthetic ones. In short, having a truly operational form is better than having a pretty one. Designers should concentrate on performance issues and, when possible, let others deal with nonperformance issues. If the designer can indicate precisely what he or she wants on the form and where and can specify the physical characteristics of the form, then an artist or professional printer can take over and deliver a fine finished product.

Printers are governed by long-established procedures and standards and precisely defined terms. If designers know the standards and terms, the chances of getting exactly what they want from a printer are very good.

Layout Aids

Form layout sheets for practically every imaginable application are available from equipment manufacturers, forms manufacturers, and printers. Do not overlook any of these sources of aids. Many printers will provide a designer with a sample book showing various type sizes and styles. Another useful device that a printer can supply is a ruler. This is used in laying out typed forms to determine how many characters can be placed in a given space. The ruler also provides direct measurement in *points* for determining what point-size type you will need.

Form Dimensions

If you have considerable freedom in determining the dimensions of a form, use a length and width that are standard in the printing industry.

Answer Space

The answer space is the space on the form that is designed to permit users to enter information. Both the size and location of answer spaces need to be considered. Answer spaces need to be big enough for a form user to comfortably enter all the necessary information.

In addition, if machines will print the entries (e.g., a typewriter), both the vertical and horizontal spacing must conform to the machine's printing characteristics. Most vertical spacing for machines is based on the standards of 6, 8, and 10 lines per inch. In horizontal spacing for machine preparation, the requirements of the equipment will put some restrictions on the design of the form.

If entries are to be handwritten, allow about 1/3 inch vertical height for each line. If vertical space is very tight on the form, this dimension can be reduced to 1/4 inch, but not smaller (Schaffer, 1980). In planning for horizontal spacing of handwritten entries, allow at least 1/6 to 1/8 inch per character or letter.

A second major consideration is that the answer space should be located relative to the question so that form users know exactly where to make each entry. A designer should put the answer space as close to the end of the question as possible. Large gaps between questions and answers are likely to lead to errors. Research has shown that, even when people are simply copying information, the bigger the gap, the greater the error (Conrad and Hull, 1967). These errors are most likely due to short-term memory limitations.

A third consideration with answer spaces concerns whether to use boxes or tics. Some designers automatically provide answer boxes, requiring one character per box, without stopping to consider the effect on human performance. Several researchers (cf. Devoe et al., 1966; Barnard and Wright, 1976; Barnard et al., 1978; Apsey, 1978) suggest that subdivided answer spaces slow people filling out a form and may possibly reduce the legibility of their responses. In fact, these studies show that writing one letter per box slowed both the writer and subsequent reader. Totally unconstrained printing (no boxes or ticks) is probably the best approach in most systems (cf. Sternberg, 1978).

Leading

Look closely at this printed page or at any printed form. Notice that there is some white space between the bottom edge of the letters of one line and the top of the letters of the next line below it. This between-line spacing is called *leading,* and on forms it should always be 2 points or more (at least 1/36 inch) (Hartley and Burnhill, 1975).

Space Between Groups

Another kind of spacing that is important in form design is the space between item groups and the space between headings and item groups. Many people feel that at least 12 points should be provided between groups of items to maintain the distinction among the groups. At least 6 points should be allowed on each side of a heading and 12 points between the head and the group.

For columns of item groups, designers should provide for item separation of at least 12 points if they do not use vertical separation lines. This spacing can be reduced slightly if a line is provided, but the 12-point minimum tends to be a good rule.

Margins

The margins of a form should provide at least 1/2 inch at the top, bottom, and sides unless a portion of the margin space is to be used for special purposes. An example of an exception would be a form designed for entries to be made in the margins for easy reference. Forms that will be bound may require considerably more than a 1/2 inch margin on the bound edge.

Typefaces

The most common typefaces used in forms design are Gothic, Italic, and Roman. Gothic type is simple, squared off, and easy to read; it has no serifs. Italic type does have serifs and a distinctive slant causes it to stand out. Roman type has serifs but does not slant.

Gothic is one of the best all-around typefaces for readability. It can be compressed more than most types without losing its legibility. This is especially important if limited space requires the use of small 6-point type. Also, Gothic is an easy typeface to read when it is capitalized. This makes it desirable for headings. Italic type is hard to read in large amounts or long lines, but it is very good when a word or phrase needs to stand out. Roman style is considered best for large quantities of text. For this reason, Roman type is commonly used for instruction sections of forms.

Type Size

The recommended type size for forms is 8 to 12 points (Tinker, 1965; Poulton, 1969). A point is 0.0138 inch. Smaller type sizes (6 points or less) can slow reading speed. It has long been established that people read lowercase print more easily than uppercase (Starch, 1914; Tinker and Paterson, 1928; Tinker, 1965; Poulton, 1968). USE UPPERCASE PRINT ONLY TO CALL ATTENTION TO CERTAIN STATEMENTS. Otherwise, use lowercase (with appropriate uppercase to begin sentences, of course) on forms.

We have covered some of the most important physical characteristics of forms. The designer should be cautioned that many of the form design guidelines just discussed have evolved over the years in the absence of good research. These general guidelines are probably acceptable for most applications; however, if a system demands an exceptionally high level of human performance, the designer may need to conduct studies to determine the best set of guidelines for that system.

Appendix D

Workplace Design

Determining Goals

The first step concerning the layout of a workplace is to determine exactly why users are interested in improving the workplace at this time. Some people are interested in trying to improve performance, which usually means increased productivity and increased accuracy. Others may be interested in trying to improve the satisfaction level. This usually refers to greater acceptance of the work and the workplace, and in some cases it simply means greater comfort. In some situations the main motivation for making changes to the workplace is to help to ensure the health and safety of the users.

Once the goals are identified, it is important to evaluate them to ensure that they are reasonable. In some cases it is nearly impossible to improve productivity by making changes to the workplace. Even when attempting to improve user comfort, it can be very difficult if recent changes already have been made or if there is no good way of measuring improved comfort.

Identifying Constraints

The second step in the workplace design process is to identify the constraints as clearly and distinctly as possible. Many times the workplace design process begins with the thought of making all changes necessary to achieve increased productivity or user acceptance, only to find out that the budget is very small. In one situation, a company specified in great detail that they wanted to improve productivity and they were willing to do all that was necessary for this to happen. However, they eventually indicated that they were not going to make any changes to existing equipment or furniture.

It is not uncommon to find situations in which the constraints on workplace design are temporary. In these cases, it is a good idea to indicate what the possibilities are if you had unlimited resources and then to give some idea of what should be done first, second, third, and so forth, until the ideal workplace is obtained by giving an idea of where you want to end up and recognizing that the constraints could change over time. This provides a goal to strive for and gives direction for making future decisions.

Keep in mind that the amount of money budgeted immediately may only be to see if anything meaningful can be done; if it is demonstrated that even minor changes are going to have a relatively large impact, this may help to provide more money in the future. The other main consideration is that some companies develop 3-, 5-, and 10-year plans. To know where you would like to be at the end of 10 years is very helpful when making decisions on workplace layout considerations.

User Profile

Once the goals are established and clear and the constraints are understood, it is time to develop a profile of typical users. To lay out a workplace for someone, we need to have a good idea of who that person is. Are they old or young? Do they see a change in the workplace layout as something to be enthusiastic about or is it perceived as just another token gesture given by management at this time? Have they just recently gone through a workplace layout project and are somewhat skeptical about the outcome? Although it does not have to be too detailed, it is important to profile the user to gain some insight into the relative impact that the new workplace layout could have.

Activity Analysis

The next consideration is to conduct an activity analysis. Here we are interested in finding exactly what users do on a daily basis. We want to know if they work mostly at their computers or spend most of their time reading. In fact, it is a good idea to interview several people who are involved in having their workplace layout changed to see if there is a common set of activities that they do and if the relative percent of time performing these activities is somewhat consistent.

Thus it is important to determine the activities performed and their frequency. When there is a sequence of repetitive activities, designers should determine the sequence in which the performance usually takes place. What is of greatest importance at this stage in the analysis is to have an exhaustive list of activities that are performed by typical users.

Selecting Equipment

Once we have a good idea of who the users are and know what kinds of activities they will be performing, it is time to determine the equipment needed to ensure that we can optimize performance. Some designers use a flow chart to indicate the order in which the activities are done. The same flow chart can be used to indicate the equipment needed at each point along the way to ensure that the performance is done in the shortest period of time and with the fewest errors.

A workplace designer needs to keep in mind that most activities will take place inside the workplace; however, some activities outside the workplace will need to be completed in order for users to successfully complete their work. This may require going to a printer outside their workplace or to a photocopy machine. If such equipment is frequently used and not located in the workplace, it should be moved close to the workplace.

Selecting Furniture

Once we understand the equipment needed for completing day-to-day activities, it is time to focus on the furniture required to house the equipment. Many times this will include a desk—usually an adjustable desk to ensure that it is at exactly the right height for the individual using it. Sometimes a special table for a CRT and keyboard will be required. Other commonly used furniture includes bookshelves, places to store a purse or other personal items, drawers, and coat racks.

After specifying equipment and furniture needs, it is a good idea to spend time finding and selecting equipment and furniture that is ergonomically designed. However, a considerable amount of equipment and furniture available on the market today is erroneously called "ergonomic" by those selling it.

It is a good idea to have two or three alternatives for each piece of equipment and furniture that has been selected. Catalogs from major distributors will help to shorten the time necessary to find the right equipment and furniture.

Workplace Layout

After the equipment and furniture have been selected, it is time to lay out the furniture inside the office. The best way to do this is to draw the office to scale and move scale models of the furniture around until a satisfactory arrangement is found. The sketches should be consistent with the individual needs of users and the usual sequence in which they perform their activities.

Some users may prefer to sit so that they can see outside a window, others may prefer to sit so that they can see people walking by the door, and others may prefer to sit so that they are facing a corner and have a minimum amount of distraction. Such alternatives should be reflected in the way recommendations are prepared for potential users.

If furniture is selected that allows considerable flexibility in moving it around, then users, after seeing the alternatives, can arrange the furniture and equipment in their office to best meet their needs. If the initial layout is not consistent with getting the work done efficiently or helping workers to feel comfortable, they should be encouraged to change it from time to time.

Keep in mind that some users are much more interested in the esthetics of the workplace layout, whereas others are much more interested in a layout that is highly functional. These personal preferences should be incorporated in the sketches provided.

Floors, Walls, and Ceiling

Once the furniture and equipment have been selected and potential layouts have been suggested, we should spend time considering the requirements for flooring, walls, and ceiling. For example, the noise level in an office area can be substantially reduced if the floor is covered with a carpet, if certain walls have sound absorption panels (or heavy wall covering), and if the ceiling is made of acoustical tile.

This is only one set of considerations; there could be others. When selecting the colors for the floors, walls, and ceilings, they should be coordinated so that the person has the feeling that the colors have been harmonized. At the very least, they should not be distracting. In some cases, particularly when using video display terminals, it is a good idea to ensure that the floors do not conduct static electricity and that the walls and ceiling are covered with subdued colors so that glare sources are kept to a minimum.

Evaluating the Environment

The next step is to conduct an environmental analysis. In most offices, the issues of greatest importance include light level, noise level, and temperature control. The light level needs to be adequate for the activities to be performed. Frequently, this is best done by providing *task lighting:* one or more table lamps are provided and can be easily directed on the material being read. This helps to reduce problems with glare and provides flexibility for the user in putting the light where it is most needed. We should be careful when using overhead fluorescent bulbs because of their reputation as a difficult glare source on CRTs. If fluorescent bulbs are used in the ceiling, special steps should be taken to ensure that reflection is not a problem.

Noise can be at least partially controlled, as indicated earlier, by ensuring that the flooring, ceiling, and walls are absorbing and not reflecting sounds. Noise can also be dealt with by using white noise, which can help to ensure privacy within an office or cubicle area.

Finally, it is a good idea to allow users to have control over the temperature level in their workplace. Sometimes this may require having a special portable heating unit; at other times it may require having special ways of cooling the area. Thermal comfort tends to be an individual issue and should be addressed in the same way that we try to ensure sufficient light in a workplace.

Workplace Testing

Once we have satisfied the items just discussed, it is a good idea to conduct a test against the original objectives. For example, if it was originally required that we increase productivity, then, prior to making any changes in the workplace, we should collect information on the existing workplace and the existing productivity level. Once changes have been made, and they may have only been made in one or two offices, designers can then track the productivity of the people involved and compare the new level of productivity with the old base-line level.

If it is comfort or greater satisfaction that is desired, an appropriate questionnaire should be administered prior to starting the workplace layout. Once two or three pilot layouts are completed, it is a good idea to again administer the questionnaire to determine if gains have taken place.

With both productivity and satisfaction, it is a good idea to readminister the test after 3 to 6 months. It is only after this period of time, after people have had an opportunity to adjust to the novelty of the new situation, that you get a good indication of whether the workplace layout has met the goals that were originally set.

Conclusion

Sometimes it requires two steps backward to eventually achieve one step forward. For example, the long-range goal may be to totally change the ceiling tile to make it more esthetically pleasing or sound absorbent and a better absorber of reflected light. However, the first step toward achieving this may be to simply paint the ceiling a darker color so that it reduces the amount of reflected light and reduces glare problems on CRT screens. Painting the ceiling is obviously not a direct step toward changing the ceiling tiles, but it is an interim step that will allow partial compliance with the original set of goals, and at some later time the ceiling can be totally removed and replaced with the ideal solution for this particular problem.

References

Abbott, R. (1983), Program design by informal English descriptions, *Communications of the ACM,* 26(11).

Akscyn, R., Yoder, E., and McCracken, D. (1988), The data model is the heart of interface design. In E. Soloway and D. Frye (eds.), *Proceedings of CHI'88 Human Factors in Computer Systems ACM,* New York, 115–120.

Alluisi, E. A., and Morgan, B. B. (1976), Engineering psychology and human performance, *Annual Review of Psychology,* 27, 305–330.

Alpert, S. R. (1991), Self-describing animated icons for human–computer interaction: A research note, *Behaviour and Information Technology,* 10(2), 149–152.

Altman, I. (1975), *The Environment and Social Behavior.* Monterey, Cal.: Brooks/Cole.

Anastasi, A. (1963), *Psychological Testing.* New York: Macmillan.

Anderson, N. S., Sobiloff, B., White, P., and Pearson, G. (1993), A foot operated PC pointer positioning device, *Proceedings of the Human Factors and Ergonomics Society 37th Annual Meeting,* 314–317.

Angiolillo, J. S., Pople, L. E., and Roberts, L. A. (1990), The evaluation of customer manuals: Initial perceptions of usability, *Proceedings of AT&T Behavioral Sciences Day,* 74–78.

Apsey, R. S. (1978), Human factors of constrained handprint for OCR. *IEEE Transactions on Systems, Man, and Cybernetics,* April, 292–296.

Asch, S. E. (1951), Effects of group pressure upon the modification and distortion of judgements. In Harold Guetzkow (ed.), *Groups, Leadership and Men.* Pittsburgh, Pa.: Carnegie Press, 177–190.

Atwood, G. E. (1969), Experimental studies of mnemonic visualization. Ph.D. dissertation, University of Oregon.

Aucella, A. F., and Ehrlich, S. F. (1986), Voice messaging enhancing the user interface based on field performance, *CHI'86 Conference Proceedings,* 156–161.

Averbach, E., and Coriell, A. S. (1961), Short-term memory in vision. *Bell System Technical Journal,* 40, 309–328.

Avons, S. E., Leiser, R. G., and Carr, D. J. (1989), Paralanguage and human–computer interaction: Identification of recorded vocal segregates, *Behavior and Information Technology,* 8(1), 13–21.

Backman, G. (1924), Korperlange und Tageszeit, *Upsala Lakar. Forhandl.,* 28, 255–282.

Backs, R. W., Walrath, L. C., and Hancock, G. A. (1987), Comparison of horizontal and vertical menu formats, *Proceedings of the Human Factors Society 31st Annual Meeting,* 715–717.

Baddeley, A. D.

——— (1976) *The Psychology of Memory.* New York: Basic Books, Inc.

——— (1992) Working memory, *Science,* 255, 556–559.

Baddeley, A. D., and Patterson, K. (1971), The relationship between long-term and short-term memory. *British Medical Bulletin,* 27, 237–242.

Baecker, R., Small, I., and Mander, R. (1991), Bringing icons to life, *Proceedings of CHI'91,* 1–6.

Bahill, A. T., and Karnavas, W. J. (1993), The perceptual illusion of baseball's rising fastball and breaking curveball, *Journal of Experimental Psychology: Human Perception and Performance,* 19(1), 3–14.

Bailey, G. D. (1993), Iterative methodology and designer training in human–computer interface design, *Proceedings of InterCHI'93,* 198.

Bailey, R. W.

——— (1972) Testing manual procedures in computer-based business information systems, *Proceedings of the 16th Annual Meeting of the Human Factors Society,* October 17–19.

——— (1973) Human error in computer-based data processing systems. *Bell Laboratories Talk Monograph.*

——— (1974) A note on the ability of experienced evaluators to determine the adequacy of a position package. Bell Laboratories Technical Report, May.

——— (1975) Handprinting errors in data systems. *Presented at the 83rd Meeting of the American Psychological Association,* August.

——— (1982) *Human Performance Engineering* (1st ed.), Prentice Hall: Upper Saddle River, N.J.

——— (1989) *Human Performance Engineering* (2nd ed.), Prentice Hall: Upper Saddle River, N.J.

——— (1991) Converting research into reality, *Proceedings of the Human Factors Society 35th Annual Meeting,* 345–349.

——— User interface update—1992, AT&T Bell Laboratories Training Course and Student Guide.

——— (1993) Performance vs. preference, *Proceedings of the Human Factors and Ergonomics Society 37th Annual Meeting,* 282–286.

Bailey, R. W., and Allen, R. W. (1991), Placement of menu choices, *Proceedings of the Human Factors Society 35th Annual Meeting,* 379–382

Bailey, R. W., and Bailey, K. N. (1992), Protoscreens: Rapid Prototyping Software, Bailey & Bailey Software Corporation, Ogden, Utah.

Bailey, R. W., and Koch, C. G. (1976), Position package test validation study (LMOS: Houston), Bell Laboratories Technical Report, August.

Bailey, R. W., and Kulp, M. J. (1971), Methodology for testing the position packages in BISCUS conversion, Bell Laboratories Technical Report, April.

Bailey, R. W., Allen, R. W., and Raiello, P. (1992), Usability testing vs. heuristic evaluation: A head-to-head comparison, *Proceedings of the Human Factors Society 36th Annual Meeting,* 409–413.

Bailey, R. W., Stemen, J. W., and Kersey, T. E. (1975), Human error in broadband operations, Bell Laboratories Technical Report, May.

Baird, J. W. (1917), The legibility of a telephone directory, *Journal of Applied Psychology,* 1, 30–37.

Ballas, J. A. (1993), Common factors in the identification of an assortment of brief everyday sounds, *Journal of Experimental Psychology: Human Perception and Performance,* 19, 250–267.

Barber, T. K. (1965), The effect of "hypnosis" on learning and recall: A methodological critique, *Journal of Clinical Psychology,* 21, 19–25.

Barnard, P. (1974), Presuppositions in active and passive questions. Paper read to the Experimental Psychology Society.

Barnard, P., and Wright, P. (1976), The effect of spaced character formats on the production and legibility of handwritten names, *Ergonomics,* 19, 81–92.

Barnard, P., Wright, P., and Wilcox, P. (1978), The effects of spatial constraints on the legibility of handwritten alphanumeric codes, *Ergonomics,* 21, 73–78.

Baron, R. A., and Bell, P. A. (1976), Aggression and heat: The influence of ambient temperature, negative affect, and a cooling drink on physical aggression, *Journal of Personality and Social Psychology,* 33, 245–255.

Baroudi, J. J., Olson, M. H., and Ives, B. (1986), An empirical study of the impact of user involvement on system usage and information satisfaction, *Communications of the ACM,* 29, 232–238.

Bartlett, F. C.

——— (1932) *Remembering.* New York: Cambridge University Press.

——— (1943) Fatigue following highly skilled work, *Proceedings of the Royal Society,* 131, 247–257.

Bates, J. (1994), The role of emotion in believable agents, *Communications of the ACM,* 37(7), 122–125.

Bayerl, J. P., Millen, D. R., and Lewis, S. H. (1988), Consistent layout of function keys and screen labels speeds user responses, *Proceedings of the Human Factors Society 32nd Annual Meeting,* 344–346.

Bednall, E. S., (1992), The effect of screen format on visual list search, *Ergonomics,* 35(4), 369–383.

Begley, S., Wright, L., Church, V., and Hager, M. (1992), Mapping the brain, *Newsweek,* April 20.

Begoray, J. A. (1990), An introduction to hypermedia issues, systems and application areas, *International Journal of Man–Machine Studies,* 33, 121–147.

Bell, C. R., Provins, K. A., and Hiorns, R. F. (1964), Visual and auditory vigilance during exposure to hot and humid conditions, *Ergonomics,* 7, 279–288.

Bell, P. A., Fisher, J. D., and Loomls, R. J. (1978), *Environmental Psychology.* Philadelphia: W. B. Saunders.

Bellantone, C. E., and Lanzetta, T. M. (1991), Works as advertised: Observations and benefits of prototyping, *Proceedings of the Human Factors Society 35th Annual Meeting,* 324–327.

Benbasat, I., Dexter, A. S., and Todd, P. (1986), An experimental program investigating color-enhanced and graphical information presentation: An integration of the findings, *Communications of the ACM,* 29(11), 1094–1105.

Benbasat, I., and Todd, P. (1993), An experimental investigation of interface design alternatives: Icon vs. text and direct manipulation vs. menus, *International Journal of Man–Machine Studies,* 38, 369–402.

Benbasat, I., and Wand, Y. (1984), Command abbreviation behavior in human–computer interaction, *Communications of the ACM,* 27(4), 376–383.

Benel, D. C. R., Ottens, D., Jr., and Horst, R. (1991), Use of an eyetracking system in the usability laboratory, *Proceedings of the Human Factors Society 35th Annual Meeting,* 461–465.

Benimoff, N. I., and Whitten, W. B. (1989), Human factors approaches to prototyping and evaluating user interfaces, *AT&T Technical Journal,* September/October, 44–55.

Beranek, L. L. (1949) *Acoustic Measurements.* New York: Wiley.

——— (1957) Revised criteria for noise in buildings, *Noise Control, 3*(1), 19.

Bergman, B. A. (1971), The effects of group size, personal space, and success–failure on physiological arousal, test performance, and questionnaire response. Ph.D. dissertation, Temple University.

Biermann, A. W., Rodman, R. D., Rubin, D. C., and Heidlage, J. F. (1985), Natural language with discrete speech as a mode for human-to-machine communication, *Communications of the ACM,* 28(6), 628–636.

Billingsley, P. (1994), The pace quickens, *SIGCHI Bulletin,* July, 8–10.

Bishu, R. R., and Zhan, P. (1992), Increasing or decreasing? The menu direction effect on user performance, *Proceedings of the Human Factors Society 36th Annual Meeting,* 326.

Blackwell, H. R.

——— (1959) Development and use of a quantitative method for specification of interior illumination levels on the basis of performance data, *Illuminating Engineering,* 55, 317–353.

——— (1964) Development of visual task evaluations for use in specifying recommended illumination levels, *Illuminating Engineering,* 59, 627–641.

——— (1972) A human factors approach to lighting recommendations and standards, *Proceedings of the 16th Annual Meeting of the Human Factors Society,* 441–449.

Blades, W. (1892), *Shakespeare and Typography.* In H. B. Wheatley, *Literary Blunders.* London: Elliot Stock, 1909. Reprinted by Gale Research Company, Detroit, 1969.

Blanchard, H. E., and Duncanson, J. (1990), AT&T's ASCII character user interface design specification (ACUIDS), *Proceedings of AT&T Behavioral Sciences Day,* June, 9–12.

Blanchard, H. E., Lewis, S. H., Ross, D., and Cataldo, G. (1993), User performance and preference for alphabetic entry from 10-key pads: Where to put Q and Z?, *Proceedings of the Human Factors and Ergonomics Society 37th Annual Meeting,* 225–228.

Blankenberger, S., and Hahn, K. (1991), Effects of icon design on human–computer interaction, *International Journal of Man–Machine Studies,* 35, 363–377.

Bliss, J. C. (1971), in T. D. Sterling (ed.), *Visual Prosthesis—The Interdisciplinary Dialogue.* New York: Academic Press, 259–263.

Bliss, J. C., Hewitt, D. V., Crane, P. K., Mansfield, P. K., and Townsend, J. T. (1966), Information available in brief tactile presentations, *Perception and Psychophysics,* 1, 272–283.

Blum, M. L., and Mintz, A. (1949), Re-examination of the accident proneness concept, *Journal of Applied Psychology,* 33(3), 195–211.

Blumenthal, A. L. (1966), Observations with self-embedded sentences, *Psychonomic Science,* 453–454.

Bodker, S., and Gronbaek, K. (1991), Cooperative prototyping: Users and designers in mutual activity, *International Journal of Man–Machine Studies,* 34, 453–478.

Boies, S. J. (1974), User behavior on an interactive computer system, *IBM Systems Journal,* 2–18.

Boies, S. J., and Gould, J. D. (1971), User performance in an interactive computer system, *Proceedings of the Fifth Annual Conference on Information Sciences and Systems,* 122.

Bolsky, M. I., and Yuhas, C. M. (1975), Performance Aids for Greater Productivity: How to Plan, Design. Organize and Produce Them, private communication.

Book, W. F. (1908), The psychology of skill: with special reference to its acquisition in typewriting. University of Montana Publications in Psychology; Bulletin No. 53, Psychological Series No. 1.

Boring, E. G. (1929), *A History of Experimental Psychology.* New York: Century Company.

Bower, G. H.

——— (1967) A multi-component theory of the memory track. In K. W. Spence and J. T. Spence (eds.), *The Psychology of Learning and Motivation: Advances in Research and Theory,* Vol. 1. New York: Academic Press, 299–325.

——— (1970) Analysis of a mnemonic device. *American Scientist,* 58, 496–510.

——— (1972) Mental imagery and associative learning. In L. W. Gregg (ed.), *Cognition in Learning and Memory.* New York: Wiley, 51–88.

Bowers, J., and Rodden, T. (1993), Exploding the interface: Experiences of a CSCW network, *Proceedings of InterCHI'93,* 255.

Brayfield, A. H., and Crockett, W. H. (1955), Employee attitudes and employee performance, *Psychological Bulletin,* 396–424.

Breisacher, P. (1971), Neuropsychological effects of air pollution, *American Behavioral Scientist,* 14, 837–864.

Brewster, S. A., Wright, P. C., and Edwards, D. N. (1993), An evaluation of earcons for use in auditory human–computer interfaces, *Proceedings of InterCHI'93,* 222.

Broadbent, D. E.

——— (1958) *Perception and Communication.* New York: Pergamon.

——— (1963) Differences and interaction between stresses. *Quarterly Journal of Experimental Psychology,* 15, 205.

——— (1977) Language and ergonomics, *Applied Ergonomics,* 8(1), 15–18.

——— (1975) The magic number seven after fifteen years, in A. Kennedy and A. Wilkes (eds.), *Studies in Long Term Memory.* New York: Wiley, 3–18.

Brown, C. M. (1988), Comparison of typing and handwriting in two finger typists, *Proceedings of the Human Factors Society 32nd Annual Meeting,* 381–385.

Brown, J. (1958), Some tests of the decay theory of immediate memory, *Quarterly Journal of Experimental Psychology,* 10, 12–21.

Bruner, J. S., and Minturn, A. L. (1955), Perceptual identification and perceptual organization, *Journal of General Psychology,* 53, 21–28.

Bryden, M. P. (1960), Tachistoscopic recognition of non-alphabetical material, *Canadian Journal of Psychology,* 14, 78–86.

Bryden, M. P., Dick, A. O., and Mewhort, D. J. K. (1968), Tachistoscopic recognition of number sequences, *Canadian Journal of Psychology,* 22, 52–59.

Buckler, A. T. (1977), A Review of the Literature on the Legibility of Alphanumerics on Electronic Displays, ADA 040625, May.

Bugelski, B. R. Kddi, E., and Segmen, J., (1968), Image as a mediator in one-trial paired-associate learning, *Journal of Experimental Psychology,* 76, 69–73.

Burchard, G., and Dragga, S. (1989), Computer-based instruction and the humanizing impulse, First Quarter, 13–18.

Burkhart, B., Hemphill, D., and Jones, S. (1994), The value of a baseline in determining design success, *CHI'94 Conference Proceedings,* 386–391.

Buros, O. K. (ed.)

——— (1974) *Tests in Print II; An Index to Tests, Test Reviews and the Literature on Specific Tests.* Gryphon.

——— (1978) *Mental Measurements Yearbook.* Gryphon.

Buros, O. K., Peace, B. A., and Matts, W. L. (eds.) (1961), *Tests in Print, a Comprehensive Bibliography of Tests for Use in Education, Psychology, and Industry.* Gryphon.

Butler, T., Domangue, J., and Felfoldy, G. (1979), Using human estimation to measure work time, private communication.

Byrne, M. C., Wood, S. D., Sukaviriya, P. N., Foley, J. D., and Kieras, D. E. (1994), Automating interface evaluation, *CHI'94 Conference Proceedings,* 232–237.

Byrne, M. D. (1993), Using icons to find documents: Simplicity is critical, *Proceedings of InterCHI'93,* 446.

Callahan, J., Hopkins, D., Weiser, M., and Shneiderman, B. (1988), An empirical comparison of pie vs. linear menus, *CHI'88 Conference Proceedings,* 95–100.

Calvin, W. H. (1994), The emergence of intelligence, *Scientific American,* October, 101–107.

Camacho, M. J., Steiner, B. A., and Berson, B. L. (1990), Icons vs. alphanumerics in pilot-vehicle interfaces, *Proceedings of the Human Factors Society 34th Annual Meeting,* 11–15.

Cann, M. T. (1990), Causes and correlates of age-related cognitive slowing: Effects of task loading and CNS arousal, *Proceedings of the Human Factors Society 34th Annual Meeting,* 149–153.

Cann, M. T., Vercruyssen, M., and Hancock, P. A. (1990), Age and the elderly internal clock: Further evidence for a fundamentally slowed CNS, *Proceedings of the Human Factors Society 34th Annual Meeting,* 1258–1262.

Capindale, R. A., and Crawford, R. G. (1990), Using a natural language interface with casual users, *International Journal of Man–Machine Studies,* 32, 341–362.

Card, S. K., Moran, T. P., and Newell, A. (1986), *The Psychology of Human–Computer Interaction.* Hillsdale, N.J.: Erlbaum.

Card, S. K., Priolli, P., and Mackinlay, J. D. (1994), The cost-of-knowledge characteristic function: Display evaluation for direct-walk dynamic information visualizations, *CHI'94 Conference Proceedings,* 238–244.

Carlson, J. R., and Hall, L. L. (1993), The impact of the design of the software control interface on user performance, *Proceedings of the 5th International Conference on Human–Computer Interaction,* 116–121.

Carmel, E., Whitaker, R. D., and George, J. F. (1993), PD and joint application design: A transatlantic comparison, *Communications of the ACM,* 36(4), 40–48.

Carr, C. (1992), PSS! Help when you need it, *Training & Development,* June, 31–38.

Carroll, J. M., and Aaronson, A. P. (1988), Learning by doing with simulated intelligent help, *Communications of the ACM,* 1988, 31(9), 1064–1079.

Carroll, J. M., and McKendree, J. (1987), Interface design issues for advice-giving expert systems, *Communications of the ACM,* 30(1), 14–31.

Carroll, J. M., and Thomas, J. C. (1988), FUN. *SIGCHI Bulletin,* 19(3), 21–24.

Carroll, J. M., Smith-Kerker, P. L., Ford, J. R., and Mazur-Rimetz, S. A. (1987–88), The minimal manual, *Human–Computer Interaction,* 3, 123–153.

Carswell, C. M. (1991), Boutique data graphics: Perspectives on using depth to embellish data displays, *Proceedings of the Human Factors Society 35th Annual Meeting,* 1532–1536.

Carter, Jr., J. A. (1986), A taxonomy of user-oriented functions, *International Journal of Man–Machine Studies,* 24, 195–292.

Casali, S. P., and Chase, J. D. (1993), The effects of physical attributes of computer interface design on novice and experienced performance of users with physical disabilities, *Human Factors and Ergonomics Society 37th Annual Meeting,* 849–853.

Cazamian, P. (1970), Round table discussion on the social factors in ergonomics, *Proceedings of Fourth International Congress on Ergonomics,* Strassburg.

Chapanis, A.

——— (1959) *Research Techniques in Human Engineering.* Baltimore, Md.: Johns Hopkins Press.

——— (1965a) *Man–Machine Engineering.* Monterey, Calif.: Brooks/Cole.

——— (1965b) Words, words, words, *Human Factors,* 7, 1–17.

——— (1975) Prelude to 2001: Exploration in human communication, *American Psychologist,* 949–961.

Chapanis, A., and Budurka, W. J. (1990), Specifying human–computer interface requirements, *Behaviour and Information Technology,* 9(6), 479–492.

Chapanis, A., and Moulden, J. V. (1990), Short-term memory for numbers, *Human Factors,* 32(2), 123–137.

Chapanis, A., Garner, W. R., and Morgan, C. T. (1949), *Applied Experimental Psychology.* New York: Wiley.

Chin, J. P. (1992), A usability and diary study assessing the effectiveness of call acceptance lists, *Proceedings of the Human Factors Society 36th Annual Meeting,* 216–220.

Chin, J. P., Diehl, V. A., and Norman, K. L. (1988), Development of an instrument measuring user satisfaction of the human–computer interface, *CHI'88 Conference Proceedings,* 213–218.

Clark, E. V. (1971), On the acquisition of the meaning of before and after, *Journal of Verbal Learning and Verbal Behavior,* 10, 266–275.

Clark, M. C., Czaja, S. J., and Weber, R. A. (1990), Older adults and daily living task profiles, *Human Factors,* 32(5), 537–549.

Clarke, A. A., and Smyth, M. G. G. (1993), A co-operative computer based on the principles of human co-operation, *International Journal of Man–Machine Studies,* 38, 3–22.

Clarke, A. C. (1968), *2001: A Space Odyssey.* New York: New American Library.

Cohen, M. (1993), Throwing, pitching and catching sound: Audio windowing models and modes, *International Journal of Human–Computer Studies,* 39, 269–304.

Cohill, A. M. (1989), The human factors design process in software development, in *Designing and Using Human-Computer Interfaces and Knowledge-Based Systems,* G. Salvendy and M. J. Smith (eds.). Amsterdam: Elsevier Science Publishers, 19–27.

Coke, E. U.

——— (1976a) Are bell system practices difficult to read? Measurement of the readability of a sample of BSPs, private communication, July.

——— (1976b) Reading rate, readability and variations in task-induced processing, *Journal of Educational Psychology,* 68, 167–173.

——— (1978) Readability and the evaluation of technical documents, private communication, October.

Coke, E. U., and Koether, M. E. (1979), The reading skills of craft and technical management employees—estimates from two samples of students, private communication, May.

Coll, R. A., Coll, J. H., and Thakur, G. (1994), Graphs and tables a four-factor experiment, *Communications of the ACM,* 37(4), 77–85.

Collins, C. C., and Bach-Y-Rita, P. (1973), Transmission of pictorial information through the skin. *Advances in Biological and Medical Physics,* 14, 285–315.

Coltheart, M. (1972), Visual information processing. In P. C. Dodwell (ed.), *New Horizons in Psychology 2.* New York: Penguin.

Condry, J. (1977), Enemies of exploration: Self-initiated versus other-initiated learning, *Journal of Personality and Social Psychology,* 35, 459–477.

Conklin, J. (1987), Hypertext: An introduction and survey, *IEEE Computer,* 20(9), 17–41.

Conrad, R.

——— (1959) Errors of immediate memory, *British Journal of Psychology,* 50(4), 349–359.

——— (1962) The design of information, *Occupational Psychology,* 36, 159–162.

——— (1967) Designing postal codes for public use, *Ergonomics,* 10(2), 233–238.

——— (1974) Acoustic confusions and immediate memory, *British Journal of Psychology,* 55, 75–84.

Conrad, R., and Hille, B. A. (1957), Memory for long telephone numbers, *Post Office Telecommunications Journal,* 19, 37–39.

Conrad, R., and Hull, A. J.

——— (1964) The preferred layout for numeral data-entry keysets, *Ergonomics.* 11, 165–173.

——— (1967) Copying alpha and numeric codes by hand, *Journal of Applied Psychology,* 51, 444–448.

——— (1968) Input modality and the serial position curve in short-term memory, *Psychonomic Science,* 10, 135–136.

Conrad, R., and Longman, D. J. A. (1965), Standard typewriter versus chord keyboard—An experimental comparison, *Ergonomics,* 8, 77–88.

Conrad, R., and Rush, M. L. (1965), On the nature of short-term memory encoding by the deaf, *Journal of Speech and Hearing Disorders,* 30, 336–343.

Cook, J., and Salvendy, G. (1989), Perception of computer dialogue personality: An exploratory study, *International Journal of Man–Machine Studies,* 31, 717–728.

Corcoran, D. W. J., Carpenter, A., Webster, J. C., and Woodhead, M. M. (1968), Comparison of training techniques for complex sound identification, *Journal of the Acoustical Society of America,* 44, 157–167.

Cornsweet, T. N. (1970), *Visual Perception.* New York: Academic Press.

Courtney, A. J. (1986), Chinese population stereotypes: Color associations, *Human Factors,* 28(1), 97–99.

Cowley, C. K., and Jones, D. M. (1992), More than meets the eye: Issues relating to the application of speech displays in human–computer interaction, *Displays: Technology and Applications,* 13(2), 69.

Crawford, R. P. (1954), *Techniques of Creative Thinking.* New York: Hawthorn.

Crockford, G. W. (1967), Heat problems and protective clothing in iron and steel works. In C. N. Davies, P. R. Davis, and F. H. Tyrer (eds.), *The Effects of Abnormal Physical Conditions at Work.* London: E. and S. Livingston.

Crosby, M. E., and Peterson, W. W. (1991), Using eye movements to classify search strategies, *Proceedings of the Human Factors Society 35th Annual Meeting,* 1476–1480.

Crossley, E. (1980), A systematic approach to creative design, *Machine Design,* March 6.

Crossman, E. R. F. W. (1959), A theory of the acquisition of speed-skill, *Ergonomics,* 2, 153–166.

Cumming, G. (1984), QWERTY and keyboard reform: The soft keyboard option, *International Journal of Man–Machine Studies,* 21, 445–450.

Curtis, B., Krasner, H., and Iscoe, N. (1988), A field study of the software design process for large systems, *Communications of the ACM,* 31(11), 1268–1287.

Cushman, W. H., Ojha, P. S., and Daniels, C. M. (1990), Usable OCR: What are the minimum performance requirements? *CHI'90 Conference Proceedings,* 145–151.

Czaja, S. J., Hammond, K., Blascovich, J. J., and Swede, H. (1989), Age related differences in learning to use a text-editing system, *Behaviour and Information Technology,* 8(4), 309–319.

Damon, A., Stoudt, H. W., and McFarland, R. A. (1966), *The Human Body in Equipment Design.* Cambridge, Mass.: Harvard University Press.

Daniels, G. S., and Churchill, E. (1952), The "average man"? WCRD-TN-53-7, Aero Medical Lab., Wright–Air Development Center, Wright–Patterson AFB, Ohio.

Danielson, R. L., and Nievergelt, J. (1975), An automatic tutor for introductory programming students. *SIGCSE Bulletin,* 7, 47–50.

Davenport, C. B., and Love, A. G. (1975), *Army Anthropology.* Washington, D. C.: U.S. Government Printing Office.

Davies, S. P., Lambert, A. J., and Findlay, J. M. (1989), The effects of the availability of menu information during command learning in a work processing application, *Behaviour and Information Technology,* 8(2), 135–144.

Davis, E. G., and Swezey, R. W. (1983), Human factors guidelines in computer graphics: A case study, *International Journal of Man–Machine Studies,* 18, 113–133.

Davis, G. A. (1973), *Psychology of Problem Solving: Theory and Practice.* New York: Basic Books.

Davis, S., and Botkin, J. (1994), A monstrous opportunity, *Training & Development,* May, 34–35.

Dearborn, W. F., Johnston, P. W., and Carmichael, L. (1951), Improving the readability of typewritten manuscripts, *Proceedings of the National Academy of Sciences,* 37, 670–672.

deCharms, R. (1968), *Personal Causation.* New York: Academic Press.

Deci, E. L. (1975), *Intrinsic Motivation.* New York: Plenum.

DeGreene, K. B. (ed.) (1970), *Systems Psychology.* New York: McGraw-Hill.

DeGroot, J., and Schwab, E. C. (1993), Understanding time-compressed speech: The effects of age and native language on the perception of audiotext and menus, *Proceedings of the Human Factors and Ergonomics Society 37th Annual Meeting,* 244.

Deininger, R. L. (1960), Human factors engineering studies of the design and use of pushbutton telephone sets, *Bell System Technical Journal,* 39(4), 995–1102.

Dennis, J. P. (1965), Some effects of vibration upon visual performance, *Journal of Applied Psychology,* 49, 245–252.

Desurvire, H., Lawrence, D., and Atwood, M. (1991), Empiricism versus judgement: Comparing user interface evaluation methods on a new telephone-based interface, *SIGCHI Bulletin,* October, 58–59.

Deutsch, D. (1973), Interference in memory between tones adjacent on the musical scale, *Journal of Experimental Psychology,* 100, 228–231.

Dever, J. J., Friend, E., Hegarty, J. A., and Rubin, J. J. (1978), Designing and presenting work procedures, private communication.

Devoe, D. B. (1967), Alternatives to handprinting in the manual entry of data, *IEEE Transactions on Human Factors in Electronics,* 21–32.

Devoe, D. B., Eisenstadt, B., and Brown, D. E. (1966), Manual input coding study, RADC-TR66-476.

Dick, A. O.

——— (1969) Relations between the sensory register and short-term storage in tachistoscopic recognition, *Journal of Experimental Psychology,* 82, 279–284.

——— (1970) Visual processing and the use of redundant information in tachistoscopic recognition, *Canadian Journal of Psychology,* 24, 133–141.

Dick, A. O., and Loader, R. (1974), Structural and functional components in the processing, organization, and utilization of tachistoscopically presented information. Technical Report 74-5, Center for Visual Science, University of Rochester.

Diggles-Buckles, V., and Vercruyssen, M. (1990), Age related slowing, S–R compatibility, and stages of information processing, *Proceedings of the Human Factors Society 34th Annual Meeting,* 154–157.

Dillon, A. (1991), Requirements for hypertext applications: The why, what and how approach, *Applied Ergonomics,* August, 258–262.

Dooley, B. B. (1974), Crowding stress: The effects of social density on men with "close" or "far" personal space. Ph.D. dissertation, University of California at Los Angeles.

Dooling, D. J., and Lachman, R. (1971), Effects of comprehension on retention of prose, *Journal of Applied Psychology,* Not available, 216–222.

Doyle, J. R. (1990), Naive users and the Lotus interface: A field study, *Behaviour and Information Technology,* 9(1), 81–89.

Dreyfus, H. L., and Dreyfus, S. E. (1979), The scope, limits and training implications of three models of aircraft emergency response behavior, ORC-79-2.

Dreyfuss, M. (1967), *The Measure of Man: Human Factors in Design* (2nd ed.). New York: Whitney Library of Design.

Drury, C. G.

——— (1987), Hand-held computers for ergonomics data collection, *Applied Ergonomics,* 18(2), 90–94.

——— (1992), A model for movement time on data-entry keyboards, *Ergonomics,* 35(2), 129–147.

Dubois, P. H. (1966), A test-dominated society: China 1115 B.C.–1905 A.D., in A. Anastasi (ed.), *Testing Problems in Perspective.* Washington, D.C.: American Council on Education.

Due, B., Jorgensen, A. H., and Nielsen, J. (1991), An observational study of user interface design practice, *Proceedings of the Human Factors Society 35th Annual Meeting,* 1219–1222.

Duffy, F., Cave, C., and Worthington, J. (eds.) (1978), *Planning Office Space.* New York: Nichols Publishing Co.

Dumais, S. T., and Landauer, T. K. (1982), Psychological investigations of natural terminology for command and query languages, in *Directions in Human/Computer Interaction,* Albert Badre and Ben Shneiderman, (eds.). Norwood, N.J.: Ablex Publishing Corporation.

Dumas, J. S. (1986), The role of human factors in software development, *Proceedings of Computer Graphics 86,* Anaheim, Calif., 530–539.

Duncan, J., and Konz, S. (1976), Legibility of LED and liquid-crystal displays, *SID Journal,* 17(4), 180–186.

Dunnett, C. W. (1972), Drug screening: The never-ending search for new and better drugs, in J. M. Tanur, *Statistics: A Guide to the Unknown.* San Francisco: Holden-Day.

Dutta, A., and Proctor, R. W. (1993), The role of feedback in learning spatially incompatible choice reaction tasks: Does it have one?, *Proceedings of the 5th International Conference on Human–Computer Interaction,* 1320–1324.

Edelson, N., and Danoff, J. (1989), Walking on an electric treadmill while performing VDT office work, *SIGCHI Bulletin,* July, 21(1), 72–76.

Edwards, A. D. N. (1988), The design of auditory interfaces for visually disabled users. In E. Soloway and D. Frye (eds.), *Proceedings of CHI '88 Human Factors in Computer Systems ACM,* New York, 83–88.

Efe, K. (1987), A proposed solution to the problem of levels in error-message generation, *Communications of the ACM,* 948–955.

Egan, D. E., and Timko, K. B. (1993), A super way to search for information, *Bellcore Exchange,* April, 15–19.

Egan, D. E., Remde, J. R., Landauer, T. K., Lochbaum, C. C., and Gomez, L. M. (1989), Behavioral evaluation and analysis of a hypertext browser, *Proceedings of CHI'89,* 205–210.

Egan, J. P. (1948), Articulation testing methods, *Laryngoscope,* 58, 955.

Egido, C., and Patterson, J. (1988), Pictures and category labels as navigational aids for catalog browsing, in E. Soloway and D. Frye (eds.), *Proceedings of CHI '88 Human Factors in Computer Systems ACM,* New York, 127–132.

Ehrenreich, S. L. (1985), Computer abbreviations: Evidence and synthesis, *Human Factors,* 27(2), 143–155.

Ekstrand, B. R. (1972), To sleep, perchance to dream (about why we forget), in C. P. Duncan, L. Sechrest, and A. W. Helton (eds.), *Human Memory: Festschrft in Honor of Benton J. Underwood.* New York: Appleton-Century-Crofts, 59–82.

Elkind, J. (1990), The incidence of disabilities in the United States, *Human Factors,* 32(4), 397–405.

Ellis, S. H. (1977), An investigation of telephone user training methods for a multiservice electronic PBX, in *Proceedings of the eighth international symposium on human factors in telecommunications.* Harlow, Essex, England. Standard Telecommunication Laboratories.

Ellis, S. H., and Coskren, R. A. (1979), New approach to customer training, *Bell Laboratories Record,* 57, 60–65.

Enderwick, T. P. (1990), Some pragmatic issues of measurement, *Proceedings of the Human Factors Society 34th Annual Meeting,* 1248–1252.

Engen, T., and Ross, B. M. (1973), Long-term memory of odors with and without verbal descriptions, *Journal of Experimental Psychology,* 100, 221–227.

Ericsson, K. A., Chase, W. G., and Faloon, S. (1980), Acquisition of a memory skill, *Science,* 208, 1181–1182.

Ertel, H. (1973), Blue is beautiful, *Time,* September.

Eschenbrenner, A. J. (1971), Effects of intermittent noise on the performance of a complex psychomotor task, *Human Factors,* 13(1), 59–63.

Farber, J. M. (1989), The AT&T user-interface architecture, *AT&T Technical Journal,* September/October, 9–16.

Faust, G. W., and O'Neal, A. F. (1977), Instructional science and the evolution of computer assisted instructional systems, *Electro Conference Record,* 34(2), 1–6.

Field, M. M., Hodge, M. H., Manley, C. W., and Sonntag, L. (1971), Guidelines for constructing human performance-based codes, private communication.

Finkelman, J. M. (1975), Effects of noise on human performance, *Sound and Vibration,* 36, 26–28.

Finn, T. A., Cambre, W. J., and Pople, L. E. (1990), Color preference in computer user interfaces: Methodological issues and experimental results, *Proceedings of AT&T Behavioral Sciences Days '90,* June, 224.

Fischer, G., Nakakoji, K., Ostwald, J., Stahl, G., and Sumner, T. (1993), Embedding computer-based critics in the contexts of design, *Proceedings of InterCHI'93,* 157.

Fisher, D. F., Monty, R. A., and Gluckberg, S. (1969), Visual confusion matrices: Fact or artifact? *Journal of Psychology,* 71, 111–125.

Fisher, D. L., and Tan, K. C. (1989), Visual displays: The highlighting paradox, *Human Factors,* 31(1), 17–30.

Fisher, D. L., Yungkurth, E. J., and Moss, S. M. (1990), Optimal menu hierarchy design: Syntax and semantics, *Human Factors,* 32(6), 665–683.

Fisher, R. A., and Yates, F. (1963), *Statistical Tables for Biological, Agricultural and Medical Research* (6th ed.). Edinburgh: Oliver and Boyd.

Fitts, P. M.

——— (1946) German applied psychology during World War II, *American Psychologist,* 1, 151–161.

——— (1966) Cognitive aspects of information processing: Set for speed versus accuracy, *Journal of Experimental Psychology,* 77, 849–857.

Fitts, P., and Posner, M. I. (1967), *Human Performance.* Belmont, Calif.: Wadsworth.

Fleishman, E. A. (1972), On the relation between abilities, learning, and human performance, *American Psychologist,* 27, 1017–1032.

Flesch, R. (1949), *The Art of Readable Writing.* New York: Collier Books.

Fletcher, J. D. (1991), Effectiveness and cost of interactive videodisc instruction in defense training and education, *Multimedia Review,* 33–42.

Foltz, P. W., Davies, S. E., and Polson, P. G. (1988), Transfer between menu systems, in E. Soloway and D. Frye (eds.), *Proceedings of CHI 88 Human Factors in Computer Systems ACM,* New York, 107–114.

Fontenelle, G. A. (1987), A contrast of guideline recommendations and Tullis' prediction model for computer displays: Should text be left-justified? *Proceedings of the Human Factors Society 31st Annual Meeting,* 1226–1228.

Ford, R. N. (1969), *Motivation through the Work Itself.* New York: American Management Association.

Fox, J. A. (1992), The effects of using a hypertext tool for selecting design guidelines, *Proceedings of the Human Factors 36th Annual Meeting,* 428–432.

Fox, W. F. (1967), Human performance in the cold, *Human Factors,* 9(3), 203–220.

Fozard, J. L., Vercruyssen, M., Reynolds, S. L., and Hancock, P. A. (1990), Longitudinal analysis of age-related slowing: BLSA reaction time data, *Proceedings of the Human Factors Society 34th Annual Meeting,* 163–167.

Francik, E. P., and Kane, R. M. (1987), Optimizing visual search and cursor movement in pull-down menus, *Proceedings of the Human Factors Society 31st Annual Meeting,* 722–726.

Franke, R. H., and Kaul, J. D. (1978), The Hawthorne experiments: first statistical interpretation, *American Sociological Review,* 43(5), 623–643.

Frankish, C., Jones, D., and Hapeshi, K. (1991), Decline in accuracy of automatic speech recognition as a function of time on task: Fatigue or voice drift?, *International Journal of Man–Machine Studies,* 36, 797–816.

Frase, L. T., and Schwartz, B. J. (1979), Typographical cues that facilitate comprehension, *Journal of Educational Psychology,* 71, 197–206.

Fraser, B. P., and Lamb, D. A. (1989), An annotated bibliography on user interface design, *SIGCHI Bulletin,* July, 21(1), 17–28.

Freedman, J. (1975), *Crowding and Behavior.* San Francisco: W. H. Freedman.

French, N. R., and Steinberg, J. C. (1947), Factors governing the intelligibility of speech sounds, *Journal of the Acoustical Society of America,* 19, 90.

Freud, S. (1901–1902), *The Psychopathology of Everyday Life.* London: Hogarth Press.

Frohlich, D. M. (1993), The history and future of direct manipulation, *Behaviour & Information Technology,* 12(6), 315–329.

Fung, L., and Ha, A. (1994), Changes in track and field performance with chronological aging, *International Journal of Aging and Human Development,* 38(2), 171–180.

Furnas, G., Landauer, T., Gomez, L., and Dumais, S. (1984), Statistical semantics: Analysis of the potential performance of keyword information systems, in *Human Factors in Computer Systems,* John Thomas and Michael Schneider (eds.). Norwood, N.J.: Ablex Publishing Corporation.

Galanter, E. (1962), Contemporary psychophysics, In R. Brown, E. Galanter, E. H. Hess, and G. Mandler, *New Directions in Psychology 1.* New York: Holt, Rinehart and Winston.

Gallagher, C. C. (1974), The human use of numbering systems, *Applied Ergonomics,* 5(4), 219–223.

Gallwey, W. T. (1974), *The Inner Game of Tennis.* Random House: New York.

Galton, F. (1883), *Inquiries into Human Faculty and Its Development.* New York: Macmillan.

Garner, W. R. (1972), The acquisition and application of knowledge: A symbiotic relation, *American Psychologist,* 941–946.

Garrett, J. W., and Kennedy, K. W. (1971), *A Collation of Anthropometry,* (Vols. I and II), AMRL-TR-68-1, AD723630.

Gaver, W. W. (1986), Auditory icons: Using sound in computer interfaces, *Human–Computer Interaction,* 2, 167–177.

Gaver, W. W., Smith, R. B., and O'Shea, T. (1991), Effective sounds in complex systems, *Proceedings of CHI'91,* 85–90.

Gawron, V. J., Drury, C. G., Czaja, S. J., and Wilkins, D. M. (1989), A taxonomy of independent variables affecting human performance, *International Journal of Man–Machine Studies,* 31, 643–672.

Geldard, F. A. (1960), Some neglected possibilities of communication, *Science,* 131, 1583.

Gentner, D. R., and Grudin, J. (1990), Why good engineers (sometimes) create bad interfaces, *CHI'90 Conference Proceedings,* 277–282.

Gignoux, M., Martin, H., and Cajgfinger, H. (1966), Troubles Cochles—Vestibulaires apres Tentative de Suicide a Laspirine, *Journal of Franc. Oto-rhinolarvng,* 15, 631–635.

Gilad, I., and Pollatschek, M. A. (1986), Layout simulation for keyboards, *Behaviour and Information Technology,* 5(3), 273–281.

Gilbert, T. (1978), *Human Competence: Engineering Worthy Performance.* New York: McGraw-Hill.

Gilbert, W. (1979), CAI notes, *USE Inc. Newsletter,* 79, 9.

Gilbreth, F. B. (1911), *Brick Laying System.* New York: Clark Publishing Co.

Gilbreth, F. B., and Gilbreth, L. M. (1917), *Applied Motion Study.* New York: Macmillan.

Gill, R., Gordon, S., Dean, S., and McGehee, D. (1991), Integrating cursor control into the computer keyboard, *Proceedings of the Human Factors Society 35th Annual Meeting,* 256–258.

Gillan, D. J., and Breeding, S. D. (1990), Designers' models of the human–computer interface, *CHI'90 Conference Proceedings,* 391–398.

Gillan, D. J., and Neary, M. (1992), A componential model of human interaction with graphs: Effects of the distances among graphical elements, *Proceedings of the Human Factors Society 36th Annual Meeting,* 365.

Gilmer, B. H. (1970), *Psychology.* New York: Harper & Row.

Gilson, E. Q., and Baddeley, A. D. (1969), Tactile short-term memory, *Quarterly Journal of Experimental Psychology,* 21, 180–184.

Gingras, B. (1987), Simplified English in maintenance manuals, *Technical Communications,* first quarter, 24–28.

Givon, M. M., and Goldman, A. (1987), Perceptual and preferential discrimination abilities in taste tests, *Journal of Applied Psychology,* 72(2), 301–306.

Glass, A. L., Holyoak, K. J., and Santa, J. L. (1979), *Cognition.* Reading, Mass.: Addison-Wesley.

Glass, D. C., and Singer, J. E., 1972, *Urban Stress.* New York: Academic Press.

Glenn, J. W. (1975), Machines you can talk to, *Machine Design,* 72–75.

Glodman-Eisler, F., and Cohen, M. (1970), Is N, P, and PN difficulty a valid criterion of transformational operations?, *Journal of Verbal Learning and Verbal Behavior,* 9, 161–166.

Glucksberg, S., and Cowan, G. N. (1970), Memory for non-attended auditory material, *Cognitive Psychology,* 1, 149–156.

Goldberg, D., and Richardson, C. (1993), Touch-typing with a stylus, *Proceedings of InterCHI'93,* 80.

Goldschmidt, E. (ed.) (1963), *The Genetics of Migrant and Isolate Populations.* Baltimore: Williams and Wilkens.

Gomez, L. M., Egan, D. E., and Bowers, C. (1986), Learning to use a text editor: Some learner characteristics that predict success, *Human–Computer Interaction,* 2, 1–23.

Gong, R., and Elkerton, J. (1990), Designing minimal documentation using a GOMS model: A usability evaluation of an engineering approach, *CHI'90 Conference Proceedings,* 99–106.

Gong, R., and Kieras, D. (1994), A validation of the GOMS model methodology in the development of a specialized, commercial software application, *CHI'94 Conference Proceedings,* April 24–28, 351–357.

Good, M., Spine, T. M., Whiteside, J., and George, P. (1986), User derived impact analysis as a tool for usability engineering, *CHI'86 Conference Proceedings,* 241–246.

Good, M. D., Whiteside, J. A., Wixon, D. R., and Jones, S. J. (1984), Building a user-derived interface, *Communications of the ACM,* 27(10), 1032–1043.

Goodwin, N. C. (1987), Functionality and usability, *Communications of the ACM,* 30(3), 229–233.

Goransson, B., Lind, M., Pettersson, E., Sandblad, B., and Schwalbe, P. (1987), The interface is often not the problem, *Proceedings of the Human Factors in Computing Systems and Graphics Interface Conference,* 133–136.

Gough, P. B. (1965), Grammatical transformations and speed of understanding, *Journal of Verbal Learning and Verbal Behavior,* 4, 107–117.

Gould, J. D. (1988), Designing for usability: The next iteration is to reduce organizational barriers, *Proceedings of the Human Factors Society 32nd Annual Meeting,* 1–9.

Gould, J. D., and Grischkowsky, N. (1984), Doing the same work with hard copy and with cathode ray tube (CRT) computer terminals, *Human Factors,* 26, 323–337.

Gould, J. D., and Lewis, C. (1985), Designing for usability: Key principles and what designers think, *Communications of the ACM,* 28(6), 300–311.

Gould, J. D., Alfaro, L., Finn, R., Haupt, B., and Minuto, A. (1987a), Reading from CRT displays can be as fast as reading from paper, *Human Factors,* 29(5), 497–517.

Gould, J. D., Alfaro, L., Finn, R., Haupt, B., Minuto, A., and Salaun, J. (1987b), Why reading was slower from CRT displays than from paper, *Proceedings of the Human Factors in Computing Systems and Graphics Interface Conference,* April, 7–11.

Gould, J. D., Alfaro, L., Barnes, V., Finn, R., Grischkowsky, N., and Minuto, A. (1987c), Reading is slower from CRT displays than from paper: Attempts to isolate a single-variable explanation, *Human Factors,* 29(3), 269–299.

Gould, J. D., Boies, S. J., Levy, S., Richards, J. T., and Schoonard, J. (1987a), The 1984 olympic message system: A test of behavioral principles of system design, *Communications of the ACM,* 30(9), 758–769.

Gould, J. D., Boies, S. J., Meluson, M., Rasamny, M., and Vosburgh, A. (1989), Entry and selection methods for specifying dates, *Human Factors,* 31(2), 199–214.

Graham, C. H. (ed.) (1965), *Vision and Visual Perception.* New York: Wiley.

Granda, R. E., Halstead-Nussloch, R., and Winters, J. M. (1990), The perceived usefulness of computer information sources: A field study, *SIGCHI Bulletin,* 21(4), 35–43.

Greaves, C., Warren, M., and Ostberg, O. (1993), Enhancing speech intelligibility using visual images, *Proceedings of the 5th International Conference on Human–Computer Interaction,* 1097–1102.

Green, H. J., and Spencer, J. P. (1969), *Drugs with Possible Ocular Side Effects.* Woburn, Mass.: Butterworths.

Greenberg, S., and Witten, I. H. (1988), How users repeat their actions on computers: Principles for design of history mechanisms. In E. Soloway and D. Frye (eds.), *Proceedings of CHI '88 Human Factors in Computer Systems ACM,* New York, 171–178.

Greene, J. M. (1972), *Psycholinguistics: Chomsky and Psychology.* New York: Penguin.

Greene, S., Gould, J. D., Boies, M. R., Rasamny, M., and Meluson, A. (1992), Entry and selection based methods of human–computer interaction, *Human Factors,* 34(1), 97–113.

Gregory, R. L. (1966), *Eye and Brain: The Psychology of Seeing.* London: World University Library.

Griffith, R. T. (1949), The minimotion typewriter keyboard, *Journal of the Franklin Institute,* 399–436.

Griffiths, I. D., and Boyce, P. R. (1971), Performance and thermal comfort, *Ergonomics,* 14, 457–468.

Grivel, F., and Candas, V. (1991), Ambient temperatures preferred by young European males and females at rest, *Ergonomics,* 34(3), 365–378.

Grudin, J.

——— (1989), The case against user interface consistency, *Communications of the ACM,* 32(10), 1164–1173.

——— (1990), The computer reaches out: The historical continuity of interface design, *CHI'90 Conference Proceedings,* 261–268.

——— (1991), Systematic sources of suboptimal interface design in large product development organizations, *Human–Computer Interaction,* 6, 147–196.

Gugerty, L. (1993), The use of analytical models in human–computer interface design, *International Journal of Man–Machine Studies,* 38(4), 625–660.

Guignard, J. C. (1965), Noise and vibration, in J. A. Gillies (ed.), *A Textbook of Aviation Physiology.* Elmsford, N.Y.: Pergamon.

Guillemette, R. A. (1989), Development and validation of a reader based documentation measure, *International Journal of Man–Machine Studies,* 30, 551–574.

Guindon, R. (1988), How to interface to advisory systems? Users request help with a very simple language, *CHI'88 Conference Proceedings,* 191–196.

Gusinde, M. (1948), *Urwaldmenschen am Ituri.* Vienna: Springer-Verlag.

Guth, S., and Eastman, A. (1955), Lighting for the forgotten man, *American Journal of Optometry,* 32, 413–421.

Haas, C. (1989), Does the medium make a difference? Two studies of writing with pen and paper with computers, *Human–Computer Interaction,* 4, 149–169.

Haas, C., and Hayes, J. R. (1986), What did I just say? Reading problems in writing with the machine, *Research in the Teaching of English,* 20(1), 22–35.

Haber, R. N., and Hershenson, M. (1973), *The Psychology of Visual Perception.* New York: Holt, Rinehart and Winston.

Hackman, G. S., and Biers, D. W. (1992), Team usability testing: Are two heads better than one? *Proceedings of the Human Factors Society 36th Annual Meeting,* 1205–1209.

Hakes, D. T., and Foss, D. J. (1970), Decision processes during sentence comprehension: Effects of surface structure reconsidered, *Perception and Psychophysics,* 413–416.

Halasz, F., and Schwartz, M. (1994), The Dexter hypertext model, *Communications of the ACM,* 37(2), 30–39.

Halgren, S. L., and Cooke, N. J. (1993), Towards ecological validity in menu research, *International Journal of Man–Machine Studies,* 39, 51–70.

Hall, E. T. (1966), *The Hidden Dimension.* New York: Doubleday.

Halstead-Nussloch, R. (1989), The design of phone-based interfaces for consumers, *Proceedings of CHI'89,* 347–352.

Hammer, W. (1972), *Handbook of System and Product Safety,* Upper Saddle River, N.J.: Prentice Hall.

Hammond, N., Long, J., Clark, I., Barnard, P., and Morton, J. (1980), Documenting human-computer mismatch in interactive systems, *Proceedings of the Ninth International Symposium on Human Factors in Telecommunications.*

Harpster, J. L. (1987), Random versus ordered menus in self-terminating menu searches, *Proceedings of the Human Factors Society 31st Annual Meeting,* 718–721.

Harris, C. S. (1980), Visual Coding and Adaptability, Hillsdale, NJ: Lawrence Erlbaum Associates.

Harris, C. S. (1980), Insight or out of sight? Two examples of perceptual plasticity in the human adult. In C. S. Harris (ed.), *Visual Coding and Adaptability.* Hillsdale, N.J.: Erlbaum.

Harris, C. S., and Shoenberger, R. W. (1970), Combined effects of noise and vibration on psychomotor performance, AMRL-TR-70-14, Wright-Patterson Air Force Base, Ohio.

Harris, H. A. (1972), *Sport in Greece and Rome.* London: Thames and Hudson.

Harris, J. R., and Harris, C. S. (1989), Through the looking glass; rapid adaptation to right–left reversal of the visual field, *Psychological Research.*

Harris, R. A. (1991), A do-it-yourself usability kit, *Journal of Technical Writing and Communication,* 21(4), 351–368.

Hartley, J., and Burnhill, P.

——— (1975) Explorations in space, *Bulletin of the British Psychological Society,* 235.

——— (1977) Fifty guide-lines for improving instructional text, *Programmed Learning and Educational Technology,* 14, 65–73.

Hartson, H. R., and Boehm-Davis, D. (1993), User interface development processes and methodologies, *Behaviour & Information Technology,* 12(2), 98–114.

Harwood, K., and Foley, P. (1987), Temporal resolution: An insight into the video display terminal problem, *Human Factors,* 29(4), 447–452.

Hayhoe, D. (1990), Sorting-based menu categories, *International Journal of Man–Machine Studies,* 33, 677–695.

Hays, W. L. (1973), *Statistics for the Social Sciences* (2nd ed.). New York: Holt, Rinehart and Winston.

Heaton, N. O. (1992), Defining usability, *Displays: Technology and Application,* 13(3), 147–150.

Hebb, D. O. (1949), *The Organization of Behavior.* New York: Wiley.

Heimstra, N. W., and Ellingstad, V. S. (1972), *Human Behavior: A Systems Approach.* Monterey, Calif.: Brooks/Cole.

Helander, M. (1989), *Handbook of Human–Computer Interaction.* Amsterdam: North Holland, 465.

Henneman, R. L., Inderrieden, M., Anderson, A., and Taylor, B. (1990), Evolutionary design of a customer activated terminal: A case study, *Proceedings of the Human Factors Society 34th Meeting,* 300–304.

Heron, W. (1957), Perception as a function of retinal locus and attention, *American Journal of Psychology, 70,* 38–48.

Herring, R. D. (1990), Evaluative methods for rapid prototypes, *Proceedings of the Human Factors Society 34th Annual Meeting,* 287–290.

Hershman, R. L., and Hillix, W. A. (1965), Data processing in typing, *Human Factors,* 7, 483–492.

Hertzberg, F. (1966), *Work and the Nature of Man.* Cleveland, Ill.: World Publishing Co.

Hertzberg, F., Mausner, B., and Snyderman, B. (1959), *The Motivation to Work* (2nd ed.). New York: Wiley.

Hewett, T. T. (1990), Evaluative methods for rapid prototypes, *Proceedings of the Human Factors Society 34th Annual Meeting,* 325–328.

Hewett, T. T., and Meadow, C. T. (1986), On designing for usability: An application of four key principles, *CHI '86 Conference Proceedings,* 247–252.

Hilgard, E. R., and Bower, G. H. (1975), *Theories of Learning.* Upper Saddle River, N. J.: Prentice Hall.

Hill, G. W., Gunn, W. A., Martin, S. L., and Schwartz, D. R. (1991), Perceived difficulty and user control in mouse usage, *Proceedings of the Human Factors Society 35th Annual Meeting,* 295–299.

Hill, J. W. (1967), Applied problems of hot work in the glass industry, in C. N. Davies, P. R. Davis, and F. H. Tyrer (eds.), *The Effects of Abnormal Physical Conditions at Work.* London: E. and S. Livingstone.

Hillman, D. J. (1985), Artificial intelligence, *Human Factors,* 27(1), 21–31.

Hirsch, R. S. (1970), Effects of standard versus alphabetical keyboard on typing performance, *Journal of Applied Psychology,* 54(6), 484–490.

Hirschleim, R., and Klein, H. K. (1989), Four paradigms of information systems development, *Communications of the ACM,* 32(10), 1199–1216.

Hoadley, E. D. (1990), Investigating the effects of color, *Communications of the ACM,* 32(2), 120–125.

Hodge, M. H., and Field, M. M. (1970), Human coding processes, *University of Georgia Report.*

Hodge, M. H., and Pennington, F. M. (1973), Some studies of word abbreviation behavior, *Journal of Experimental Psychology,* 98, 350–361.

Hollan, J. D. (1992), Visualizing information, *Bellcore Exchange,* July/August, 13–17.

Holleran, P. A. (1991), A methodological note on pitfalls in usability testing, *Behaviour & Information Technology,* 10(5), 345–357.

Holtzblatt, K., and Beyer, H. (1993), Making customer-centered design work for teams, *Communications of the ACM,* 36(10), 93–103.

Horgan, J. (1994), Experiments reveal links between memory and sleep, *Scientific American,* 32–33.

Howard, C., O'Boyle, M. W., Eastman, V., Andre, T., and Motoyama, T. (1991), The relative effectiveness of symbols and words to convey photocopier functions, *Applied Ergonomics,* 22(4), 218–224.

Howe, J. A. (1970), *Introduction to Human Memory.* New York: Harper & Row.

Howell, W. C., and Kreidler, D. L. (1963), Information processing under contradictory instructional sets, *Journal of Experimental Psychology,* 65, 39–46.

Howes, A. (1994), A model of the acquisition of menu knowledge by exploration, *CHI'94 Conference Proceedings,* 445–451.

Howes, D. H. (1957), On the relation between the intelligibility and frequency of occurrence of English words, *Journal of the Acoustical Society of America,* 29, 296.

Hoyos, C. G., Gstalter, H., Strube, V., and Zang, B. (1987), Software design with the rapid prototyping approach: A survey and some empirical results, in G. Salvendy (ed.), *Cognitive Engineering in the Design of Human–Computer Interaction and Expert Systems.* Elsevier: Amsterdam, 329–340.

Hsia, Y., and Graham, C. H. (1965), *Color blindness,* in C. H. Graham (ed.), *Vision and Visual Perception.* New York: Wiley.

Hsu, S.-H., and Wu, S.-P. (1991), An investigation for determining the optimum length of chopsticks, *Applied Ergonomics,* 22(6), 395–400.

Hull, A. J.

——— (1973) A letter-digit matrix of auditory confusions, *British Journal of Psychology,* 64(4), 579–585.

——— (1975) Nine codes: A comparative evaluation of human performance with some numeric, alpha and alphanumeric coding systems, *Ergonomics,* 18, 567–576.

Human Factors Society (1988), American National Standard for Human Factors Engineering of Visual Display Terminal Workstations, ANSI/HFS 100, Santa Monica, Calif.

Hunt, J. McV. (1965), Intrinsic motivation and its role in psychological development, in D. Levine (ed.), *Nebraska Symposium on Motivation,* vol. 13. Lincoln: University of Nebraska Press.

Hunt, S., and Schur, A. (1991), The user oriented evaluation process: A process for preserving user needs during iterative system test and evaluation, *Proceeding of the Human Factors Society 35th Annual Meeting,* 1336–1339.

Hunter, I. M. L. (1957), *Memory: Facts and Fallacies.* New York: Penguin.

Huuhtanen, P., Leino, T., Miemela, T., and Shola, K. (1993), Mastering the changes in information technology: A follow-up study of insurance tasks, *Proceedings of the 5th International Conference on Human–Computer Interaction,* 703–708.

Imada, A. S. (1990), Ergonomics: Influencing management behaviour, *Ergonomics,* 33(5), 621–628.

Israelski, E. W. (1977), *Human Factors Handbook for Telecommunications Product Design,* private communication.

Ives, B., and Olson, M. H. (1984), User involvement and MIS success: A review of research, *Management Science,* 30(5), 586–603.

Jackson, J. C., and Roske-Hofstrand, R. J. (1989), Circling: A method of mouse-based selection without button presses, *Proceedings of CHI'89,* 161–166.

Jacob, R. J. K. (1990), What you look at is what you get: Eye movement-based interaction techniques, *CHI'90 Conference Proceedings,* 11–18.

Jacob, V. S., Gaultney, L. D., and Salvendy, G. (1986), Strategies and biases in human decision-making and their implications for expert systems, *Behaviour and Information Technology,* 5(2), 119–140.

Jahnke, J. (1963), *Journal of Verbal Learning and Verbal Behavior.* New York: Academic Press.

James, W. (1914), *Habit.* New York: Henry Holt and Co.

Jamison, D., Suppes, P., and Wells, S. (1974), The effectiveness of alternative instructional media: A survey, *Review of Educational Research,* 44(1), 1–67.

Janosky, B., Smith, P. J., and Hildreth, C. (1986), Online literary catalog systems: An analysis of user errors, *International Journal of Man–Machine Studies,* 25, 573–592.

Jarvenpaa, S. L., and Dickson, G. W. (1988), Graphics and managerial decision making: Research based guidelines, *Communications of the ACM,* 31(6), 764–774.

Jarvis, J. F. (1966), A case of unilateral permanent deafness following acetyl salicylic acid, *Journal of Laryngology,* 80, 318–320.

Jeffries, R., Miller, J. R., Wharton, C., and Uyeda, K. M. (1990), User interface evaluation in the real world: A comparison of four techniques, *Proceedings of CHI'91,* 119–124.

Jellinek, H. D., and Card, S. K. (1990), Powermice and user performance, *CHI'90 Conference Proceedings,* 213–220.

Jenkins, J. G., and Dallenbach, K. M. (1924), Obliviscence during sleep and waking, *American Journal of Psychology,* 35, 605–612.

Jerrams-Smith, J. (1987), An expert system within a supportive interface for UNIX, *Behaviour and Information Technology,* 6(1), 37–41.

John, B. E. (1980), Thoughts on System Design, Analysis, and Human/Machine Allocation, private communication.

——— (1990), Extensions of GOMS analysis to expert performance requiring perception of dynamic visual and auditory information, *CHI'90 Conference Proceedings,* 107–115.

Johnsgard, T. (1994), Fitts' law with a virtual reality glove and a mouse: Effects of gain, *Graphics Interface'94,* 8–15.

Johnsgard, T. J., Page, S. R., Wilson, R. D., and Zeno, R. J. (1995), A comparison of graphical user interface widgets for various tasks, *Proceedings of the Human Factors and Ergonomics Society 39th Annual Meeting.*

Johnson, H., and Johnson, P. (1989), Integrating task analysis into system design: Surveying designer's needs, *Ergonomics,* 32(11), 1451–1467.

Johnson, R. E. (1970), Recall of prose as a function of the structural importance of the linguistic units, *Journal of Verbal Learning and Verbal Behavior,* 9, 12–20.

Johnson, W., Jellinek, H., Klotz, L., Jr., Rao, R., and Card, S. (1993), Bridging the paper and electronic worlds: The paper user interface, *Proceedings of InterCHI'93,* 507.

Johnston, J. C., and McClelland, J. L. (1974), Perception of files in words: Seek not and ye shall find, *Science,* 184, 1192–1194.

Jones, E. M., and Munger, S. J. (1969), The development of performance-based coding principles, *American Institutes for Research Reports.*

Jones, S. (1968), *Design of Instructions.* London: Her Majesty's Stationary Office.

Jonides, J. (1968), Toward a model of attention shifts, in A. L. Glass, K. J. Holyoak, and J. L. Santa, *Cognition.* Reading, Mass.: Addison-Wesley.

Jorgensen, A. H. (1990), Thinking aloud in user interface design: A method promoting cognitive ergonomics, *Ergonomics,* 33(4), 501–507.

Joyce, R., and Gupta, G. (1990), Identity authentication based on keystroke latencies, *Communications of the ACM,* February, 33(2), 168–176.

Joyce, R. P., Chenzoff, A. P., Mulligan, J. F., and Mallory, W. J. (1973), *Fully Proceduralized Job Performance Aids,* Vols. I, II, and III, AFHRL-TR-73-43.

Kabbash, P., Mackenzie, I. S., and Buxton, W. (1993), Human performance using computer input devices in the preferred and non-preferred hands, *Proceedings of InterCHI'93,* 474.

Kacmar, C. J., and Carey, J. M. (1991), Assessing the usability of icons in user interfaces, *Behaviour & Information Technology,* 10(6), 443–457.

Kantorowitz, E., and Sudarsky, O. (1989), The adaptable user interface, *Communications of the ACM,* 32(11), 1352–1358.

Kantowitz, B. H. (1989), The role of human information processing models in system development, *Proceedings of the Human Factors Society 33rd Annual Meeting,* 1059–178.

Kaplan, R. (1974), Some methods and strategies in the prediction of preference, in E. H. Zube, J. G. Fabos, and R. O. Brush (eds.), *Landscape Assessment: Values, Perceptions, and Resources.* Stroudsburg, Pa.: Dowden, Hutchinson, and Ross.

Kaplan, S. (1974), An informal model for the prediction of preference, in E. H. Zube, J. G. Fabos, and R. O. Brush (eds.), *Landscape Assessment: Values, Perceptions, and Resources.* Stroudsburg, Pa.: Dowden, Hutchinson, and Ross.

Karat, C. M.

——— (1989), Iterative usability testing of a security application, *Proceedings of the Human Factors Society 33rd Annual Meeting,* 273–277.

——— (1990), Cost–benefit analysis of usability engineering techniques, *Proceedings of the Human Factors Society 34th Annual Meeting,* 839–843.

——— (1992), Cost-justifying human factors support on software development projects, *Human Factors Bulletin,* 35(11), 1–4.

Karat, C. M., Campbell, R., and Fiegel, T. (1992), Comparison of empirical testing and walkthrough methods in user interface evaluation, *CHI'92 Conference Proceedings,* 397–404.

Karat, J., McDonald, J. E., and Anderson, M. (1986), A comparison of menu selection techniques: Touch panel, mouse and keyboard, *International Journal of Man–Machine Studies,* 25, 73–88.

Karlin, J. E. (1977), The changing and expanding role of human factors in telecommunication engineering at Bell Laboratories, in *Proceedings of the Eighth International Symposium on Human Factors in Telecommunications.* Harlow, Essex, England: Standard Telecommunication Laboratories.

Katz, P. (1937), *Animals and Men.* Longmans Green.

Kaufman, J. E. (1972), *IES Lighting Handbook.* New York: Illuminating Engineering Society.

Kay, H. (1951), Learning of a serial task by different age groups, *Quarterly Journal of Experimental Psychology,* 3, 166–183.

Kazan, B. (1976), Electroluminescent displays, *Proceedings of the S.I.D.,* 17, 23–29.

Kearsley, G. (1985), Automation in training and education, *Human Factors,* 27(1), 61–74.

Keele, S. W. (1968), Movement control in skilled motor performance, *Psychological Bulletin,* 70(6), 387–403.

——— (1973), *Attention and Human Performance.* Pacific Palisades, Calif.: Goodyear Publishing Co.

Keele, S. W., and Chase, W. G. (1967), Short-term visual storage, *Perception and Psychophysics,* 2, 383–386.

Keister, R. S. (1989), The content of help screens: Users versus developers, *Proceedings of the Human Factors Society 33rd Annual Meeting,* 390–393.

Kelley, J. F., and Ukelson, J. (1992), COAS: Combined object–action selection: A human factors experiment, *Proceedings of the Human Factors Society 36th Annual Meeting,* 316.

Kemeny, J. G. (1979), *The Accident at Three Mile Island,* Report of the President's Commission. Washington, D.C.: Government Printing Office.

Kempton, W. (1986), Two theories of home heat control, *Cognitive Science,* 10, 75–90.

Kemske, F., and Weingarten, N. J. (1989), A memo from the future: Data training in 1999, *Data Training,* November, 16–21.

Kennedy, K. W. (1964), The human body in equipment design: Reach capability of the USAF population, AMRL-TDR-64-59.

Keyson, D. K., and Parsons, K. C. (1990), Designing the user interface using rapid prototyping, *Applied Ergonomics,* 21(3), 207–211.

Kherumian, R., and Pickford, R. W. (1959), *Heredite et Frequence des Anomalies Congenitales du Sens Chromatique.* Paris: Vigot Freres.

Kieras, D. E., and Bovair, S. (1984), The role of a mental model in learning to operate a device, *Cognitive Science,* 8, 255–273.

Kieras, D. E., and Polson, P. G. (1985), An approach to the formal analysis of user complexity, *International Journal of Man–Machine Studies,* 22, 365–394.

Kincaid, J. P., Fishburne, R. P., Rogers, R. L., and Chissom, B. S. (1975), Derivation of new readability formulas (Automated Readability Index, Fog Count, and Flesch Reading Ease Formula) for Navy enlisted personnel, Naval Training Command Research Branch Report 8-75, February.

Kincaid, J. P., Thomas, M., Strain, K., Couret, I., and Bryden, K. (1990), Controlled English for international technical communication, *Proceedings of the Human Factors Society 34th Annual Meeting,* 815–819.

Klare, C. R., Mabry, J. E., and Gustafson, L. M. (1955), The relationship of patterning (underlining) to immediate retention and to acceptability of technical material, *Journal of Applied Psychology, 39,* 40–42.

Klare, G. R.

——— (1975) Assessing readability, *Reading Research Quarterly,* 10, 62–102.

——— (1975) A second look at the validity of readability formulas. Invited address at the annual meeting of the National Reading Conference, St. Petersburg, Florida.

Klare, G. R., Nichols, W. H., and Shuford, E. M. (1957), The relationship of typographical arrangement to the learning of technical material, *Journal of Applied Psychology,* 41(1), 41–45.

Klemmer, E. T. (1969), Grouping of printed digits for manual entry, *Human Factors,* 11(4), 397–400.

Klemmer, E. T., and Lockhead, G. R. (1962), Productivity and errors in two keying tasks: A field study, *Journal of Applied Psychology,* 46, 401–408.

Kling, J. W., and Riggs, L. A. (1971), *Experimental Psychology* (Chapter 7: The Chemical Senses). New York: Holt, Rinehart and Winston.

Knox, S. T., Bailey, W. A., and Lynch, E. F. (1989), Directed dialogue protocols: Verbal data for user interface design, *Proceedings of CHI'89,* 283–287.

Kohler, I.

——— (1962) Experiments with goggles, *Scientific American,* 206(5), 62–72.

——— (1964) The formation and transformation of the perceptual world, *Psychological Issues,* 3(4), Monograph 12.

Kolodner, J. L. (1983), Toward an understanding of the role of experience in the evolution from novice to expert, *International Journal of Man–Machine Studies,* 19, 497–518.

Kornhauser, A. (1965), *Mental Health of the Industrial Worker.* New York: Wiley.

Kotsonis, M. E., and Lehder, D. Z. (1990), If we had our way: Organizational alternatives for behavioral scientists, *Human Factors Society 34th Annual Meeting* (poster session).

Koved, L., and Shneiderman, B. (1986), Embedded menus: Selecting items in context, *Communications of the ACM,* 29(4), 312–318.

Kreifeldt, J. G., and Levine, S. L. (1989), Reduced keyboard designs using disambiguation, *Proceedings of the Human Factors Society 33rd Annual Meeting,* 441–444.

Kroemer, K. H. E.

——— (1972) Human engineering the keyboard, *Human Factors,* 14(1), 51–63.

——— (1975) Muscle strength as a criterion in control design for diverse populations, in A. Chapanis (ed.), *Ethnic Variables in Human Factors Engineering.* Baltimore, Md.: Johns Hopkins University Press.

——— (1970) Human strength: Terminology, measurement and interpretation of data, *Human Factors,* 12, 297–313.

——— (1974) Designing for muscular strength of various populations, AMRL-TR-72-46, Wright–Patterson Air Force Base: Aerospace Medical Research Laboratory.

Kruk, R. S., and Muter, P. (1984), Reading of continuous text on video screens, *Human Factors,* 26(3), 339–345.

Kryter, K. D.

——— (1970) *The Effects of Noise on Man.* New York: Academic Press.

——— (1972) Speech communication, in H. P. Van Cott and R. G. Kinkade, *Human Engineering Guide to Equipment Design.* Washington, D.C.: U.S. Government Printing Office.

Kryter, K. D., and Whitman, E. C. (1965), Some comparisons between rhyme and PB—word intelligibility tests, *Journal of the Acoustical Society of America,* 39, 1146.

Kubovy, M., and Howard, F. P. (1976), Persistence of a pitch segregating echoic memory, *Journal of Experimental Psychology: Human Perception and Performance,* 2, 531–537.

Kucera, H., and Francis, W. N. (1967), *Computational Analysis of Present-day American English.* Providence, R.I.: Brown University Press.

Kulp, M. J. (1976), The effects of position practice readability level on performance, private communication.

Lachnit, H., and Pieper, W. (1990), Speed and accuracy effects of fingers and dexterity in 5-choice reaction tasks, *Ergonomics,* 33(12), 1443–1454.

Lalomia, M. J., and Happ, A. J. (1987), The effective use of color for text on the IBM 5153 color display, *Proceedings of the Human Factors Society 31st Annual Meeting,* 1091–1095.

Lamb, L. (1987), Documentation: Fitting it to the job, *Data Training,* February.

Lang, K., Graesser, A. C., and Hemphill, D. D. (1991), Understanding errors in human computer interaction, *Proceedings of CHI'91* (poster session).

Langan, L. M., and Watkins, S. M. (1987), Pressure of menswear on the neck in relation to visual performance, *Human Factors,* 29(1), 67–71.

Lanning, E. D. (1967), *Peru before the Incas.* Upper Saddle River, N.J.: Prentice Hall.

Larkin, J., McDermott, J., Simon, D. P., and Simon, H. A. (1980), Expert and novice performance in solving physics problems, *Science,* 208, 1335–1342.

Lashley, K. S. (1951), The problem of serial order in behavior, in L. A. Jeffress (ed.), *Cerebral Mechanisms in Behavior.* New York: Wiley.

Lawton, M. P. (1990), Aging and performance of home tasks, *Human Factors,* 32(5), 527–536.

Lazonder, W., and vanderMeij, H. (1993), The minimal manual: Is less really more?, *International Journal of Man–Machine Studies,* 39, 729–752.

Leahy, M., and Hix, D. (1990), Effects of touch screen target location on user accuracy, *Proceedings of the Human Factors Society 34th Annual Meeting,* 370–374.

Lederer, A. L., and Prasad, J. (1992), Nine management guidelines for better cost estimating, *Communications of the ACM,* 35(2), 51–59.

Lee, E., MacGregor, J., and Lam, N. (1986), Keyword-menu retrieval: An effective alternative to menu indexes, *Ergonomics,* 29(1), 115–130.

Lehner, P. E., Rook, F. W., and Adelman, L. (1984), Mental models and cooperative problem solving with expert systems, *National Technical Information Service,* No. AD-A147 843.

Lepper, M. R., and Greene, D. (1979), *The Hidden Costs of Reward.* Morristown, N.J.: Erlbaum.

Lewis, J. R.

——— (1993), Problem discovery in usability studies: A model based on the binomial probability formula, *Proceedings of the 5th International Conference on Human–Computer Interaction,* 666–671.

——— (1994a), Sample sizes for usability studies, *Human Factors,* 36(2), 368–378.

——— (1994b), A critical review of human factors studies of split keyboards through 1993, *IBM Technical Report.*

Lin, R., and Kreifeldt, J. G. (1992), Understanding the image functions for icon design, *Proceedings of the Human Factors Society 36th Annual Meeting,* 341.

Lineberry, C. S., Jr. (1977), When to develop aids for on-the-job use and when to provide instruction, *Improving Human Performance: A Research Quarterly,* 6, 87–92.

Linvill, J. G., and Bliss, J. C. (1966), A direct translation reading aid for the blind, *Proceedings of the IEEE,* 54, 40–51.

Locke, J. L., and Locke, V. L. (1971), Deaf children's phonetic, visual, and dactylic coding in a grapheme recall task, *Journal of Experimental Psychology,* 89, 142–146.

Loeb, M., and Holding, D. H. (1975), Backward interference by tones or noise in pitch perception as a function of practice, *Perception and Psychophysics,* 18, 205–208.

Loftus, E. F., and Loftus, G. R. (1980), On the permanence of stored information in the human brain, *American Psychologist,* 409–420.

Loftus, E. F., Freedman, J. L., and Loftus, G. R. (1970), Retrieval of words from subordinate and supraordinate categories in semantic hierarchies, *Psychonomic Science,* 235–236.

Lohse, J. (1990), A cognitive model for the perception and understanding of graphics, *Proceedings of CHI'91,* 137–144.

Lombaers, J. H. M. (1990), Supermarketing ergonomics, *Ergonomics,* 33(5), 541–545.

Lorayne, H. (1976), *Good Memory—Successful Student!* New York: Stein and Day.

Lovesey, E. S. (1970), The multi-axis vibration environment and man, *Ergonomics,* 1(5), 258–261.

Luchins, A. S. (1942), Mechanization in problem solving—the effect of Einstellung, *Psychological Monographs,* 54(6, whole no. 248).

Lundell, J., and Notess, M. (1991), Human factors in software development: Models, techniques and outcomes, *Proceedings of CHI'91,* 145–151.

Luria, A. R. (1968), *The Mind of a Mnemonist.* New York: Basic Books.

MacDonald, N. H., Frase, L. T., Gingrich, P. S., and Keenan, S. A.

——— (1980) Writer's Workbench: Computer Programs for Text Editing and Assessment, private communication.

——— (1982) The Writer's Workbench: Computer Aids for Text Analysis, *IEEE Transactions on Communications.*

MacDonald, N. H., Keenan, S. A., Gingrich, P. S., Fox, M. L., Frase, L. T., and Collymore, J. L. (1981), Writer's Workbench: Computer Aids for Writing, private communication.

MacDonald-Ross, M. (1977), *Graphics in Text: A Bibliography.* Institute of Educational Technology, Open University; Graphics in texts, in L. S. Shulman (ed.), *Review of Research in Education,* Vol. 5, Hasca, Ill.: Reacock, 1977.

MacGregor, J. N. (1987), Short-term memory capacity: Limitation or optimization? *Psychological Review,* 94(1), 107–108.

MacGregor, J. N., and Lee, E. S. (1987), Performance and preference in videotex menu retrieval: A review of the empirical literature, *Behaviour and Information Technology,* 6(1), 43–68.

MacGregor, R. C., Hasan, H., and Liao, H. T. (1993), The effect of user characteristics on interface choice, *Proceedings of the 5th International Conference on Human–Computer Interaction,* 570–574.

Mack, R., and Lang, K. (1989), A benchmark comparison of mouse and touch interface techniques for an intelligent workstation windowing environment, *Proceedings of the Human Factors Society 33rd Annual Meeting,* 325–329.

MacKay, W. E. (1991), Triggers and barriers to customizing software, *Proceedings of CHI'91,* 153–160.

MacKenzie, I. S. (1992), Fitts' law as a research and design tool in human–computer interaction, *Human–Computer Interaction,* 7, 91–139.

MacKenzie, I. S., and Ware, C. (1993), Lag as a determinant of human performance in interactive systems, *Proceedings of InterCHI'93,* 488.

Mackworth, J. F. (1963), The duration of the visual image, *Canadian Journal of Psychology,* 17, 62–81.

MacNeilage, P. F. (1964), Typing errors as clues to serial ordering mechanisms in language behavior, *Language and Speech,* 7, 144–159.

MacNeilage, P. F., and MacNeilage, L. A. (1973), Central processes controlling speech production during sleep and waking, in F. J. McGuigan (ed.), *The Psychophysiology of Thinking.* New York: Academic Press.

Maes, P. (1994), Agents that reduce work and information overload, *Communications of the ACM,* 37(7), 31–40.

Mager, R. F.

——— (1973) On the other hand, *Improving Human Performance: A Research Quarterly,* 2, 77–88.

——— (1975) *Preparing Instructional Objectives.* Belmont, Calif.: Fearon Publishers.

Mager, R. F., and Beach, K. M. (1967), *Developing Vocational Instruction.* Belmont, Calif.: Fearon.

Mager, R. F., and Pipe, P. (1970), *Analyzing Performance Problems.* Belmont, Calif.: Fearon Publishers.

Maguire, M., and Dillon, A. (1993), Usability measurement: Its practical value to the computer industry, *Proceedings of InterCHI'93,* 145.

Mahoney, F., and Lyday, N. (1984), Design is what counts in computer-based training, *Training and Development Journal,* July, 40–41.

Malde, B. (1986), How a real-life system stands up to the commandments, *Behavior and Information Technology,* 5(1), 81–87.

Malone, T. W.

——— (1980) What makes things fun to learn? Palo Alto, Calif.: *Xerox Research Center Technical Report.*

——— (1984), Heuristics for designing enjoyable user interfaces: Lessons from computer games, in *Human Factors in Computer Systems,* John Thomas and Michael Schneider (eds.). Norwood, N.J.: Aldex Publishing Corporation.

Mantei, M. M., and Teorey, T. J. (1988), Cost/benefit analysis for incorporating human factors in the software lifecycle, *Communications of the ACM,* 31(4), 428–439.

Margolis, B. L., and Kroes, W. H. (1975), *The Human Side of Accident Prevention: Psychological Concepts and Principles Which Bear on Industrial Safety.* Springfield, Ill.: Charles C Thomas.

Marics, M. A. (1990), How do you enter "D'Anzi-Quist" using a telephone keypad, *Proceedings of the Human Factors Society 34th Annual Meeting,* 208–211.

Marshall, C., McManus, B., and Prail, A. (1990), Usability of product X: Lessons from a real product, *Behaviour and Information Technology,* 10(1), 243–253.

Martin, C. R.

——— (1974) Contribution of position level testing to BISCUS/FACS system success, private communication.

——— (1975) The BISCUS/FACS position package testing program, private communication.

——— (1979) Human performance testing in the design of computer-based systems, private communication.

Martin, G. L. (1989), The utility of speech input in user–computer interfaces, *International Journal of Man–Machine Studies,* 30, 355–375.

Martin, G. L., and Corl, K. G. (1986), System response time effects on user productivity, *Behaviour and Information Technology,* 5(1), 3–13.

Martin, J. (1973), *Design of Man–Computer Dialogues.* Upper Saddle River, N.J.: Prentice Hall.

Maslow, A. H.

——— (1954) *Motivation and Personality.* New York: Harper.

——— (1970) *Motivation and Personality* (2nd ed.). New York: Harper & Row.

Mason, M. V. (1986), Adaptive command prompting in an on-line documentation system, *International Journal of Man–Machine Studies,* 25, 33–51.

Massaro, D. W.

——— (1972) Preperceptual images, processing time and perceptual units in auditory perception, *Psychological Review,* 79, 124–145.

——— (1975) *Experimental Psychology and Information Processing.* Chicago: Rand McNally College Publishing.

Matthews, M. L., and Mertins, K. (1989), Visual performance and subjective discomfort in prolonged viewing of chromatic displays, *Human Factors,* 31(3), 259–271.

Mayhew, D. J. (1990), Cost-justifying human factors support: A framework, *Proceedings of the Human Factors Society 34th Annual Meeting,* 834–838.

Mayzner, M. S., and Tresselt, M. E. (1965), Tables of single-letter and digram frequency counts for various word-length and letter-position combinations, *Psychonomic Monograph Supplements,* 1(2), 13–32.

Mayzner, M. S., Tresselt, M. E., and Wolin, B. R. (1965), Tables of trigram frequency counts for various word-length and letter-position combinations, *Psychonomic Monograph Supplements,* 1(3).

McArthur, B. N. (1965), Accuracy of source data: Human error in hand transcription, AD 623157.

McClelland, L. A. (1974), Crowding and social stress. Unpublished Ph.D. dissertation, University of Michigan.

McDonald, J. E., Molander, M. E, and Noel, R. W. (1988), Color-coding categories in menus, in E. Soloway and D. Frye (eds.), *Proceedings of the CHI'88 Human Factors in Computer Systems,* New York, 101–106.

McGregor, D. (1960), *The Human Side of Enterprise.* New York: McGraw-Hill.

Medsker, K. L. (1979), CAI/CMI, private communication.

Mehlenbacher, B., Duffy, T. M., and Palmer, J. (1989), Finding information on a menu: Linking menu organization to the user's goals, *Human–Computer Interaction,* 4, 231–251.

Meister, D. (1976), *Behavioral Foundations of System Development.* New York: Wiley.

Melder, K., and Hix, D. (1991), Task-based modeling of the user population of a complex interactive system domain, *Proceedings of the Human Factors Society 35th Annual Meeting,* 430–433.

Merikle, P. M., and Coltheart, M. (1972), Selective forward masking, *Canadian Journal of Psychology,* 26, 296–302.

Merikle, P. M., Coltheart, M., and Lowe, D. G. (1971), On the selective effects of a patterned masking stimulus, *Canadian Journal of Psychology,* 25, 264–279.

Merwin, D. H., Dyre, B. P., Humphrey, D. G., Grimes, J., and Larish, J. F. (1990), The impact of icons and visual effects on learning computer databases, *Proceedings of the Human Factors Society 34th Annual Meeting,* 424–428.

Mewhort, D. J. K., and Cornett, S. (1972), Scanning and the familiarity effect in tachistoscopic recognition, *Canadian Journal of Psychology,* 26, 181–189.

Meyer, J., Shinar, D., and Leiser, D. (1990), Time estimation of computer "Wait" message displays, *Proceedings of the Human Factors Society 34th Annual Meeting,* 360–364.

Michaels, S. E. (1971), Qwerty versus alphabetic keyboards as a function of typing skill, *Human Factors,* 13(5), 419–426.

Milgram, S. (1965), Some conditions of obedience and disobedience to authority, *Human Relations,* 18, 57–76.

Miller, G. A.

——— (1962) Some psychological studies of grammar, *American Psychologist,* 17, 748–762.

——— (1956) The magical number seven, plus or minus two: Some limits on our capacity for processing information, *Psychological Review,* 63, 81–97.

Miller, G. A., and Nicely, P. E. (1955), An analysis of perceptual confusions among some English consonants, *Journal of the Acoustical Society of America,* 27, 338.

Miller, G. A., Galanter, E., and Pribram, K. H. (1960), *Plans and the Structure of Behavior.* New York: Holt, Rinehart and Winston.

Miller, L. A., and Thomas, J. C. (1977), Behavioral issues in the use of interactive systems, *International Journal of Man–Machine Studies,* 9, 509–536.

Miller, R. B.

——— (1968) Response time in man–computer conversational transactions, *AFIPS Conference Proceedings* (Fall Joint Computer Conference). Washington, D.C.: Thompson Book Co., 1968.

——— (1954) Survey of human engineering needs in maintenance of ground electronics equipment, Contract AF 30(602) 24, 274.

Milner, B., and Taylor, L. (1972), Right-hemisphere superiority in tactile pattern recognition after cerebral commisurotomy: Evidence for non-verbal memory, *Neuropsychologia,* 10, 1–15.

Mohageg, M. F. (1989), Differences in performance and preference for object-oriented vs. bit-mapped graphics interfaces, *Proceedings of the Human Factors Society 33rd Annual Meeting,* 385–389.

Molich, R., and Nielson, J. (1990), Improving a human–computer dialogue, *Communications of the ACM,* 32(3), 338–348.

Montanelli, R. G. (1977), Using CAI to teach introductory computer programming, *ACM SIGCUE Bulletin,* 11(1), 14–22.

Morehouse, L. E., and Gross, L. (1977), *Maximum Performance.* New York: Pocket Books.

Morgan, C. T., Cook, J. S., Chapanis, A., and Lund, M. W. (eds.) (1963), *Human Engineering Guide to Equipment Design.* New York: McGraw-Hill.

Morland, V. D. (1983), Human factors guidelines for terminal interface design, *Communications of the ACM,* 26(7), 484–494.

Moroney, W. F., and Reising, J. (1992), Subjects in human factors: Just who are they?, *Proceedings of the Human Factors Society 36th Annual Meeting,* 1227–1231.

Morse, A., and Reynolds, G. (1993), Overcoming current growth limits in UI development, *Communications of the ACM,* April, 72.

Muller, G. E., and Pilzecker, A. (1900), An experimental contribution to teaching the mind, *Zeitschrift fur Psychology,* 212–216.

Muller, M. J. (1991), PICTIVE—An exploration in participatory design, *Proceedings of CHI'91,* 225–231.

Mulligan, R. M., Altom, M. W., and Simkin, D. K. (1991), User interface design in the trenches: Some tips on shooting from the hip, *Proceedings of CHI'91,* 232–236.

Munsterberg, H. (1913), *Psychology and Industrial Efficiency.* Boston: Houghton Mifflin.

Murch, G. M. (1985), Colour graphics: Blessing or ballyhoo? *Computer Graphics Forum,* 4, 127–135.

Murdock, B. B., and Von Saal, W. (1967), Transpositions in short-term memory, *Journal of Experimental Psychology,* 74(1), 137–143.

Murray, B. S., and McDaid, E. (1993), Visualizing and representing knowledge for the end user: A review, *International Journal of Man–Machine Studies,* 38, 23–49.

Murrell, K. F. H. (1965), *Ergonomics: Man in His Working Environment.* London: Chapman and Hall.

Muter, P., and Maurutto, P. (1991), Reading and skimming from computer screens and books: The paperless office revisited, *Behaviour & Information Technology,* 10(4), 257–266.

Muter, P., Latremouille, S. A., Treurniet, W. C., and Beam, P. (1982), Extended reading of continuous text on television screens, *Human Factors,* 24, 501–508.

Myers, B. A., and Rosson, M. B. (1992), Survey on user interface programming, *CHI'92 Conference Proceedings,* 195–202.

Mynatt, E. D., and Weber, G. (1994), Nonvisual presentation of graphical user interfaces: Contrasting two approaches, *CHI'94 Conference Proceedings,* 166–172.

Napier, H. A., Lane, D. M., Batsell, R. R., and Guandango, N. S. (1989), Impact of a restricted natural language interface on ease of learning and productivity, *Communications of the ACM,* 32(10), 1190–1198.

Nass, C., Steuer, J., and Tauber, E. R. (1994), Computers are social actors, *CHI'94 Conference Proceedings,* 72–78.

National Research Council Committee on Undersea Warfare (1949). *Human Factors in Undersea Warfare.* Washington, D.C.

Neerinex, M., and deGreef, P. (1993), How to aid non-experts, *Proceedings of InterCHI'93,* 165–170.

Neisser, U. (1976), *Cognition and Reality: Principles and Implications of Cognitive Psychology.* San Francisco: W. H. Freeman.

Neisser, U., and Weene, P. (1963), A note on human recognition of hand-printed characters, *Information and Control,* 3(2).

Nelson, B. C., and Smith, T. J. (1990), User interaction with maintenance information: A performance analysis of hypertext versus hard copy formats, *Proceedings of the Human Factors Society 34th Annual Meeting,* 229–233.

Nelson-Denny Reading Test Form D (1982). Boston: Houghton Mifflin.

Newell, A. (1971), Speech understanding systems, Report AFOST-TR-72-0142. Pittsburgh, Pa.: Carnegie-Mellon University.

Newman, R. W., and White, R. M. (1951), Reference Anthropometry of Army Men, Report No. 180. Lawrence, Mass.: Quartermaster Climatic Research Laboratory.

Nielsen, J.

——— (1989) Usability engineering at a discount, in *Designing and Using Human–Computer Interfaces and Knowledge-based Systems,* G. Salvendy and M. J. Smith (Eds.). Amsterdam: Elsevier, 394–401.

——— (1990) Traditional dialogue design applied to modern user interfaces, *Communications of the ACM,* 33(10), 109–118.

——— (1992) Finding usability problems through heuristic evaluation, *CHI'92 Conference Proceedings,* 373–380.

——— (1993) Noncommand user interfaces, *Communications of the ACM,* April, 82.

——— (1994) Enhancing the explanatory power of usability heuristics, *CHI'94 Conference Proceedings,* 152–155.

Nielsen, J., and Landauer, T. K. (1993), A mathematical model of the finding of usability problems, *Proceedings of InterCHI'93,* 206–213.

Nielsen, J., and Levy, J. (1994), Measuring usability preference vs. performance, *Communications of the ACM,* 37(4), 67–75.

Nielsen, J., and Molich, R. (1990), Heuristic evaluation of user interfaces, *CHI'90 Conference Proceedings,* 249–256.

Nielsen, J., and Phillips, V. L. (1993), Estimating the relative usability of two interfaces: Heuristic, formal and empirical methods compared, *Proceedings of InterCHI'93,* 214.

Nielsen, J., and Schaefer, L. (1993), Sound effects as an interface element for older users, *Behaviour & Information Technology,* 12(4), 208–215.

Nilsen, E., Jong, H., Olson, J. S., Biolsi, K., Rueter, H., and Mutter, S. (1993), The growth of software skill: A longitudinal look at learning and performance, *Proceedings of InterCHI'93,* 149.

Nolan, P. R. (1989), Designing screen icons: Ranking and matching studies, *Proceedings of the Human Factors Society 33rd Annual Meeting,* 380–384.

Norcio, A. F., and Stanley, J. (1989), Adaptive human–computer interfaces: A literature survey and perspective, *IEEE Transactions on Systems, Man and Cybernetics,* 19(2), 339–408.

Nordhausen, B., Chignell, M. H., and Waterworth, J. (1991), Comparison of manual and automated linking in hypertext engineering, *Proceedings of the Human Factors Society 35th Annual Meeting,* 310–314.

Norman, D. A.

——— (1980) Post-Freudian slips, *Psychology Today,* 42–50, April.

——— (1984) Stages and levels in human–machine interaction, *International Journal of Man–Machine Studies,* 21, 365–375.

Norman, K. L., and Chin, J. P. (1989), The menu metaphor: Food for thought, *Behaviour and Information Technology,* 8(2) 125–134.

Norman, K. L., Weldon, L. J., and Shneiderman, B. (1986), Cognitive layouts of windows and multiple screens for user interfaces, *International Journal of Man–Machine Studies,* 25, 229–248.

Norman, R. J., and Nunamaker, J. F. (1989), CASE productivity perceptions of software engineering professionals, *Communications of the ACM,* 32(9), 1102–1108.

North, A. J., and Jenkins, L. B. (1951), Reading speed and comprehension as a function of typography, *Journal of Applied Psychology,* 35(4), 225–228.

Noyes, J. (1983), The QWERTY keyboard: A review, *International Journal of Man–Machine Studies,* 18, 265–281.

Noyes, J. M., and Frankish, C. R. (1989), A review of speech recognition applications in the office, *Behaviour and Information Technology,* 8(6), 475–486.

O'Day, V. L., and Jeffries, R. (1993), Orienteering in an information landscape: How information seekers get from here to there, *Proceedings of InterCHI'93,* 438.

Ogawa, K. (1993), The role of design guidelines in assisting the interface design, *Proceedings of the 5th International Conference on Human–Computer Interaction,* 272–276.

Ohnemus, K. R., and Biers, D. W. (1993), Retrospective versus concurrent thinking-out-loud in usability testing, *Proceedings of the Human Factors Society 37th Annual Meeting,* 1127–1131.

Olsen, D. R., Foley, J. D., Hudson, S. E., Miller, J., and Myers, B. (1993), Research directions for user interface software tools, *Behaviour & Information Technology,* 12(2), 80–97.

Open Software Foundation (1990), *OSF/Motif Style Guide.* Upper Saddle River, N.J.: Prentice Hall.

Oppermann, R. (1994), Adaptively supported adaptability, *International Journal of Human–Computer Studies,* 40, 455–472.

Osborn, A. F. (1963), *Applied Imagination* (3rd ed.). New York: Scribner.

Osga, G. A. (1991), Using enlarged target area and constant visual feedback to aid cursor pointing tasks, *Proceedings of the Human Factors Society 35th Annual Meeting,* 369–373.

Owsowitz, S., and Sweetland, A. (1965), Factors affecting coding errors, Rand Corporation Technical Report.

Paap, K. R., and Roske-Hofstrand, R. J. (1986), The optimal number of menu options per panel, *Human Factors,* 28(4), 377–385.

Page, S. R. (1993), Selecting colors for dialog boxes and buttons in a text interface, *Proceedings of the 5th International Conference on Human–Computer Interaction,* 208–213.

Palerno, D. S. (1973), More about less: a study of comprehension, *Journal of Verbal Learning and Verbal Behavior,* 211–221.

Parsons, H. M.

——— (1974) What happened at Hawthorne? *Science,* 183, 922–932.

——— (1972) *Man–Machine System Experiments* (Chapter 3: Forebears). Baltimore, Md.: Johns Hopkins Press.

Parsons, M. (1972), Environmental design, *Human Factors,* 14, 369–482.

Parsons, S. O., Eckert, S. K., and Seminara, J. L. (1978), Human factors design practices for nuclear power plant control rooms, *Proceedings of the Human Factors Society 22nd Annual Meeting.*

Patrick, A., and Whalen, T. (1989), Conversational hypertext, *Proceedings of CHI'89,* 289–292.

Paxton, A. L., and Turner, E. J. (1984), The application of human factors to the needs of the novice computer user, *International Journal of Man–Machine Studies,* 20, 137–156.

Pearson, G., and Weiser, M. (1988), Exploratory evaluation of a planar foot-operated cursor-positioning device, in E. Soloway and D. Frye (eds.), *Proceedings of CHI'88 Human Factors in Computer Systems,* New York, 13–18.

Penfield, W.

——— (1958) *The Excitable Cortex in Conscious Man.* Liverpool: Liverpool University Press.

——— (1969) Consciousness, memory, and man's conditioned reflexes, in K. Pribam (ed.), *On the Biology of Learning.* New York: Harcourt Brace Jovanovitch.

Penfield, W., and Perot, P. (1963), The brain's record of auditory and visual experiences, *Brain,* 86, 595–696.

Pepler, R. (1972), Thermal comfort of students in climate controlled and non-climate controlled schools, *ASHRAE Transactions,* 78, 97–109.

Perkins, R., Blatt, L. A., Workman, D., and Ehrlich, S. F. (1989), Iterative tutorial design in the product development cycle, *Proceedings of the Human Factors Society 33rd Annual Meeting,* 268–272.

Perlman, G. (1991), The HCI bibliography project, *SIGCHI Bulletin,* 15–20.

Petajian, E., Bishoff, B., and Bodoff, D. (1988), An improved automatic lipreading system to enhance speech recognition, in E. Soloway and D. Frye (eds.), *Proceedings of CHI '88 Human Factors in Computer Systems ACM,* New York, 19–25.

Pew, R. W.

——— (1966) Acquisition of hierarchical control over the temporal organization of a skill, *Journal of Experimental Psychology,* 71, 764–771.

——— (1969) The speed–accuracy operating characteristic, *Acta Psychologica,* 30, 16–26.

Piaget, J. (1952), *The Origins of Intelligence in Children.* New York: International Universities Press.

Pierce, J. R., and Karlin, J. E. (1957), Reading rates and the information rate of a human channel, *Bell System Technical Journal,* 36, 497–516.

Pierson, W. R. (1961), Body size and speed, *Research Quarterly,* 32, 197–200.

Pirow, P. C. (1994), Six ages of man as determined by athletic performance, *International Journal of Aging and Human Development,* 38(3), 237–277.

Plaisant, C., and Sears, A. (1992), Touchscreens interfaces for alphanumeric data entry, *Proceedings of the Human Factors Society 36th Annual Meeting,* 293.

Pollack, I. (1959), Message uncertainty and message reception, *Journal of the Acoustical Society of America,* 31, 1500–1508.

Poller, M. F., Friend, E., Hegarty, J. A., Rubin, J. J., and Denver, J. J. (1981), *Handbook for Writing Procedures.* Indianapolis, Ind.: Western Electric Publication Center.

Polson, P. G., and Lewis, C., (1992), Cognitive walkthroughs: A method for theory-based evaluation of user interfaces, *International Journal of Man–Machine Studies,* 36, 741–773.

Porter, L. W., Lawler, E. E., and Hackman, J. R. (1975), *Behavior in Organizations.* New York: McGraw-Hill.

Posner, M. I. (1973), *Cognition: An Introduction.* Glenview, Ill.: Scott, Foresman.

Potter, R. L., Weldon, L. J., and Shneiderman, B. (1988), Improving the accuracy of touch screens: An experimental evaluation of three strategies, in E. Soloway and D. Frye (eds.), *Proceedings of CHI '88 Human Factors in Computer Systems ACM,* New York, 27–32.

Potter, S. S., Cook, R. I., Woods, D. D., and McDonald, J. S. (1990), The role of human factors guidelines in designing usable systems: A case study of operating room equipment, *Proceedings of the Human Factors Society 34th Annual Meeting,* 392–395.

Poulton, E. C.

——— (1968) Rate of comprehension of an existing teleprinter output and of possible alternatives, *Journal of Applied Psychology,* 16–21.

——— (1969) Skimming lists of food ingredients printed in different sizes, *Journal of Applied Psychology,* 55–58.

——— (1972) *Environment and Human Efficiency.* Springfield, Ill.: Charles C. Thomas.

Poulton, E. C., Hunt, J. C. R., Mumford, J. C., and Poulton, J. (1975), Mechanical disturbance produced by steady and gusty winds of moderate strength: Skilled performance and semantic assessments, *Ergonomics,* 18, 651–673.

Provins, K. A., and Bell, C. R. (1970), Effects of heat stress on the performance of two tasks running concurrently, *Journal of Experimental Psychology,* 85, 40–44.

Quetelet, A. (1870), *Anthropometrie.* Brussels: C. Munquardt.

Raalte, J. A. (1992), CRT's for high-definition television, *Information Display,* January, 6–10.

Raban, A. (1988), Word processing learning techniques and user learning preferences, *SIGCHI Bulletin,* October, 20(2), 83–87.

Rabbitt, P. (1990), Age, IQ and awareness, and recall of errors, *Ergonomics,* 33(10/11), 1291–1305.

Ramsey, H. R., and Atwood, M. E. (1979), *Human Factors in Computer Systems: A Review of the Literature.* Englewood, Colo.: Science Applications.

Reaux, R. A., and Williges, R. C. (1988), Effects of level of abstraction and presentation media on usability of user system interface guidelines, *Proceedings of the Human Factors Society 32nd Annual Meeting,* 330–334.

Reder, L. M., and Anderson, J. R. (1979), A comparison of texts and their summaries: Memorial consequences, Carnegie-Mellon University Technical Report.

Reed, P. (1993), HFES/ANSI software user interface standardization: Critical issues, *Proceedings of the 5th International Conference on Human–Computer Interaction,* 268–271.

Reicher, G. M. (1969), Perceptual recognition as a function of meaningfulness of stimulus material, *Journal of Experimental Psychology,* 81, 275–280.

Reynolds, S. L. (1991), Longitudinal analysis of age changes in speed of behavior, *Proceedings of the Human Factors Society 35th Annual Meeting,* 198–202.

Rieman, J. (1993), The diary study: A workplace-oriented research tool to guide laboratory efforts, *Proceedings of InterCHI'93,* 321–326.

Riggs, L. A. (1971), Vision, in J. W. Kling and L. A. Riggs, *Experimental Psychology.* New York: Holt, Rinehart and Winston.

Rissland, E. L. (1984), Ingredients of intelligent user interfaces, *International Journal of Man–Machine Studies,* 21, 377–388.

Roberts, D. F. (1975), Population differences in dimensions, their genetic basis and their relevance to practical problems of design, in A. Chapanis (ed.), *Ethnic Variables in Human Factors Engineering.* Baltimore, Md.: Johns Hopkins University Press.

Robertson, G. G., Card, S. K., and Mackinlay, J. D. (1993), Information visualization using 3D interactive animation, *Communications of the ACM,* April, 56.

Roethliesberger, F. J., and Dickson, W. J., (1939), *Management and the Worker—An Account of a Research Program Conducted by the Western Electric Company, Hawthorne Works, Chicago.* Cambridge: Harvard University Press.

Romero, H. A., Ostrom, L. T., and Wilhelmsen, C. A. (1993), What difference can the data make?, *Human Factors and Ergonomics Society 37th Annual Meeting,* 841–845.

Ronco, P. G., Hanson, J. A., Raben, W. W., and Samuels, I. A. (1966), *Characteristics of Technical Reports That Affect Reader Behavior: A Review of the Literature.* National Science Foundation. Washington, D.C.: Government Printing Office.

Rothkopf, E. Z. (1963), Some observations on predicting instructional effectiveness by simple inspection, *Journal of Programmed Instruction,* 2, 18–20.

Rotter, J. B. (1966), Generalized expectancies for internal versus external control of reinforcement, *Psychologieal Monographs,* 80, 1–28.

Rowley, D. E. (1994), Usability testing in the field: Bringing the laboratory to the user, *CHI'94 Conference Proceedings,* 252–257.

Rudd, J., and Isensee, S. (1991), Twenty-two tips for a happier, healthier prototype, *Proceedings of the Human Factors Society 35th Annual Meeting,* 328–331.

Rudnicky, A. I., Hauptmann, A. G., and Lee, K. F. (1994), Survey of current speech technology, *Communications of the ACM,* 37(3), 52–57.

Rullo, J. C., and McDonald, L. B. (1990), Factors related to skill degradation and their implications for refresher training, *Proceedings of the Human Factors Society 34th Annual Meeting,* 1396–1399.

Rumelhart, D. E., Widrow, B., and Lehr, M. A. (1994), The basic ideas in neural networks, *Communications of the ACM,* 37(3), 87–92.

Russo, P., and Boor, S. (1993), How fluent is your interface? Designing for international users, *Proceedings of InterCHI'93,* 342.

Sacks, O.

——— (1987), *The Man Who Mistook His Wife for a Hat.* New York: Harper & Row.

——— (1995), *An Anthropologist on Mars.* New York: Knopf.

Saja, A. D. (1985), The cognitive model: an approach to designing the human computer interface, *SIGCHI Bulletin of the ACM,* 16(3), 36–40.

Sakitt, B. (1975), Locus of short-term visual storage, *Science,* 190, 1318–1319.

Salthouse, T. A. (1990), Influence of experience on age differences in cognitive functioning, *Human Factors,* 32(5), 551–569.

Sanders, L. (1976), *The Tomorrow File.* New York: Berkley Publishing.

Santhanam, R., and Wiedenbeck, S. (1993), Neither novice nor expert: The discretionary user of software, *International Journal of Man–Machine Studies,* 38, 201–229.

Santon, P. J., Baltzr, A. J., Badre, A. N., Henneman, R. L., and Miller, M. S. (1992), On handwriting recognition system performance: Some experimental results, *Proceedings of the Human Factors Society 36th Annual Meeting,* 283.

Schaffer, E. (1980), Writing space on forms, private communication.

Schaffer, L. H., and Hardwick, J. (1969), Errors and error detection in typing, *Quarterly Journal of Experimental Psychology,* 21, 209–213.

Scheibe, K. E., Shaver, P. R., and Carrier, S. C. (1967), Color association values and response interference on variants of the Stroop test, *Acta Psychological,* 28, 286–295.

Schriver, K. A. (1986), Designing Computer Documentation: A Review of the Relevant Literature. Pittsburgh, Pa.: Carnegie Mellon University.

Schweighardt, M. F. (1990), Using the context of interactions to adapt to users, *Proceedings of the Human Factors Society 34th Annual Meeting,* 346–350.

Sears, A., and Shneiderman, B. (1991), High precision touchscreens: Design strategies and comparisons with a mouse, *International Journal of Man–Machine Studies,* 34, 593–613.

Sears, A., Revis, D., Swatski, J., Crittenden, R., and Shneiderman, B. (1993), Investigating touchscreen typing: The effect of keyboard size on typing speed, *Behaviour & Information Technology,* 12(1), 17–22.

Sebrechts, M. M., and Marsh, R. L. (1989), Components of computer skill acquisition: Some reservations about mental models and discovery learning, in *Designing and Using Human Computer Interfaces and Knowledge-based Systems,* G. Salvendy and M. J. Smith (eds.). Amsterdam: Elsevier, 168–173.

Seibel, R. (1964), Data entry through chord parallel devices, *Human Factors,* 6, 189–192.

Seyer, P. (1989), Performance improvement with hypertext, *Performance and Instruction,* February, 22–28.

Seymour, P. J. (1965), A study of the relationship between the communication skills and a selected set of predicators and of the relationship among the communication skills. Ph.D. dissertation, University of Minnesota.

Shackel, B. (ed.) (1979), *Man/Computer Communication,* Vol. 1. Maidenhead, Berkshire, England: Infotech International Ltd.

Shackel, B., Alty, J. L., and Reid, P. (1992), Hilites—The information service for the world HCI community, *SIGCHI Bulletin,* 40–46.

Sheldon, K. M. (1989), Moby Dick 2.1, *Byte,* July, 344.

Shepard, R. N., and Sheenan, M. M. (1965), Immediate recall of numbers containing a familiar prefix or postfix, *Perceptual and Motor Skills,* 21, 263–273.

Shinar, D., and Stern, H. I. (1987), Alternative option selection methods in menu-driven computer programs, *Human Factors,* 29(4), 453–459.

Shneiderman, B.

——— (1982), System message design: Guidelines and experimental results, in *Directions in Human/Computer Interactions,* Albert Badre and Ben Shneiderman (eds.). Norwood, N.J.: Ablex Publishing Corporation.

——— (1987), *Designing the User Interface: Strategies for Effective Human–Computer Interaction.* Reading, Mass.: Addison-Wesley.

——— (1988), We can design better interfaces: A review of human–computer interaction styles, *Ergonomics,* 31(5), 699–710.

Shulman, L. S., Loupe, M. J., and Piper, R. M. (1968), Studies of the inquiry process: Inquiry patterns of students in teacher-training programs. East Lansing: Educational Publications Services, Michigan State University.

Silbiger, H. R. (1973), Speech Communication, private communication.

Silvern, L. C. (1970), Training: Man–man and man–machine communications, in K. B. DeGreene (ed.), *Systems Psychology.* New York: McGraw-Hill.

Simonton, K., Vercruyssen, M., and Hashizume, K. (1991), Effects of posture on reaction time: Influence of gender and practice, *Proceedings of the Human Factors Society 35th Annual Meeting,* 768–771.

Singleton, W. T. (1974), *Man–Machine Systems.* New York: Penguin Books.

Skinner, B. F. (1953), *Science and Human Behavior.* New York: Free Press.

Slobin, D. (1966), Grammatical transformations and sentence comprehension in childhood and adulthood, *Journal of Verbal Learning and Verbal Behavior,* 5, 219–227.

Smelcer, J. B., and Walker, N. (1990), Transfer across computer command menus, *Proceedings of the Human Factors Society 34th Annual Meeting,* 439–443.

Smith, D. B. D. (1990), Human factors and aging: An overview of research needs and application opportunities, *Human Factors,* 32(5), 509–526.

Smith, E. A., and Kincaid, P. (1970), Derivation and validation of the automated readability index for use with technical materials, *Human Factors,* 12, 457–464.

Smith, J. B., and Weiss, S. F. (1988), An overview of hypertext, *Communications of the ACM,* July.

Smith, L. D. (1943), *Cryptography: The Science of Secret Writing.* New York: W. W. Norton.

Smith, S. L. (1986), Standards versus guidelines for designing user interface software, *Behavior and Information Technology,* 5(1), 47–61.

Smith, S. L., and Mosier, J. N. (1984), Design guidelines for user system interface software, Mitre Corporation Technical Report. Bedford, Mass.: Mitre Corp.

Smith, S. L., and Mosier, J. N. (1986), *Guidelines for Designing User Interface Software.* Bedford, Mass.: Mitre Corp.

Smither, J. A. (1992), The processing of synthetic speech by older and younger adults, *Proceedings of the Human Factors Society 36th Annual Meeting,* 190.

Snyder, H. L. (1980), Human visual performance and flat panel display image quality, AD A092685.

Snyder, H. L., Decker, J. J., Lloyd, C. J. C., and Dye, C. (1990), Effect of image polarity on VDT task performance, *Proceedings of the Human Factors Society 34th Annual Meeting,* 1447–1451.

Somberg, B. L., (1987), A comparison of rule-based and positionally constant arrangements of computer menu items, *Proceedings of the Human Factors in Computing Systems and Graphics Interface Conference,* April, 255–260.

Sonntag, L. (1971), Designing human-oriented codes, *Bell Laboratories Record,* February, 43–49.

Soth, M. W. (1976), The human work module: A structural entity for personnel subsystem design, private communication.

Sparrow, J. A. (1989), Graphical displays in information systems: Some data properties influencing the effectiveness of alternative forms, *Behaviour and Information Technology,* 8(1), 43–56.

Sperling, G. (1960), The information available in brief visual presentations, *Psychological Monographs,* 74.

Spoto, C. G., and Babu, A. J. G. (1989), Highlighting in alphanumeric displays: The efficacy of monochrome methods, *Proceedings of the Human Factors Society 33rd Annual Meeting,* 370.

Staggers, N., and Norcio, A. F., Mental models: Concepts for human–computer interaction research, *International Journal of Man–Machine Studies,* 38, 587–605.

Staples, L. (1994), Vision and vitality, *SIGCHI Bulletin,* July, 11–12.

Starch, D. (1914), *Advertising.* Chicago: Scott, Foresman.

Starker, I., and Bolt, R. A. (1990), A gaze-responsive self-disclosing display, *CHI'90 Conference Proceedings,* April, 3–9.

Steinmetz, C. S. (1967), The Evolution of Training, in *Training and Development Handbook.* New York: McGraw-Hill.

Sternberg, S. (1978), Character delimiters are a liability in data-entry forms, private communication.

Stewart, T. F. M.

——— (1976), The specialist user, *Proceedings of NATO Advanced Study Institute,* Mati, Greece.

——— (1992), The role of HCI standards in relation to the directive, *Displays: Technology of Applications,* 13(3), 125–134.

Stewart, T. F. M., Damodaran, L., and Eason, K. D. (1976), Interface problems in man–computer interaction, in E. Humford and H. Sackman (eds.), *Human Choice and Computers.* New York: American Elsevier.

Sticht, T. G. (ed.)(1975), *Reading for Working: A Functional Literacy Anthology.* Alexandria, Va.: Human Resources Research Organization.

Stokols, D. (1972), On the distinction between density and crowding: Some implications for future research, *Psychological Review,* 79(3), 275–278.

Stramler, J. H., Jr. (1992), The dictionary for human factor/ergonomics: A significant reference work in human factors, *Proceedings of the Human Factors Society 36th Annual Meeting,* 544.

Strong, E. P. (1956), A comparative experiment in simplified keyboard retraining and standard keyboard supplementary training. Washington, D.C.: General Services Administration.

Strunk, W., and White, E. B. (1972), *The Elements of Style.* New York: Macmillan.

Sweeney, M., Maguire, M., and Shackel, B. (1993), Evaluating user–computer interaction: A framework, *International Journal of Man–Machine Studies,* 38(4), 689–712.

Swensson, R. G. (1972), The elusive tradeoff: Speed vs. accuracy in visual discrimination tasks, *Perception and Psychophysics,* 12, 16–32.

Swink, J. R. (1966), Intersensory comparisons of reaction time using an electropulse tactile stimulus, *Human Factors,* 8(2), 143–145.

Takeuchi, A., and Nagao, K. (1993), Communicative facial displays as a new conversational modality, *Proceedings of InterCHI'93,* 187.

Tanaka, T., Fukumoto, T., Yamamoto, S., and Noro, K. (1988), The effects of VDT work on urinary excretion of catecholamines, *Ergonomics,* 31(12), 1753–1763.

Tannehill, R. E. (1974), *Job Enrichment.* Chicago: Dartnell Corporation.

Tannenbaum, J. J. (1977), The MAP process, private communication.

Taylor, F. W.

——— (1911) *The Principles of Scientific Management.* New York: Harper.

——— (1947) *Principles of Scientific Management.* New York: Harper and Brothers.

——— (1957) Psychology and the design of machines, *American Psychologist,* 12, 249–258.

Taylor, J. M., and Murch, G. M. (1986), The effective use of color in computer graphics applications, *Computer Graphics'86 Conference Proceedings,* 515–521.

Terrace, H. S. (1963), Errorless transfer of a discrimination across two continua, *Journal of the Experimental Analysis of Behavior,* 6(2), 223–232.

Teubner, A. L., and Vaske, J. J. (1988), Monitoring computer users' behaviour in office environments, *Behaviour and Information Technology,* 7(1), 67–78.

Thacker, P., Tullis, T. S., and Babu, A. J. G. (1987), Application of Tullis' visual search model to highlighted and non-highlighted tabular displays, *Proceedings of the Human Factors Society,* 1221–1225.

Thimbley, H. (1982), Character level ambiguity: Consequences for user interface design, *International Journal of Man–Machine Studies,* 16, 211–225.

Thomas, M., Jaffe, G., Kincaid, J. P., and Stees, Y. (1992), Learning to use simplified English: A preliminary study, *Technical Communication,* first quarter, 69–73.

Thomas, R. E., Butterfield, R. K., Hool, J. N., and Herrick, R. T. (1993), Effects of exercise on carpal tunnel syndrome symptoms, *Applied Ergonomics,* 24(2), 101–108.

Thompson, E. T. (1979), How to write clearly, *Rolling Stone,* November 15.

Thomsen, C. T. (1990), The business instrument panel: A new paradigm for interacting with financial data, *CHI'90 Conference Proceedings,* 161–166.

Thorell, L. G., and Smith, W. J. (1990), *Using Computer Color Effectively: An Illustrated Reference.* Prentice Hall: Upper Saddle River, N.J.

Thorpe, C. E., and Rowland, G. E. (1965), The effect of natural grouping of numerals on short-term memory, *Human Factors,* 7, 38–44.

Tichauer, E. R. (1962), The effects of climate on working efficiency, *Impetus,* Australia, 1(5), 24–31.

Tinker, M. A.

——— (1943) Illumination intensities for reading newspaper type, *Journal of Education Psychology.* 34, 247–250.

——— (1963) *Legibility of Print.* Ames, Iowa: Iowa State University Press.

——— (1965) *Bases for Effective Reading.* Minneapolis: University of Minnesota Press, 136–141.

Tinker, M. A., and Paterson, D. G. (1928), Influence of type form on speed reading, *Journal of Applied Psychology,* 359–368.

Travis, D., Stewart, T. F. M., and Mackay, C. (1992), Evaluating image quality, *Displays: Technology and Application,* 13(3), 139–146.

Travis, D. S. (1990), Applying visual psychophysics to user interface design, *Behavior and Information Technology,* 9(5), 425–438.

Triplett, N. (1897), The dynamogenic factors in pacemaking and competition, *American Journal of Psychology,* 9, 507–533.

Trollip, S. R. (1979), The evaluation of a complex computer-based flight procedures trainer, *Human Factors,* 21(1), 47–54.

Trumbly, J. E. (1994), Productivity gains via an adaptive user interface: An empirical analysis, *International Journal of Human–Computer Studies,* 40, 63–81.

Tufte, E. R. (1983), *The Visual Display of Quantitative Information.* Cheshire, Conn.: Graphics Press.

Tullis, T. S.

——— (1983) The formatting of alphanumeric displays: a review and analysis, *Human Factors,* 25(6), 657–682.

——— (1993) Is user interface design just common sense?, *Proceedings of the 5th International Conference on Human–Computer Interaction,* 9–14.

Tullis, T. S., and Kodimer, M. L. (1992), A comparison of direct-manipulation, selection, and data-entry techniques for reordering fields in a table, *Proceedings of the Human Factors Society 36th Annual Meeting,* 298.

Tulving, E., and Thomson, D. M. (1973), Encoding specificity and retrieval processes in episodic memory, *Psychological Review,* 80, 352–373.

Turn, R. (1974), The use of speech for man–computer communication, Report R-1386-ARPA. Santa Monica, Calif.: Rand Corporation, January.

Turner, A. N., and Lawrence, P. R. (1965), *Industrial Jobs and the Worker.* Boston: Harvard Graduate School of Business Administration.

Ullsperger, P., Metz, A. M., and Gille, H. G. (1988), The p300 component of the event-related brain potential and mental effort, *Ergonomics,* 31(8), 1127–1137.

Urlings, I. J. M., Nuboer, I. D., and Dul, J. (1990), A method for changing the attitudes and behavior of management and employees to stimulate the implementation of ergonomic improvements, *Ergonomics,* 33(5), 629–637.

Van Cott, H. P., and Chapanis, A. (1972), Human engineering tests and evaluation, in H. P. Van Cott and R. G. Kinkade (eds.), *Human Engineering Guide to Equipment Design.* Washington, D.C.: U.S. Government Printing Office.

Van Cott, H. P., and Kinkade, R. G. (eds.) (1972), *Human Engineering Guide to Equipment Design.* Washington, D.C.: U.S. Government Printing Office.

Vercruyssen, M., Carlton, B. L., and Diggles-Buckles, V. (1989), Aging, reaction time, and stages of information processing, *Proceedings of the Human Factors Society 33rd Annual Meeting,* 174–178.

Virzi, R. A.

——— (1990) Streamlining the design process: Running fewer subjects, *Proceedings of the Human Factors Society 34th Annual Meeting,* 291–294.

——— (1992) Refining the test phase of usability evaluation: How many subjects is enough?, *Human Factors,* 34(4), 457–468.

Virzi, R. A., Sorce, J. F., and Herbert, L. B. (1993), A comparison of three usability evaluation methods: Heuristic, think-aloud, and performance testing, *Human Factors and Ergonomics Society 37th Annual Meeting,* 309–313.

Von Papstein, P., and Frese, M. (1988), Transferring skills from training to the actual work situation: The role of task application knowledge, action styles and job decision latitude, in E. Soloway and D. Frye (eds.), *Proceedings of CHI '88 Human Factors in Computer Systems ACM,* New York, 55–60.

Von Wright, J. M. (1957), A note on the role of "guidance" in learning, *British Journal of Psychology,* 48, 133–137.

Vroom, V. H. (1964), *Work and Motivation.* New York: Wiley.

Walker, J. H., Sproull, L., and Subramani, R. (1994), Using a human face in an interface, *CHI'94 Conference Proceedings,* 85–91.

Walker, M. A. (1989), Natural language in a desktop environment, in *Designing and Using Human Computer Interfaces and Knowledge-based Systems,* G. Salvendy and M. J. Smith (eds.). Amsterdam: Elsevier, 502–509.

Walker, N., and Smelcer, J. B. (1990), A comparison of selection times from walking and pull-down menus, *CHI'90 Conference Proceedings,* 221–225.

Walker, N., Smelcher, J. B., and Nilsen, E. (1991), Optimizing speed and accuracy of menu selection: A comparison of walking and pull-down menus, *International Journal of Man–Machine Studies,* 35, 871–890.

Wallas, G. (1926), *The Art of Thought.* New York: Harcourt, Brace, and World.

Wang, E. M. Y., Shahnavaz, H., Hedman, L., Papadopoulos, K., and Watkison, N. (1993), A usability evaluation of text and speech redundant help messages on a reader interface, *Proceedings of the 5th International Conference on Human–Computer Interaction,* 724–729.

Wargo, M. J. (1967), Human operator response speed, frequency, and flexibility: A review and analysis, *Human Factors,* 9(3), 221–238.

Warner, H. D. (1969), Effects of intermittent noise on human target detection, *Human Factors,* 11(3), 245–250.

Warren, R. E.

——— (1974) Norms of restricted color association, *Bulletin of the Psychonomic Society,* 4, 37–38.

——— (1980) Color associations of alarm and status labels, private communication.

Wason, P. C. (1961), Response to affirmative and negative binary statements, *British Journal of Psychology,* 133–142.

Wason, P. C., and Johnson-Laird, P. N. (1972), *Psychology of Reasoning: Structure and Content.* London: Batsford.

Weiner, B., and Kukla, A. (1970), An attributional analysis of achievement motivation, *Journal of Personality and Social Psychology,* 15, 1–20.

Weiss, D. S. (1980), The effects of text segmentation on reading. Ph.D. dissertation, University of Toronto.

Weisstein, N., and Harris, C. S. (1974), Visual detection of line segments: An object-superiority effect, *Science,* 186, 752–755.

Welch, J. R. (1980), Automatic speech recognition—putting it to work in industry, *Computer.*

Welford, A. T.

——— (1960) The measurement of sensory-motor performance: Survey and reappraisal of twelve years progress, *Ergonomics,* 3, 189–230.

——— (1976) *Skilled Performance: Perceptual and Motor Skills.* Glenview, Ill.: Scott, Foresman.

West, L. J. (1967), Vision and kinethesis in the acquisition of typewriting skill, *Journal of Applied Psychology,* 51, 161–166.

Weston, G. F. (1975), Plasma panel displays, *Journal of Physics E: Scientific Instruments,* 8, 981–991.

Weston, H. C. (1949), *Slight Light and Efficiency.* London: Lewis.

Wheatley, D. M., and Unwin, A. W. (1972), *The Algorithm Writer's Guide.* London: Longman Group.

Wheatley, H. B. (1883), *Literary Blunders.* London: Elliot Stock. Reprinted by Gale Research Company, Detroit, 1969.

White, R. W. (1959), Motivation recountered: The concept of competence, *Psychological Review,* 66, 297–333.

Whitefield, A., Wilson, F., and Dowell, J. (1991), A framework for human factors evaluation, *Behaviour and Information Technology,* 10(1), 65–79.

Wickelgren, W. A.

——— (1965) Acoustic similarity and intrusion errors in short-term memory, *Journal of Experimental Psychology,* 70, 102–108.

——— (1966) Associative intrusions in short-term recall, *Journal of Experimental Psychology,* 72(6), 853–858.

——— (1969) Auditory or articulatory coding in verbal short-term memory, *Psychological Review,* 76, 232–235.

Widrow, B., Rumelhart, D. E., and Lehr, M. A. (1994), Neural networks: Applications in industry, business and science, *Communications of the ACM,* 37(3), 93–105.

Williams, C. M. (1973), System response time: A study of users' tolerance, *IBM Technical Report,* 217–232.

Williams, J. D., Swenson, J. S., Hegarty, J. A., and Tullis, T. S. (1977), The effects of mean CSS response time and task type on operator performance in an interactive computer system, private communication.

Williams, J. R. (1988), The effects of case and spacing on menu option search time, *Proceedings of the Human Factors Society 32nd Annual Meeting,* 341–343.

Wilson, J., and Rosenberg, D. (1988), Rapid prototyping for user interface design, in M. Helander (ed.), *Handbook of Human–Computer Interaction,* Amsterdam, North-Holland: 859–867.

Winkler, R. L. (1972), *Introduction to Bayesian Inference and Decision.* New York: Holt, Rinehart and Winston.

Wohlwill, J. F.

——— (1974) Human adaption to levels of environmental stimulation, *Human Ecology,* 2, 127–147.

——— (1976) Environmental aesthetics: The environment as a source of affect, in I. Altman and J. F. Wohlwill (eds.), *Human Behavior and Environment: Advances in Theory and Research,* Vol. 1. New York: Plenum.

Wolf, C. G. (1988), A comparative study of gestural and keyboard interfaces, *Proceedings of the Human Factors Society 32nd Annual Meeting,* 273–277.

——— (1992), A comparative study of gestural, keyboard, and mouse interfaces, *Behaviour & Information Technology,* 11(1), 13–23.

Wolf, C. G, Glasser, A. R., Fujisaki, T., and Watson, T. J. (1991), An evaluation of recognition accuracy for discrete and run-on writing, *Proceedings of the Human Factors Society 35th Annual Meeting,* 359–363.

Woodson, W. E., and Conover, D. W. (1964), *Human Engineering Guide for Equipment Designers* (2nd ed.). Berkeley, Calif.: University of California Press.

Woodworth, R. S.

——— (1899) The accuracy of voluntary movement, *Psychological Monographs,* 3 (whole no. 13).

——— (1938) *Experimental Psychology.* New York: Holt.

Wooldridge, S. (1974), *Computer Input Design.* New York: Petrocelli Books.

Work in America (1974), Report of a Special Task Force to the Secretary of Health, Education, and Welfare. Cambridge, Mass.: The MIT Press.

Wright, P.

——— (1969) Transformations and the understanding of sentences, *Language and Speech,* 12, 156–166.

——— (1977) Presenting technical information: A survey of research findings, *Instructional Science,* 6, 93–134.

Wright, P., and Barnard, P. (1975), Just fill in this form—A review for designers, *Applied Ergonomics,* 6(4), 213–220.

Wright, P., and Lickorish, A. (1984), Ease of annotation in proofreading tasks, *Behaviour and Information Technology,* 3, 185–194.

Wright, P., and Reid, F. (1973), Written information: Some alternatives to prose for expressing the outcomes of complex contingencies, *Journal of Applied Psychology,* 57(2), 160–166.

Wright, P., Lickorish, A., and Milroy, R. (1994), Remembering while mousing: The cognitive costs of mouse clicks, *SIGCHI Bulletin,* 41–46.

Wright, P. C., and Monk, A. F. (1991), A cost-effective evaluation method for use by designers, *International Journal of Man–Machine Studies,* 35, 891–912.

Wright, R. B. (1992), Method bias and concurrent verbal protocol in software usability testing, *Proceedings of the Human Factors Society 36th Annual Meeting,* 1220–1224.

Wulff, J. J., and Berry, P. C. (1962), Aids to job performance, in R. M. Gagne and A. W. Melton, *Psychological Principles in System Development.* New York: Holt, Rinehart and Winston.

Wyatt, S., and Langdon, J. N. (1932), *Inspection Process in Industry.* London: Her Majesty's Stationery Office, Industrial Health Board, Report 3.

Wyndham, C. (1969), Adaption to heat and cold, *Environmental Research,* 2, 442–469.

Wyon, D. P. (1970), Studies of children under imposed noise and heat stress, *Ergonomics,* 13, 598–612.

Yamamoto, S., and Kuto, Y. (1992), A method of evaluating VDT screen layout by eye movement analysis, *Ergonomics,* 35(5/6), 591–606.

Yerkes, R. M., and Dodson, J. D. (1908), The relation of strength of stimulus to rapidity of habit formation, *Journal of Comparative Neurology and Psychology,* 18, 459–482.

Young, R. M. (1981), The machine inside the machine: Users' models of pocket calculators, *International Journal of Man–Machine Studies,* 15, 51–85.

Young, R. M., Green, T. R. G., and Simon, T. (1989), Programmable user models for predictive evaluation of interface designs, *Proceedings of CHI'89,* 15–19.

Zinke, R. A. (1977), A study of the application of personnel subsystem development technology in the design of the BISCUS/FACS Conversion System, private communication.

Zipf, G. (1935), *The Psycho-biology of Language.* Boston: Houghton Mifflin.

Index